BEEF CATTLE

A. L. Neumann,

Head of Department and Professor of Animal Science,
New Mexico State University

Roscoe R. Snapp,

Late Professor of Animal Science,
University of Illinois

Sixth Edition

John Wiley & Sons, Inc.,
New York · London · Sydney · Toronto

This book is dedicated to my hundreds of
undergraduate and graduate students,
now working around the world,
who have been my inspiration during
some thirty years of teaching.

PREFACE

The decade of the 1960s has seen costs of beef production, especially non-feed costs, spiral to all-time highs. The costs of land, buildings, feedlot equipment, farming tools, and taxes and credit have increased much faster than have the sale prices of breeding, feeder, or fed cattle. Lower feed costs and more efficient rations and feeding and breeding programs have tended to offset some of the increases in operating costs. Reductions in labor requirements, due to mechanization and overall efficiencies resulting from the increased size of farms, ranches, and feedlots, have further assisted in making it possible for the beef cattle enterprise—the most important single source of cash farm income—to survive profitably in the face of increasing pressures from rising costs.

The above financial picture of the current status of the beef cattle enterprise is responsible for some shifts in emphasis in this, the sixth, edition of *Beef Cattle*. The financial risks and the profit potential of the various cattle production programs are given more attention. Efficiency of production, especially with respect to increasing percent calf crop weaned, and improved weaning weights, feedlot gain, and feed conversion, have been given more space and attention in this revision, because I believe that new scientific knowledge is now available which, if universally applied at the farm, ranch, or feedlot level, will enable cattlemen to compete more successfully with producers of the other species of meat animals.

Not overlooked are the possibilities for improving the acceptability of the beef carcass by all those concerned, including packer, retailer, and consumer. Conventional selection programs and mating systems are treated at some length, as is the now generally accepted possibility of introducing non-British breeds in a crossbreeding system to maximize heterotic response and combine production traits in a manner to produce a superior product at lower cost.

A change was made in organization that, it is hoped, will tend to bring students with varying degrees of prior preparation and beef cattle background nearer to a common level. Four review chapters are placed at the opening of the text so as to prepare all students for what is to follow. A notable example is the chapter dealing with beef cattle breeding, by

Dr. K. E. Gregory, now director of the United States Meat Animal Research Center at Clay Center, Nebraska, whose assistance is gratefully acknowledged. This chapter illustrates the change in teaching material in this text, which completes the transformation of the approach to the beef cattle industry from one of tradition and husbandry to one of science.

The suggestions given by instructors of the beef cattle production courses in our universities and colleges have always been and will continue to be welcomed and appreciated. Researchers whose findings are cited have made an important contribution to this revision, as have the extension livestock specialists who have translated these findings into practical recommendations for producers. For the cooperation of these three groups of very important people I am much indebted.

I further wish to acknowledge the generosity of a large number of farm journal editors, government agency personnel, and breed association representatives who have so kindly granted permission to use their charts, photographs, and other material in this new edition. Without such assistance this revision would have been more difficult. I have attempted to cite the source of all such materials and to give credit to the proper individuals or institutions. If I have failed to give recognition for help and materials received, it has been wholly unintentional and is deeply regretted.

Las Cruces, New Mexico   *A. L. Neumann*

October 1968

CONTENTS

PART V

SPECIALIZED BEEF CATTLE PROGRAMS   571

PART VI

SPECIAL PROBLEMS IN BEEF PRODUCTION   609

BEEF CATTLE

Part I

ORGANIZATION OF THE BEEF CATTLE ENTERPRISE

Chapter I

PROGRAMS and AREAS of BEEF PRODUCTION

Beef cattle production differs from production of most other kinds of livestock in that the operation is frequently divided into several distinct steps or phases. It is possible to carry on all these phases on a single farm or ranch as successive steps of a continuous process. More often, however, one or two phases are carried on to the exclusion of the others, not only on individual farms, but also in agricultural regions. In addition there are two or three highly specialized forms of beef production that differ so much in management methods from those commonly followed that they deserve special mention.

BEEF PRODUCTION PROGRAMS

COMMERCIAL COW-CALF PROGRAM

The first and most fundamental step in the beef enterprise is the production of a baby calf and raising it to weaning age. The calf, so to speak, is the raw material from which the finished beast is eventually made. The breeding herds in which calves are produced need little grain or other finishing feeds. Consequently, the raising of beef calves is generally, though not always, confined to those sections that have an abundance of comparatively low-carrying-capacity grazing land. Thus most of the important breeding centers are located either in regions that receive insufficient rainfall to produce cultivated crops or in hilly areas where the land is too rolling to be farmed to advantage. Aside from its effect on crops grown in a given area, climate further plays an important role in determining the location of the cow-calf enterprise. The southern and southwestern states have a decided advantage over those farther north. Because of the shorter winters in these regions the calves are ordinarily born 4 to 6 weeks earlier than in the north, and they may even be born in the fall, making them larger and heavier when they are marketed the following fall.

The raising of beef calves is most successful on comparatively large farms or ranches, because it requires little more labor to care for a herd of 75 to 100 cows than for a herd of 10 to 20. So long as the United States has extensive areas of government-controlled land or of privately owned grazing lands ere large herds are cared for at the rate of only one man for about 250 cows, it will be difficult for the farmer who operates a quarter section of high-priced tillable farm land to compete in the breeding of commercial beef calves for the open market as his principal enterprise. This does not mean that a breeding herd may have no place on a great many farms. It does mean that the breeding of commercial cattle must often be secondary in importance to some other enterprise that more efficiently utilizes the available land and labor. On such farms the breeding herd is a sort of by-product plant that uses the otherwise unmarketable products of the farm.

Commercial—that is, not purebred—cow herds can be grouped into four broad categories, depending on system of land management, available feeds and pastures, and the method of marketing the calf crop:

1. There are the large spectacular herds that sometimes contain a thousand cows or more, operated on ranches located principally in the Rocky Mountain region. These ranches usually include some deeded land situated along rivers or streams where the winter feed supply of hay or silage, if needed, is produced on irrigated meadows and cropland. The remainder of the ranch usually consists of extensive acreages of low-carrying-capacity, government-controlled land, such as Bureau of Land Management land, national parks, or forests, that may be near or adjacent to the deeded land. The rancher has grazing privileges or permits for a given number of cows for the summer grazing season. The calf crop is sold either as calves at weaning time in the fall or as yearlings the following fall after having spent another grazing season on the range. Little if any finishing is done on these ranches because grain is not grown to any extent.

2. Then there are herds varying in size from 30 to 50 cows to very large herds operated on ranches usually owned by the rancher or leased from private owners. Most of these herds are found in the Plains and the Pacific Coast regions. The operation of these ranches varies considerably, depending on whether they are located in the northern or southern portion of these regions and on the feed-producing capabilities of the soil. Although

most of these ranches sell either calves or yearlings to other ranchers or to feeders, some may feed out their own production. This practice is followed by more and more of the ranchers in the Southwest who are growing dryland grain sorghums or who have access to irrigation water for the production of feed grains. Still other ranchers contract-feed their calves and yearlings in feedlots in the vicinity.

3. Farm-sized herds of 20 to 100 cows are typically operated on farms in the Corn Belt and adjacent regions. On these farms commercial cows may or may not be a secondary enterprise. The beef cow herd often has replaced a dairy herd, and sometimes tillable land incapable of maintaining high yields of cash crops has been seeded down to improved pasture which is utilized by a cow herd. In other instances, cows may utilize only the permanent pasture and aftermath and other cash-crop residues such as straw, cornstalk fields, and corn cobs. The calves from these herds are usually either sold in the fall or spring or are fed out on the farms where they are produced. The large increase in beef cow numbers in the North Atlantic states consists of herds that fall in this category.

4. A fourth type of operation consists of herds varying in size from a few to several hundred cows, typically found in the Cotton Belt and Gulf Coast regions. These regions are favorable to the cow-calf program because long-season grazing is possible on pastures now occupying acres once depleted by erosion and continuous production of cotton, peanuts, sweet potatoes, or corn. Winter oats, fescue, and long-season grasses such as orchard grass make it possible to reduce costly winter feeding. The climate is such that winter or very early spring calving is common, so that calves are weaned and sold earlier than are calves from the other areas. Herds that do not market their production as fat slaughter calves or do not feed out their calves as baby beeves usually sell them as stocker or feeder calves rather than as yearlings. Much of the increase in finishing cattle that has occurred in the Arizona-California region is based on the calves and light yearlings produced in these herds.

Successful commercial cow-calf operations usually have several things in common regardless of category:

1. Relatively low investment in land required per cow.
2. Maximum utilization of pasture and low-sale-value roughages.
3. Minimum outlay for supplemental feed.
4. Low labor costs.
5. High-percentage calf crops of high-quality, heavyweight calves.
6. Minimal losses from diseases and parasites.

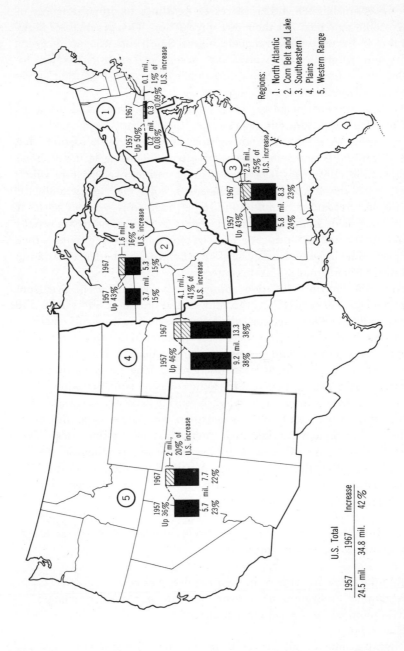

FIG. 1. Beef cows on farms and ranches, by regions, in 1957, and 1967. Cow numbers increased steadily throughout the country during this decade. (From U.S.D.A. statistics.)

Regions:
1. North Atlantic
2. Corn Belt and Lake
3. Southeastern
4. Plains
5. Western Range

	U.S. Total	Increase
1957	1967	
24.5 mil.	34.8 mil.	42%

Region 1:
0.1 mil., 1% of U.S. increase
1967 0.3 0.09%
1957 0.2 mil. 0.08%
Up 50%

Region 2:
1.6 mil., 16% of U.S. increase
1967 5.3 15%
1957 3.7 mil. 15%
Up 43%

Region 3:
2.5 mil., 25% of U.S. increase
1967 8.3 23%
1957 5.8 mil. 24%
Up 43%

Region 4:
4.1 mil., 41% of U.S. increase
1967 13.3 38%
1957 9.2 mil. 38%
Up 46%

Region 5:
2 mil., 20% of U.S. increase
1967 7.7 22%
1957 5.7 mil. 23%
Up 36%

6

The largest numbers of beef cows are found in the western and plains states, but Fig. 1 shows the extent of the increase in cow numbers in the eastern half of the country, notably in the southeastern states.

STOCKER PROGRAM

A stocker is a young animal that is being fed and cared for in a way that promotes growth rather than improved condition. Stockers or stock cattle are of two kinds: heifers that are intended for the breeding herd, and steers and heifers that are intended for the market as feeders or for finishing by the present owner. With both kinds of stockers the owner's main objective is to produce as much gain as is economical, consistent with normal growth and development. Necessarily, then, stockers are handled only by farmers or ranchers who have quantities of nonfinishing feed in the form of pasture or harvested roughage such as hay, straw, fodder, and silage.

Because stock heifers for breeding purposes are in demand principally in the breeding centers where they were produced, few animals of this class are to be found outside such areas. In general, they are managed in much the same way as the breeding herd.

FIG. 2. A top-quality commercial cow herd with calves, on typical mountain foothill range in the Western Range area. (The Record Stockman.)

FIG. 3. Stocker steer calves being wintered on legume-grass silage in Indiana. These calves were born and raised to weaning age on a western ranch but will spend a year on this Corn Belt farm before being marketed as finished steers. (Corn Belt Farm Dailies.)

Stockers intended for the market, however, may be grown out in the region where they were bred and reared by allowing them to graze grassland like that used by their mothers. Or they may be shipped soon after being weaned, either to grazing areas that are not fully stocked with cows and young calves or to grain-growing sections where they ultimately will be finished. In grain sections their feed consists mainly of the aftermath of meadows, legume pasture crops grown in the regular farm rotation, stalk fields, oat straw, legume hay, and silage. Many Corn Belt feeders make a practice of buying their cattle as calves or yearlings in the fall and carrying them on such feeds through the winter or for a full year before putting them into the feedlot. In this way the by-products of grain farming are utilized. Any undesirable or unthrifty cattle are weeded out before the use of expensive feeds is begun and, probably most important, the cattle are bought when market conditions are especially favorable to the buyer.

Stock cattle may be laid in at almost any time of year on a well-diversified farm or by a commercial feedlot operator. Hence an order may be placed in the hands of a commission firm or dealer with instructions to buy when the next big "bargain day" occurs. On the other hand, cattle that are to go directly into the feedlot must be bought within a rather short period if the plans for feeding and marketing are not to be disarranged. Feeder cattle often sell unusually high at the time the feeder wishes to start his feeding operations. Had he bought his animals 3 to 6 months earlier when the market was depressed and carried them along on rough feed until he really needed them, he might have saved $1 to $5 per hundredweight in the cost price.

Stockers are seldom carried longer than a year before they are placed in the feedlot or sold to some other cattleman for further development or finishing. An exception is yearlings that are kept over in the fall to utilize winter wheat pasture in the Southwest. In the spring these steers, now 2-year-olds, are usually sold as fleshy feeders for immediate finishing. The famous Flint Hills section of eastern Kansas and the Osage country in Oklahoma are still utilized to some extent by 2-year-old stockers, but this program is rapidly giving way to the commercial cow program in these sections. In the Corn Belt it is more common to keep yearling stockers only through the grazing season of summer and fall or during the fall and winter months when stalk fields, oat straw, silage, and other coarse roughages are available. Commercial feedlot operators may feed a stocker or "warm-up" ration for only 50 to 75 days before shifting to higher-energy finishing rations.

THE FINISHING PROGRAM

The finishing or fattening program consists of feeding thrifty calves and well-grown-out older cattle, mainly yearlings, a ration that is moderately high to very high in energy, until they are sufficiently conditioned or finished to yield a carcass with enough fat to be palatable and acceptable to the consumer. Such rations must contain considerable low-fiber, high-energy feeds such as farm grains, molasses, protein concentrates, or waste animal fats. Roughages are less important in this program but nevertheless must be available economically, whether they are homegrown or purchased.

Corn (*Zea mays*) and the sorghums make up the major portion of the high-energy feeds. Naturally, then, the Corn Belt and the sorghum-growing areas of western Kansas, the Texas Panhandle, and parts of western

Oklahoma and eastern New Mexico are important cattle-finishing areas. Other important finishing sections are found in the irrigated valleys of north central Colorado and the Platte Valley in Nebraska, where the cattle are fed large quantities of corn silage, alfalfa, and sugarbeet by-products, plus shipped-in grains. The Northwest, including Montana and Idaho, where quantities of barley and wheat are being grown and fed to finishing cattle, is developing rapidly in this industry.

Other important cattle-finishing centers are in the productive Mississippi Delta area in the South, in parts of Florida where large quantities of citrus by-products are available, and in Louisiana where blackstrap molasses, a by-product of the sugar industry, is abundant.

For the two decades following World War II the fastest-growing area of all was in California and Arizona where the large-scale feedlot with completely mechanized feed-handling facilities made for efficient operation. These large lots are near the fastest-growing population centers of the United States and find a ready market for their product. They have a favorable climate for cattle finishing and are fairly close to sources of feeder cattle and grain. (Some inconsistencies in the axiom that fat cattle and grain farms go together will be discussed later.) During the mid-1960s the sorghum belt in the Texas Panhandle experienced a similar growth period, and only some uncertainty about the supply of irrigation water for sorghum production dims its future expansion.

In sections of the Corn Belt, where the land is somewhat rolling and considerable improved pasture is available, a common method of finishing

FIG. 4. Choice yearling steers on a finishing ration in northwestern Iowa. Shelter belt and board fence in background are sufficient protection from prevailing winds. (American Hereford Association.)

cattle is to self-feed corn on grass during the summer and fall months. Such sections often produce a small percentage of their supply of feeders; the balance is purchased from the western or southern states. In the Corn Belt proper, however, where most of the land is tillable, cattle are finished almost entirely on harvested feeds and the feeding is done in drylot or confinement and usually during the winter months. A common practice is to buy, in the fall, yearling steers or heifers or heifer calves that are ready to go directly into the feedlot. They are accustomed to a full feed of grain as rapidly as possible and by April or May are usually carrying sufficient finish to grade high enough to suit the majority of consumers. Cattle handled in this way interfere very little with the growing of crops. They arrive at the farm in the fall about the time the corn is harvested and leave in the spring before the busy season begins. On the other hand, many of the steer calves finished in the Corn Belt are carried well into the summer and fall, either in drylot or on pasture. In the summer drylot finishing program there is wider use of "green chop"—rotation pasture forage that is chopped and fed fresh daily along with the concentrate portion of the ration. Haylage, an ensiled product made from the same type of crop but intermediate in moisture content between hay and silage, is also finding favor.

The finishing of cattle for market has several important advantages in addition to the direct financial profit. One benefit is the fertility that remains on the farm in the manure. Farms where cattle have been fed for a number of years are much more productive, as a rule, than adjoining farms where grain and hay have been sold. Cattle feeding makes use of damaged grain and hay that would contribute little to the farm income if sold for cash. Farmers located at a distance from shipping points can greatly reduce their marketing expenses by marketing their farm products through cattle rather than in the form of hay and grain. Another source of income from feeding cattle is the gain made by hogs that are kept in the feedlot to utilize the partially digested grain found in the cattle manure. As already mentioned, cattle-finishing programs also provide a good market for an almost innumerable list of by-product feeds.

Cattle feeding, like other industries that derive their profits from buying raw materials and selling them later in the form of a finished product, involves some speculative risk. This risk is necessarily high when both cattle and feed are purchased, and there is always the temptation to expand or contract operations according to the prospect of favorable or unfavorable prices for finished cattle in the future. However, in a feeding program that is adapted to marketing the feeds grown on a given farm, the risk is probably no greater than that connected with a number of other major farm enterprises.

THE BABY-BEEF PROGRAM

Strictly choice or better, fat, young, cattle, varying in age from 8 to 15 months and weighing 650 to 950 pounds, are called baby beeves. The finishing of baby beeves is a highly specialized form of beef production; to finish animals at so early an age there must be maximum use of palatable and highly nutritious feedstuffs. Baby-beef production is carried on throughout the country wherever grain is grown, but the program is naturally concentrated in areas such as the fringes of the Corn Belt where grain is grown in large quantities and permanent pastures also play an important role.

The production of baby beef implies the breeding, rearing, and finishing of the calves on the same farm. Since only calves of strictly beef type

FIG. 5. Homebred baby beeves. 16 months of age and almost ready for market as 1,000-pound slaughter cattle. These steers spent their entire lives on this Corn Belt farm. Note effective homemade feeding equipment. (Corn Belt Farm Dailies.)

show a pronounced tendency to finish at so early an age, the breeding herd itself must demonstrate the characteristics that signify early maturity. Dairy crosses are usually unsatisfactory, because these cattle tend to grow rather than finish at the desired weights for baby beef.

Calves are provided with creep feed as soon as they are old enough to eat, which is usually at 4 to 6 weeks of age. Roughage has a minor place in the finishing ration because the calves have little capacity for quantities of roughage after consuming the desired amount of grain. However, the cow herd will utilize the available roughage materials.

Western calves carrying considerable flesh and showing excellent breeding can be finished into acceptable baby beeves if carefully handled and fed for 6 to 7 months, especially if they have previously been accustomed to grain from a creep feeder.

THE FAT-CALF PROGRAM

Much of the increase in beef cow numbers since World War II in the southeastern states and the Corn Belt can be attributed to the financial success achieved by those farmers who adapted the cow-calf program to a set of circumstances peculiar to these areas. In the gradual shift from dairying, large numbers of dairy cows of all breeds were bred to beef bulls. The resulting calves made very rapid gains because of the extra milk supplied by their dams and because of the longer lactation period. Calves weighing upward of 600 pounds in slaughter condition were not uncommon at 8 to 9 months of age, and this increase was often made without grain feeding. Traditionally, consumers in the South and Southeast preferred beef with less finish than was desired by the majority of consumers elsewhere. Therefore, consumer acceptance of this young milk-fed beef in the regions where it was produced was sufficient to ensure acceptable selling prices. Additional weight and condition were being added by creep feeding and by earlier calving, with many fall calves being dropped by cows on winter oat and wheat pastures.

The future of this program is now at the crossroads. As income levels have risen in the belt extending from the southeastern coastal regions to Texas, and as more people from other sections of the country, with their traditional food habits, have moved into this area, the demand for beef has shifted from calf or young, grass-fat beef to more mature, fed beef. This has caused lower prices for fat calves. Furthermore, with successive use of the third or fourth beef bull on the dairy-bred cows and their female descendants found in the area, the milk production levels have dropped. As a result the condition or finish of the weaner calf has been greatly

FIG. 6. Native grade cows with their 600-pound milk-fat calves ready for slaughter. These calves, sired by a registered Hereford bull, were dropped in January and February and sold for slaughter at weaning time in October. (University of Kentucky.)

reduced, generally to below desired slaughter finish. The improved conformation caused by the increase in beef breeding with each generation has placed these calves into a competitive situation with western calves for feeding in the Corn Belt feedlots. Therefore they are no longer available for slaughter as fat calves, and the fat-calf program may be on the way out, although there is likely to be some demand for fat-calf beef for some time to come.

Calves in this program should by all means be marketed as slaughter calves, right off the cows or "off the teat," even if creep feeding is required to insure adequate finish. Once these calves lose their milk bloom, they sell at a distinct disadvantage as stockers or feeders. With age, their defects in beef conformation become more pronounced, and the final selling price after feeding in the manner usually used on straight beef-bred feeders is considerably lower. Brahman bulls, which are commonly used in the Gulf Coast region, sire calves that fit into the fat-calf program very well, although western feedlots are quite competitive for these calves as feeders when the fat-calf market drops.

Fewer and fewer cows of the dual-purpose breeds such as Milking

Shorthorn and Red Poll are milked and handled as dairy cattle. The more common practice is for most, if not all, of the cows to nurse their own calves until they are 8 to 10 months old. Mature cows of good milking ability can easily suckle two calves, and extra calves are often purchased or calves from cows that are being milked are transferred to foster mothers. If labor is available, calves are often housed in drylot, apart from the cows, and turned in to nurse twice daily. In this method of managing dual-purpose cows the calves readily learn to eat grain, and the result is an ideal slaughter calf at weaning time, with extraordinary weights combined with enough beef type to satisfy the most discriminating buyer. When research now under way with the newest breed importation, the Charolais, is more conclusive, this program may well be the one where they will fit best.

THE PUREBRED PROGRAM

The breeding of purebred or registered cattle is a highly specialized form of beef production, requiring a relatively large amount of capital for foundation cattle and equipment and a great deal of husbandry skill and sound judgment on the part of the manager before success is possible.

FIG. 7. A purebred Angus herd on a productive Iowa farm. Purebreds, if high enough in quality, may be used to increase volume of business without increasing numbers. (American Angus Association.)

This phase of cattle breeding is better suited to men of considerable experience than to beginners. This does not rule out the 4-H Club or Future Farmers of America youngster who starts a heifer project and conducts it under the supervision of an experienced father or other adult leader. In this instance immediate financial gain is not so much the goal; rather, the training involved in the conduct of the project is uppermost. Indeed, a large number of successful herds were started from just such a project.

The opportunities offered the breeder of purebred cattle are almost unlimited. Honor, fame, and large financial rewards are all within possibility of realization. In the United States the purebred cattle business, large though it is, still represents only a small part of the entire industry, and there is great progress yet to be made in expanding it. It is estimated that only about 3 percent of all beef females in the country are registered.

In the western states, where cattle are the principal source of income, ranchers have long recognized the value of purebred stock and have been willing to pay from $500 to $1500 per head for young purebred bulls in order to produce grade calves and yearlings for market. Farmers in the Corn Belt and elsewhere, who maintain small breeding herds as a minor enterprise, have possibly been slower to recognize the importance of superior breeding. They have now learned, however—mainly from their swine-breeding friends—that superior performance-tested bulls are a good investment, and in some respects these farmers are today leading many of the western breeders.

Despite the need for large numbers of purebred bulls to serve as sires in both purebred and commercial herds, most purebred herds at present are not sufficiently high in quality to warrant saving more than one-half to two-thirds of the bull calves dropped. Castration of mediocre calves gives the purebred breeder a source of feeder steers that can contribute to his income as well as raise the quality of commercial feeder cattle. Purebred bulls and surplus breeding females must sell for substantially more than market prices if the purebred program is to be profitable.

Although purebred beef cattle have increased noticeably in the Corn Belt and the southeastern states in recent years, they have not kept pace with the increase of commercial beef cows in these areas. Considering that probably not more than half of the grade beef cows of these regions are now bred to purebred bulls and that probably a similar proportion of the purebred beef bulls now used in grade herds are of mediocre quality, the future demand for young purebred bulls with production records or with progeny-tested sires appears unusually bright.

No single part of the country any longer holds a monopoly on good beef bulls. Excellent purebred herds of all the major breeds are found

Table 1
Membership and Registration Data for Selected Beef Breed Associations (1966)

	Hereford[a]	Angus	Polled Hereford	Charolais	Shorthorn and Polled Shorthorn	Brahman
Membership	75,149	48,231	30,000	3,414	3,700	1,004
Registrations	467,573	389,141	165,500	39,478	39,207	15,281
Transfers	297,857	324,397	121,000	29,408	26,476	10,337

Registrations, rank by states

	Hereford[a]	Angus	Polled Hereford	Charolais	Shorthorn and Polled Shorthorn	Brahman
1	Texas	Missouri	Texas	Texas	Iowa	Texas
2	Kansas	Texas	Missouri	Missouri	Illinois	Florida
3	Oklahoma	Iowa	Kentucky	Kansas	Kansas	Louisiana
4	Nebraska	Oklahoma	Illinois	Oklahoma	Missouri	Arkansas
5	Montana	Illinois	Tennessee	Florida	Nebraska	Mississippi
6	Missouri	Tennessee	Oklahoma	Iowa	Indiana	North Carolina
7	South Dakota	Kentucky	Kansas	South Dakota	North Dakota	Arizona
8	Colorado	Nebraska	Mississippi	Nebraska	Texas	Georgia
9	North Dakota	Kansas	Arkansas	Illinois	South Dakota	Nevada
10	California	Montana	Alabama	Colorado	Oklahoma	Alabama

[a] Numbers include data on both horned and polled Herefords from the American Hereford Association.

throughout the country, and the day is past when commercial cowmen had to drive hundreds of miles to find good bulls. There is considerable evidence that bulls produced in the immediate area of a farm or ranch, from cows and bulls adapted to the prevailing environmental conditions, will work out more satisfactorily for the buyer than will bulls shipped in from a distance.

AREAS OF BEEF PRODUCTION

There are five rather well defined areas of beef production in the United States. Each differs from the others in the extent to which beef production is practiced and in the relative importance of the various programs that have been discussed above. These five areas may be divided into still smaller regions, each noted for a particular phase of cattle raising or for a distinctive method of handling cattle. These smaller subdivisions will not be discussed in detail here because they are largely of local interest.

THE RANGE REGIONS

The Range regions are contained roughly within the area lying west of the 100th meridian. This area may be subdivided into the Plains, Rocky Mountain, and Pacific Coast regions, and certain smaller sections of more or less local importance.

In the Range regions the breeding of calves and the growing out of young cattle on grass are the dominant phases of beef production. Because of the great amount of grazing land, much of which is still in public domain, this region is particularly well suited for these extensive, rather than intensive, forms of cattle raising. Most of the cattle produced in the Range regions are sold as stockers and feeders for further development or finishing.

Because beef production is a major enterprise with a majority of the ranchers of this area, the cattle are well bred and are handled according to modern methods of ranch management.

FIG. 8. A commercial cow herd is the major enterprise on this Oregon ranch on the western slope of the Rocky Mountains. Yearling replacement heifers, being bred to calve as 2-year-olds, may be seen in the herd. (The Record Stockman.)

Formerly most of the purebred bulls used in improving the type and quality of the range cattle were purchased from Corn Belt breeders, but at present a large percentage of the bulls used in the Range area are bred there. In fact, breeding good bulls for sale to commercial ranchers, as well as breeding superior young bulls and heifers to go into other pure-bred herds, is now a highly important phase of beef production in the Range states.

Until the years following World War II, the ranchers in the Range regions finished few cattle, preferring to ship most of their calves and yearlings to the Corn Belt. This practice has changed appreciably because of four principal factors:

1. Higher freight rates on shipment of live cattle to Corn Belt markets and of dressed beef back to consuming centers in the West.

Table 2
Relative Growth in Beef Cow and Total Cattle Numbers in Different Regions and Type-of-Farming Areas[a]

Region and Type-of-Farming Area	1957 (1,000 head)	1967 (1,000 head)	1967 as Percentage of 1957	Rank in Increase Type-of-Farming Area	Region
Beef Cows					
North Atlantic region	**194**	**337**	**174**		**1**
Lake	500	775	155	2	
Corn Belt	3,160	4,533	143	5	
Lake and Corn Belt region	**3,660**	**5,308**	**145**		**3**
Appalachian	1,460	2,711	186	1	
Southeastern	2,158	2,721	126	9	
Delta	2,223	2,883	129	8	
Southeastern region	**5,841**	**8,315**	**142**		**4**
Northern Plains	4,230	6,109	144	4	
Southern Plains	4,908	7,193	147	3	
Plains region	**9,138**	**13,302**	**146**		**2**
Rocky Mountain	4,145	5,583	135	6	
Pacific Coast	1,556	2,068	133	7	
Western Range region	**5,701**	**7,651**	**134**		**5**
United States total	**24,534**	**34,592**[b]	**141**		

Table 2 (Continued)

Region and Type-of-Farming Area	1957 (1,000 head)	1967 (1,000 head)	1967 as Percentage of 1957	Rank in Increase Type-of-Farming Area	Region
All Cattle and Calves					
North Atlantic region	**6,019**	**5,081**	**84**		**5**
Lake	10,030	9,741	87	9	
Corn Belt	18,638	19,819	106	7	
Corn Belt and Lake region	**28,668**	**29,560**	**103**		**4**
Appalachian	6,570	7,589	115	5	
Southeastern	5,614	6,156	110	6	
Delta	5,947	5,867	99	8	
Southeastern region	**18,131**	**19,612**	**108**		**3**
Northern Plains	13,029	18,553	142	1	
Southern Plains	10,693	15,065	141	2	
Plains region	**23,722**	**33,618**	**142**		**1**
Rocky Mountain	10,018	11,585	116	4	
Pacific Coast	6,174	7,791	126	3	
Western Range region	**16,192**	**19,376**	**120**		**2**
United States total	**92,860**	**108,491**[c]	**117**		

[a] Compiled from U.S.D.A., *Agricultural Statistics*, 1957 and 1967.
[b] Includes 2 and 87 thousand in Alaska and Hawaii respectively.
[c] Includes 8 and 236 thousand in Alaska and Hawaii respectively.

2. Substantial disproportionate increases in consumer population in the western states.

3. Simultaneous increase of irrigation facilities and development of high-yielding grain sorghum varieties that may be combine-harvested.

4. Development of large-scale feedlot operations.

California, western Texas, Arizona, and Washington show especially large increases in numbers of beef cattle finished, and this can be said about other more or less isolated spots throughout the area. By the mid-1960s the western states were feeding about ten times more cattle than just before World War II. At times up to one-third of all cattle on feed may be found in the western states today, demonstrating the real competition that Corn Belt feeders are getting from the West.

Although the West has seen spectacular increases in cattle feeding, it is highly possible that further increases will be smaller for a number of reasons. Continuing rise in freight rates for both grain and feeder cattle being shipped into the West, shortages of suitable roughages, competition for space and land from other industries, and labor shortages may be important deterrents to further expansion. If the present high level of beef consumption per capita is to be maintained, a large portion of the increased numbers of fed cattle must come from the Corn Belt and the grain sorghum–winter wheat belt found in western Kansas, western Oklahoma, the Texas Panhandle, and eastern New Mexico.

THE CORN BELT AND ADJACENT REGION

This area is noted for its broad stretches of prairie land and exceedingly fertile soil, making it a region devoted largely to the growing of crops. Land is high-priced, and farms are comparatively small to medium-sized. They are largely used for general farming, with the growing of grain as the major enterprise. Intensive rather than extensive methods of feed or crop production prevail.

Beef production under such conditions necessarily assumes a quite different role from that in the Range areas where grass is the main crop to be sold. Beef production in the Corn Belt is conducted according to intensive management methods. The finishing of western-grown, and, in increasing numbers, southern-grown cattle, either in drylot or on a limited amount of pasture, represents the typical method of handling cattle in the Corn Belt and adjoining region. As a rule the cattle are bought in the fall at public stockyards, feeder cattle auctions, or direct from the range. Yearling steers are commonly allowed to run on pastures and stalk fields for 3 to 4 weeks, after which they are confined in drylots and placed on a finishing ration. Normally 5 to 6 months are required to make satisfactory market beasts out of these yearlings, by which time they will at least grade choice as carcasses on the rail. As discussed in greater detail elsewhere, yearling heifers are fed for shorter periods than are steers and are hurried along faster on more concentrated rations.

Summer feeding on grass of overwintered calves is commonly practiced in sections that are somewhat rolling though tillable and therefore have large areas in both rotation and permanent pasture. Because thin cattle are scarce in the spring, steers that are to be fed on pasture are often bought in the fall and carried through the winter as stockers on such feeds as hay, silage, and a low level of concentrates, which tend to produce growth rather than fat. Thus two phases of beef production are represented

FIG. 9. Prime, long-fed steers that have been on full feed in a Corn Belt feedlot for about 10 months after having been bought in the Nebraska Sandhills as weanling calves the previous fall. (American Angus Association.)

in such a method of management. In the level areas where pasture crops are less important, calves may be finished out in drylot during the summer after being wintered on stocker rations.

Cow-calf operations in the Corn Belt, although less numerous than at the beginning of the twentieth century, are more important than is commonly supposed. Most of the native calves are raised and fed in small groups of 10 to 100 head, making the breeding industry less spectacular than feedlot operations involving larger droves of shipped-in feeders. Nevertheless, fully one-half of the cattle and calves slaughtered within the Corn Belt are native cattle that were bred in the region. Many of these cattle are discarded dairy cows and veal calves, but a large number are young cattle of beef breeding that are marketed as baby beeves or as finished yearlings.

Another aspect of breeding operations in the Corn Belt is the raising of purebred cattle, for which this region enjoys particularly favorable conditions. The abundant feed supply promotes low production costs, and the proximity of the Range and Southeastern areas insures a broad market for all surplus breeding stock.

THE APPALACHIAN AND GREAT LAKES REGIONS

These regions are noted for their rolling topography, which in places has a mountainous character. Permanent pastures and woodlands occupy a considerable portion of the total area and, as always in broken regions, there are wide variations in the character and fertility of the soil. In certain sections there are extensive outcrops of limestone where the best bluegrass of the country is found. Grass grown on such a soil is exceedingly nutritious and is prized by cattle grazers. These areas are usually stocked with older steers, which finish satisfactorily during the summer on this grass alone. In other parts of these regions the soil is quite thin. These sections are used mainly by breeding herds or by young stocker cattle that have been purchased in the West or South. At the end of the grazing season these stocker cattle are either finished on feeds grown on the bottomland or are wintered on hay or silage and held over for another summer.

Despite the predominance of dairy cattle in the Appalachian and Great

FIG. 10. One of the many new cow herds in Virginia, being used to utilize the improved pastures that have displaced cash crops from these rolling hills. (American Angus Association.)

Lakes regions, beef cattle numbers have increased noticeably there since the mid-1930s. Especially is this true in Kentucky, Tennessee, Virginia, West Virginia, Minnesota, Wisconsin, and Michigan. Many plantations and farms in these states, which formerly produced principally dairy products, truck, cash grain crops, or tobacco, are now largely in grass utilized by beef cattle. Many excellent purebred beef herds have been established and are a source of improved breeding stock for the areas' rapidly expanding beef industry.

The Appalachian and Great Lakes states are better suited for the production of grass than for any other agricultural product because of the rolling topography, making them natural beef cattle areas. Further increases in beef cattle population may be expected there as existing pastures are improved and as more land is reclaimed from scrubby timber or is taken out of cultivation and used for pasture and hay.

THE SOUTHEASTERN REGION

Before the mid-1930s the Southeastern region was not an important beef-producing area. As a rule, few cattle were found on each farm and often these were noticeably lacking in type and quality. This is no longer true, although there are several factors that handicap beef production in this area.

Rainfall is highly variable in this region, and short droughts are rather common. The result is a rather unreliable pasture situation. Winter feed supplies, unless provided in the form of winter pastures, are sometimes rather expensive and poor in quality as a consequence of the uncertain moisture supply. Once-productive, tillable soils used for growing cotton and other cash crops are sometimes badly eroded and depleted of soil nutrients. Consequently, either forage production is low or investment in fertilizers is high. Wet winter weather in parts of the area makes grazing of winter oats and tame pastures difficult during some years and adds to the uncertainty of the winter feed supply.

Offsetting these disadvantages are the relatively low investment in land and buildings, the short, mild winters and long growing seasons, and the multitude of forage and pasture crops that can be grown successfully. The increased production of improved Bermuda grass strains, winter oats, corn, the newer hybrid Sudans, and hybrid grain sorghums of both the forage and grain types, is making this region a real competitor as a cattle-producing area. Proof of the superior adaptation of the Brahman breed and its crosses in the hotter, more humid portions of this area is responsible for much of the increase in cattle numbers. The application of soil conservation and

FIG. 11. Many cotton plantations and farms in the Gulf Coast region have been sodded with Coastal Bermuda and other adapted perennial grasses. Brahman cattle or Brahman crosses are popular in this region because of their adaptation to the environment. (National Cottonseed Products Association, Inc.)

soil-building practices that were encouraged by government crop control programs, the improvement of pastures, and the production of better adapted forage crops come in for their share of credit.

It will be surprising if this region does not continue its rapid expansion in all types of beef production programs. The South is the logical place to look for much of the increase in beef cow numbers that must occur if the growing demands of an ever-increasing population are to be met.

THE BEEF CATTLE CYCLE

A study of the fluctuations in cattle population in the United States over a series of years shows that these fluctuations have not been haphazard but have followed a rather regular order. Periods of large cattle numbers have occurred about every 14 to 16 years. Also, periods of unusually low numbers of cattle have been separated by about the same interval. (See Fig. 12.) Since price is largely a function of supply, it follows that approximately the same amount of time—that is, about 15 years—intervenes between periods of very high prices or between periods of very low prices. This time interval, called the cattle cycle, is much used by agricultural statisticians and market forecasters in predicting the supply and price of cattle in the immediate future. A livestock marketing specialist has described the evolution of a typical cattle cycle as follows:

> Briefly, a typical cycle begins with an increased demand for breeding stock to expand herds. Prices of breeding stock soar and the producing (cow-and-calf) enterprise becomes especially profitable. As cows, heifers, and calves are held back, only steers are marketed in large numbers for slaughter. Later when calves from enlarged breeding herds reach maturity, total slaughter increases. Prices break, often severely. Declines are sharpest for breeding stock, and least for high grade fed cattle. The producing enterprise becomes relatively unprofitable, more cows are slaughtered, and a scramble ensues to expand the feeding business. Both cow and calf slaughter are larger, cow herds are reduced, and the calf crop becomes smaller. Ultimately total slaughter decreases and prices turn upward, initiating a new cycle.

Having located our place on the cycle with reference to the last peak or low spot, we need only construct that part of the curve that represents the next 5 or 6 years to see what long-term supply and price changes

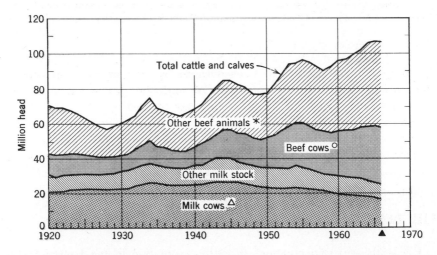

FIG. 12. Cycles in beef cattle numbers, 1920–1966. It will be noted that "peaks" in cattle population have occurred about every 15 to 16 years. △ Cows and heifers 2 years and older for milk; ○ 2 years and older not for milk; * heifers and calves not for milk and all steers and bulls; ▲ preliminary. (U.S.D.A.)

are likely to occur during that period. However, it should be noted that cattle prices are not entirely determined by the cattle population, but also by the number of cattle marketed and by the consumer demand for beef. Consequently, any event that changes either of these factors has far more effect on cattle prices than does the total number of cattle in the country. For example, the severe droughts experienced in the Range and Corn Belt states during 1934 and 1936 resulted in a sharp reduction in cattle numbers after only 5 or 6 years of expansion instead of the usual 7 or 8. The decline was abnormally severe but was of short duration because of the onset of World War II. Thus the cattle cycle may be greatly disturbed by unforeseen abnormal conditions. It should not be followed rigidly in planning a cattle enterprise on an individual farm but should be regarded as a rough guide. There are indications that the number of years required for the buildup to the peak of the cattle cycle has become less with each cycle—that is, the buildup appears to be occurring at a faster rate than formerly.

Chapter 2

The BEEF CATTLE ENTERPRISE

The rancher usually does not have to be concerned with the question of whether he should be running cattle on his spread. He knows that range livestock, either breeding herd or young stock, is his program, and his only major decision, in certain parts of the country, is whether a combination of cattle and sheep, or possibly sheep alone, might be more profitable. The farmer, on the other hand, may first have to wrestle with the problem of whether he should engage in any livestock enterprise at all or whether he should give all his attention to commercial crop production. If he decides to divide his time between grain and animal production, the question of whether to cast his lot with cattle, swine, or sheep becomes a real one.

FIG. 13. Purebred Polled Hereford heifers making use of Nebraska Sandhills range. The owner of this grass could have chosen from among several beef cattle programs, but crop farming is out of the question in this area because of limited rainfall and soil type. (Western Livestock Journal.)

Beef cattle have some advantages over other kinds of livestock, and some of these advantages are of sufficient importance to warrant analysis and discussion. It should be noted that some of the data to be cited in support of certain statements are based on all cattle, both beef and dairy. Unfortunately, statistics are not always available for an accurate separation of these two types, but certain generalizations can be made with respect to the advantages of beef cattle, as briefly discussed below.

CATTLE UTILIZE LARGE QUANTITIES OF ROUGHAGE

Any system of general farming produces a large quantity of coarse, low-grade roughages that have a low market value. In the common 4-year Corn Belt rotation of corn, corn, small grain, and legumes, approximately 2 tons of roughage are produced for every ton of grain harvested. The roughage resulting from the corn crop is practically unsalable, and the small-grain straw and legume hay are often hard to dispose of at worthwhile prices, especially if damaged by rain or the presence of weeds. All these

Table 3
Source of All Feed Fed Different Classes of Livestock (1963)[a]

Item	Beef Cattle (%)	Dairy Cattle (%)	Swine (%)	Sheep and Goats (%)	Poultry (%)	Other Livestock (%)
Concentrates						
Corn	9.6	14.5	70.8	3.2	38.3	34.4
Sorghum grain	3.0	1.6	6.8	0.8	11.0	1.9
Other grains	2.1	6.8	6.2	1.3	11.5	12.5
High-protein feed	4.0	5.1	11.9	5.1	29.7	6.1
Other by-product feed	1.6	3.9	—	—	7.6	9.4
Total	20.3	31.9	95.7	10.4	98.1	64.3
Roughages						
Hay	12.8	30.0	—	5.3	—	13.8
Silage and stover	5.9	12.0	—	3.4	—	—
Pasture	61.0	26.1	4.3	80.9	1.9	21.9
Total	79.7	68.1	4.3	89.6	1.9	35.7
Overall total	100.0	100.0	100.0	100.0	100.0	100.0

[a] Compiled from U.S.D.A. statistics.

roughages, properly fed, are good feed for beef cattle and will, on the average, return more to the farmer when fed than when sold on the market.

Large quantities of corn and sorghum stover are being wasted which, if properly stored and fed or even grazed, would support millions of beef cows and stocker cattle during the winter months. The annual production of straws and stover in the United States is at least 250 million tons. When this amount is compared with the total production of wild and tame hay, namely 100 million tons, the importance of the roughages resulting from grain production is obvious. In the corn and sorghum production areas, where a system of general farming prevails, stover and straw constitute an even greater percentage of the total roughage supply than for the country as a whole, but comparatively little of this roughage is used efficiently in these areas.

A more efficient utilization of all farm products is one of the important problems of the general farmer. With respect to the economic use of these coarse roughages, beef cattle offer a usually satisfactory solution. Mature beef cows can be maintained effectively on rations composed of roughage alone, whereas steers being finished for the market may consume from 10 to 300 percent as much roughage as grain, depending on the level of grain being fed. Illinois studies summarizing the cost account data obtained from a typical cattle feeder in western Illinois for a 7-year period show that the ratio of roughage to concentrates fed was 141 to 100 (including straw used for bedding). In California and Arizona, roughages comprise a much smaller portion of the rations fed feeder cattle, but this is usually because roughage is in short supply and high-priced. In other words, the situation is reversed from that in the more traditional feeding areas where roughage supplies are more economical and where, in fact, cattle are often being fed in order to provide a market for the vast quantities of roughage available.

CATTLE UTILIZE PASTURE CROPS

The importance of pastures and their utilization by beef cattle is taken for granted in the ranching sections of the country, but the importance of pasture crops in the nonranching areas is often overlooked. In the major general farming areas of the country—the Corn Belt and the Middle Atlantic states—grain farming, pastures, and livestock production are highly developed. Beef cattle make it possible for the permanent and rotation pastures to contribute a fair share of income on the farms in this vast area. In other parts of the country real efforts are being made to reclaim and rebuild soils that no longer support cash crops. However, without beef

FIG. 14. Yearling steers on alfalfa, clover, timothy rotation pasture. Beef cattle provide the farmer with a strong incentive for staying with a good rotation cropping system, because cattle often make it possible for pasture crops to produce returns equal to those from cash crops. (University of Illinois.)

cattle and other ruminants to convert the grass and roughage produced on these lands into income, such reclamation is economically unfeasible.

Pastures necessarily occupy a large fraction of the total farming area in this country. Permanent pastures have by no means disappeared from the Corn Belt and other farming areas, nor is it likely that they ever will, because even in the Corn Belt much of the land is better suited for pasturage than for anything else. Furthermore, farmers have found that a certain area of permanent grassland is almost indispensable to good animal management and over a series of years is likely to prove as profitable per acre as any other part of the farm. In addition to this permanent pasture there are acres of temporary pasture crops such as winter small grains, timothy, red and sweet clover, Sudan grass, and various mixtures of grasses and legumes, mostly grown as a part of the regular farm rotation. It is not surprising, then, that the total pasture area is considerable, even in the major farming states.

Just what percentage of this great pasture acreage is grazed by beef cattle can only be estimated. In some sections, such as central Missouri, southwestern Wisconsin, and southern Iowa, beef cattle are by far the most important kinds of livestock kept. Around Chicago, Omaha, Min-

Table 4

Importance of Pasture in General Farming Areas[a]

	Average Pasture Acreage per Farm			Percentage of Total Farm Area in Pasture		
	Cropland Used Only for Pasture (acres)	Woodland and Other Pasture (acres)	Total Pasture (acres)	Cropland Used Only for Pasture (%)	Woodland and Other Pasture (%)	Total Pasture (%)
Illinois	11	27	38	7	18	25
Indiana	10	24	34	10	23	33
Iowa	7	40	47	4	25	29
Missouri	21	52	73	15	35	50
Ohio	8	28	36	9	28	37
Middle Atlantic states	4	29	33	4	29	33
East North Central states	10	30	40	8	25	33
West North Central states	10	101	111	4	37	41
United States	8	98	106	4	51	55

[a] Compiled from U.S. Census of Agriculture.

neapolis, St. Paul, and other large cities dairy cattle are important. Making due allowance for the pasture used by hogs, sheep, and dairy cattle, about 60 to 70 percent of the total pasture area of the North Central states is utilized by beef cattle. This is equivalent to approximately 100 million acres, or 30 percent of the total area in farms.

CATTLE FURNISH A HOME MARKET FOR GRAIN AND HAY

Excepting wheat, comparatively little of the immense production of grain and hay is used other than as feed for farm animals. Most of these feed materials are used on the farms where they are produced. Of the quantity sold, the larger portion is bought by feeders who do not raise enough for their own needs. Even in the heart of the Corn Belt more than 75 percent of the grain and hay is fed on the farms where it is grown.

Beef cattle undeniably constitute an important home market for these farm-grown feedstuffs. For the United States as a whole, beef cattle consume 15 percent of the 4-billion-bushel annual corn crop. For the Corn Belt section this percentage is much higher, as corn is the principal feed used in the extensive finishing operations prevalent in this region. A still larger percentage of the nation's sorghum crop is fed to cattle.

The demand for corn and sorghum for cattle feeding has an effect on the grain market that can hardly be overestimated. It is far more important than the total demand for commercial uses and export combined. Since the livestock industry, including the beef cattle enterprises, provides the principal outlet or market for feed grains, protein concentrates, and salable roughages, it is not surprising that the price of feed crops is closely related to the prices received by feeders and farmers for their livestock. This is conclusively shown in Fig. 15. Only in situations where heavy demand for export purposes may bolster prices above their value for feed do feed prices increase above what they are worth when converted to slaughter livestock. Another exception may be when government price support programs keep the normal supply and demand situation from functioning.

The commercial production of alfalfa in many of the irrigated valleys such as those of the Platte River in Nebraska, the Rio Grande in Colorado

Table 5
Percentage of Feeds Consumed by Different Classes of Livestock (1963)[a]

Item	Corn[b]	Sorghum Grain	Other Grain	High-Protein Feed	Hay	Silage and Stover	Pasture
Beef cattle	15.6	35.2	15.5	19.3	37.3	42.3	68.1
Dairy cattle	15.4	12.3	33.5	16.1	57.4	55.6	19.1
Swine	47.1	16.7	18.9	23.3	—	—	2.0
Sheep and goats	0.5	0.9	0.9	2.2	1.3	—	7.9
Poultry	15.8	32.7	21.9	36.2	—	—	0.5
Other livestock	5.6	2.2	9.3	2.9	4.0	2.1	2.5
Total	100.0	100.0	100.0	100.0	100.0	100.0	100.0

[a] Compiled from U.S.D.A. statistics.
[b] Includes the corn in the silage fed.

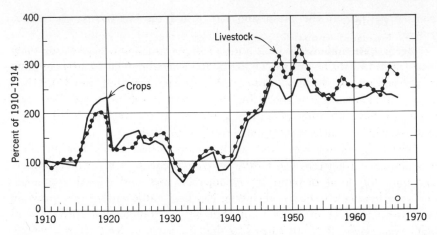

FIG. 15. Relationship of prices received for all crops and for livestock and animal products. ○ January–August average. (U.S.D.A.)

and New Mexico, or the Imperial Valley in California depends in large part on the cattle-finishing enterprise for its market. Dehydrated alfalfa has become such an important part of most commercially prepared fortified protein concentrates that it can almost be said that as the fat-cattle market goes, so goes the alfalfa market. The commercial production of corn and forage sorghums, for sale as silage or as green chop to local cattle feeders, is becoming an important enterprise in such areas as Weld County, Colorado, and the Texas-Oklahoma Panhandle where cattle feeding is more highly concentrated than in any other part of the country. Naturally the price received depends on the profitability of the cattle-feeding enterprise at a given time.

BEEF CATTLE UTILIZE CASH-CROP BY-PRODUCTS

The oilseed meals resulting from the processing of soybeans, cottonseed, and flax are the major components of the protein concentrates fed to beef cattle. Wheat bran, blackstrap molasses, beet pulp, citrus pulp, and cottonseed hulls are other important by-products of the milling and processing of farm crops that are used in beef cattle rations.

Farmers receive higher prices for crops from which these by-products are made because of the large quantity of such by-products being consumed by beef cattle and other farm animals. The total by-product production of at least 10 million acres is fed to beef cattle.

Table 6
**Estimated Annual Grain and Protein Concentrate
Requirement for Beef Cattle
(Feed Year October 1, 1962, through September 30, 1963)**[a]

Class of Beef Cattle	Number (1,000)	Grains and Mill Feeds		Protein Concentrate	
		Required per Head (lb)	Total Requirement (1,000 tons)	Required per Head (lb)	Total Requirement (1,000 tons)
Breeding cattle					
Cows	29,000	50	725	100	1,450
Yearling heifers	6,100	100	305	125	381
Heifer calves	9,000	200	900	100	450
Bulls	1,300	300	195	150	98
Stockers and feeders					
Yearling steers	10,250	2,600	13,325	250	1,281
Slaughter heifers	4,000	2,100	4,200	210	420
Slaughter calves	3,250	300	488	100	162
Stocker calves	5,600	100	280	125	350
Dairy steers	3,000	1,000	1,500	100	150
Total			21,918		4,742[b]

[a] Estimates made by Feed Survey Committee of the American Feed Manufacturers' Association.
[b] Protein concentrate expressed as soybean oil meal equivalent based on 40 percent crude-protein content.

BEEF CATTLE REQUIRE A SMALL INVESTMENT IN BUILDINGS AND EQUIPMENT

The average investment in buildings and equipment per hundred dollars invested in livestock is generally lower for beef cattle than for other farm animals. The investment for buildings and other shelter for the cow-calf enterprise may range from no shelter at all to $15 to $50 per cow, depending on the section of the country. The increasing cost of labor has necessitated an increase in the investment in laborsaving equipment such as silo unloaders, auger bunks, and self-feeders, but labor costs are lower for beef cattle than for other livestock enterprises, as will be discussed later.

Greater use of pole-type barns and open sheds instead of conventional barns, or the elimination of barns and sheds entirely, along with cheaper bunker-type or trench silos, is lowering the relative cost of buildings and equipment still further.

Although expensive barns, mechanical feed- and litter-handling equipment, extensive corrals, and the like are convenient and desirable from a laborsaving standpoint, they are not indispensable. Some of the best known feeders of market-topping steers, and the owners of some of the better purebred beef cattle herds, use equipment that is no more expensive than that found on the average well-kept livestock farm.

BEEF CATTLE UTILIZE LABOR EFFICIENTLY

Beef cattle require little labor compared with dairy cattle and hogs or with cultivated crops occupying an equal area of land. Of the labor that is required, the larger part is needed during the winter and early spring when there is little demand for labor in the fields. Beef cattle thus tend to distribute the labor requirements of the farm throughout the year by (1) utilizing labor that otherwise would be unemployed during the

Table 7
Monthly Expenditure of Labor in Feeding Beef Cattle[a]

	Labor Expended	
	For Cattle (hr)	For Total Farm (hr)
January	164	511
February	144	496
March	132	718
April	79	916
May	31	1,000
June	—	1,133
July	—	1,370
August	—	889
September	1	934
October	13	815
November	78	825
December	129	648
Total labor per year	771	10,255
Labor November 1 to April 1	647	3,198
Percent of labor coming in fall and winter	83.8	31.2

[a] Illinois Bulletin 261.

winter and (2) lessening the summer demand for labor by using part of the tillable land for hay and pasture. A farm that would require the employment of an extra man during the spring and summer under a grain system is frequently operated by the owner alone under a livestock system in which beef cattle predominate. Moreover, the cattle furnish many hours of profitable employment for the owner and his tractor during the winter, which without cattle might not be profitably utilized.

Table 7 shows the average monthly labor expenditure on a 295-acre Corn Belt farm where 40 to 79 cattle were fed annually over a 7-year period. Had cattle feeding not been practiced, it is likely that only a fraction of the 647 hours of labor expended during the winter would have gone into remunerative enterprises. The returns realized from this labor, therefore, may be counted as an almost clear gain to the farm income. In large commercial feedlots it is not uncommon for the labor requirement to be no more than one man per 2,000-head capacity.

DEATH LOSS IS LOW IN BEEF CATTLE

Beef cattle, unlike most other livestock, are subject to few ailments and diseases that are likely to result in death. Compared with sheep and hogs, beef cattle have an unusually low death rate. No beef cattle disease is at all comparable with hog cholera in mortality rate, nor are cattle an easy prey to internal parasites, as are sheep. Losses among calves and yearlings are, of course, higher than for mature cattle, but even here they are far below those for young pigs and lambs.

Beef cattle are not, as a rule, seriously injured by improper methods of feeding. They do not go off feed easily, and usually show rapid recovery from the effects of improper rations or poor management methods. Two important exceptions are bloat and founder, discussed in other chapters. These points are of special importance to the ambitious young farmer whose experience with livestock is limited.

BEEF IS THE MOST POPULAR MEAT

A continuing increase in beef consumption is ample evidence of the overwhelming popularity of beef. Table 8, which shows consumption trends for meats and poultry, indicates that people are eating more total red meats than ever before and that the consumption of poultry has also increased. Even though dressed poultry usually sells for from one-third to

Table 8
Per Capita Meat Consumption in the United States[a]

	Pounds Consumed per Person					
Year	Beef	Veal	Lamb and Mutton	Pork	Total Red Meat	Poultry[b]
1910–1919 average	63.0	6.7	6.2	64.0	139.9	14.4
1950	63.4	8.0	4.0	69.2	144.6	24.7
1955	82.0	9.4	4.6	66.8	162.8	26.3
1960	85.2	6.2	4.8	65.2	161.4	34.3
1965	100.5	5.2	3.7	58.6	168.0	40.3

[a] U.S.D.A.-ERS.
[b] Not considered to be red meat, but shown for comparison purposes.

one-half the price of beef per pound, it is not being used as a substitute for beef. The average consumer in California uses about 125 pounds of beef per year, suggesting that the ceiling on total beef consumption has not yet been reached in the United States. The average beef consumption in several other countries substantially exceeds that of the United States, as shown in Table 9, further illustrating the potential of the demand for beef in America.

Table 9
Worldwide Per Capita Meat Consumption (1964)[a]

Country	Beef and Veal (lb)	Pork (lb)	Lamb and Mutton (lb)	Total (lb)
New Zealand	95	36	95	226
Uruguay	167	19	35	221
Australia	108	23	88	219
Argentina	149	17	11	177
United States	106	65	4	175
Canada	89	55	3	147
France	70	61	6	142
USSR	30	20	8	60
Mexico	24	12	3	39
Japan	5	7	2	15

[a] U.S.D.A., *Livestock and Meat Situation*, 1966.

BEEF CATTLE MANURE IS A VALUABLE BY-PRODUCT

One of the important advantages enjoyed by the farmer who markets his crops through animals is the conservation of the fertility of his soil. Animals retain only a small part of the plant food elements contained in the feeds consumed, returning the greater part to the soil in the manure produced. (In beef cattle, for instance, the retention values for nitrogen, phosphorus, potassium, and organic matter are 25, 15, 10, and 70 percent respectively.) As a result the livestock farmer is able to maintain his land in a high state of fertility with the use of less commercial fertilizer than is needed by the farmer who sells his grain outright.

An estimated 1.5 billion tons of manure are produced annually by the livestock on the farms and ranches of the United States. The value of this manure in terms of the increased crop yield that would result if it were completely recovered and carefully used is enormous. The amount of manure produced by feeder cattle of various ages is shown in Table 10.

Table 10
Manure Obtained from Cattle Fed on Paved Floor
in Open Shed and Adjoining Paved Lot[a]

			Average Manure per Head	
	Number of Lots Averaged	Average Days Fed	Total Period (tons)	Per Month (tons)
Calves				
Full-fed with silage	29	229	1.82	0.23
Full-fed with dry roughage	10	235	2.18	0.30
Full-fed with ear corn silage	8	231	2.16	0.28
Wintered without grain	5	132	1.59	0.36
Yearlings				
Fed over 140 days	4	150	2.24	0.45
Fed under 101 days	11	91	1.52	0.51
Dry beef cows	4	134	2.91	0.66

[a] Illinois Agricultural Experiment Station, Mimeographed Reports.

DISADVANTAGES OF THE
BEEF CATTLE ENTERPRISE

The disadvantages of the beef enterprise should also be listed and discussed. Fortunately the list is short but knowledge concerning these points is important in the choice of a farm or ranch enterprise.

A SPECULATIVE RISK IS INVOLVED IN STOCKER AND FEEDER PROGRAMS

In periods of rapidly declining prices there is a real possibility that purchased calves will sell a year later, as feeder yearlings or as finished steers, for considerably less per hundredweight than the original cost. This of course applies only directly to the original weight bought, and this situation occurs only occasionally, but the possibility must be considered. In some of the finishing programs, especially those involving either heavy or low-grade feeders, the selling price may also be below the actual cost of the gains—that is, after conversion to cattle gains the feeds fed to such cattle may sell for less than their market value.

ANNUAL OUTLAY OF CAPITAL IS HIGH IN STOCKER AND FINISHING PROGRAMS

Most cattle feeders must buy their stockers or feeders and, since such cattle often remain on the farm or in the feedlot for from 6 months to a year, it is necessary annually to invest as much as one-third to two-thirds of the final value of the finished cattle, aside from the investment in feed. Commercial feedlots may feed two or three droves per year, hence the interest charges on cattle on feed go on throughout the year. This requires a rather large supply of either available cash or credit. In any case, interest charges on this large investment must be reckoned with.

BEEF CATTLE PROGRAMS REQUIRE LABOR AND MANAGEMENT OF ABOVE-AVERAGE QUALITY

In order to reduce costs and losses from disease and death, and in order to utilize the latest research findings in the feeding and management of beef cattle, skilled labor and scientific knowledge are required. This is possibly even more true if the buying and selling of cattle are not delegated to a commission man or order buyer especially trained for the job. Knowing when fed cattle are ready to sell to best advantage will prevent the costly gains that result when cattle are held beyond the point where they are finished for their grade. Equally important is the ability to choose the proper feeding program to utilize the available feed supply, and to adapt the program to take advantage of seasonal trends in supply of feed and both feeders and fed cattle. The importance of such decisions means that exceptional managerial ability is essential to success. The alternative is to leave these managerial decisions to professional help, which is usually available on some sort of fee basis.

BEEF CATTLE PRODUCTION IS A YEAR-ROUND OPERATION

Except for a few short-term feeding programs, the production of beef cattle is a confining enterprise. For best results daily inspection of cattle is desirable, and, of course, in practically all of the finishing programs daily chore work is necessary. True, the total hours required may be low, especially if laborsaving feeding systems are used, but still someone must be on hand. In ranching areas, where pastures are large and sometimes hard to reach, daily inspection is unfeasible and yet the cattle cannot be long neglected. Lengthy vacations and frequent weekend absences are not conducive to success with beef cattle.

Part II

PRINCIPLES OF BREEDING, REPRODUCTION, AND FEEDING

Chapter 3

PRINCIPLES of BEEF CATTLE BREEDING

Keith E. Gregory[*]

The beef cattle industry in the United States is composed of several segments: (1) the purebred breeder or seedstock producer, (2) the commercial producer, (3) the feeder, (4) the packer, and (5) the retailer.

The purebred breeder of beef cattle maintains seedstock herds to provide bulls for the commercial producer. The commercial producer provides feeder stock to the feeder, who in turn provides the packer with finished beef cattle ready for slaughter. The packer slaughters the cattle and provides the retailer with either dressed carcasses or wholesale cuts from these carcasses. The retailer breaks down the dressed carcasses or wholesale cuts into retail cuts, trimmed and packaged suitably for his customers, the consumers.

There is an interdependence among these segments because each affects cost of production or desirability of product, or both. Both desirability and price of product are reflected in changes in consumption or use. Level of consumption is important to all segments. Consumption of beef depends primarily on how much it costs the consumers relative to other food items and on how well they like it. The profits to be realized by all segments of the beef cattle industry depend on continued improvement in both productive efficiency and carcass desirability.

Only traits that contribute to productive efficiency and carcass desirability are of economic importance to the beef cattle industry. These traits, frequently referred to as performance traits, are (1) reproductive perfor-

---

[*] Director, U.S. Meat Animal Research Center, Clay Center, Nebraska.

mance or fertility, (2) mothering or nursing ability, (3) rate of gain, (4) economy of gain, (5) longevity, and (6) carcass merit.

RESPONSIBILITY OF THE PUREBRED BREEDER OR SEEDSTOCK PRODUCER

The opportunity for genetically improving traits of economic value rests primarily with the purebred breeder or in seedstock herds. Most of the opportunity for selection in beef cattle is among bulls. Level of performance in commercial beef cattle populations is determined primarily by the bulls available to commercial herds from the purebred segment of the industry. In order to fulfill his responsibility to the other segments of the beef cattle industry, the purebred breeder or seedstock producer should have a working knowledge of genetics or the science of heredity, along with an appreciation of all traits of economic importance to the industry. In addition, he should understand the procedures for measuring or evaluating differences in these traits and be able to develop effective breeding practices for making genetic improvement in them.

THE BASIS FOR GENETIC IMPROVEMENT

Differences among animals result from the hereditary or genetic differences transmitted by their parents and the environmental differences in which they are developed. With minor exceptions each animal receives half its inheritance from its sire and half from its dam. The units of inheritance are known as genes and are carried on threadlike material, present in all cells of the body, called chromosomes. Cattle have 30 pairs of chromosomes. The chromosomes and genes are paired, each gene being at a particular place on a specific chromosome pair. There are thousands of pairs of genes in each animal, and one member of each pair comes from each parent. All cells in an animal's body have essentially the same makeup of chromosomes and genes.

The ovaries of females and the testicles of males produce the reproductive cells, which contain only one member of each chromosome pair, and it is purely a matter of chance which gene from each pair goes to each reproductive cell. In this halving process a sample half of each parent's inheritance goes to each reproductive cell, meaning that the genetic potentialities of an individual are determined at fertilization. The pairing of chromosomes restores the full complement when a reproductive cell from the male fertilizes a reproductive cell from the female, keeping the number of chromosomes constant over countless generations. Since the half of each of its parents' inheritance that each reproductive cell receives is strictly a matter of chance, some reproductive cells will contain more desirable genes for economically important traits than will others. This results in a superior individual and offers the opportunity for selection. The chance segregation in the production of reproductive cells and recombination upon fertilization is the cause of genetic differences among offspring of the same parents.

The genetic merit of a large number of offspring will average that of their parents. However, some individuals will be genetically superior to the average of their parents and an approximately equal number will be inferior. Those that are superior provide the opportunity for selection and genetic improvement. The basis for genetic improvement is differential reproduction, which is accomplished by permitting some animals to leave a greater number of offspring than others or some to leave offspring while others do not. This is what happens when selection is practiced.

Genes vary greatly in their effects. Some traits are controlled primarily by a single pair of genes, whereas other traits are affected by many genes. Examples of traits controlled primarily by a single pair of genes are dwarfism and color. Most of the economically important traits—carcass characteristics, growth rate, feed efficiency, and mothering ability—are affected by many genes. The thousands of genes present make countless combinations possible in any animal. Genes are too small to be individually identified, and their presence is evident only in outward effects such as differences in growth rate, feed efficiency, and conformation.

In traits controlled by a single pair of genes, one member of the pair must be dominant. The dominant gene has the capacity for covering or masking the effect of the other member of the pair, which is referred to as the recessive gene. For example, the gene for polled is dominant and masks the gene for horns when both are present. In another example, the gene for dwarfism is recessive to the gene for normal appearance. Thus, if N represents the gene for normal appearance and n represents the gene for dwarfism, individuals with the "genetic makeup" of NN and Nn are normal in appearance, but Nn individuals carry the gene for dwarfism and transmit this gene to approximately half their offspring. Dwarfs (nn individuals) can result from mating normal-appearing parents if each carries the gene for dwarfism ($Nn \times Nn$). Mating normal-appearing individuals that carry the dwarf gene ($Nn \times Nn$) results in noncarriers (NN), carriers (Nn), and dwarfs (nn) in a $1:2:1$ ratio.

Among animals, all differences that are not genetic are classified as environmental. Even though every attempt may be made to provide a uniform environment, there are still random environmental differences among animals. For example, identical twins are exactly alike in their genetic makeup but differ in their performance because of random or chance environmental differences. All animals are not at exactly the same place at the same time, grazing the same area, and exposed to the same environmental elements. Some members of a group may contact infectious organisms while others do not. Another example might be injury to the udder of a cow, which would reduce her milk production and result in decreased weaning weight of her calf. There are many random environmental factors that may affect some members of a group and not others and thus affect the expression of differences in economically important traits.

GENE FREQUENCY

The objective of selection for any performance trait is to increase in the population the number or frequency of desirable genes affecting that trait. This is accomplished by selecting animals that are above the herd average in genetic merit.

Differential reproduction is the basis for change in gene frequency and genetic improvement. Culling animals that are poor in economically important traits reduces the frequency of undesirable genes in a herd if the culled animals are replaced by animals that are superior in those traits and thus have a high percentage of desirable genes. Differential reproduction is the basis for continuous improvement in livestock, for the increase of desirable genes in one generation is added to those of the previous generation and the improvement tends to be permanent.

Gene frequency refers to the percentage of the available locations that a particular gene occupies in a herd or population. Since genes are paired in each animal, gene frequency includes both members of each pair and ranges from 0.00 to 1.00. For example, if a herd is free of dwarfism (NN), frequency of the dwarf gene in the herd is 0.00 and frequency of the gene for normal condition is 1.00—that is, it occupies every potential location. Conversely, in a herd of dwarfs (nn), frequency of the dwarf gene is 1.00 and frequency of the gene for normal condition is 0.00. In a herd where all animals carry the dwarf gene (Nn), the frequency is 0.5 for both genes; thus, combined frequencies of both members of a gene pair is 1.00.

KINDS OF GENETIC VARIATION

Genetic variation is caused by either additive or nonadditive gene effects. Many genes are involved in the expression of each performance trait, and when these genes produce their effects in a manner comparable to adding block upon block, as in construction of a building, their effects are said to be additive. Selection aims to increase the frequency of desirable genes that produce additive effects. The part of the total variation, genetic and environmental, due to additive effects is called heritability.

In nonadditive genetic variation, specific combinations of genes produce special effects as a result of being present together. When specific combinations of genes produce a favorable effect, the genetic variation is referred to as hybrid vigor or heterosis.

Traits vary in the degree to which they are controlled by these two kinds of genetic variation. For traits where most of the genetic variation is additive and where it is large compared with the environmental variation, selection based on differences in individual performance will be effective. For traits where most of the genetic variation is nonadditive, selection based on differences in individual performance will be relatively ineffective. For the latter type of trait, the breeding program must be designed to make use of specific crosses that produce favorable gene combinations. This involves crossing lines or breeds to obtain favorable combinations of genes for the expression of these traits.

A knowledge of the relative amounts of additive and nonadditive genetic variation that affect each economically important trait is fundamental to the development of an effective breeding program.

FACTORS AFFECTING RATE OF IMPROVEMENT FROM SELECTION

The factors affecting rate of improvement from selection are (1) heritability, (2) selection differential, (3) genetic association among the traits, and (4) generation interval.

HERITABILITY

Heritability is the proportion of the differences between animals, measured or observed, that are transmitted to the offspring. Thus it is the proportion of the total variation that is due to additive gene effects. The higher the heritability for any trait, the greater the rate of genetic improvement or the more effective selection will be for that trait. For traits of equal economic value, those with high heritability should receive more attention in selection than those with low heritability. Every attempt should be made to provide all animals from which selections are made with as

nearly the same environment as possible, which will result in a larger proportion of the observed differences among individuals being genetic and
will increase the effectiveness of selection. It is important to adjust for
known environmental differences before making selections if the environmental factors can be evaluated. Adjustments can be made for differences
in age, age of dam, season of calving, and sex.

The average heritability estimates for some of the economically important traits of beef cattle are presented in Table 11. Of the total difference
between the selected individuals and the average of the population from

Table 11
Heritability Estimates of Some
Economically Important Traits

Trait	Heritability (%)
Calving interval (fertility)	10
Birth weight	40
Weaning weight	30
Cow maternal ability	40
Feedlot gain	45
Pasture gain	30
Efficiency of gain	40
Final feedlot weight	60
Conformation score	
Weaning	25
Slaughter	40
Carcass traits	
Carcass grade	40
Rib-eye area	70
Tenderness	60
Fat thickness	45
Retail product (%)	30
Retail product (lb)	65
Cancer-eye susceptibility	30

which they were selected, the percentage indicated in the table for each
trait is actually transmitted to the offspring. For example, if the selected
bulls and heifers were 30 pounds above herd average in weaning weight
(selection differential), their progeny would be expected to average 9
pounds heavier than if no selection had been practiced for this trait
($30\% \times 30 = 9$).

These heritability estimates were obtained under carefully controlled
environmental conditions from a large number of research herds, and ad-

justments were made for known major environmental sources of variation. The heritability of any trait can be expected to vary slightly in different herds, depending on the genetic variability present and the uniformity of environment. However, estimates from different research herds have been reasonably consistent. The heritability estimates in Table 11 probably represent average expectations for many herds, provided that the general environment is similar for all cattle within the herd. These estimates indicate that selection should be reasonably effective for most performance traits. But since these traits vary both in heritability and economic importance the rate of improvement in them and the emphasis that they should receive will also vary considerably.

SELECTION DIFFERENTIAL

Selection differential is the difference between the selected individuals and the average of all animals from which they were selected. Selection differential is determined by the proportion of progeny needed for replacements, the number of traits considered in selection, and the differences that exist among the animals in a herd. If the average weaning weight of a herd is 450 pounds, and the individuals retained for breeding average 480 pounds, the selection differential is 30 pounds.

In beef cattle there are some rather severe limitations on selection differentials possible for the various traits. The relatively low reproductive rate of beef cattle usually necessitates keeping approximately 40 percent of the females for replacements to maintain the herd, and an even higher percentage to expand it. Most of the opportunity for selection is among the bulls because a smaller percentage of the bulls must be saved for replacement. Increasing the number of traits selected for reduces the opportunity for selection for any one trait, making it important to select only for those traits of economic value that are heritable. Every effort should be made to get the maximum selection differentials possible for the traits of greatest economic importance and of highest heritability, ignoring traits that have little or no bearing on either efficiency of production or desirability of product.

GENETIC ASSOCIATION AMONG TRAITS

A genetic correlation among traits results when genes favorable for the expression of one trait tend to be either favorable or unfavorable for the expression of another trait. Genetic correlations may be either positive or negative. If the association is favorable, the rate of improvement in total merit is increased; conversely, if a genetic antagonism exists among traits, the rate of improvement from selection is reduced.

Available information indicates a favorable association between rate and efficiency of gain and between growth rate in different periods. The only major unfavorable genetic association that has been reported among traits of economic value in beef cattle is a positive genetic correlation between outside fat thickness and marbling score. This means that when marbling, an important determinant of carcass grade, is selected for, excessive outside fat may also result.

GENERATION INTERVAL

The fourth major factor that influences rate of improvement from selection is the generation interval—that is, the average age of all parents when their progeny are born. Generation interval averages approximately 4.5 to 6 years in most beef cattle herds.

The progress made per generation in any trait is equal to the superiority of the selected individuals above the population average from which they came (selection differential), multiplied by the heritability of the trait. This can be put on a yearly basis by dividing by the average length of generation. For example:

$$\text{annual progress for a trait} = \frac{\text{heritability} \times \text{selection differential}}{\text{generation interval}}$$

If heritability of yearling weight is 50 percent, if the selected individuals (males and females) are 50 pounds heavier than the average of all animals, and if the generation interval is 5 years, then the rate of improvement per year in yearling weight would be $(0.50 \times 50)/5$, or 5 pounds. It is evident that progress can be greater when the generation interval is shortened, which can be accomplished by vigorous culling of cows on the basis of production, but it also means that the herd would contain more young or nonproductive-aged cattle.

METHODS OF SELECTION

Selection may be based on (1) pedigree information, (2) individual performance information (mass selection), (3) progeny test or family performance information, or (4) a combination of all three.

Pedigree information is most useful in selecting among young animals before their own or their progeny's performance is known. Pedigree information may also be used in selecting for characters that are measured late in life, such as longevity and resistance to cancer eye, or for traits expressed only in one sex, such as mothering or nursing ability (selecting bulls out of cows that have produced calves with a high average weaning weight). Pedigree information should be given less attention after information is available on an individual's own performance or that of its progeny.

Selection on an individual's own performance (mass selection) will result in most rapid improvement when the trait is one with a high heritability—for example, growth rate. The advantage of selecting on individual performance is that it permits a rapid turnover of generations, shortening the generation interval.

Use of progeny test information results in the most accurate selection if the progeny test is adequate. Progeny tests are most needed in selecting for carcass traits (if good indicators are not available in the live animal), for sex-limited traits such as mothering ability (where individual performance information is not available on bulls), and for traits with low heritabilities. Progeny test information is more accurate than pedigree information and individual performance, provided the progeny test is extensive. Disadvantages are the less intense culling possible because of the small proportion of animals that can be adequately progeny-tested, the longer generation interval required to obtain progeny test information, and the decreased accuracy as compared to individual performance if not enough progeny are tested or if they are improperly evaluated.

All three types of information—pedigree, individual performance, and progeny test—should be used in selecting beef cattle. In using pedigree information, only the closest relatives should receive much consideration, because the more distant relatives can influence the individual's heredity only through the close relatives, the sire and dam. Information on the poor-performing ancestors should be considered along with the good-performing ones because, on the average, the influence of all grandparents on the heredity of the individual is the same, and both parents equally influence the heredity of an individual.

A good policy is to make initial selections on the basis of pedigree and individual performance information and let the extent to which a bull is used in a herd in later years be determined on the basis of progeny test information.

TYPES OF SELECTION

The three types of selection are (1) tandem selection, (2) selection based on independent culling levels, and (3) selection based on an index of net merit.

Tandem selection is selection for one trait at a time. When the desired level of performance is reached in a given trait, a second trait is given primary emphasis, and so on. This is the least effective of the three types the second most effective type of selection but it has one disadvantage: is that by selecting for only one trait at a time, some animals that are at the same time poor in other traits will be retained.

Independent culling levels require that an animal reach specific levels of performance in each trait before it is kept for replacement. This is the second most effective types of selection but it has one disadvantage: in requiring specific levels of performance in all traits, it does not allow for slightly substandard performance in one trait to be offset by superior performance in another.

Selection based on an index of net merit gives weight to the various traits in proportion to their relative economic importance and their heritability, and recognizes the genetic association, if any, among the various traits. The use of the index or some modification of it is the preferred type for most herds. It allows slightly substandard performance in one trait to be offset by outstanding performance in another. Also, by giving additional weight to traits of higher heritability or greater economic importance, there can be greater improvement in net merit.

Differences in heritability of traits should be considered in selection because, obviously, if a trait has extremely low heritability, little genetic improvement in it can be expected, and emphasizing it will reduce the emphasis that can be put on traits with higher heritability that will give a greater response to selection.

Although increasing the number of traits reduces the selection differential for any one trait, it results in more rapid improvement in total genetic merit or net worth. Average reduction in progress in each trait as a result of considering several traits is approximately $1/\sqrt{n}$, where n is the number of traits selected for. For example, if four genetically independent traits are involved in selection, the selection differential for each will be approximately half what it would have been if only one trait were involved ($1/\sqrt{4} = \frac{1}{2}$). This is based on the assumption that there are no genetic associations, favorable or unfavorable, among the four traits. It is obvious that considering all heritable, economically important traits simultaneously will result in faster improvement in genetic merit involving all traits.

Relative rates of improvement in some traits of economic value with different selection intensities or different percentages saved for breeding are considered in Table 12. These estimates are based on phenotypic evaluation (visibly obvious through performance) for the traits indicated, and assume that the percentage saved and used produce progeny that have an opportunity to be selected for the next generation—that is, the selected

Table 12

Estimates of Potential Progress in Ten Years when Different Intensities of Mass Selection Are Practiced for Specific Traits[a]

	Percentage of Bulls Saved					Assumptions
	1	10	20	50	70	
Weaning weight and no other traits (lb)	41.6	30.6	26.4	19.2	15.6	50% of heifers saved
Weaning weight and 1 other trait (lb)	29.4	21.6	18.7	13.6	11.0	$h^2 = 0.3$ in both sexes[b]
Weaning weight and 2 other traits (lb)	24.0	17.7	15.2	11.1	9.0	SD = 40 lb in both sexes[c]
Weaning weight and 3 other traits (lb)	20.8	15.3	13.2	9.6	7.8	
Postweaning daily gains and no other traits (lb)	0.44	0.30	0.25	0.17	0.12	50% of heifers saved
Postweaning daily gains and 1 other trait (lb)	0.31	0.21	0.18	0.12	0.08	$h^2 = 0.5$ in bulls
Postweaning daily gains and 2 other traits (lb)	0.25	0.17	0.14	0.10	0.07	$h^2 = 0.3$ in heifers
Postweaning daily gains and 3 other traits (lb)	0.22	0.15	0.12	0.08	0.06	SD = 0.29 lb in bulls
						SD = 0.20 lb in heifers
Yearling weight and no other traits (lb)	147.4	103.2	86.4	57.6	43.2	50% of heifers saved
Yearling weight and 1 other trait (lb)	104.2	73.0	61.1	40.7	30.5	$h^2 = 0.6$ in bulls
Yearling weight and 2 other traits (lb)	85.0	59.5	49.8	33.2	24.9	$h^2 = 0.4$ in heifers
Yearling weight and 3 other traits (lb)	73.7	51.6	43.2	28.8	21.6	SD = 80 lb in bulls
						SD = 60 lb in heifers
Yearling conformation score and no other traits (units)	1.39	1.02	0.88	0.64	0.52	50% of heifers saved
Yearling conformation score and 1 other trait (units)	0.98	0.72	0.62	0.45	0.37	$h^2 = 0.4$ in both sexes
Yearling conformation score and 2 other traits (units)	0.80	0.59	0.51	0.37	0.30	SD = 1 unit in both sexes
Yearling conformation score and 3 other traits (units)	0.70	0.51	0.44	0.32	0.26	

[a] Assumes that selection will be only for the criteria indicated, that when selection is for more than one trait, each trait is given equal emphasis, and that the traits are inherited independently. Generation interval is 5 years.

[b] h^2 = heritability.

[c] SD = standard deviation; it is an estimate of variation.

bulls from each generation are sired by bulls selected by the same criteria in the previous generation. Table 12 shows the advantages of saving bulls from among the top and selecting only for traits that have real economic value. Obviously, a closed-herd system must be used for the above conditions to prevail.

MATING SYSTEMS

There are five fundamental types of mating systems: (1) random mating, (2) inbreeding, (3) outbreeding, (4) assortative mating, and (5) disassortative mating.

Random mating is the mating of individuals without regard to similarity of pedigree or similarity of performance.

Inbreeding is the mating of individuals that are more closely related than the average of the breed or population. Linebreeding is a form of inbreeding, and refers to the mating of individuals so that the relationship to a particular individual is either maintained or increased. This method automatically results in some inbreeding because related individuals must be mated to accomplish it.

Outbreeding is the mating of individuals that are less closely related than the average of the breed or population. The term outcrossing is also used to mean outbreeding when matings are made within a breed. Crossbreeding is a form of outbreeding.

Assortative mating is the mating of individuals that are more alike in performance traits than the average of the herd or group.

Disassortative mating is the mating of individuals that are less alike in performance traits than the average of the herd or group.

Inbreeding and outbreeding refer to similarity of pedigree of relationship, and assortative and disassortative mating refer to phenotypic resemblance.

Inbreeding adversely affects most performance traits or results in some reduction in general vigor. However, herds of reasonable size, where several sires are used, can be maintained closed to outside breeding for relatively long periods without any appreciable increase in inbreeding or decline in performance associated with inbreeding.

Within a closed herd where the mating is random as far as relationship is concerned, the rate of increase in inbreeding per generation is $\frac{1}{8}m$ + $\frac{1}{8}f$, where m is the total number of males used in each generation and f is the total number of females in the herd in each generation. Thus in a 100-cow herd where 4 sires are used per generation with 100 cows in the herd per generation, the increase in inbreeding per generation is $\frac{1}{8}(4)$ +

$\frac{1}{8}(100) = \frac{1}{32} + \frac{1}{800} = 0.031 + 0.0012 = 0.0322$, or 3.22 percent per generation. If generation interval is 5 years, 15 years on such a program would result in a herd with average inbreeding of 9.66 percent. This is not a rapid rate of inbreeding compared, for example, with the mating of half brothers and sisters which results in offspring that are 12.5 percent inbred. Offspring of sire-daughter, son-dam, and full brother and sister matings are 25 percent inbred.

Sire numbers per generation are of paramount importance in affecting rate of inbreeding. The rate of inbreeding can be reduced by deliberately avoiding close matings such as sire-daughter and half brother and sister. Whereas linebreeding will result in some loss of vigor, if the animal to which a herd is being linebred is truly outstanding the increase in performance as a result of intensifying the genes of an outstanding individual may more than offset any decline in performance due to inbreeding. Rigid selection accompanying linebreeding should also reduce some of the undesirable effects of inbreeding. When inbred or linebred herds are outcrossed, the loss of vigor that accompanies inbreeding is more than restored.

Linebreeding and inbreeding make the individuals in a herd more alike genetically and thus more uniform in their transmitting ability. A major advantage of linebreeding and inbreeding is that a breeder knows his own herd better than he knows someone else's and is likely to do a more effective job of selecting from within his herd. The effectiveness of linebreeding depends primarily on the genetic merit of the animal to which the linebreeding is directed.

Many breeders fear the consequences of inbreeding because it intensifies what is already present in the herd, including both bad and good traits. If an undesirable trait is present, inbreeding will tend to bring it to light, but the genes responsible for the undesirable effect were already present. For example, if genes responsible for dwarfism are present in a population, inbreeding may increase the number of dwarf calves born, but it is not the cause of dwarfism. Inbreeding may be used to determine the presence of undesirable genes in a herd and, if accompanied by rigid selection, may effectively reduce their frequency.

The main disadvantage of linebreeding and inbreeding is that the foundation animals may not be truly superior. A genetic defect in the foundation animals can by chance rise to a high frequency and greatly interfere with the breeding program and materially reduce the value of the herd regardless of its genetic merit for major performance traits. Because it reduces genetic variation, inbreeding results in decreased heritabilities, and selection on individual performance is less effective. Since inbreeding makes individuals more alike in their genetic makeup, it increases the effectiveness of family selection.

Linebreeding and inbreeding should be practiced only in herds of outstanding genetic merit. The herds should be large enough so that the rate of inbreeding will be slow enough to provide opportunity for selection before genetic variation is reduced to the point where selection is not effective. All commercial producers and purebred breeders with small herds or herds of only average genetic merit should avoid linebreeding and inbreeding.

Outbreeding or outcrossing is recommended for all commercial producers and for secondary seedstock herds. Close matings should be avoided, but owners of secondary seedstock herds may profitably secure bulls from linebred herds. If sources of linebred bulls are changed periodically for use in secondary herds, the system is still outbreeding. If it becomes necessary to outcross linebred herds to correct a deficiency, breeders may find it advantageous to outcross to other linebred herds that are particularly outstanding in the trait that needs improvement. After such an outcross it may be desirable to resume a program of linebreeding.

Many breeders practice both assortative and disassortative mating. Assortative mating is practiced when superior cows are mated to superior bulls or when the poorer cows are mated to the unproved or less highly regarded sires. Disassortative mating is practiced when a breeder attempts to make "corrective matings"—that is, by mating cows that are mediocre or poor in one trait to bulls considered superior or outstanding in that trait. Assortative mating results in increased genetic variation in a herd, while disassortative mating tends to reduce the genetic variation in a herd.

CROSSBREEDING FOR HETEROSIS
IN COMMERCIAL PRODUCTION

Heterosis or hybrid vigor is the result of nonadditive gene effects and is the difference in performance between breed crosses and the average of the parental breeds or groups used in the cross. Heterosis results from favorable combinations of genes or groups of genes brought about by specific crosses. Utilization of heterosis necessitates crossbreeding, a form of outbreeding.

Commercial utilization of heterosis depends on crossing breeds or groups that result in generally favorable genetic combinations. Heterosis is used extensively in commercial swine and poultry production, and evidence is accumulating to indicate that heterosis is of appreciable economic importance in beef cattle. Although the results vary from different experiments and from crosses of different breeds, they generally show a heterotic effect on early postnatal mortality, preweaning and postweaning growth

FIG. 16. These crossbred calves produced in Wyoming demonstrate the increased growth rate and muscling that usually result from crossing Charolais bulls and Angus cows. The heifer mates to these steers should make excellent mother cows. (Western Livestock Journal.)

rate, age at puberty, and in fertility and mothering ability of crossbred cows. The heterotic effect on feed efficiency has been small.

The level of heterosis is inversely proportional to heritability or additive genetic variation. Thus in traits of highest heritability such as postweaning growth rate and feed efficiency the level of heterosis is relatively low, whereas in traits with low heritability such as livability and fertility the heterotic effects are great.

Research and actual practice indicate that many of the advantages to be gained from a systematic crossing program result only if the crossbred of F_1 females are kept and used as mother cows. This is because of the relatively high levels of heterosis on fertility and mothering ability and the economic importance of these traits.

Preliminary results indicate that a systematic crossbreeding program may result in a 15 to 20 percent increase in pounds of calf weaned per cow bred over a program of straightbreeding. This involves the cumulative heterotic effects on fertility, mothering ability, and preweaning growth rate.

The following provides a summary of the results from an extensive crossbreeding experiment involving the Hereford, Angus, and Shorthorn breeds, conducted at the Fort Robinson Beef Cattle Research Station, a U.S. Department of Agriculture station in northwestern Nebraska. These results are typical of those obtained from other experiments and are presented here because this represents the largest and most complete experiment of the kind ever conducted.

In the first phase of this experiment the three straightbreds and all reciprocal crosses among them were produced. These studies included a total of 751 calves from four calf crops sired by 16 Hereford, 17 Angus, and 16 Shorthorn bulls. Heterosis or hybrid vigor was evaluated by comparing the crossbreds with the average of the straightbreds. Crossbreds and straightbreds were sired by the same bulls and were out of comparable cows. These studies involved an evaluation of the effects of hybrid vigor on embryo survival, postnatal mortality, birth weight, preweaning growth, weaning weight, weaning conformation score, postweaning growth rate and yearling weight of heifers developed under two management programs, age and weight at first heat of heifers developed under two management programs, postweaning growth rate and yearling weight of steers on a growing-finishing ration, postweaning feed efficiency of steers on a growing-finishing ration, slaughter grade of steers, and detailed information on carcass characteristics of steers involving complete cutout data on one side of each carcass.

The effects of hybrid vigor were significant for most of the economic traits evaluated. A 3 percent greater calf crop was weaned in the crossbred than in the straightbred calves because of differences in early postnatal mortality. The heterotic effect on 200-day weight was 24 pounds in heifers and 16 pounds in steers. The heterotic effect on postweaning growth rate of heifers on a low level of feeding was greater than in steers on a growing-finishing ration. The magnitude of the heterotic effect on growth rate was related to level of feeding and age. That is, heterosis or hybrid vigor tended to decrease with increasing age after approximately 1 year and was greatest on a restricted feed intake when comparing heifers with steers. The heterotic effect was 50 pounds on 550-day weight of heifers and 29 pounds on 452-day weight of steers. The heterotic effect on carcass weight at 452 days was 23 pounds for steers. Heterotic effects on age at first heat of heifers were 41 and 35 days for low and moderate levels of feeding, respectively. After adjusting age at puberty for the effects of average preweaning and postweaning daily gains, approximately one-half to three-fourths of the heterotic effect on age at puberty remained. Thus there was a heterotic effect on age at puberty independent of its effects through increased average daily gains.

The advantage of the crossbred steers in feed efficiency was small.

The crossbred steers produced carcasses with slightly more finish when killed at the same age. However, when adjustments were made for the effects of weight, there was no difference in carcass composition. Thus had the steers been slaughtered at the same weight, the composition of the carcasses would have been the same.

In net merit (value of the boneless, closely trimmed retail meat, adjusted for quality grade, minus feed costs from weaning to slaughter) the advantage of the crossbred steers over the straightbred steers was $8.81 per carcass. This net merit difference is among the steers that lived to slaughter. The 3 percent advantage for the crossbreds in calf crop weaned was not involved in computing this difference.

For growth, feed efficiency, and carcass traits the heterotic effect was greater in the Hereford-Angus and Hereford-Shorthorn combinations than in the Angus-Shorthorn combination, while for age and weight at puberty the heterotic effect was greatest for the Hereford $\times$ Shorthorn and reciprocal cross. In evaluating all traits for the effects of heterosis, it can be concluded that heterosis results in an increased rate of maturity.

The second phase of this experiment is in progress and involves the evaluation of the effects of hybrid vigor on fertility and mothering ability. That is, straightbred cows of the three breeds are being compared with their crossbred half sisters when both are bred to the same purebred bulls. For the five years 1963 through 1967 on which data have been collected, the advantage of the crossbred cows has been 17, 6, 10, —3, and 11 percent, respectively, for calf crop weaned and 17, 31, 20, 22, and 27 pounds, respectively, in average weaning weight of calves at 200 days. The results of heterotic effects on cow performance traits (fertility and mothering ability) should be regarded as preliminary because data are still being collected from this phase of the experiment.

Several questions are to be answered as to the most effective procedures for the utilization of heterosis in commercial beef production, involving which breeds to use and the number of breeds that may be used in a rotation of sires program for the maintenance of heterosis.

The effective utilization of heterosis in practice depends on using as breeding stock animals that are superior in their own performance in the traits that have high heritability. Not only is it essential to use superior purebred bulls, but it is necessary to find specific bulls that combine most favorably with the female herd.

Some problems are inherent in the utilization of heterosis. One is the overlap of generations among females in the herd. The percentage of different lines or breeds represented in the females will vary, because only approximately 15 or 20 percent of the female herd is replaced each year and several different age and breeding groups are present in a herd at a given time. More than one source of bulls must be in service at a time if bulls are used that combine most favorably with the specific females from the different breeding groups, necessitating more than one breeding pasture.

While perhaps none of these problems is insurmountable, relatively large herds that can use several breeding pastures are indicated, so that females of different crosses can be separated and bred to appropriate sires. However, if artificial insemination is used, this is not a major problem.

GENETIC ENVIRONMENTAL INTERACTIONS

Genetic environmental interactions refer to the extent to which the same genes contribute to superior performance in different environments. For example, assume two groups or lines of cattle, A and B. If line A is superior in environment 1 but line B is superior in environment 2, a genetic environmental interaction exists. The extent to which cattle superior in one environment maintain that superiority over a wide range of environments is not known. There has been little research to evaluate the importance of genetic environmental interactions on performance traits of beef cattle, and little is known about the range in adaptability of different kinds and types of beef cattle to variations in climate and environment. Research on the adaptability of different types of cattle to different systems of production is now under way.

Beef cattle provide a means of utilizing the feed resources over a wide range of environments and in various types of production programs, ranging from the lush improved pastures of the Corn Belt and Southeastern regions to the sparsely vegetated desert ranges of some of the western states. There is basis to question whether the genetic makeup capable of the maximum response in one environment is the same as the one capable of maximum response in another. For example, is the same cow capable of using equally well the feed resources in widely varying environments in terms of calf weaning weight? To obtain maximum use of available feed resources, the genetic makeup best adapted to each situation is the one toward which specific breeding programs should be directed, and this involves an understanding of genetic environmental interactions and selection for adaptability to specific climatic conditions and production programs.

Among the evidence that suggests the importance of adaptation in beef cattle is the superior performance in some traits of cattle possessing some Brahman breeding under the subtropical conditions of the Gulf Coast region. Yet even in this region, the more the environmental conditions are improved, the less the advantage of the Brahman breeding in these traits. In the more temperate regions, cattle with some Brahman breeding do not seem so well adapted as cattle with British ancestry. More specific questions involving the matter of adaptation relate to specific production programs and practices.

In the beef cattle industry the breeding stock are often moved to conditions greatly different from those in which they and their parents were selected. Such movement also characterizes feeder and replacement cattle for commercial production. The most effective use of the feed resources of this country necessitates movement of many feeder cattle. Perhaps little can be done in this situation, but the extensive movement of breeding stock, particularly herd bulls, in the purebred segment of the industry is undesirable if adaptation is of major importance. Until the importance of genetic environmental interactions is fully evaluated, breeders would do well to select breeding stock under environmental conditions comparable to those under which their progeny are expected to perform.

USE OF RECORDS

Record of performance is the systematic measurement of traits of economic value and the use of these records in selection, with the aim of finding the genetically superior individuals in all economically important traits so that they may be used for breeding. Records increase a breeder's knowledge of differences between animals and thus increase the accuracy of his selections.

The preferred measurements are those that give most accurately the breeding value or genetic merit of an animal relative to the others in a herd. Research on beef cattle breeding has demonstrated that appreciable genetic improvement can be made in most economically important traits by selection on the basis of differences in individual performance, as indicated by the estimates presented in Table 12. Such research has involved methods of measuring these traits and estimating their heritability, and developing selection procedures for traits that contribute to both productive efficiency and carcass merit. The systematic measurement of differences among animals in the economically valuable traits, the recording of these measurements, and using the records in selection will increase the rate of genetic improvement.

Performance records of animals should be adjusted to eliminate known environmental differences between animals, so that genetic differences will be a larger part of the total differences measured or observed. Adjustments should be made for differences in age, sex, age of dam, and any other "environmental" variable that can be measured or evaluated. Because any increase in environmental variation tends to obscure genetic differences

and decrease the effectiveness of selection, every precaution should be taken
to measure economically important traits as accurately as possible. For
example, an effort should be made to equalize fill in animals before they
are weighed, because errors in weighing decrease the accuracy of selection.
Fill can be equalized somewhat by removing water and feed for 12 hours
before weighing and by recording more than one weight. This applies to
both initial and final weights.

Record of performance is useful primarily to provide a basis for comparing cattle handled alike within a herd and not for comparing differences
between herds, because there are apt to be large environmental differences
between herds due to location, management, and nutrition. It is difficult
to adjust accurately for these differences, making the evaluation of genetic
differences extremely difficult, even though genetic differences between
herds do exist.

Average weaning weights of 500 pounds may be realistic in some environments and in some production programs, whereas 350-pound weaning
weights may be reasonable under more adverse conditions. Yet, beef cattle
may provide the most desirable means of utilizing the land under both
conditions. Furthermore, the genetic merit of a herd weaning 350-pound
calves may be equal or even superior to that of a herd weaning 500-pound
calves. Standards of performance expressed as deviations from individual

FIG. 17. Gain-testing yearling bulls on grass measures performance under environmental conditions similar to those in which their progeny will be expected to
perform well. (Western Livestock Journal.)

herd or group averages are advisable for making comparisons with a herd, but comparisons between herds based on minimum standards of performance can be undesirable and misleading.

Minimum standards of performance for the various production and carcass traits have been considered in some record of performance programs. Because of the variation in environmental conditions and production programs, standards involving between-herd comparisons may tend to recognize herds carried under superior environmental conditions rather than those that are genetically superior.

Comparing animals within a herd that are subject to different environmental conditions, such as having part of the calves on nurse cows or other variations in feeding and management, is as objectionable as comparing the records of different herds. If variations in treatments exist, comparisons should be restricted to animals treated alike unless appropriate adjustments can be made for treatment effects.

All economically important traits that are heritable should be evaluated for all animals in a herd. An effective record of performance program should be compatible with practical management regimes. Cattle should be evaluated under the approximate environmental conditions in which their progeny are expected to perform.

From the standpoint of genetic improvement for the entire beef cattle industry, record of performance will have greatest impact in purebred or seedstock herds. Commercial producers can use records of performance to cull cows, to select replacement heifers, and to evaluate bulls on their progeny's performance where progeny groups are kept under comparable conditions. Since approximately 40 percent of all heifers must be saved for replacements just to maintain a herd, opportunity for selection among females is limited.

Commercial producers can also make effective use of performance records by selecting bulls on the basis of records from purebred or seedstock herds that are on a systematic record of performance program. In selecting herd bulls from their own herds as well as from other breeders' herds, purebred breeders should evaluate prospects on the basis of their records as compared with the herd average. Over a period of time the inherent productivity of any herd depends largely on the genetic merit of the bulls used.

The goals in record of performance are not greatly different from those that have always been sought by progressive breeders. The principal differences lie in a systematic record-keeping program and the use of these records in making selections. Record of performance up to slaughter requires no new or additional facilities except a scale and forms for keeping records.

The principal features of a good record of performance program are:

1. All animals are given equal opportunity.

2. Systematic, written records are kept of all economic traits on all animals.

3. Records are adjusted for known sources of variation such as age of dam, age of calf, and sex.

4. Records are used in selecting replacement stock and in culling poor producers.

5. Nutritional program and management practices are practical and compatible with those where progeny of the herd are expected to perform and are uniform for the entire herd.

Space does not permit sufficient detail to provide guidance for an individual record of performance program. Extension county agents can supply guides and record blanks prepared by their extension livestock specialists for this purpose. The various purebred associations also provide guidebooks and record forms designed to meet their special needs. Methods differ slightly in different areas, and breeders are advised to adopt those generally in use in their areas.

Relative emphasis put on the different traits may vary in different herds, but the attention given each trait should be based primarily on its heritability and economic importance to the entire beef cattle industry. Keeping records does not change what an animal will transmit. Records must be used to locate and use the genetically superior individuals if genetic improvement is to be accomplished.

MAJOR PERFORMANCE TRAITS OF BEEF CATTLE

All traits of economic value should be considered in selecting beef cattle. The major traits influencing productive efficiency of highly desirable beef cattle are (1) reproductive performance or fertility, (2) mothering or nursing abilty, (3) growth rate, (4) efficiency of gain, (5) longevity, (6) carcass merit, and (7) conformation.

REPRODUCTIVE PERFORMANCE OR FERTILITY

A high level of reproductive performance or fertility is fundamental for making genetic improvement in beef cattle, because increased calf crops decrease the percentage that must be saved for replacement and thus increase the selection differential. Efficient cow-calf operations are fundamental to an efficient industry, and no single factor in commercial cow operations has a greater bearing on production costs than does calf crop. With cows composing a higher percentage of the total beef cattle population, fertility is an increasingly important trait from the standpoint of total industry efficiency. Both the male and the female should be considered in selecting for fertility, as reduced calf crops can be the result of sterility or partial sterility of either.

Reproductive performance or fertility is a complex trait. Many random or chance environmental factors affect fertility from the time a cow is turned with a bull until her calf is normally weaned, so that fertility in any given year reveals little of the real genetic differences among cows. Better measures of fertility are needed for both cows and bulls.

Because reproductive performance is important to efficient production, it commands attention in a breeding program, even though research results indicate that heritability is low and rate of improvement will be slow. There are reported instances where close culling for fertility has improved calf crops. In purebred herds, consideration should be given to culling open cows if they are below average in previous production, and all cows open in successive years regardless of production, assuming that no reproductive disease problems exist. Herd bulls should be selected from cows with good fertility records, and should be sired by bulls of high fertility and show high fertility themselves as measured by their ability to settle cows.

In herds where reduced calf crops are a problem, close attention to feeding, disease control, and management practices is definitely indicated. Reproductive diseases markedly influence fertility. Level of feeding—particularly level of energy, vitamin A, protein, and phosphorus—is important. Management of bulls, bull-cow ratio, size of pastures, and distribution of water may be related to whether cows conceive during the normal breeding season.

Recording birth weight is optional in a record of performance program. Knowing the birth weight provides a more accurate measure of gain from birth to weaning. Direct selection for heavier birth weights seems undesirable because of the increased likelihood of calving difficulty. Selection for traits that are of major economic importance should favor selection toward the optimum birth weight. Because of a high positive genetic correlation between birth weight and postnatal gain, information on birth weight by sire progeny groups may be useful in deciding which sires to use for their second breeding season, since often this is the only progeny information available at the time a decision must be made.

NURSING OR MOTHERING ABILITY

Weaning weight of calf is used as a measure of mothering ability. The calf's own genetic impulse for growth is confounded with mothering ability by this procedure, but this is not a serious handicap, as half of the growth impulse of the calf is transmitted by the dam. The ability to wean heavy, vigorous calves is necessary for efficient cow-calf operations. With the trend of marketing cattle at younger ages, weaning age represents a higher proportion of total age at market time and increases the relative importance of weaning weight. Increasing the pounds of calf produced per cow increases efficiency because certain fixed costs such as veterinary, labor, and bull service are on a per head basis. Feed costs for cows seem to be rather closely related to size of cow, but faster calf gains decrease feed requirements per unit of gain among cattle of the same breed.

It is emphasized that the objective is to increase weaning weight relative to mature cow size. Thus pounds of calf achieved for each unit of cow weight maintained may be a good measure of efficiency of operations from the standpoint of returns per unit of feed.

Selection of bulls and replacement heifers with heavy weaning weights relative to the herd average will lead to genetic improvement in mothering ability. Selection for increased weaning weight is practiced not only for mothering ability but for the calf's own ability to grow. Research indicates that selection among cows for mothering ability should be reasonably effective. This can be accomplished by selecting cows on the basis of their calves' weaning weights, since cows that wean calves heavier than the herd average in one year are more apt to produce calves heavier than average in succeeding years.

Differences in mothering ability can be evaluated about as accurately on the basis of 112-day calf weights as on the conventional weaning age of approximately 200 days. If calves are creep-fed, 112-day calf weights are perhaps preferable. Adjusting for differences in age of dam, sex of calf, and age of calf is necessary, as these factors influence weaning weight. In adjusting for differences in calf ages it is recommended that average daily gain from birth to weaning be used for each calf (subtract constant or actual birth weight, calculate average daily gain, and adjust to standard age for the group).

Mothering ability of cows may be compared within groups of the same sex of calf and within ages of cows if numbers are large. This avoids an adjustment for differences in sex of calf and age of dam. The most accurate adjustment factors for sex of calf and age of dam are those developed in the herd in which they are used, provided the data are not biased and the herd is large enough for reliable estimates to be made. Adjustment factors for smaller herds should be developed from herds with similar management regimes. Records are more accurate where the calving season is relatively restricted so that major differences in age and seasonal influences are avoided. Since weaning weight is used as a measure of mothering ability, it is important that all calves be treated the same for such things as creep feeding, so that the major variable is difference in nursing ability of the cows.

GROWTH RATE

Growth rate is important because of its high association with economy of gain and its relation to fixed costs, such as veterinary, buildings, grazing fees, and labor, that tend to be on a per head or per unit of time basis. In most instances differences in growth rate have been measured in time-constant, postweaning feeding tests, and results indicate that differences in growth rate can be appraised rather accurately in this manner. A postweaning period of at least 140 days is required to measure differences in growth rate. This minimum length is based on rather uniform initial weights, condition, age, and previous treatments. Final weight at 12 to 18 months (standardized for age differences) is probably a better measure of genetic differences in growth rate than any individual component of final weight (i.e., birth weight, preweaning gains, and postweaning gains).

Final weight at a standard age of 18 months seems to be a good measure of growth rate and it fits the management programs of many purebred herds. Bulls can be carried on a relatively low level of concentrate feeding (4 to 5 pounds of concentrates plus full feed of roughage) their first winter and fed at a higher level of concentrate either on grass or in drylot during their yearling summer. By this procedure bulls are developed at a high enough level of feeding and over a long enough period for genetic differences in growth rate to be expressed, and a good appraisal of growth can be made. Bulls handled in this manner are in good sale condition at a desirable age and season. Postweaning gains are measured for approximately 350 days and gains made in this period can be added to 200-day weaning weight, appropriately adjusted for age of dam, to arrive at an adjusted 550-day weight.

Final weight and grade at somewhere near normal market age for a high percentage of slaughter cattle seems to be of most interest on an industrywide basis. The use of postweaning gain alone as a measure of growth could foster poor milking ability because of compensatory gains, in that a poor feed supply in one period tends to be followed by a period of increased rate of gain.

An alternate program for measuring growth rate in bulls is to feed at a higher level and for a shorter period immediately after weaning. Bulls may be put on feed when they are weaned and full-fed for 5 to 6 months on a ration of from approximately equal parts of concentrates and roughage to two parts concentrates and one part roughage. In this program an adjusted final weight at 365 days can be used as a measure of differences in growth rate. For example, adjusted 365-day weight may be obtained by adding the gain made in a 165-day postweaning period to 200-day weaning weight, appropriately adjusted for age of dam. The postweaning feeding period may be intermediate to the two described above—for example, it may be 252 days with an adjusted final weight of 452 days computed and used as a basis for selection.

Research results indicate that a reasonably high level of feeding is desirable to appraise differences in growth rate most accurately. If a lower level of feeding is used, the period for measuring differences in growth rate should be longer. However, it is recommended that a relatively low level of feeding, promoting gains of 0.75 to 1 pound per day, be used for heifers during their first winter. Research results indicate that full-feeding a high-concentrate ration during the first winter may interfere with reproductive performance and mothering ability. Because a high percentage of heifers must be kept for replacements, there is little opportunity to select among heifers for differences in growth rate. Hence, from this standpoint, little can be gained from the heavy feeding of heifers.

In selecting heifer replacements for differences in growth rate it is suggested that long yearling age (approximately 18 months) be used, with adjustments in the same manner suggested for bulls (by adding the gain made after weaning to weaning weight, adjusted to a constant age, and appropriately adjusted for age of dam). This assumes that heifers are carried at a relatively low level of feeding during their first winter. If heifers are bred as yearlings, it may be desirable to make selections prior to 15 months of age. This can be done effectively with a 252-day postweaning period and adjusting final weights to 452 days.

The relation of growth rate to differences in composition of gain is of great importance. For example, a 600-pound carcass with 30 percent fat trim will yield approximately the same amount of edible meat as a 470-pound carcass with 10.5 percent fat trim. Such differences in fat trim have been observed in carcasses of the same quality grade. In considering differences in total gain it seems appropriate to be concerned with differences in composition of gain. Increased rate of gain is of little value if the additional gain is due to fat rather than muscle growth.

EFFICIENCY OF GAIN

Efficiency of gain is one of the traits of greatest economic importance in beef cattle. Efficiency of gain is difficult to estimate because it requires individual feeding and adjustments for differences in weight, as increased weight is associated with higher feed requirements per unit of gain.

Present information indicates that genetic improvement can be made in efficiency of gain by selecting for it through rate of gain, because the fast gainers will also be efficient gainers. It is therefore recommended that breeders depend on differences in rate of gain as an indicator of efficiency of gain rather than incur the added expense of individual feeding. However, if a breeder desires to feed individually and adjust the records for differences in weight in order to measure differences in efficiency of gain, this is more accurate.

LONGEVITY

The longer animals remain productive in a herd, the fewer replacements will be needed, and thus the costs of growing out replacements to productive age will be reduced. However, the longer an animal remains in a herd, the longer will be the generation interval, which may reduce the rate of genetic improvement from selection. Breeders of purebred cattle or seedstock herds should be concerned with making genetic improvement in longevity, so that commercial beef cattle populations will be productive at older ages. Yet a fairly rapid turnover of generations in purebred herds is desirable for making a maximum rate of genetic improvement in other traits of economic value.

With the trend toward marketing cattle at younger ages and somewhat lighter weights, a higher percentage of the beef cattle population must be cows in order to produce the same amount of beef. This higher proportion of cows tends to make longevity of greater economic importance from an industrywide standpoint. Longevity in bulls is important because it decreases the annual cost of bull service.

The major factors affecting longevity—or, more important, number of years spent in the breeding herd—are sterility, unsoundness of feet and legs, serious eye diseases such as cancer eye, udder troubles, and unsound mouth. Research shows that susceptibility to cancer eye is heritable, and selection against it should be reasonably effective; however, it is a trait that can be measured only late in life.

Selection for longevity must be confined primarily to indicators such as structural soundness and to pedigree information—that is, selection of close relatives of individuals that have had a long productive life. There is a certain amount of automatic selection for fertility and longevity, because animals that remain in a herd long enough to produce a large number of offspring tend to have a larger number saved for replacements.

CARCASS MERIT

Carcass merit is of fundamental importance to the beef cattle industry, because desirability of product together with price is the major factor affecting consumption. In selecting for improved carcass merit, the factors that contribute to carcass desirability and their relative importance must be known. In selecting among breeding animals the conformation items indicative of desirable carcass traits, and ways of measuring or evaluating differences in these traits in live cattle, must be available.

Research in many states indicates that the American public desires beef with a high percentage of lean as compared to fat and bone, but that the lean must be tender, flavorful, and juicy. The maximum muscle development is desired in the portions yielding the more preferred and higher-priced cuts—the back, loin, rump, and round.

In grading carcasses, meat quality is determined by marbling, texture, color, and firmness in relation to maturity. Research indicates that marbling—that is, finely dispersed fat within the muscle—contributes to juiciness and flavor, but its relation to tenderness is low. Among carcasses from beef breeds that are similar in other respects, marbling is the major factor that contributes to quality grade.

Large differences in marbling occur in cattle fed and managed in the same manner, and research indicates that part of this variation is due to genetic differences. There are no known reliable indicators in the live animal for predicting marbling. The length of time on feed and the energy content of the ration seem to be the best guides for predicting marbling in live steers.

Outside fat is related to marbling when extreme variations in outside fat and marbling are considered. However, the relation between outside fat and marbling at the top end of the carcass grades (prime, choice, and good) is low among cattle that have been fed similarly. The amount of outside fat on a carcass is not a factor in determining quality grade.

Thickness of muscling as measured by rib-eye area is one of the more highly heritable characteristics in beef cattle; therefore, selection for muscling should be effective.

Youthfulness is probably one of the best indicators of tenderness; therefore, selecting animals that will reach desirable market weights and grades at young ages should result in more tender carcasses. Additionally, research indicates that tenderness is a trait with a fairly high heritability; thus selecting for both rate of gain and for tenderness should accelerate improvement for this trait.

As mentioned in discussing growth rate, difference in composition of carcasses (relationship of fat to lean) is a major factor influencing difference in value. Considerable variation in fat trim has been observed in carcasses of the same quality grade. The choice carcass, on the average, has about 20 percent of fat trim upon breakdown to retail cuts left with approximately 0.375 to 0.5 inch of outside fat, with the range being from less than 10 percent to more than 30 percent. Thus a 470-pound carcass with 10.5 percent fat trim will yield as much edible meat as a 600-pound carcass with 30 percent fat trim. The value of the fat trim is negligible in today's market.

The two primary factors that determine differences in the real value of carcasses are the yield of boneless, retail-trimmed cuts and the quality grade of the carcass, and appraisal of differences in carcass merit should include these two factors. The Livestock Division, Consumer and Marketing Service, U.S. Department of Agriculture, has developed a system of grading beef carcasses that describes carcass differences in these two factors. In this system one grade describes differences in quality of the meat and a separate grade describes differences in estimated yield of boneless, retail-trimmed cuts from the round, loin, rib, and chuck. Estimated yield of such cuts as a percentage of carcass weight has been referred to as "cutability." These four wholesale cuts represent approximately 80 percent of the value of a carcass, and the relation between yield of boneless, retail-trimmed cuts from the round, loin, rib, and chuck and from the remainder of the carcass is high. Since differences in cutability and differences in quality grade are the primary factors that determine differences in carcass value, a carcass merit index combining these two variables is appropriate for ranking carcasses on the basis of value and for ranking sires on the basis of the carcass value of their offspring.

Studies conducted by the U.S. Department of Agriculture have shown that the yield from the round, loin, rib, and chuck can be predicted rather accurately by using fat thickness at the twelfth rib, rib-eye area at the twelfth rib, percentage of kidney and pelvic fat of the carcass, and carcass weight. Their prediction equation is as follows: estimated percentage of boneless, retail-trimmed cuts from round, loin, rib, and chuck (cutability) = 52.56 — 4.95 (thickness measurement of fat over rib eye, twelfth rib, in inches) — 1.06 (percentage of kidney fat) + 0.682 (area of rib eye, twelfth rib, in square inches) — 0.008 (carcass weight, in pounds). For example, the computations for a 600-pound carcass with a rib-eye area at twelfth rib of 10 square inches, 0.6-inch fat at twelfth rib, and

3.5 percent of kidney fat would be: cutability $= 52.56 - 4.95(0.6) -$
$1.06(3.5) + 0.682(10) - 0.008(600) = 52.56 - 2.97 - 3.71 + 6.82$
$- 4.80 = 47.90$ percent.

In evaluating animals for carcass merit it is desirable to combine cutability and quality grade of the carcass into a single index that will describe
differences in carcass value. This can be done by combining differences
in cutability and differences in quality grade of the carcass into an index
of carcass merit.

At recent price levels of prime, choice, and good grades, a 2 percent
change in cutability has approximately the same effect on value as a change
of one full U.S.D.A. grade in carcass quality. This can be expressed as
2 percent cutability equals 1 quality grade. An index describing this relation
would be:

$$I = \frac{\text{cutability}}{2} + \frac{\text{quality grade}}{1}$$

Since it is desirable to use carcass quality grade to the nearest one-third
grade, the correct relation between the values of cutability and grade can
be obtained by using a descending scale with one unit change equated
to each one-third of a grade and dividing the quality grade component
of the index by 3. Thus the index would be:

$$I = \frac{\text{cutability}}{2} + \frac{\text{quality grade}}{3}$$

To simplify the index: $I = 0.5$ cutability $+ 0.33$ quality grade. This index
can be further simplified by multiplying by a factor of 2: $I =$ cutability $+$
0.66 quality grade. For ease of computation it can be rounded to: $I =$ cutability $+ 0.7$ quality grade, with quality grade expressed to the nearest one-
third of a grade.

Carcass quality grade may be coded to a numerical scale for computing
a carcass merit index. Differences among carcasses in the index reflect
differences in their real value, if both differences in cutability and carcass
grade are weighted according to their relative economic values.

The following numerical values for carcass quality grades are suggested
for use in computing indexes:

Carcass Grade	Numerical Value	Carcass Grade	Numerical Value
High prime	52	High good	46
Average prime	51	Average good	45
Low prime	50	Low good	44
High choice	49	High standard	43
Average choice	48	Average standard	42
Low choice	47	Low standard	41

Table 13

Application of an Index to Evaluate and Rank Carcass Merit

		Carcass Quality Grade				
Carcass Number	Cutability (%)	U.S.D.A. Grade	Number Code	Grade × 0.7	Index	Rank in Value
1	50	Average choice	48	33.6	83.6	5
2	47	Low prime	50	35.0	82.0	6
3	51	Average prime	51	35.7	86.7	1
4	47	High good	46	32.2	79.2	8
5	53	Low choice	47	32.9	85.9	2
6	53	Average good	45	31.5	84.5	4
7	46	High good	46	32.2	78.2	10
8	47	Average choice	48	33.6	80.6	7
9	44	High choice	49	34.3	78.3	9
10	54	Low good	44	30.8	84.8	3

Using the index, I = cutability $+ 0.7$ quality grade, the example in Table 13, involving 10 carcasses, shows how such an index works in actual practice. Any descending code scale for quality grade may be used, provided one unit change is equated to one-third of a grade. Thus 17, 16, and 15 may be used for high, average, and low prime, respectively, with a comparable descending scale for the lower grades.

If ages of the cattle are known, growth rate can be included in the index by adjusting carcass weight to an age-constant basis. Liveweight adjusted for differences in age and multiplied by dressing percentage provides an estimate of carcass weight on an age-constant basis. Carcass weight (adjusted for age) multiplied by the carcass merit index yields an index of value of a carcass or an estimate of its worth including growth rate, cutability, and quality of meat.

Obtaining an index of value on the progeny of several different sires should provide a logical basis for ranking them on their most probable genetic worth for several economically important traits. The information is relatively easy to obtain.

While the suggested index may be improved with additional information, the principle is basic—that is, differences in value are determined by differences in amount of salable meat and differences in quality of meat.

In considering cattle from the same herd or breeding group, as in a breeding program, that are fed and managed alike, weight at a constant age is by far the most important factor affecting pounds of edible meat at a constant age. In fact, research shows that differences in weight at a constant age account for approximately 90 percent of the variation in pounds of edible meat at a constant age. These results may be interpreted to mean that in selecting cattle from the same herd, variations in growth rate should be given much more attention than variations in finish if pounds of edible meat at a constant age are a primary objective.

CONFORMATION AND ITS EVALUATION

Performance traits other than carcass merit and structural soundness should be measured directly or through the indicators that have been discussed rather than through conformation. Conformation is a performance trait to the extent that it contributes to carcass merit and longevity. Basically, the important conformation items are structural soundness, which may contribute to longevity, and beefiness (thickness or natural fleshing or muscling), particularly in the regions of cuts that contribute most to carcass value—back, loin, rump, and round.

Research is in progress to develop new tools to measure differences in fat and muscling in live beef cattle. It is recommended that breeders use the best current procedures for evaluating differences in the major items of conformation. The word "major" is emphasized and refers only to items of conformation that contribute to carcass merit and longevity— that is, correct skeletal structure or structural soundness, beefiness or thickness of natural fleshing, particularly in the regions of the high-priced cuts, and optimum finish at a relatively young age.

In evaluating differences in conformation it is recommended that a score at weaning and another at the time of final weight (12 to 18 months of age) be obtained. The weaning score is probably of less value than final score; therefore, the greatest emphasis should be placed on the final conformation score. Since 12 to 18 months is somewhere near normal market age for a high percentage of slaughter cattle, it should help to guard against producing the "wrong kind" of cattle—those that mature either too early or too late.

Size or weight is a measure of growth rate and should not be considered in evaluating conformation. However, it is difficult to score conformation completely independent of growth, since a thrifty, growthy animal that has been doing well naturally looks better than one that has not done so well, even though both may be basically the same in the major items of conformation.

A scoring system may be simple or considerably detailed, including independent scores of each of the major items of conformation. One with greater detail helps to point out both the items of conformation that are good and those that are deficient. A simple system tends only to group animals of approximately equal desirability from a conformation standpoint, without indicating items in which they are deficient or superior. Each breeder should use a systematic scoring system, choosing for himself whether to use a simple or more complex one. Breed association herd classification programs are designed to evaluate conformation in detail.

Research indicates that differences in outside fat and differences in thickness of muscling can be appraised, but with somewhat limited accuracy, by subjective or "eyeball" evaluation in live cattle. Quality of the meat, as determined by marbling, color, texture, and firmness in relation to age, and the amount of edible meat produced per unit of carcass weight are the primary factors that determine real differences in carcass value. The shape of muscling has some effect on desirability when the carcass is broken into retail cuts. Also, thickness of muscling does affect yield of retail-trimmed cuts.

Among cattle that have been fed alike there is little relation between outside fat and marbling. Because marbling is a primary factor in determining carcass quality grade, a major objective in assessing differences in the indicators of carcass merit in live breeding cattle is to evaluate differences in the amount of lean relative to fat. Shape of muscling should also be considered. The other major factor to be considered in conformation evaluation is structural soundness, which is indicative of longevity. Differences in thickness of muscling or natural fleshing can be appraised to best advantage in the areas where the least amount of outside fat is normally present, namely the outside of the round and the forearm.

Research is in progress to develop techniques for objectively measuring differences in outside fat and muscling in live cattle. Until such techniques are perfected, a subjective score to reflect differences in these traits is recommended. In scoring for differences in fat thickness, an optimum of 0.3 to 0.5 inch is desired in slaughter steers and heifers. At yearling age, bulls have approximately 0.2 to 0.3 inch less outside fat than steers developed in a comparable manner. Thus it seems that one need not be greatly concerned if bulls have less than the amount of outside fat that is optimum for steers, because most slaughter steers and heifers have appreciably more than the optimum amount. One of the real opportunities for reducing outside fat in slaughter cattle is to select bulls that have the minimum amount.

In bulls developed alike it seems reasonable to give independent scores for differences in (1) structural soundness, (2) thickness of natural fleshing or muscling, and (3) outside fat. A muscle score reflects differences in thickness of muscling in relation to length of long bones. Thus muscle score usually reflects weight in relation to height. With two animals of the same weight but differing in height, the animal with less height will ordinarily receive the higher score for muscling. Preliminary information indicates that mature size may be highly associated with long-bone length at yearling age. Thus in two animals weighing the same as yearlings but differing in height, the one with the greater height may be expected to have a heavier mature weight and a lower muscling score at yearling age.

On the basis of these preliminary results, selection for heavy weights at yearling age, along with a high muscling score, should result in a growth curve with rapid early growth without excessive mature size. Thus the use of a muscling score at yearling age may be a factor in affecting mature size. Yearling weight should be given major attention because of its great economic importance, and indications are that near-maximum yearling weight may be obtained along with a high muscling score at yearling age.

RECORDS INTERPRETATION

The systematic collection of records on economic traits, the making of appropriate adjustments of these records, and the development of records summaries provide the basis for effective selection for genetic improvement. However, the effectiveness of a systematic record-of-performance program depends entirely on the extent to which records are used in making selection decisions.

It was emphasized earlier that most of the opportunity for selection should be used on traits of greatest economic value that have heritabilities high enough to effect a reasonable rate of improvement. Increasing the number of traits reduces the amount of selection that can be practiced for any one trait. Fertility is a trait of great economic importance but it is low in heritability; thus rate of improvement in fertility is expected to be slow. On the other hand, yearling weight is a trait of great economic importance with high heritability.

As already indicated, yearling weight at a constant age accounts for more than 90 percent of the variation in pounds of boneless, trimmed retail cuts at a constant age. Thus yearling weight adjusted to constant age is a trait that should receive major attention in selection. While the economic value of conformation score at yearling age is not so well documented as yearling weight, it may be of considerable economic importance if it is based primarily on thickness of natural fleshing or muscling and on structural soundness.

Another trait of major economic importance is mothering ability, because heavy weaning weights are basic to an efficient beef cattle industry. It is suggested that selections be based primarily on appropriately adjusted yearling weight and yearling conformation score based on thickness of muscling and structural soundness, along with some attention to weaning weight.

It is evident that adjusted yearling weight is composed of weaning weight and postweaning gains. Thus selection for yearling weight automatically gives attention to weaning weight.

Table 14 provides an example of the basic records necessary for making effective selections. These are some of the records on the surplus bulls offered for sale in 1967 at the Fort Robinson Beef Cattle Research Station in a cooperative research project of the U.S. Department of Agriculture and the Nebraska Agricultural Experiment Station.

Table 14, in abbreviated form, provides weight and conformation score information on all bulls from the entire calf crop, the bulls offered for sale and their sires, including sire progeny records, and individual information on two particular sires, the averages of their entire progeny, and individual information on their sons in the sale. Weight ratios are used in the information, and they provide a basis for making comparisons. Complete information supplied at the sale included records of the performance of the dams of the sale bulls, consisting of weaning weights and scores.

Among the many items of interest in Table 14 are:

1. The difference between the average weights and scores of the bulls offered for sale and the average of all bulls from the entire calf crop.

2. The difference in weights and scores among the bulls offered for sale.

3. The difference among the sire progeny groups in weight ratios and scores.

In reviewing these records it should be kept in mind that the sale bulls were a selected group—that is, above average in their adjusted 452-day weight—and, as evidenced by their sires' records on their own performance, their sires were a highly selected group. It is obvious that about half of the sires will be below the average weight ratio of 100 in the performance of their progeny in yearling weight ratio.

If heritability were perfect, or 1.0, the sires would rank the same in their progeny averages as they do in their own performance. This is not the case, as the heritability of adjusted yearling weight is only about 60 percent.

The same kind of summary illustrated in Table 14 is appropriate for selecting heifers, because the basic procedures and criteria should be the same in both sexes.

Table 14
Record of Performance for Yearling Bulls, Fort Robinson Beef Cattle Research Station, 1967

Lot Number	Sire and Tattoo of Sons	200-Day Performance				452-Day Performance			
		Weight[a] (lb)	Weight Ratio[b]	(205)[d]	Score[c]	Weight[a] (lb)	Weight Ratio[b]	(205)[d]	Score[c]
	Average all bulls	452	100	(205)[d]	12.3	1,001	100	(205)[d]	12.5
	Average sale bulls	476	105	(45)	12.5	1,064	106	(45)	12.8
	Nebraska Comet 2135		102		14		103		15
	Progeny average 1965–66		97	(44)	12.5		97	(23)	12.7
8	6098	508	112		13	1,084	108		14
9	6147	484	107		13	1,084	105		13
10	6212	489	108		14	1,078	108		13
11	6287	511	113		14	1,057	106		12
12	6432	464	103		12	1,052	105		13
	Nebraska Comet 1109		107		14		103		16
	Progeny average 1964–65–66		101	(62)	12.8		103	(32)	13.2
13	6011	452	100		12	1,073	107		12
14	6138	483	107		12	1,095	109		14
15	6195	468	104		13	1,037	104		13
16	6350	454	100		13	1,063	106		14
17	6352	512	113		13	1,069	107		13

[a] 200-day weights are adjusted for age of dam; 452-day weights are the 200-day weights plus the gain during the 252 days following weaning.
[b] Weight ratios for 200- and 452-day weights are computed by dividing the bull's 200- or 452-day weight by the average weight of all bulls. A ratio of 110 means that an animal is 10 percent above average.
[c] Scores follow the recommendations of the U.S. Beef Cattle Records Committee: 17-16-15 = fancy; 14-13-12 = choice, and so on.
[d] The figures in parentheses indicate the number of animals involved in computing the average ratios. The progeny averages for 200-day weight include both bulls and heifers; the 452-day averages are for bulls only.

CENTRAL TESTING STATIONS

Central testing stations are locations where animals are assembled from many herds to evaluate differences in some performance traits under uniform conditions. Present and potential uses of central testing stations include (1) estimating genetic differences between herds or between sire progenies in gaining ability, grade, finishing ability, and carcass characteristics; (2) determining the gaining ability, grade, and finishing ability of potential sires as compared with similar animals from other herds; (3) determining gaining ability, grade, and finishing ability under comparable conditions of bulls being readied for sale to commercial producers; and (4) as an educational tool to acquaint breeders with performance testing.

In setting up a central testing station, its objectives should be clearly defined and procedures designed to accomplish the objectives. Because objectives and procedures vary with location, only general principles will be discussed here.

In beef cattle, nutritional level at one stage of life usually has carryover effects on performance at later stages. A poor feed supply in one period tends to be followed by a period of increased or compensatory gain when rations are improved. Conversely, a higher than normal plane of nutrition, such as that provided by creep-feeding, is likely to be followed by a period of subnormal gains on a normal feeding regime.

Because pretest levels of nutrition and management usually differ from farm to farm or ranch to ranch, performance at a central testing station is influenced by pretest environment. From one standpoint this is a serious disadvantage of central testing stations, as part of the observed differences at a station will be due to pretest conditions. It is almost always impossible to estimate the importance of these effects, but carryover herd environmental effects are less important than herd differences due to environment when all animals have been fed for a comparable period in the herds in which they were produced. If this is considered, central testing stations minimize herd environmental effects.

Bull buyers must decide from which herds to buy bulls and which bull or bulls to buy within a herd. If the bulls are raised and fed entirely on the farm or ranch where dropped, the buyer has the difficult task of deciding how much of the apparent superiority or inferiority of bulls in a specific herd is the result of feeding and herdsmanship rather than heredity. If the bulls have spent part of their lives under standard conditions that minimize these effects, the buyer's task is easier, whether he is buying commercial bulls or herd sires for a purebred herd.

Similarly, if progeny test groups of steers from different herds are being fed out to determine the transmitting ability of the sires for growth rate, feed efficiency, and carcass characteristics, sire comparisons are more accurate if all progeny are fed under standard conditions for the final feeding period.

Central tests have limited use for estimating genetic differences among herds. The larger the herd size, the greater the number needed to adequately sample the herd. The precision of the tests is greatly improved if five to eight progeny of each of two or more sires from each herd are tested each year. This permits assessment of within-herd differences to compare with between-herd differences. Furthermore, there should be an adequate sample of animals from each herd on test or little real information on herd differences will be accumulated.

If central testing stations are used to estimate genetic differences between herds, it is recommended that samples of those completing the evaluation be used in topcross comparisons in commercial herds, so that additional traits can be measured and the precision can be increased. If the purpose is to evaluate individual potential sires, the number tested per head or per sire is of no importance; but between-herd comparisons should be discouraged if numbers from each herd are small. Preferably, bulls should be entered in this type of test only if they meet rigid qualifications for preweaning rate of gain and conformation score.

If the purpose of the testing station is solely to develop bulls and make objective performance information available to prospective buyers, a service especially valuable to small breeders, the number of bulls per herd or per sire is immaterial. To be most useful, however, large numbers should be fed at a single location, giving buyers an adequate number from which to choose. This is possible if commercial-type feedlots are used.

Influences of pretest environment on test performance can probably never be eliminated, but they can be minimized. Animals should be used whose pretest treatment was similar, and should be grouped within relatively narrow age ranges. Animals for a given test should be delivered to the station on a specific date and should undergo an adjustment period of 14 to 84 days on the test ration before beginning the official test. The test should run for an adequate length of time—140 to 182 days if a high-concentrate ration is used the entire time and longer if the ration is high in roughage.

Influences of pretest environment can be minimized in appraisal of results if the final reports include both pretest and test gains. If test gain alone is used, cattle on a suboptimum pretest feeding level that did not permit full expression of their inherent ability to grow are likely to compensate with inflated test gains. Using both pretest and test gains avoids labeling an unduly high test gain as the animal's real gaining ability. This can be done by either averaging pretest and test gains or, if test starts immediately after weaning, computing a final weight as a standard weaning weight (e.g., 200 days) plus test gain. The animal's entire life must be accounted for; "loafing periods" of unequal length, which tend to influence subsequent gains, should not be omitted.

The problem of compensatory gain is not limited to central testing stations. Within a herd the inherently fast-gaining calf whose mother was a poor milker is likely to have a low weaning weight with a correspondingly inflated postweaning gain. Comparisons, whether between herds at test stations or within a herd on an individual farm, should consider both preweaning and postweaning gains.

Central testing stations are most valuable if it is recognized that they can evaluate only a limited number of traits and that at best they are only one phase of a complete performance evaluation program. A primary measure of their effectiveness should be the impact they have on increased herd testing for all economically important traits. Central testing stations can cause difficulty in the maintenance of herd health, but proper precautions can minimize this problem.

HEREDITARY DEFECTS OF BEEF CATTLE

A large number of hereditary defects of possible economic importance have been reported in all breeds of beef cattle and also among the dairy breeds. Perhaps the hereditary defect most widely known is "snorter" dwarfism, which occurred at troublesome frequencies in some herds in the late 1940s and early 1950s. Discrimination against lines of breeding known to carry this defect has reduced its frequency. Snorter dwarfism, like most other hereditary defects, is inherited as a simple recessive—that is, it is caused by a single pair of genes that must be present together before the trait is expressed—and thus results from the mating of parents that both carry the defective gene.

Other types of hereditary dwarfism are due to different genes. "Comprest" dwarfism seems to result from a gene with incomplete dominance, meaning that the carrier individuals are comprest and an extreme type of dwarfism segregates from comprest $\times$ comprest matings. The comprest condition in Herefords and the "compact" condition in Shorthorns are probably due to the same gene. Snorter dwarfism has been authentically reported in both the Angus and Hereford breeds, and longheaded dwarfism, which is also inherited as a simple recessive, has been reported in the Angus breed.

The most practical means of testing a bull for a *specific* defective recessive gene is to breed him to 16 females that are known carriers of the gene. To determine if a bull is a carrier of *any* genetic defect, the most appropriate test is on his own daughters. On the average, half of the daughters of a bull with a defective gene will be carriers of that gene. After 30 to 35 matings of a bull to his daughters without the occurrence of some genetic defect, one can be reasonably sure that the bull does not carry a genetic defect noticeable in the offspring. Table 15 shows the percentage of bulls that are carriers of a hereditary defect inherited as a simple recessive that will not be detected with different numbers and kinds of test matings.

Although many genetic defects are present in all breeds of beef and dairy cattle and in all classes of farm livestock and probably cannot be eliminated, it is possible to curb their effect. Increased frequency of genetic defects in a breed or population can be explained by a gene's producing an effect in carriers that causes them to be preferred to noncarriers; or by chance, a defective gene may happen to be present in a line of breeding that is favored and used extensively by the industry. Seedstock producers are in a position to keep these defective genes from becoming a problem to commercial producers by closely observing operations and by realistically approaching a solution once a problem arises. This may require careful screening of herd bulls by progeny testing and the prompt elimination of those proved to be carriers of a defective gene.

If an abnormal calf is born in a herd, the breeder should establish the most probable cause of the abnormality, which can only be done by complete records. A limited number of developmental abnormalities may occur that do not have a genetic basis. If a breeder decides that an abnormality has a hereditary basis, he should breed away from the source of the trouble, so that minimum damage will result to his herd and to others. This may be done by outcrossing to a linebred herd, after a careful study of the outcross so that the same or an equally undesirable defect will not be introduced. Another method involves the progeny testing of bulls from his own herd, to insure that future herd bulls are not carriers of the gene responsible for the defect. The latter procedure may be indicated if the genetic merit of the herd is particularly high. If only a small percentage of the animals in a herd are possible carriers of the genetic defect, the best course is to eliminate those animals from the herd, provided that their genetic merit is not superior to the remainder of the herd.

Although it is unwise to use sons of bulls or cows known to be carriers of defect-producing genes without first progeny-testing the sons, discrimination against lines of breeding involving animals several generations removed from a known carrier is unjustified. Only one-half of the progeny of a carrier bull will be carriers when the bull is mated to cows that are noncarriers. Thus it seems more reasonable to handle such situations on an indi-

Table 15
Testing Bulls for Hereditary Defects Inherited as Simple Recessives

Number of Matings	Percentage of Carrier Bulls That Will Not Be Detected When Mated	
	To Known Carriers of a Specific Defect (%)	To Own Daughters or to Unselected Daughters of Known Carriers of a Specific Defect (%)
5	23.73	51.29
6	17.80	44.88
7	13.35	39.27
8	10.01	34.36
9	7.51	30.06
10	5.63	26.30
11	4.22	23.01
12	3.16	20.13
13	2.37	17.61
14	1.78	15.41
15	1.34	13.48
16	1.00	11.80
17		10.32
18		9.03
19		7.90
20		6.91
21		6.05
22		5.29
23		4.63
24		4.05
25		3.54
26		3.10
27		2.71
28		2.37
29		2.07
30		1.81
31		1.58
32		1.38
33		1.21
34		1.06
35		0.93

vidual herd or bull basis than to discriminate against other herds descended from similar lines of breeding if they are not directly incriminated.

DEVELOPING BREEDING PLANS

Attaining the maximum rate of genetic improvement in all traits of economic value in beef cattle requires a clear perspective of objectives and a planned breeding program for accomplishing them. The different segments of the industry are interested in various specific traits but the breeder must be aware of the demands of the entire industry. The commercial producer wants cows that have long productive lives and wean a high percentage calf crop of heavy, high-grading calves. The feeder demands rapid and efficient feedlot gains, and the packer and retailer are interested in cattle that will produce high-grading carcasses with a minimum of excess fat and the maximum yield of closely trimmed retail cuts from the wholesale cuts of greatest value.

The heritability, genetic association with other traits, and relative economic importance determine the attention each trait should receive in selection. Traits vary in heritability and economic value. The greater the number of traits selected for, the smaller the selection differential will be for any one trait. Traits of low heritability respond less to selection than do traits of high heritability. The opportunity for selection should be used for traits that will result in the maximum genetic progress for the traits of greatest economic value. Little can be gained and much can be lost by paying too much attention to traits of little economic value and traits of low heritability. While there are genetic differences between herds, evidence indicates that the large differences in feed resources and management programs between herds make it extremely difficult to compare the records of different herds. Thus comparisons should be among animals in the same herd.

Rate of improvement in most economically important traits of beef cattle is relatively slow primarily because of the inherently low reproductive rate, the large number of traits of economic value, and the long generation interval. The low reproductive rate makes it necessary to keep a high percentage of the offspring, especially females, as replacements, and the large number of desired traits limits the selection that can be practiced for any one trait. However, most of the economically important traits have reasonably high heritability, fertility being the most notable exception. Though improvement is slow, it tends to be permanent, cumulative from year to year, and transmitted to future generations. Over a period of 15 to 20 years, production in a herd or breed that has been subjected to systematic selection for all economically important traits should be noticeably higher than in those where no such effort was made.

A systematic record of performance program with selection based on differences in records is basic to any planned breeding program. The choice of breeding plans involves many considerations. In seedstock herds, if genetic merit is already high, if the herd is large, if it is not particularly deficient in some trait of major economic importance, and if it is relatively free of hereditary defects, a closed herd program of linebreeding may be desirable. One advantage of a closed herd is that the breeder knows the differences in performance of his own cattle better than another breeder's and can better evaluate differences in their most probable genetic worth.

In large herds where a relatively large number of sires are used in each generation, a closed herd can be maintained for extremely long periods without any appreciable increase in inbreeding if an attempt is made to avoid the mating of close relatives such as sire-daughter, full brother–sister, and half brother–sister. In herds where as many as 8 to 10 sires are used per generation, the decrease in performance and the reduction in genetic variation as a result of inbreeding will hardly be noticeable.

If the herd is not large and only a small number of sires are used in each generation, the level of inbreeding will increase more rapidly, and performance and genetic variation in the herd will decrease. Decrease in genetic variation decreases the effectiveness of selection.

If the genetic merit of the herd is already high, it will be difficult to bring in an outcross that is genetically superior to some individuals in the herd. Also, in herds that are relatively free of genetic defects, the chance of increasing this problem with introductions of outside bulls is greater.

Whenever a genetic defect is troublesome in a herd, or when performance in some economically important trait is particularly low, perhaps an outcross is indicated. Although the outcross should be selected to correct the deficiency, the other traits of economic value should also receive major consideration. Minimum sacrifice in other traits is a primary objective when bringing in an outcross to correct some deficiency. Outcrossing for any reason in herds of superior genetic merit should be done only on a cautious and systematic basis, and only herds known to be outstanding in the trait of major interest and superior in all traits should be considered. Perhaps this may be another linebred herd. Certainly, records are as fundamental in making selections for outcrossing as they are in making selections from within the herd. After selecting the outcross, a comparison with sires in the herd in a properly conducted progeny test is desirable before extensive use is made of the sires brought into a herd for outcrossing.

In herds that are only average or below in genetic merit, an outcrossing program may logically be the one of choice. However, since most of the opportunity for selection in beef cattle is in the bulls used, records in the herds where bulls are selected should be helpful in locating individuals that have superior genetic merit. Securing outcross sires from linebred herds is desirable. Smaller commercial herds should follow a program of outcrossing.

Pedigree, individual performance, and progeny test information all have a place in a constructive breeding program. Young sires should be initially selected on the basis of pedigree and individual performance data. The extent to which they are used in a herd will depend on their rank with other sires based on progeny test information. After progeny test information is available, it should be used in making decisions among sires, remembering that an increase in generation interval is involved. However, in herds of superior genetic merit, where increased accuracy is of fundamental importance, there is justification for using progeny test information more extensively than in herds only average or below average in genetic merit.

Because generation interval affects rate of improvement from selection, it should be kept relatively short. If a bull is truly superior, he should sire sons that have genetic merit surpassing his own, when he is bred to cows comparable in genetic merit to the population that produced him. The problem is one of devising an evaluation program based on use of records that aid in locating such sons. Perhaps one handicap to continued improvement in some herds is the extensive use of an old sire without sufficient attention to locating sons to replace him. When the old bull passes out of the picture, the herd is left without sires that are superior to him. Continued improvement depends on use of herd bulls that are superior to the ones used in the previous generation.

Before a bull is used extensively in a herd, as would be the case with artificial insemination, it may be desirable to progeny-test him for genetic defects on 30 to 35 of his daughters. If a herd is following a linebreeding program, it should be determined that the bull to which the linebreeding is directed is not a carrier of genetic defects, and known carriers should be discarded.

The breeding of cattle of truly superior genetic merit is a great challenge. Many decisions must be made on breeding plans, and in selecting herd bulls and replacement females. One difficult decision seems to be "dropping" a bull that still has a good market for his progeny, even though the breeder may have determined that the bull is not contributing to the accomplishment of his goals and may be inferior to others in the herd.

The more a breeder knows about the animals in his herd and the more clearly he understands his objectives, the more frequently he should make correct decisions. Success in breeding superior beef cattle, like success in other ventures, depends primarily on the utility of the goals and the accuracy of decisions while working toward the goals. A complete record of performance program provides the basis for making correct decisions.

Goals can be attained only by those who have the objectivity to keep them in perspective and the dedication to remain steadfast in achieving them. They can be accomplished only by a planned breeding program based on the systematic use of records for selection on all traits of economic value.

The contributions to genetic improvement have been made and will continue to be made by those who have chosen goals based on utility. The successful breeders have not been faddists, but have exercised common sense and good judgment with the long-term outlook in mind.

Chapter 4

REPRODUCTION and MATING

The percent calf crop weaned is the most important single factor in determining profit or loss in the cow-calf program. The term "percent calf crop weaned," as used in this book, refers to the number of calves weaned from all the cows and heifers of breeding age in the herd at breeding time. Calculating percent calf crop weaned from the cows that calved does not present the true picture of the reproductive performance of a herd.

It is reliably reported that the percent calf crop weaned is as low as 60 percent in most of the herds in certain areas, and not over 90 percent in the best of herds throughout the country over a period of years. The national average is estimated at 80 percent. Table 16 illustrates the effect of percent calf crop weaned on the cost of each calf weaned under varying annual cow costs. It is obvious that an understanding of the physiology of reproduction in beef cattle is extremely important.

THE REPRODUCTIVE ORGANS OF THE COW

The reproductive system of the cow consists of the genital tract and certain related organs. The genital tract may be described as a tube extending from the posterior end of the body, forward into the body cavity. This tube varies considerably in size and shape throughout its length and, inasmuch as each part performs a special function in the process of reproduction, each part is spoken of as an *organ* of the reproductive system. In accordance with such a definition the following reproductive organs exist in the cow:

The Vulva. The vulva is the exterior opening of the female genital tract. It consists of two *labia*, or lips, which close the opening of the tract, and of an internal chamber just within the labia called the *vulvar cavity*. Into this cavity opens the *urethra,* the duct from the bladder.

Table 16
Influence of Percent Calf Crop Weaned, Weaning Weight, and Weaner Calf Grade on Calf Costs and Gross Return[a]

Percent Calf Crop	Weaning Weight (lb)	Pounds Calf Weaned per Cow	Cost per cwt Calf Weaned ($)	Value of Calf Production per Cow ($)		
				Fancy[b]	Choice[b]	Good[b]
90	550	495	20.20	148.50	141.00	128.70
90	500	450	22.20	135.00	128.25	117.00
90	450	405	24.70	121.50	115.40	105.30
90	400	360	27.75	108.00	102.60	93.60
80	550	440	22.75	132.00	125.40	114.40
80	500	400	25.00	120.00	114.00	104.00
80	450	360	27.75	108.00	102.60	93.60
80	400	320	31.25	96.00	91.20	83.20

[a] Assumes a $100 cost for keeping a cow for 1 year.
[b] Prices of $30, $28.50, and $26 are assumed for fancy, choice, and good grade calves, respectively.

The Vagina. The portion of the tract just forward of the vulvar cavity is called the vagina. It is about 10 inches long and lies immediately below the colon. Its principal functions are to receive the penis during service and to afford a passage for the calf from the uterus to the outside of the cow's body at birth.

The Cervix. This is a constriction in the genital canal that marks the division between the vagina and the uterus. It is also called the *os uteri,* or neck of the uterus, and is really a cone-shaped part of that organ which projects back into the forward end of the vagina. During estrus or "heat" and at the time of calving the cervix is much dilated, but normally it is contracted so as to close the uterus. If, at time of service, whether natural or by artificial insemination, the cervix is closed so tightly as to make the passage of sperm cells into the uterus unlikely, conception of course will not occur. During pregnancy the cervix is tightly closed and is sealed against the entrance of bacteria from the vagina by a plug of mucus secreted by the mucous membrane of this region.

The Uterus. The uterus is the portion of the genital tract that is designed to retain and nourish the embryo or fetus between the time of fertilization and calving. The uterus consists of a main portion or *body,* lying just beyond the cervix, and two branches or *horns* at its forward end. In the cow the body of the uterus is relatively small, whereas the

FIG. 18. The first goal of every cattle breeder should be the production of a healthy, vigorous calf, normally born and well started by a mother cow of sufficiently good inherent milk production to raise a calf for every cow in the herd. (American Hereford Association.)

horns are long and large. For a short way they extend forward nearly parallel to each other, then spiral outward, terminating near the lower wall of the vagina. The horns of the uterus are held in place by a tough, elastic membrane called the *broad ligament,* which forms a connection between the uterus and the abdominal walls.

    The Oviducts. The end of each horn of the uterus narrows down to a threadlike tubule, the oviduct or Fallopian tube, leading toward the ovary. There are two oviducts, just as there are two ovaries and two horns. Although the ovaries are located quite near the end of the uterus, the oviducts are so tortuous that the total length of each is some 5 or 6 inches. At its outer extremity the oviduct broadens to form a funnel-shaped opening called the *infundibulum,* into which the ripened egg migrates when liberated

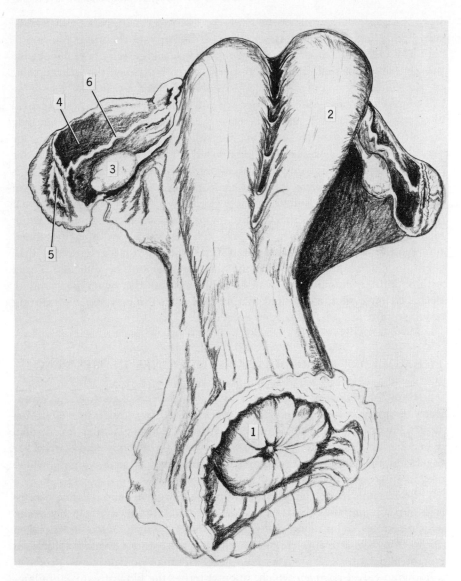

FIG. 19. The reproductive system of the cow. 1, cervix or os uteri; 2, right horn of the uterus; 3, ovary; 4, broad ligament; 5, the infundibulum of the oviduct; 6, the oviduct. (After Williams in *Diseases of the Genital Organs of Animals.*)

from the ovary. The walls of the oviduct are covered with cilia or threadlike projections, which facilitate the passage of the egg into the uterine horn.

The Ovaries. The ovaries are groups of specialized cells, actually outside the genital tract, that produce, at fairly regular periods, the *ova* or eggs, as the female sex cells are called. In the cow the ovaries lie loosely in the body cavity alongside the forward part of the vagina. They are oval in shape, about 1 inch in diameter around the thickest part. They may be felt with comparative ease by inserting the hand into the colon for about 15 inches by way of the rectum. Each ovary consists of a cluster of small egg sacs, probably several thousand in number, and each sac is called a *Graafian follicle*. Every female is born with a large number of Graafian follicles and still others develop throughout her lifetime. Each follicle contains an egg that eventually is theoretically capable of being fertilized and growing into a calf. The follicles remain in an undeveloped stage until puberty when, one at a time and at 3-week intervals, they begin to enlarge through an increase in the amount of follicular liquid within, until eventually the wall is ruptured and the ovum is liberated. Probably fewer than a hundred ova are liberated through ovulation during the life of a cow.

THE ROLE OF REPRODUCTIVE HORMONES IN THE COW

From physiological studies it is known that reproduction is a complex series of processes that are carefully synchronized by substances called *hormones.* Hormones may be defined as chemical substances that are formed by endocrine (ductless) glands in one part of the body and carried by the blood or lymph to another part of the body or to an organ, where they modify the activity of that organ.

Briefly, the role of hormones in reproduction is as follows: The process is begun by the tiny *pituitary gland,* situated at the base of the brain, which secretes a hormone called *gonadotropin* which, upon being taken up by the bloodstream, stimulates the growth of a Graafian follicle in one of the ovaries. Now this enlarging Graafian follicle itself produces a hormone called *estrogen* which, upon entering the bloodstream, stimulates the secretion of mucus in the vagina and acts upon the central nervous system to cause the cow to show signs of heat. Upon the rupture of the Graafian follicle and release of the egg, the production of estrogen ceases, and a substance known as *corpus luteum* is formed in the ruptured follicle from which the egg has just been released. The corpus luteum secretes a third hormone called *progesterone,* which (1) inhibits the production of gonadotropin by the pituitary, thereby preventing the development of another Graafian follicle, and (2) prepares the mucous lining of the uterus to receive and nourish the fertilized egg. However, if the cow was not served or if conception did not occur, the corpus luteum begins to degenerate in about 15 days and the secretion of progesterone is stopped, whereupon the pituitary gland again begins the production of gonadotropin to start the process all over again.

Should the cow be served during estrus and the ovum be fertilized, the corpus luteum remains in the ovary throughout pregnancy and continues to produce progesterone, which prevents the development of Graafian follicles and thereby the occurrence of estrus. Normally the corpus luteum atrophies within about 6 weeks following calving, and preparation for the resumption of the estrus cycle takes place.

Occasionally the corpus luteum fails to atrophy at the normal time after calving, inducing temporary sterility. Such a "retained" corpus luteum is commonly termed an *ovarian cyst*. If the cyst is expelled from the ovary, which a veterinarian or an experienced herdsman can bring about by working through the rectum, the cow usually begins estrus within a few days and can then be served. The forced removal of an ovarian cyst may be a poor practice, however, as cows prone to this condition are apt to be chronic "problem breeders." Culling them will help ensure a larger calf crop and a shorter calving season.

ESTRUS AND CONCEPTION

From the standpoint of the practical cattleman, the reproduction process begins with the first heat period of the young heifer. Ordinarily this condition, known as puberty, is first observed in heifers soon after they are 1 year old, although some heifers reach puberty while still nursing and thus are in danger, if a bull is running with the herd, of being bred when they are as young as 6 to 8 months of age. Liberal feeding appears to hasten the advent of puberty and scanty feeding tends to retard it. Apparently Brahman females or Brahman crossbreds and some of the newer breeds containing some Brahman blood reach sexual maturity at a much older age than heifers of the British breeds.

A few hours after the end of estrus, an egg is liberated by the rupture of a Graafian follicle, thereby presenting the conditions for conception and pregnancy. The interval between the end of estrus and ovulation varies from 6 to 26 hours (Table 17), whereas the time required for the sperm to travel from the vagina to the oviducts, where fertilization normally occurs, is only a few minutes. Consequently it is advisable to delay breeding until toward the close of the heat period, or even a few hours afterward

Table 17
Time Interval Between Estrus and Ovulation in Beef Cows[a]

End of Estrus		Time of Ovulation		Interval Between End of Estrus and Ovulation	
Hour of Day	Number of Cows	Hour of Day	Number of Cows	Interval in Hours	Number of Cows
4– 5 P.M.	2	10–11 P.M.	1	1– 2	0
6– 7 P.M.	7	12– 1 A.M.	0	3– 4	0
8– 9 P.M.	7	2– 3 A.M.	0	5– 6	2
10–11 P.M.	7	4– 5 A.M.	0	7– 8	1
12– 1 A.M.	10	6– 7 A.M.	1	9–10	0
2– 3 A.M.	4	8– 9 A.M.	7	11–12	8
4– 5 A.M.	2	10–11 A.M.	5	13–14	8
6– 7 A.M.	0	12– 1 P.M.	5	15–16	11
8– 9 A.M.	0	2– 3 P.M.	11	17–18	6
10–11 A.M.	1	4– 5 P.M.	5	19–20	1
		6– 7 P.M.	1	21–22	2
		8– 9 P.M.	2	26	1
Total observations	40		38		40
Average time estrus ended, 10:30 P.M.		Average ovulation time, 1 P.M.		Average interval, 14.6 hours	

[a] Adapted from *Journal of Animal Science*, 1:192.

if hand breeding is practiced, in order to favor the presence of strong, vigorous sperm in the oviduct at the time the egg is liberated.

Because the duration of estrus in cattle is short, seldom exceeding 18 hours, if natural mating is attempted breeding should not be delayed because the cow may soon be out of standing heat. In studies made at the Michigan station with beef cows, 4 P.M. was the earliest and 10:30 P.M. the average time of day at which estrus ended, indicating that cows that are to receive only a single service should be bred in midafternoon. If two services are possible, one should be made soon after the cow is observed to be in heat, usually in the morning, and another in the evening. Artificial breeding is most likely to be successful if insemination is made on the morning of the day after estrus. If hand mating in a breeding chute is practiced, bulls can be trained to serve cows the morning following estrus. Young bulls used in pasture mating may often breed cows only once, during the early part of estrus, and their semen may not remain viable long enough to ensure fertilization.

New research data suggest that much embryonic mortality or early abortion may be caused by the death of an embryo that resulted from fertilization by old or senile sperm cells. This could be the result of breeding too early in the heat period or, in artificial insemination, of using old (unfrozen) semen. As shown in Table 17, ovulation occurs, on the average, 14.6 hours after the end of estrus, although it may occur still later. Delaying artificial breeding until late in the day after estrus may very well result in conception, but can also result in embryonic mortality, in this case caused by an old or senile ovum rather than semen.

SIGNS OF ESTRUS

Cows vary greatly in their behavior during estrus. The condition can usually be detected by extreme nervousness in the animal and by her attempts to mount other members of the herd, which in turn often mount her. Examination usually discloses a noticeable swelling of the vulva, which often appears slightly inflamed in light-skinned animals.

As a rule, estrus is accompanied by a slight mucous discharge. Rarely is there any loss of blood until a day or two after estrus, when a slight bloody discharge or "menstruation" sometimes occurs. Many herdsmen believe that the appearance of blood on the genitals, hips, or tail following breeding is an indication that conception has not occurred, but this theory is refuted by reliable research.

RECURRENCE OF ESTRUS

Unless fertilization takes place, heat periods normally recur at intervals of approximately 3 weeks. There is some variation in length of the estrus cycle even in the same individual. Usually, however, it is seldom shorter than 18 or longer than 21 days. Cows that are to be bred should be closely observed at least twice a day during the third week following their last heat period.

Estrus, of course, does not normally appear during pregnancy. Its occurrence at this time is often followed by abortion, although there are numerous instances on record where pregnant cows have been served in apparently normal heat periods with no ill effect whatever on the embryo. Usually estrus reappears about 6 weeks after calving, but few cowmen breed their cows back so soon. The common practice is to wait until about 2 or 3 months after calving, so that the calves will be born at the desired time in the calving cycle. Experience in artificial insemination in dairy cattle suggests that it may even be harmful to breed a cow on her first estrus following calving.

Range-area cowmen, especially those in the semi-arid regions, are more concerned with getting cows bred so that they will have a calving interval of not over 12 months than with getting them bred back too soon. As mentioned in Chapter 8, inadequate protein and energy in the period immediately following calving causes delayed recurrence of estrus and therefore makes the calving of range cows on a 12-month schedule more difficult.

AGE AT WHICH TO BREED HEIFERS

Inasmuch as the process of reproduction, and especially lactation, imposes a heavy tax upon the mother, heifers should not be bred until they are reasonably mature. Nature apparently provides that the growth of the fetus and care of the young shall take precedence over everything else, even over the requirements of the mother's body for maintenance and growth. Whether delayed growth of the young cow, brought on by too early calving, is later resumed depends on the level of feed given the heifer after calving and during lactation.

There is much evidence that gestation is less apt to stunt immature heifers than is lactation. This seems reasonable in view of the fact that the newborn calf contains only about 15 pounds of protein and 3 pounds of fat, whereas about 65 pounds of protein, 70 pounds of fat, and 90 pounds of carbohydrates are in the total milk production of the young mother during the first 4 months of lactation. The milk produced also contains many times more calcium, phosphorus, and other minerals than are present in the newborn calf. Thus it is obvious that the average daily demands on the mother during lactation are several times greater than those during gestation. This has considerable importance in the accidental pregnancy of an immature heifer. If the heifer is not herself permitted to raise her calf but is allowed to dry up, there may not be a serious loss of size in the heifer. Ranchers experiencing adverse range conditions often early-wean the calves from 2-year-old heifers when the calves are 4 to 5 months old, thereby reducing the deleterious effects of lactation.

Since a majority of farmers and ranchers want their calves to be born during not more than a 2-to-3-month period in the spring, replacement heifers must be bred to drop their first calves very close to either their second or their third birthdays. There is much difference of opinion as to which is the better age for first calving. Several experiments and surveys have been made, particularly in the range states, to study this question. Although the results do not agree in all respects, experiments such as the

Table 18
Production Records of Cows, Bred First to Calve at Two and Three Years of Age, Through Fourteen Years of Age[a]

Age at First Calving	Two-Year-Old	Three-Year-Old
Number of females started on test, fall 1948	60	60
Number remaining, March 1962	23	22
Reasons for removal from test		
Open or failing to calve in two successive years	16	16
Cancer eye	9	6
Spoiled udder	4	5
Crippled	2	2
Disease	1	2
Hardware disease	2	1
Accidental	0	2
Unknown	2	3
Heifers assisted at first calving	28	1
Average mature body weight, fall 1956 (lb)	1,148	1,178
Total number of calves weaned	533	482
Number of calves weaned per cow year	0.80	0.71
Total percent calf crop weaned	86.7	85.2
Average weaning weight, all calves (lb)	476	485
Average weaning weight, minus two-year-old calf (lb)	482	485
Cow cost per cwt calf weaned ($)	10.33	11.34
Extra pounds of calf weaned	330	—

[a] Oklahoma MP-67:69.

one summarized in Tables 18 and 19 tend to justify the following statements regarding heifers bred as yearlings to calve as 2-year-olds:

1. Size at first breeding is more important than age, with a minimum weight of 600 pounds in average flesh being desirable.

2. Total number of calves weaned and total weaned calf weight during a cow's lifetime favor breeding as yearlings if the heifer receives assistance, when needed, during calving.

Table 19
Summary of 8.5 Years' Results in Study of Beef Cows Wintered at Different Levels (1948–1956)[a]

Age at First Calving	Two-Year-Olds			Three-Year-Olds		
Lot Number	1	3	5	2	4	6
Level of Winter Supplement Fed[b]	Low	Medium	High	Low	Medium	High
Number of cows at start of experiment	15	15	15	15	15	15
Number remaining on test November 1956	14	14	10	14	12	13
Average weight changes of cows on test (lb)						
Initial weight 10/29/48	473	471	476	476	461	470
Average winter weight loss	−108	−98	−63	−112	−97	−67
Average summer gain	188	185	147	198	179	160
Final weight 10/30/56	1,103	1,165	1,165	1,182	1,128	1,223
Calf production records at 8.5 years of age						
Heifers assisted at first calving	6	8	4	—	—	1
Calves lost at first calving	1	1	2	—	—	2
Total number of calves weaned	91	93	75	82	71	73
Percent calf crop weaned[c]	93	95	89	97	89	90
Total number of calves weaned per cow	6.44	6.58	6.03	5.79	5.13	5.15
Average calving date	3/14	3/9	3/8	3/15	3/4	3/5
Average calf weights (lb)						
At birth (sex-corrected)	76	76	77	76	76	78
At weaning (age- and sex-corrected)	480	472	471	495	474	492
Total feed, pasture and mineral cost per cow ($)	224.18	299.66	402.54	224.18	299.66	402.54
Cow cost per cwt calf weaned ($)	7.26	9.65	14.16	7.83	12.31	15.87

[a] Oklahoma MP, 48:46.
[b] Supplements fed: low level, 1 pound cottonseed meal pellets; medium level, 2.5 pounds cottonseed meal pellets; high level, 2.5 pounds cottonseed meal pellets and 3 pounds oats.
[c] Based on number of cows bred to calve each year. Calf losses not due to experimental treatment were not charged against the lot.

3. When heifers are bred as yearlings, maturity is delayed 3 or 4 years, and they may never reach full mature size, especially if supplemental winter feed is inadequate.

4. Average weaning weights of the first two or three calves produced by heifers bred as yearlings will be slightly lower than weights of calves produced by heifers bred first as 2-year-olds.

5. Cow cost per 100 pounds of weaned calf favors breeding heifers to calve first as 2-year-olds.

6. More heifers will need assistance at first calving if bred to calve as 2-year-olds rather than as 3-year-olds.

7. Using small, refined bulls on yearling heifers is worthwhile if pasture calving is practiced and if experienced assistance is unavailable during the calving season, because the resulting calves tend to be somewhat smaller at birth. Using Angus bulls on first-calf Hereford heifers may be helpful, but does not necessarily insure against calving difficulty.

8. Level of supplemental winter feed has more effect on the weight of the young cow than on the weight of the weaned calf or percent calf crop weaned, provided that minerals, carotene, and protein are adequate in all cases.

9. The feed or pasture available during the nursing period affects weaning weight of the calf more than does the age at which its mother is first bred.

10. The number of open cows among second-calf heifers that calved first as 2-year-olds can be reduced by supplemental feeding during the breeding season when they are nursing their first calves.

The choice of age at which first to breed heifers thus depends on the ration available during the first winter as heifer calves, the level of management available during the calving season, and the quality and quantity of summer pasture available during the lactation period. Small ranch herds and farm herds tend to practice breeding as yearlings, whereas large spreads generally (but not always) breed heifers first as 2-year-olds. More information concerning the role of nutrition as it affects first-calf heifers is found in Chapter 8.

THE BREEDING SEASON

The time of the breeding season, of course, depends on when the farmer or rancher would most like the calves to be born. As the average gestation period is about 283 days, mating should begin approximately 9 months and 10 days before the earliest date on which the calves are wanted. Although it is highly desirable to have all calves born as close together as possible, it will be found that, because of delay of the estrus periods of some cows following calving and failure of others to conceive from the first service, the period of calving, even in the better-managed herds, usually extends over 2 or 3 months. Any greater irregularity in the span of calving time is usually regarded as unsatisfactory under ordinary circumstances, especially in a commercial herd. There are instances where low rainfall causes poor range conditions, hence the nutritional status of lactating cows is so poor that longer, or even year-round, breeding seasons are justifiable to insure any calves at all.

Usually bulls should be removed from the pastures within 4 months or less after the beginning of the breeding season. Leaving the bulls with the cows for 65 days will give almost every cow in the herd three opportunities to conceive, which should suffice if the bulls are fertile and active throughout the breeding season. A few cows will not begin cycling soon enough to have several opportunities to be served. Except in unusual circumstances, such as drought or injury, these late-cycling cows should be culled. Pregnancy tests in the fall will reveal open or nonpregnant cows, which should be sold for slaughter at weaning time or later if some cheap gains can be put onto the dry cow or if prices are almost certain to strengthen substantially during the holding period.

SPRING AND FALL CALVES

When possible, the commercial cattleman tries to have his calves born at a time when range and weather conditions are most favorable. This usually means that they should be born either in the spring, after the cold weather of winter but before the heat and flies of summer, or in the fall before cold weather sets in. The exact calendar dates depend somewhat on the latitude and, in the western states, upon the altitude as well. Table 20 shows the results of one study of the effect of birth month on survival, growth, and weaning weight of beef calves. By far the greater number of the calves of the country are born in the spring. However, some farmers, especially in the central and southern states, find it more advantageous to have the calves dropped in the fall or even during winter. Below are the principal advantages claimed for spring- and fall-born calves, respectively:

Advantages of Spring Calves
1. Dry cows can be wintered more cheaply.
2. Calves are of good age by wintertime and can better withstand cold weather.

Table 20
Effect of Month of Birth on Survival, Growth, and Weaning Weight of Beef Calves[a]

Month	Number of Cows	Number of Calves	Percentage of Calf Crop	Average Age Weaned (days)	Average Weaning Weight (lb)	Average Daily Gain (lb)	Percetage on Totals
January	130	128	98.5	275	558	1.72	1.24
February	714	676	94.7	250	503	1.69	6.52
March	3,625	3,474	95.8	221	466	1.75	33.53
April	4,366	4,183	95.8	198	433	1.79	40.37
May	1,548	1,447	93.5	180	402	1.79	13.96
June	335	326	97.3	163	374	1.81	3.15
July	20	20	100.0	130	355	2.10	0.20
September	11	10	91	336	596	1.53	0.10
October	16	15	93.7	284	517	1.61	0.13
December	90	83	92.2	304	557	1.57	0.80

[a] Charles R. Kyd, Missouri Extension Service. Information to the author.

3. Cows milk better while on grass than they do on dry winter feed.

4. Labor is saved when cows and calves run together on pasture.

5. Calves may be sold either at weaning time with no wintering or as yearlings with only one wintering.

6. Cows are bred while on pasture, when they are most likely to conceive.

7. A smaller investment in shelter and equipment is required.

8. Condition of cows at calving time is easier to control, because some winter feeding is usually practiced and this can be regulated to prevent unnecessary weight gain. Fall-calving cows are sometimes quite fat and will have more calving difficulties.

Advantages of Fall Calves

1. Cows usually are in better physical condition, nutritionally speaking, for calving in the fall than in the spring because of good summer grazing conditions; hence the calves are likely to be stronger.

2. Young calves escape the severe heat and flies of midsummer.

3. At weaning in the springtime the calves may be turned on grass to be sold in the fall as short yearlings with a sale weight at least 200 pounds heavier than in a spring-dropped calf.

4. Cows that freshen in the fall or early winter lactate longer than those that freshen in the spring. Spring grass stimulates milk flow in cows

FIG. 20. Spring calves born on clean pasture are rarely affected by diseases or parasites. (Denver and Rio Grande Western Railroad.)

that calved the preceding fall, whereas changing to dry feed in the fall tends to diminish milk flow in cows that calved the preceding spring.

5. Cattle numbers are at their peak during the winter season when labor for their care is more apt to be available.

6. Cows are bred in the winter when hand mating or artificial insemination can conveniently be used.

7. Creep-feeding, when practiced, is more conveniently managed.

8. Calves are weaned at a more favorable marketing time, whether sold as stockers or as fat slaughter calves.

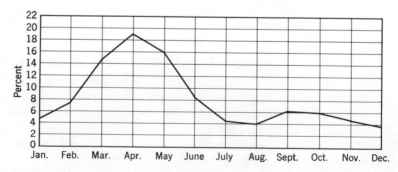

FIG. 21. Births of calves by months. Nearly 50 percent of all beef calves are born during March, April, and May. (U.S.D.A.)

The advantages of spring calves have the most weight under extensive rather than intensive methods of cattle production. Fall-born calves, on the other hand, are particularly well suited to farms where beef cattle are only one of several enterprises that contribute to the income from the farm. Farms once equipped for dairying may well utilize their equipment and buildings with a fall-calving program. In regions where winter small grain pastures are important, a fall-calving system fits extremely well because it enables the cow herd to utilize such pastures to best advantage.

Calving in very early spring, such as in January or February, while seeming to be a compromise that would have the advantage of both seasons of calving, actually has most of the disadvantages instead. Feed costs are high, calf death losses are apt to be high, labor requirements are increased, and if calves are sold off the cows in the early fall they are often just heavy enough to be discriminated against as to price. The result may well be less net income rather than the hoped-for increase.

In some of the Range area, notably the Sandhills region in Nebraska, the practice of splitting the calving season into a spring and a fall season is increasing. Apparently this is being done because the ranchers believe

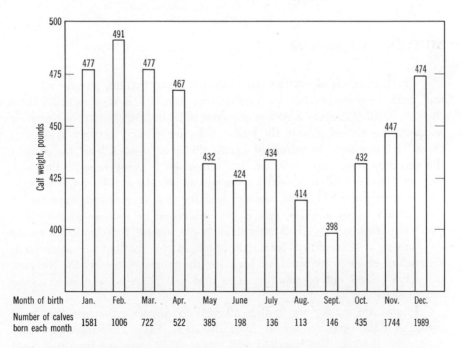

FIG. 22. Average 205-day weight of calves born in different months in South Texas. (Texas A. and M. University, Animal Science Department.)

they can sell more pounds of calf per cow with this combination. The fall-dropped calves are usually sold the following fall, weighing 600 to 700 pounds. In marketing circles these calves are called "calf yearlings." Another reason some give for this practice is that the ranchers select from the fall-dropped calves their heifer replacements for the main herd, which calves in the spring. This procedure enables them to compromise on the breeding date for the first-calf heifers, since they will be calving at 2.5 years of age. The replacements for the fall-calving herd come from the spring-dropped calves for the same reason. Naturally this system is adaptable only to herds large enough to make dividing the herd practical.

Researchers and extension service personnel in most states have collected data similar to those shown in Fig. 22. The data represent almost 25,000 weaner calves weighed in Webb County in South Texas near the Mexican border. Undoubtedly the extreme heat of the summer in this region accounts for the light weaning weights of the calves dropped between May and September. Corresponding data from other areas will probably differ in some respects, so that breeders in other sections should use data collected for their own regions, when possible, in choosing the ideal calving season.

METHODS OF MATING

Three methods of mating are followed: hand mating, pasture mating, and artificial insemination. In hand mating the bull is kept separate from the cow herd. Whenever a cow is observed to be in heat during the breeding season, she is turned in with the bull, where she remains until she is served; or, as is often done in purebred herds, she is led into a level lot or into the breeding chute and held or tied while she is being served. The bull may or may not be managed by an attendant. As a rule, only a single service is permitted in hand mating, and the cow is removed immediately after service. Two matings, one in the afternoon or evening and another the following morning, will almost certainly ensure that egg and sperm are both in best condition for fertilization. If two matings are attempted, a breeding chute is required, at least for the morning service, for the cow will have passed out of standing heat and will not readily submit to the second service.

In pasture mating the bull is allowed to run with the breeding herd throughout the breeding season. Data shown in Table 21 demonstrate that most cows that are easily settled will settle either from the first or second service. The table also shows that even after six estrus periods 27 percent

Table 21
Calculated Estrus Period at Which Cows Conceived During a Breeding Season with Pasture Matings[a]

Estrus Period (20-Day Basis)	Number of Cows Conceiving	Percentage of Herd			
		100% Fertility		73% Fertility	
		For Period	Cumulative	For Period	Cumulative
First	295	52	52	38	38
Second	155	28	80	20	58
Third	61	11	91	8	66
Fourth	30	5	96	4	70
Fifth	19	3	99	2	72
Sixth	3	1	100	1	73
Total: 120 days (4 months)	563	100		73	

[a] *Journal of Animal Science*, 3:156. (Data from U.S.D.A. Experiment Station, Jeanerette, Louisiana.)

of the cows were still not settled. These cows doubtless would be difficult to settle under any system of mating and should be culled.

Pasture mating saves the labor of inspecting the herd each day for cows that are in heat and driving them to the breeding pen for service. Moreover, it precludes the possibility of a cow's "going by" unbred because the herdsman failed to detect her estrus. The objections to pasture mating are:

1. All cows that run in the same pasture are bred by the same bull, or by any one of several bulls if more than one bull is present. This makes it impossible to be certain of the sire of each calf. This information is essential in herds on performance test or in purebred herds.

2. Bulls wear themselves out by repeatedly serving a cow while she remains in heat. Sometimes as many as six or more services are performed by several bulls, and should another cow be in heat the next, or even the same, day she might not be served.

3. If two or more cows are in heat at the same time in a single-sire herd, the bull may give all his attention to one of them, allowing the others to go unbred.

There is little doubt that in a herd of commercial cattle the disadvantages of pasture mating are more than balanced by the saving of labor

FIG. 23. Pasture mating of 2-year-old heifers. Labor requirements are low and conception rates are high in healthy cattle bred on lush spring pasture. (American Hereford Association.)

and the greater certainty of getting all cows in calf. With purebred cattle, however, pasture mating is less widely used. Here, certain knowledge as to whether a cow has been served, the particular bull performing the service, and the date on which it occurred is so important, especially for sale cattle, that many purebred breeders resort to pasture breeding only to pick up any cows that were missed during hand breeding. Pasture mating is also used to some extent in large purebred establishments where bulls and pastures are numerous enough to permit dividing the herd into lots of 15 to 20 cows, each with a bull at the head.

ARTIFICIAL INSEMINATION

Artificial breeding has not been so widely practiced by beef cattle breeders as it has been by dairymen. This is changing rapidly, and only the matter of a generally reduced conception rate and the additional work involved are keeping the practice from becoming more widely accepted by beef cattlemen. Performance-tested beef bulls are now found in all important AI (artificial insemination) bull studs, and owners of small commercial herds should by all means avail themselves of this service if practicable. Of the 7.7 million cows bred by artificial insemination in 1965 in the United States, 0.5 million were beef cows. This figure represented about 2 percent of the total beef cow population.

Perhaps the most important reason for the limited use of artificial insemination by the practical cattleman is the difficulty of determining when beef cows are in heat, and the problems associated with sorting them out and restraining them for breeding. Beef cows that run together day after day are much less inclined to ride one another than are dairy cows that are penned at night and turned together after being milked in the morning. Because the breeding season for beef cows usually comes at the time of year when they are on pasture, often on a remote part of the farm or ranch, it is difficult not only to determine each day which cows should be bred but to separate these cows from the herd and drive them to the barn or corral to be inseminated. The daily sorting out of cows suspected of being in heat, driving them to the corral, confining them in the squeeze gate, and returning them to pasture cause a great deal of undesirable disturbance to both cows and young calves. Confinement or lot feeding of open cows is feasible in many instances and eliminates some of the objections mentioned. Use of vasectomized bulls in pastures or in drylot will help locate cows that are in heat. Such a bull is incapable of inseminating a cow because the vas deferens has been surgically cut, prohibiting the passage of semen from testicle to penis.

Added to the disadvantages mentioned is the fact that sometimes a relatively low percentage of cows have conceived after two and even three inseminations. Usually after 2 or 3 months of artificial breeding a cleanup bull is turned with the herd to settle the cows that are still open. As a result the next year's calves show much variation in age and weight, thereby complicating their feeding and management.

Artificial insemination is of great value in large purebred herds because it permits the mating of an outstanding sire to many more cows than he could handle by either hand or pasture breeding. Bulls owned in partnership can be used in several herds, often located hundreds of miles apart, by resorting to artificial insemination. Purebred breeders should know the rules of their respective breed associations before using this method of breeding, however. As this is being written, most beef breed associations permit only two owners of a given bull, and only one association places no restrictions on numbers of breeders using one sire.

Artificial insemination can be employed to advantage in prolonging the usefulness of valuable sires which, because of accidents or advanced age, can no longer perform natural service. The development of a technique for freezing semen so that it can be stored almost indefinitely in a "semen bank" is one of the great advances in commercial animal husbandry. Semen from outstanding sires can thus be used to inseminate outstanding cows long after the sire's death, and enough semen can be stored to produce literally thousands of offspring. Here again, breed association rules should be known. Most purebred associations will not permit the registry of calves produced by stored semen from a bull no longer living.

SYNCHRONIZATION OF ESTRUS

If all cows to be bred in a herd could be bred artificially in a 1- or 2-day period, and if a high conception rate could be assured without undue cost, artificial insemination would be readily accepted by the majority of cattlemen. Some promising research is under way on a workable method for achieving this goal.

Synthetic progesterone-like compounds, which have properties similar to those of the hormone produced by the corpus luteum of a pregnant cow, can be fed at low levels, causing cessation of heat in cows and heifers just as occurs during pregnancy. This makes it possible to halt the estrus cycles of all cows in a herd regardless of when their cycles have been occurring, and then cause the cycles to begin again simultaneously in all of the cows. All cows and heifers that are to be bred are fed the material in a small amount of supplement for 18 days, then the feeding of the compound is discontinued. Within 24 to 48 hours, 90 percent of the cows will be in heat. The best conception rate is being obtained by breeding all cows on both the second and third days following withdrawal of the hormone. Conception rates as high as 85 percent have been reported, although rates of 50 to 60 percent are more common. Once synchronized, the cows will all continue to come in heat at about the same time. Synchronization will never have a place where natural breeding is practiced because too much bull-power would be required at one time. Even when artificial insemination is used, it can be a problem to provide enough cleanup bulls for natural breeding of the still-open cows.

Cows with calves less than 2 months old should not be synchronized until they are recycling; otherwise the chances of settling them will be lessened because normal ovulation will not follow the artificially induced estrus. Getting a uniform consumption of hormone is difficult but absolutely essential. Research on use of injections of still other hormones, administered near the end of the quiet period to induce normal ovulation upon heat, holds promise of increasing the conception rate. Also, work is under way on use of injected or implanted progesterone-like compounds to replace the feeding method most commonly used at present.

If all of the problems associated with this practice can be solved, the number of bulls required in commercial beef herds could be reduced to the extent that purebred breeders who rely on sales to such breeders would be in serious financial difficulty.

NUMBER OF COWS PER BULL

The number of cows that can be successfully bred by a single bull during a short, intensive breeding season depends on, first, the age of the bull and, second, the manner in which the cows and bull are handled. A yearling bull may be allowed an occasional service but in no case should he be mated with more than 12 or 15 cows during a breeding season of 2 or 3 months. Two-year-old bulls are capable of caring for 25 to 30 cows and a bull 3 years old or over can handle 40 to 50 cows if hand mating is practiced. If pasture mating is followed, these figures should be reduced by about one-third. In no case should a bull under 15 months of age be allowed to run with his cows. Instead, hand mating should be used and only a single service allowed per cow.

As the size of the herd is increased, more bulls should be provided in relation to the number of cows, because there is always a tendency for a large herd to break up into small droves of 10 to 50 cows each.

Table 22
Effect of Bull-Cow Ratio on Length of Calving Period[a]

| | | | | | Length of Calving Period | | | |
| | | | | | Minimum Period | | Maximum Period | |
Cows per Bull	Number of Herds	Number of Cows	Average Calf Crop (%)	Average (days)	Number of Cows in Herd	Days	Number of Cows in Herd	Days
20 or less	5	756	95.4	77	51	53	455	99
21–30	36	3,682	94.5	102	101	41	110	219
31–40	12	1,223	93.1	118	34	46	80	212
Over 40	12	2,027	93.6	132	82	73	110	243

[a] Compiled from Mimeographed Reports of Kansas Beef Production Contest, 1946–1950 inclusive, Kansas Agricultural Extension Service.

There should be enough bulls in the herd to make sure that there is little chance of any such drove of cows remaining long without a bull. Sometimes bulls running together in one pasture tend to bunch up, and it may be necessary to break up this tendency by riding the pasture daily for a while.

Table 22, compiled from published records of the Kansas Beef Production Contest, indicates that having more than 25 cows per bull is likely to result in a calving period that extends over 4 to 6 months or even longer, if the bulls are left with the herd until all the cows are settled. If they are removed earlier, some of the cows will not have been bred and will produce no calves the following year.

EVALUATION OF BULLS FOR BREEDING SOUNDNESS

A discussion of the physiology of the reproductive organs of the bull has been omitted, only because such information is perhaps less essential to the breeder or rancher than is similar information with respect to the cow. Most bulls are normal and fertile, and few anatomical abnormalities

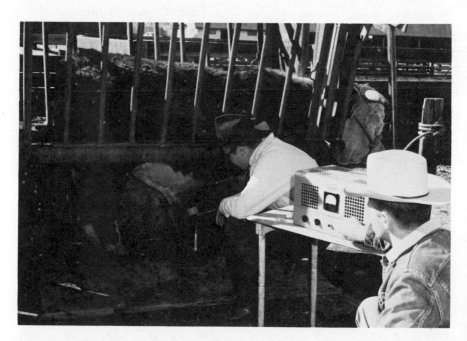

FIG. 24. Collection of a semen sample by the use of an electro-ejaculator—a technique best left to a trained expert—is only one step of several required in fertility-testing a bull. (Animal Science Department, New Mexico State University.)

are seen. True, in small herds or single-sire herds an unsound bull may cause complete loss of a calf crop if commonsense observations are not made during the breeding season. A survey conducted in Texas found that approximately 16 percent of 1,369 bulls, representing all breeds and ages and all seasons of breeding, were placed in either the "questionable" or the "cull" category on the basis of semen evaluation—that is, examination for presence of a sufficient number of live, motile sperm cells. Such tests are indicative of the bull's soundness at the time of testing but do not necessarily reflect his ability to settle cows during a 60-to-90-day breeding season. Some bulls that produce a good semen sample may not have enough sex drive or libido to breed the cows, or they may be unable to follow cows or serve them because of leg or foot problems. On the other hand, the collection of a usable sample of semen is not always possible, and errors in testing can creep in by this means also.

In general, fertility testing of bulls should be left to a veterinarian or to specialists in physiology of reproduction who are technically trained in this procedure. Most small-scale breeders and even most ranchers cannot justify the cost of the specialized equipment needed to make the test themselves, nor will they be apt to examine enough bulls to become expert to the extent that the test will be reliable. Further information on the subject of evaluation of bulls for breeding soundness can be obtained from textbooks and bulletins devoted exclusively to the subject.

Chapter 5

PREGNANCY, PARTURITION,

and CARE of the CALF

Estrus normally precedes ovulation by several hours; consequently the spermatozoa or sperm cells of the male usually have sufficient time to reach the oviducts, where they await the liberation of the ripened egg. Normally fertilization occurs in the upper end of the oviduct, just below the infundibulum, although there may be times when it does not occur until the egg reaches the uterus.

The length of time required for the fertilized ovum to move down the Fallopian tube to the horn of the uterus is not definitely known, but probably is about 10 days. During this time the fertilized ovum, now known as a zygote, is undergoing division or segmentation. While increasing little if any in size, the egg by successive cleavage divides first into 2, then 4, 8, 16, 32, and so forth, segments, finally reaching the morula or "mulberry" stage. During this migratory stage the zygote depends on the yolk of the egg for nourishment.

Soon after reaching the uterus the zygote becomes greatly enlarged by the absorption of fluids, and segmentation proceeds at a rapid rate. Also the cells begin to exhibit marked differences in size and shape, first assuming the appearance of well-defined layers; later, the differentiation of cells and tissues to form the different systems of organs takes place.

For the first 5 or 6 days inside the uterus the segmented zygote lies free within the uterine cavity, but shortly thereafter it becomes implanted—that is, attached to the wall of the uterus. This implantation not only protects the embryo from sudden and violent displacement, but also affords a method for the transfer of nutritive material from the mother to the young, and the transfer of waste products from the embryo to the mother, thereby making possible growth and develement. As soon as this exchange of materials begins to take place, the embryo is called a fetus.

THE FETAL MEMBRANES

The fetal membranes consist of three separate structures or parts: the *chorion,* the *amnion,* and the *allantois.* The chorion is the outer membrane and lies close to the mucous membrane of the uterus. The surface of the chorion is much greater than that of the impregnated horn. Consequently, it may extend into the nonpregnant horn, as well as into the body of the uterus. It has many blood vessels leading into the *placenta,* discussed below.

The amnion is a membrane that begins at the navel and surrounds the fetus like a sac, enclosing it entirely. It contains a liquid that protects the fetus from external injury. In the cow there are about 6 or 7 quarts of this liquid, called the *amniotic fluid.* During parturition the amniotic fluid serves to lubricate the vagina, thus aiding in the expulsion of the fetus. If the amnion does not break until the late stages of parturition and the birth canal does not benefit from this lubrication, birth may be slowed or made more difficult.

The allantois is a large membranous sac, between the chorion and the amnion, containing the fetal urine. The urine enters the allantois through a tube from the fetal bladder, called the *urachus.* All of these membranes, taken together, constitute the fetal membranes or "afterbirth."

THE PLACENTA

The placenta is the portion of the fetal membranes that unites the mother and the fetus. Although there is no direct vascular connection, the blood vessels of each lie quite close together, so that an interchange of materials can be made through their extremely thin but extensive walls. The capillaries of the fetal membranes, especially those of the allantois, which penetrate the chorion, become imbedded in the mucous walls of the uterus where they are in contact with the capillaries of the uterus.

This penetration of the capillaries of the fetal membranes into the walls of the uterus is by no means general over the entire surface of the impregnated horn. Rather, such contact is made only at specialized points on the uterine wall, which are known as cotyledons, illustrated in Fig. 25. These cotyledons are small prominences resembling scars or warts in the nonpregnant cow, somewhat oblong in shape, with their long axis at right angles to the long axis of the uterine horn. Because there are 40 to 60 cotyledons in each horn, the cow is said to have a multiple placenta.

During pregnancy these cotyledons greatly enlarge, and numerous fol-

FIG. 25. Cotyledon of cow, showing relation of the maternal and fetal circulations: *u,* uterus; *Ch,* chorion; C^1, maternal, and C^2, fetal, portion of cotyledon. (After drawing by Colin.)

licles or depressions are developed on each one's surface. Into these follicles the villi of the chorion and other portions of the fetal membrane are inserted, making an extensive and extremely close attachment between the fetus and the mother. These groups of villi of the fetal membranes are called the *fetal cotyledons*. Since each fetal cotyledon surrounds and "dovetails" into a maternal cotyledon, it follows that the fetal and maternal cotyledons are present in equal numbers. Between the cotyledons the chorion is free of the walls of the uterus.

THE UMBILICAL CORD

The membranous portion that unites the fetus and the placenta is the *umbilical cord*. Practical stockmen usually refer to it simply as the navel. Its sheath is composed of amniotic membrane within which are found two umbilical arteries, two umbilical veins, and the urachus, the tube leading from the urinary bladder into the allantois. Interspersed between these vessels is a gelatinous mass called the *Whartonian gelatin*. The umbilical arteries carry blood from the fetus to the capillaries of the fetal placenta, while the veins carry it back. These blood vessels have strong muscular walls that contract forcefully when ruptured at calving time, thereby preventing excessive bleeding from the calf's navel. The umbilical arteries are only loosely attached to the umbilical opening in the calf, and their ends retract within the calf's abdominal cavity upon being ruptured at birth. This retraction prevents the entrance of disease-producing bacteria through the severed vessels and also serves to check the flow of blood. The umbilical veins, however, are attached firmly to the umbilical opening or ring and do not retract when ruptured but remain open for a time and are occasionally the avenue for infectious bacteria. Treatment of the navel with iodine at birth is recommended to prevent bacterial infection.

POSITION OF THE FETUS IN THE UTERUS

In the early stages of development the embryo floats freely in the amniotic fluid, occupying no distinct position. With the growth of the fetus, however, it becomes fixed in position, usually with the anterior end, or head, toward the cervix. As the end of gestation approaches, the weight of the fetus causes it to rest upon its side on the floor of the cow's abdomen. As the cow's rumen or paunch occupies the entire left side, the fetus must arrange itself on the right.

MULTIPLE PREGNANCY

The cow usually is uniparous; more than one fetus is seldom formed in the uterus in a given pregnancy. However, twin calves occur occasionally, and triplets and quadruplets are not unknown. When only one fetus is present, it usually occupies one horn of the uterus; with twins, each horn usually contains a fetus. As the fetal membranes of calves are large and extensive, filling, even in a single birth, nearly the entire uterus, those of twins are almost certain to be in close contact and to become fused. If the fetal membranes of twins of opposite sexes are fused and a more or less common circulatory system is established, development of the reproductive organs of the female fetus is arrested. Apparently the hormones of the male are dominant over those of the female; or, as seems more likely, the male sex cells are the first to appear in the development of the two fetuses. Heifer calves born twin with bulls are almost always sterile because of the imperfect development of their reproductive organs. Such heifers are called *freemartins*.

SIGNS OF PREGNANCY

The gestation period of the cow is approximately 283 days but varies with breed. Angus cows have shorter gestation periods by about one week, whereas Brahman cows have periods about one week longer. It is believed these differences in gestation length are largely responsible for the differences in calf birth weights seen among the various breeds.

Long before the end of the gestation period certain changes are observable in the pregnant female that indicate the existence of the developing fetus. Because a diagnosis of pregnancy is often of great importance, every cattleman would like to know how to determine, as early as possible, whether his cows are pregnant.

Unfortunately none of the signs of pregnancy that can be observed by the layman is infallible during the first half of the gestation period. However, there are certain changes generally observed in pregnant cows that give good reason for suspecting pregnancy in any bred female that exhibits them. As gestation progresses, more obvious signs of pregnancy appear, although even these are sometimes misleading owing to the presence of certain diseases that produce changes similar to those caused by a developing fetus. Only direct examination made after the second month of gestation discloses beyond doubt the presence or absence of pregnancy. Among the many signs of pregnancy are the following important ones:

1. Cessation of estrus or heat. After seeing a bull serve a cow that has an identifying number, the cowman may easily make a record of it and determine when estrus should again occur if the cow fails to conceive. If pasture breeding is practiced, the experienced cowman checks the herd periodically during the breeding season to see if the cows are "passing over." If identifiable cows do not return or show estrus, he can be assured that the cows are being settled.

2. A noticeable enlargement of abdomen and udder. An enlargement of the abdomen is usually a good sign but not necessarily foolproof. As parturition approaches, the udder fills and teats firm up. First-calf heifers usually exhibit udder development sooner than mature cows do.

3. "Pregnancy testing" by internal examination. Manual examination or palpation of the reproductive tract by way of the rectum and colon may be made to verify pregnancy beyond a doubt. Experienced persons can detect a fetus as young as 2 months by this method, but examination

after the third month is more reliable. The effectiveness of rectal palpation as a pregnancy test is possible because the uterus and ovaries lie just beneath the colon and can easily be felt through the wall of the large gut. In early pregnancy (2 to 6 months) the presence of the fetus can be felt beneath the floor of the colon. When gently pressed by the hand, the fetus slips away as though it were floating in a liquid, which of course it is, and it returns immediately to its original position when pressure is released. As the gestation period advances, the uterus is pulled down within the abdominal cavity owing to the weight of the fetus and accompanying fluids. In this stage it is usually impossible for the examiner to feel the fetus. However, careful exploration will disclose a large, thick, firmly stretched band—the posterior end of the uterus—which passes downward and forward into the abdominal cavity. The blood vessels supplying the fetus will be enlarged enough so that they can also be felt at this stage.

IMPORTANCE OF PREGNANCY TESTING

Examination for pregnancy as a routine practice at the end of the breeding season can be an important tool in the efficient operation of a beef cow herd, because carrying nonpregnant cows for a full year without any return is one of the largest drains on profits. Table 23 shows the breeding

Table 23
Effect of Pregnancy Testing on Percent Calf Crop Dropped in a Colorado Herd[a]

| | | | | Percent Calf Crop Dropped | |
Year	Cows Examined	Number Found Open	Percent Open	Without Pregnancy Testing[b]	With Pregnancy Testing[c]
1952	343	62	18.1	80.2	98.6
1953	352	24	6.7	92.4	99.1
1954	406	22	5.4	92.9	98.2
1955	469	28	6.0	92.1	97.9
1956	539	94[d]	17.4	82.0	99.3

[a] Personal communication to the author from Dr. Lloyd C. Faulkner, Colorado State University.
[b] Calculated on the basis of number of breeding-age females exposed to the bull.
[c] Based on the number of cows remaining in the herd after selling the cows declared open upon pregnancy testing.
[d] Not all the open females were sold in 1956 owing to the large number found to be open, but calving percentage is calculated as if they had been sold.

and calving record of a grade herd in Colorado over a 5-year period. All females, including breeding-age heifers, were examined for pregnancy in the fall before winter feeding started and, except for the last year, all open females were sold. Prompt disposal of the open females increased the annual net return of the remaining cows by approximately $8 per head by reducing the total winter feed bill. Experimental work has indicated, although not conclusively, that heritability plays a role in regularity of breeding, and vigorous culling of slow breeders or nonbreeders has been shown to increase the breeding efficiency of a herd materially.

DETERMINATION OF CALVING AND BREEDING DATE

It is often desirable to know approximately the date on which a cow will calve in order to be prepared to give special attention should it be needed. If the date of service or the date the bull was turned in with a herd of cows is known, one need only count back 3 months and forward 10 days to determine the approximate calving date or the beginning of the calving season. For example, if a cow is served on July 21, counting back 3 months to April 21 and forward 10 days gives an approximate calving date of May 1.

It is often desirable to have calves dropped shortly after a certain date—for instance, if cattle are to be fitted for show, in which case full-aged animals may have a slight advantage over short-aged ones. The junior calf class, for example, usually consists of calves born after January 1 of the year they are shown, and it is obvious that a calf born on January 3 would be further along in development than one born on February 20. To determine when to start breeding so as to have the calves bunched after a certain base date, count forward 3 months and back 10 days from the base date. Because of the natural variation in length of gestation period, going back only 5 days instead of the usual 10 will give a safety factor. Thus, if it is desirable for calves to be born on or soon after May 1, determine the breeding date by counting forward 3 months to August 1, then back 10 or 5 days to July 20 or 25, depending on how safe one wishes to be. Gestation tables, found in many publications, are of course handy to use but may not always be available when needed; thus the above rules of thumb are a convenient substitute.

CARE OF PREGNANT COWS

If possible, pregnant cows should be separated from the rest of the herd to avoid injury and possible abortion from riding, butting, and fighting.

A good plan, where adequate lots and pastures are available, is to remove all bred cows from the herd at about the middle of the gestation period.

The pregnant cow needs exercise and, except during cold or stormy weather, there is no better place for her than out of doors where she can move about freely. A windbreak and a dry place in which to lie are ideal when available, but probably the majority of commercial beef cows do not enjoy such luxuries.

SIGNS OF PARTURITION

During the last few days of the gestation period certain changes in the pregnant animal will indicate to the experienced observer that parturition is not long away. The more important signs are the following:

1. A relaxation of the pelvic (sacrosciatic) ligaments that permits the muscles of the rump to drop inward, causing a noticeable falling away or sinking about the tailhead and pinbones, and a general softening or loosening of the flesh in this region.
2. An enlargement and thickening of the vulva, which appears swollen and somewhat inflamed.
3. A noticeable distention of the teats and an enlargement of the udder, as well as an abrupt change in its secretion, from a watery material to thick, milky colostrum.

These changes usually begin some 3 or 4 weeks before birth occurs, but become more and more pronounced as the time of parturition draws nearer. As a rule, they appear sooner in heifers than in older cows, and the latter sometimes calve with little or no "notice."

LABOR

The act of birth is accomplished through much exertion on the part of the mother. The first real sign of labor is a noticeable uneasiness on the part of the animal several hours before calving. The cow will, if possible, leave the herd and move to a more isolated part of the pasture. She turns her head, glancing nervously to the rear, and frequently lies down and gets up at short intervals.

The preliminary contractions of the uterus, which undoubtedly are responsible for the uneasiness described, are extremely important, because they move the fetus from a lateral, recumbent position on the floor of the abdomen to a longitudinal, upright attitude immediately in front of the pelvic girdle, from which position it can be easily expelled. The uterine contractions also bring about a dilation or enlargement of the cervix, through which the fetus must pass. In fact this restriction, the cervix, which normally separates the vagina and the uterus, is practically obliterated, and the two organs form one continuous passage to the exterior. Enlargement of the cervix is largely caused by the pressure exerted by the uterine walls upon the fetal fluids. These fluids within their elastic membranes are forced into the rear part of the uterus and, by transmitting the pressure from the contracting uterus equally in all directions, serve as an elastic dilator, first of the cervix and later of the parts beyond. Assisting the uterine muscles in this and the subsequent steps of parturition are the large muscles of the abdominal walls, as well as the diaphragm. These muscles, contracting in unison with those of the uterus at intervals of 1 to 2 minutes, exert a tremendous force upon the fetus. Normally they bring about its expulsion, unassisted, within 30 or 40 minutes after the contractions first begin.

The effect of this great and rhythmic pressure upon the fetus is to cause it and its enveloping membranes to be moved through the dilated cervix into the vagina. The chorion or outer membrane, being firmly attached to the uterine walls except near the fundus, does not withstand much displacement and is soon ruptured by the increasing pressure. This step in the birth process is highly essential to the well-being of the young, as the continuation of close contact between the chorion and uterus maintains the oxygen supply and nourishment of the fetus throughout labor. When the chorion ruptures, the allantois with its liquid contents—that is, the fetal urine—is forced into the vagina to appear at the vulva, where gradually it forms a semitransparent, balloonlike sac holding 1 to 2 pints of fluid. This sac is called the first water bag. Usually it increases in size until it ruptures from the pressure of the liquid within.

Following the escape of the allantoic fluid, the pressure exerted by the uterus and the abdominal muscles is applied directly to the fetus, now suspended only in the amniotic fluid. Should this pressure be applied before the fetus is in proper position or before the genital passages are sufficiently enlarged, birth will be more difficult. Therefore, the longer the water bag remains intact the better.

Within minutes following the rupture of the allantoic sac or first water bag, the amniotic membrane appears with the contained fetus. This membrane is glistening-white and contains a rather viscid, slimy, opalescent liquid. Within the membrane a portion of the fetus, usually the front feet, can plainly be seen. With each labor contraction more and more of the amniotic liquid is forced out to form a sac, the second water bag, which eventually bursts, thereby lubricating the genital passage.

The rupture of the amniotic membrane is followed by violent strainings on the part of the cow, which soon force the head and then the shoulders of the fetus through the pelvic canal. At this stage the cow often rises to her feet, and the calf completely emerges under the influence of gravity.

RENDERING ASSISTANCE

Assistance should not be given except when it is actually necessary. Some herdsmen and even veterinarians are prone to rush in and "take" the calf by force at the first appearance of the emerging fetus. This practice is to be condemned, because there is likely to be injury to both cow and calf in the form of torn membranes and strained ligaments. If no progress has been made toward delivery within about 2 hours after the beginning of labor, a careful examination should be made, by a veterinarian if possible, to determine whether an abnormal presentation is causing the delay.

While unnecessary aid is to be discouraged, one should not make the

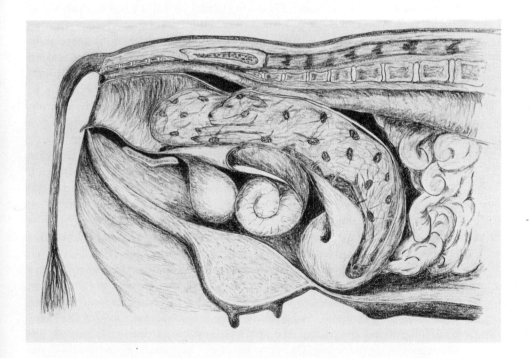

FIG. 26. The normal position of the fetus at the time of birth. (After Skelett.)

opposite mistake of permitting the cow to labor until she is completely exhausted before aid is furnished. Occasionally a calf has an abnormally large head or unusually heavy shoulders or hips that greatly delay or totally halt its passage through the pelvic cavity, even in a normal presentation. Anything other than a presentation where both front feet and the head appear almost simultaneously is considered abnormal. In a posterior presentation (hind feet first), the hips of the calf frequently cause trouble, mainly because the small size of the allantoic sac in the region of the hindquarters results in insufficient dilation of the cervix.

In either event assistance should be given by fastening small ropes or chains well above the calf's pasterns, to avoid injuring the soft hoofs, and pulling backward and downward each time the cow labors. No more force should be exerted than is necessary to overcome the obstruction. Under no condition should traction be applied except when the cow labors, unless she is exhausted to the point where she refuses to labor. In that event the calf must be removed entirely by traction. Use of mechanical calf-pullers by anyone but an experienced person usually proves fatal to the calf and results in permanent injury to the cow.

So long as there is no tension on the umbilical cord there is no cause for hastening the act of birth. However, when the calf's head has advanced as far as the eyes, the cord is pressed against the floor of the cow's pelvis in such a way that the placental circulation is jeopardized. At this point it is highly advisable to rupture the amnion, if it is still intact, to permit respiration and to prevent the calf from drowning in the amniotic fluid. Because respiration is unlikely to be effective so long as the chest is in the viselike grasp of the genital passage, birth should be hastened by traction in such situations. In a posterior presentation there is always considerable danger of the calf's suffocating through rupture or strangulation of the umbilical cord, and assistance in hastening parturition is more likely to be necessary in posterior than in anterior births. A posterior presentation can usually be anticipated if the feet first appear with the pads or soles of the hoofs topside.

Some other abnormal presentations occasionally encountered are: a breech presentation—that is, a posterior presentation with the hind feet forward rather than backward; one or both front feet turned back in an otherwise normal presentation; one with the front feet normally presented, but the head turned back. In all these situations the calf must be pushed back into the abdominal cavity and the limbs or head straightened. This is difficult to do because of the counteracting force of the labor contractions and because of the space limitations. Drugs or hormones are sometimes used to cause the cow to cease her straining and allow the corrections to be made. Generally only experienced herdsmen or veterinarians should attempt these adjustments, but when such help is unavailable an amateur will have to make the best of the situation.

BIRTH BY CAESAREAN SECTION

In some instances where valuable cows—for instance, purebreds that have been fitted for show—can obviously not deliver a calf, even with assistance, an operation known as a Caesarean section is performed. This surgery must be done by a veterinarian or it is almost certain to end in loss of both calf and cow. As already mentioned, excessive condition or fat can interfere with even a normal birth; if an abnormal presentation is found in such an overfinished animal, a Caesarean operation may be the only solution.

It consists of removal of the calf through a large incision in the right side just above the flank. Because the opening must be made through the skin, several membranes and muscle layers, and finally the wall of the uterus itself, it is obvious why only a skilled surgeon can do the work. The cow usually is immobilized for the operation by spinal anesthesia or a spinal block.

If the surgery is skillfully done and infection is held to a minimum, the cow should be left in good condition for future calvings. Usually the calf is saved if the operation takes place at the normal calving time or before prolonged labor has occurred. Ordinarily a Caesarean section is not attempted more than a few days in advance of what is believed will be the normal calving time.

CARE OF THE NEWBORN CALF

After the calf is born, all membranes should be cleared from its nostrils to facilitate breathing. Normally the mother cow takes care of this process. If parturition has taken a long time and it is suspected that the calf breathed while still inside the cow, all amniotic liquid should hastily be cleared from the nasal passages and throat to prevent drowning. This can usually be accomplished by holding the calf head-downward for an instant to permit the material to drain out of the lungs and air passages. Artificial respiration should then be given at once and continued until breathing is established. Many calves that appear to be dead can be revived by this method.

Because the umbilical cord is relatively short (12 to 15 inches), it always ruptures during the act of birth—usually about the time the forequarters pass through the vulva. If sanitary precautions have been taken in preparing clean quarters for the parturient cow, or if the cow has calved on clean pasture, there is little likelihood of any trouble from navel infection. However, some cattlemen make a practice of applying either tincture of iodine or formalin to the navel stump to destroy any pus-forming bacteria that may be present or the even more dangerous soilborne tetanus organisms found on some farms. Others dust the stump with antiseptic powder to hasten its drying and sloughing off.

As soon as possible after birth, the cow and calf should be left alone. The cow apparently derives satisfaction from drying her calf by licking it, and this may be nature's form of artificial respiration. Also it apparently stimulates the functioning of the calf's circulatory system.

Usually the calf stands and nurses of its own accord. The taking of nourishment a few minutes after birth is neither necessary nor advisable. If at the end of 5 or 6 hours the calf has not nursed, however, it should be given assistance in finding the udder. Seldom is more than one lesson necessary. Most authorities agree that the colostrum, or first milk, acts as a mild laxative, but perhaps its most important function is to provide protective antibodies that reduce the danger from various respiratory and gastrointestinal infections.

It is important that young animals receive plenty of sunshine and exercise, which are necessary for health and rapid growth. Calves born on pasture naturally get sufficient exercise in following their mothers back and forth over the pasture. Fall- and winter-born calves in the nonrange area do not have this advantage and should be provided with a good-sized well-drained lot in which to scamper and run on fine days when the ground is dry or frozen. In addition to this lot, a small paved corral on the sheltered side of the barn is ideal as a sunning place when the larger lot is wet and muddy.

EXPULSION OF THE PLACENTA

After the calf is born, the outer fetal membranes and the chorion are still attached to the walls of the uterus by means of the cotyledons. With parturition, however, the exchange of nutritive elements between the maternal and fetal placentae immediately stops, and there is a shrinkage of the villi of the cotyledons. The attachments between the chorion and the uterus are thus loosened, a process that is greatly hastened and facilitated by the contractions of the uterus, and through which the uterus returns to its normal size. One by one the fetal cotyledons separate from the cotyledons of the uterus, and the freed membranes are gradually forced out through the vulva.

RETENTION OF THE PLACENTA

Normally the fetal membranes are expelled 2 to 6 hours after parturition. If they remain longer than 24 hours, it is likely that an abnormal condition exists. Cattle are more susceptible to this condition than are other species of domestic animals, and it is not uncommon to encounter herds in which nearly 20 percent of the cows are troubled with this problem during some calving seasons.

There are various possible causes of retained placenta. In all probability many cases result from an infection that causes inflammation and enlargement of the maternal cotyledons. Such infection may have been present before parturition, in association with contagious abortion and other reproductive diseases (see Chapter 27), or it may have occurred while the cow was given assistance during labor. Also, retention is likely to accompany a failure of the uterine walls to contract promptly because of general weakness on the part of the animal or its exhaustion from an especially long labor. Cows that are thin and half-starved, as well as cows that are in high show condition, are likely to be troubled with retained placenta. A deficiency of carotene, or of vitamin A itself, in the ration of the pregnant cow has been strongly linked with retained placenta. Such a deficiency is most apt to occur in cows being wintered on cured grass, without supplementation, following one or more seasons of severe drought.

A retained placenta is an ideal medium for the development of putrefying bacteria. An almost infallible signal of retained placenta is a stringlike portion of the unexpelled membranes hanging from the vulva, where it comes in contact with the tail and hindquarters of the cow. It of course becomes heavily laden with all sorts of bacteria, which quickly spread into the interior. Decomposition and putrefaction begin in a remarkably short time. Except in cold weather, an obnoxious odor, warning that attention is urgent, appears within 48 hours after parturition.

REMOVING RETAINED PLACENTA

There is considerable difference of opinion among veterinarians as to the proper time to remove a retained placenta. Some recommend removal as soon as 24 hours after calving. Others advocate waiting another day in order that the attachments may be partly loosened by the process of decomposition. Still other authorities advise postponement until the fourth or fifth day after calving to permit the uterus to contract sufficiently so that the operator can easily reach all parts of the affected horn. Some do not recommend removal at all, preferring to treat the condition with sulfa drugs or antibiotics, thereby helping the cow combat the resultant infection.

If possible the services of a veterinarian or experienced herdsman should be obtained when removal of a retained placenta is attempted. The inexperienced operator who tries to carry out printed instructions is likely to seriously damage the walls of the uterus, especially the cotyledons. Moreover, there is danger of the layman's introducing harmful bacteria into the uterus, as well as infecting himself through a cut or abrasion on the arm. Few laymen realize the importance of observing sanitary precautions throughout the operation. Others go to the opposite extreme and use strong antiseptic solutions that scar and deaden the tender membranes of the cow's genital tract. Under no condition should the layman attempt to remove a retained placenta until he has seen the operation performed several times by trained practitioners and has learned from experience and observation the exact method of procedure. The average cattleman does better to send for a qualified person to perform the operation. To do otherwise will almost certainly result in the sterility of the animal. After the placenta has been forcibly removed, treatment with recommended antibiotics or sulfa drugs usually prevents infection.

OCCURRENCE OF ESTRUS AFTER PARTURITION

Cows usually do not show signs of estrus until some 6 or 8 weeks after parturition. Some cows go even longer before coming in heat. Although the calving dates of some cows can be advanced a few weeks each year, relying on this practice to bring about earlier calving in an entire herd of cows usually ends in disappointment. Late calvers should be culled or the average calving date may become still later. As previously mentioned, late-dropped heifer calves should ordinarily not be retained for breeding purposes, because regularity of breeding may be heritable. For the same reason such heifers should not be sold to other breeders.

SIZE OF THE CALF CROP

The term "calf crop" refers to the percentage of cows in the breeding herd that produce calves of weaning age in a given 12-month period. This percentage often varies widely between different herds and in the same herd in different years. The more important factors that determine size of the calf crop are:

1. Small percentage of pregnancies resulting from (a) failure to detect cows in heat if hand mating is practiced, (b) cows not recycling before the end of the breeding season, (c) impaired breeding qualities in the bull, (d) too many cows per bull, and (e) poor semen or technique in artificial insemination.

2. Abortion. When more than 5 or 6 percent of the cows abort, the abortion is probably a contagious form.

3. Loss of calves during or soon after birth. Abnormal presentations, exposure, scours, and congenital weakness frequently account for the loss of 10 to 15 percent of calves born.

4. Loss of calves over 1 month old but under weaning age. The percentage of loss here is small, perhaps less than 1 percent. Pneumonia, bloat, and various accidents claim an occasional calf during this period of life.

Table 24
Production Record of Breeding Herd of U.S.D.A. Range Experiment Station, Miles City, Montana[a]

		Percentage of All Cows in Herd				
	Number of Cows	Cows		Calves		
Year		Open	Pregnant	Normal Birth	Weaned	Loss to Weaning[b]
1925	106	14.2	85.8	84.0	81.1	5.5
1926	104	26.9	72.1	72.1	69.2	4.0
1927	126	9.5	90.5	77.0	71.4	21.1
1928	161	14.9	85.1	80.7	78.3	6.8
1929	168	22.0	78.0	75.0	74.4	4.6
1930	186	8.6	91.4	86.6	85.5	6.5
1931	234	11.5	88.5	87.2	85.9	2.9
1932	291	16.2	83.8	81.8	79.7	4.9
1933	331	11.8	88.2	86.7	84.3	4.1
1934	353	14.7	85.3	83.6	81.6	4.3
1935	366	23.0	77.0	71.9	70.8	8.2
1936	314	14.6	85.4	81.8	79.9	6.3
1937	233	31.8	68.2	66.1	64.4	5.7
1938	290	9.7	90.3	88.3	86.2	4.6
1939	305	6.6	93.4	92.5	90.5	3.2
1940	351	8.5	91.5	89.2	86.9	5.0
1941	388	12.1	87.9	87.1	84.5	3.8
1942	446	12.8	87.2	85.9	83.4	4.4
Total	4,753	Av. 14.4	85.6	83.1	81.0	5.4

[a] Baker and Quesenberry, U.S.D.A., *Journal of Animal Science*, 3:80.
[b] Based on number of pregnant cows.

With so many possible adverse factors it is not surprising that a herd of 10 or more cows seldom escapes all of them. In other words, a perfect calf crop (100 percent) is seldom realized. Instead, the crop is likely to be somewhere between 70 and 95 percent, as shown in Table 24.

In general, the size of the calf crop varies inversely with the size of the herd, because of the smaller amount of individual attention given to the animals of larger herds. However, studies by the U.S. Department of Agriculture, both in the Corn Belt and Range areas, indicate that good care and management on the part of the owner are much more important in determining the percentage of calves raised than is the size of the breeding herd or even the ratio of bulls to cows. In many instances great variation existed in the size of the calf crop on practically adjoining ranches "with no perceptible difference in range, feed, water facilities, quality of animals, or animal losses."

Undoubtedly such differences are largely due to careful culling of nonbreeding cows, the amount of attention given to the conditioning of bulls before the breeding season, the amount of time spent in systematic inspection of the herd for cows in heat, and the attention given the cows during both breeding and calving periods. Nevertheless, even with the same system of management, much variation in the percentage of calves raised was found on the same farm or ranch from one year to another.

Weather conditions, particularly as they affect the feed supply of the cow herd and the exposure to which the young calves are subjected during the 2 or 3 weeks following birth, are largely responsible for these yearly

Table 25

Influence of Size of Calf Crop on Net Cost per Calf[a]
(North Central Texas Ranches)

Percent Calf Crop (by Groups)	Number of Ranches in Each Group	Number of Calves	Net Cost per Calf (4-Year Average)	Percent of Total Calves
30–40	1	590	$43.92	1.4
40–50	2	585	49.21	1.3
50–60	14	11,880	36.15	27.4
60–70	18	8,020	38.88	18.5
70–80	22	10,848	28.03	25.6
80–90	22	9,749	25.46	22.5
90–100	5	1,695	22.15	3.9
Totals and averages	84	43,367	$31.95	100.0

[a] California Bulletin 458.

Table 26
Percentage of Calves Raised in Corn Belt Herds[a]

State	Number of Farms	Average Number of Cows per Farm	Average Percent Calf Crop
Indiana	6	16.5	96.2
Illinois	13	21.0	89.3
Missouri	33	22.1	90.2
Iowa	76	31.1	86.5
Minnesota	12	21.7	85.0
South Dakota	6	30.2	86.4
Kansas	46	50.4	81.5
Nebraska	38	27.0	78.9
Total	230	31.5	84.9

[a] U.S.D.A. Report 111.

variations. These can be overcome to a great extent by providing sufficient emergency feed supplies and adequate shelter facilities to meet such emergencies if winter calving is practiced.

The influence of size of calf crop on the net cost of raising calves to weaning age is well illustrated in Table 25. It should be noted that 68 percent of these Texas ranches realized a calf crop under 80 percent. In sharp contrast to these figures are the records of more than 100 small breeders in the North Central states, many of whom weaned calves from 90 percent or more of their cows (Table 26). In view of the gradual decrease in size of calf crop from east to west, it seems probable that feed conditions and available shelter were the chief causes for the variations noted. The data in both of these tables were obtained several years ago when many breeders, especially large ranchers, gave less attention to their herds than they do now.

Chapter 6

PRINCIPLES of FEEDING BEEF CATTLE

Except for the increasingly large number of feeder cattle that are fed high-concentrate rations in drylot, beef cattle, being ruminant animals, are usually found on farms and ranches that produce large quantities of harvested roughage and/or pasture. The prevalence of roughage may result from the operator's choice, from necessity because perpetually low rainfall makes growing of other crops difficult, or from erosion or soil infertility.

Ruminant or cud-chewing animals have specially adapted digestive systems that enable them to utilize roughages or feeds that contain comparatively high levels of crude fiber or cellulose and related compounds. An understanding of these adaptations is valuable in determining feed or nutrient requirements of beef cattle, and knowledge of nutrient requirements is essential for proper ration formulation.[1]

SIGNIFICANT FEATURES OF RUMINANT NUTRITION

Monogastric animals—those having one simple stomach, such as the pig and man—have a relatively low-capacity alimentary tract consisting of stomach, small and large intestines, and accessory glands. In these animals digestion is largely enzymatic in nature and little provision is made for digesting roughages; therefore, their diet must consist mainly of concentrates or feeds low in crude fiber. In contrast, ruminant animals have compound

[1] The nutrient requirements of beef cattle, established through worldwide research, are regularly reviewed by the Sub-Committee on Beef Cattle Nutrition of the National Research Council, of which the author is a member. Recommendations published by the Sub-Committee in its 1963 revision of *Nutrient Requirements of Beef Cattle* have been used for reference in this chapter.

132

stomachs and a much more complex digestive system, and much remains to be learned about their anatomy and function.

Using tools such as the artificial rumen and various fistulae (e.g., esophageal, rumen, and abomasal) in the live animal, researchers are shedding much light on the so-called darkest spot in animal nutrition, the rumen. By means of fistulae or semipermanent surgical openings, samples of feeds in various stages of digestion may be withdrawn at intervals to follow the progress of physical and chemical changes. Also, specific chemical substances may be introduced through these openings to study their effects on the digestive process.

The most successful cattle feeders today are those who know and take advantage of the following unique characteristics of the ruminant animal:

The Four-Compartment Stomach. The *rumen or* paunch, the first compartment, constitutes about 80 percent of the total stomach capacity in adult cattle and may hold up to 50 or 60 gallons. Connected with the paunch are the second and third compartments, the *reticulum* or honeycomb and the *omasum* or manyplies, which constitute 5 and 7 or 8 percent of the total stomach capacity, respectively, in mature animals. All three of these compartments have a common opening or passageway called the *esophageal groove,* through which materials may pass freely. The function of the reticulum is not well understood, but it is known that the omasum is the site where much water is absorbed from the paunch contents before it passes into the fourth compartment, the *abomasum* or true stomach. The abomasum holds about 7 to 8 percent of the total stomach contents, and its function is similar to that of the stomach in monogastric animals.

Symbiotic Microorganisms of the Rumen. The rumen provides an ideal environment as to temperature, moisture, and nutrient supply for microbial life, and literally billions of bacteria—up to 100 billion per gram of dried rumen contents—and somewhat fewer protozoa live in the rumen, to the mutual benefit of both the microorganisms and the host animal, the ruminant. This mutual benefit or support is known as symbiosis. The breakdown of cellulose and related compounds by the enzymes produced by these microorganisms accounts for the higher feeding value of roughages when fed to ruminants.

Volatile fatty acids (VFA) are produced as a result of the microbial fermentation of carbohydrates in the rumen. The most important of these acids, in terms of amount produced, are acetic, propionic, and butyric, in that order. It is estimated that up to 80 percent of the digestible carbohydrate portion of beef cattle rations may be converted to these acids and absorbed directly from the rumen, to be metabolized and used as energy in meeting maintenance requirements or stored as fat. Thus fatty acids are a major source of energy in ruminant rations as contrasted with the situation in monogastric animals, where carbohydrates are largely absorbed as glucose after digestion in the stomach and small intestine.

Recent research in Great Britain and in the United States has shown that if the level of acetic acid resulting from the fermentation of carbohydrates in the rumen can be reduced while the level of propionic acid is increased, the energy of the ration will be more efficiently used. This shift of the acetic-propionic acid ratio in favor of propionic acid reduces the energy losses that occur in metabolism at the cellular level. Promising leads have been uncovered that may make it possible quantitatively to affect the VFA levels.

Bacterial Synthesis of Protein in the Rumen. As bacteria and the other microorganisms living in the rumen multiply, they build or synthesize the protein required for the next generation of organisms, using whatever source of nitrogen is available from the feeds consumed by the host. The bacterial protein thus synthesized in the rumen undergoes digestion in the abomasum or true stomach and intestine, and is absorbed by the host in the form of amino acids, regardless of the source or quality of the protein or nitrogen originally consumed in the ration.

Thus microorganisms play a significant role in protein as well as carbohydrate utilization in the ruminant. By contrast, the nonruminant animal has specific requirements for about 10 of the amino acids, the building blocks of protein. These specific amino acids are called the essential amino acids, and their balance—that is, presence in required proportions—in a feed protein determines the *quality* of the protein for the nonruminant.

The microflora of the rumen are not specific in their requirements as to source of nitrogen or protein; therefore, low-quality proteins are well utilized by the ruminant. Furthermore, nonprotein nitrogenous compounds, such as urea, ammonium salts, and amides, can make up a substantial portion of the total nitrogen or protein requirement of the ruminant. Such nonprotein nitrogenous compounds as urea are almost always a cheaper source of nitrogen, per pound of protein equivalent, than are the usual protein concentrates, and thus it is possible to formulate more economical protein supplements for cattle than for nonruminant animals. A majority of the commercially prepared protein concentrates designed for beef cattle now contain some urea, generally about 3 percent, or about 8 percent protein equivalent. More about urea and protein equivalent is found in Chapter 16.

Bacterial Synthesis of the B-Complex Vitamins in the Rumen. Just as the microflora of the rumen are able to synthesize protein, they also synthesize many of the vitamins required by the host. Once the rumen becomes functional and a bacterial population is established, B-complex vitamins such as riboflavin, niacin, pyridoxine, biotin, folic acid, and

B_{12} are synthesized in the rumen at a rate sufficient to meet the needs of the host. The rate of synthesis of some of these vitamins can be altered by varying the level of certain nutrients in the ration. For example, if cobalt, one of the essential mineral elements, is deficient in the ration, vitamin B_{12} synthesis is too low for maximum performance on the part of the host. Vitamin K, one of the fat-soluble vitamins, is also synthesized in the rumen, but such important fat-soluble vitamins as A, D, and E must be supplied in the ration.

Important Role of Saliva. The salivary glands of cattle are located beneath the tongue and produce 15 to 20 gallons of saliva per day. This saliva, a highly alkaline material, is mainly secreted during rumination or cud-chewing, but some appears during the act of eating. The saliva serves several functions but perhaps the most important is that of buffering or neutralizing the large amount of acid produced in the rumen by bacterial enzymatic fermentation.

When the total production of volatile fatty acids is extremely high, as often happens when high-concentrate rations are used in cattle-finishing programs, not enough saliva is secreted to neutralize the increased acid production. The resulting condition is known as rumen parakeratosis or an inflammation and keratinization of the hairlike papillae of the rumen wall. It is believed that this condition interferes with absorption of nutrients through the rumen wall, and that anything which would increase saliva flow, such as feeding coarse roughage even in small amounts (as little as 2 pounds of hay daily) to ensure rumination, would prevent the problem. The addition of buffering agents such as sodium bicarbonate has not been highly successful.

Saliva also contains urea, which is continuously recycled and thus provides the rumen microflora with another nitrogen source for the synthesis of protein. Because much of the salivary urea was originally absorbed as ammonia from the rumen, the recycling serves to increase the utilization of dietary nitrogen.

Ruminant saliva contains minute quantities of certain enzymes, but they are considerably less important than are enzymes in nonruminant digestion.

FACTORS AFFECTING MICROBIAL ACTIVITY IN THE RUMEN

A more thorough understanding of the relationship between rumen microflora and the host animal has led to the concept that, in order to feed ruminants adequately or, specifically, to feed beef cattle, the nutrient requirements of the microflora should first be met. It is quite well established that the bacterial flora of the rumen consist of at least several dozen forms and that the relative distribution as well as total numbers present can be altered by changes in the ration. Certain forms are known to predominate if finishing rations that are high in readily available carbohydrates are being fed, whereas still other forms prevail when cattle are grazing lush pastures. Changing from one type of ration to another results in bacterial population shifts, but these shifts take place rather slowly. As a result, digestive disturbances may occur if the type of ration is changed too suddenly. Undoubtedly this fact partly explains the poor performance, often amounting to actual weight loss, that results when cattle are shifted from dry wintering rations to succulent spring pastures.

Protein or nitrogen level in the ration has a marked effect on the total digestibility of the dry matter of ruminant rations, and especially on the crude fiber. This level is extremely important, because differences in crude-fiber digestibility have an influence on digestibility of remaining nutrients in the ration that may be encased within the cell walls of the fibrous portion of the plant.

Including a small amount of readily available carbohydrate such as ground corn or molasses in high-roughage rations has been reported to increase rumen microfloral activity and thereby increase the feeding value of high-fiber roughages. If large amounts of such easily digestible carbohydrates are fed, as in finishing rations, the reverse happens—that is, crude-fiber digestibility is reduced—but in this kind of ration the crude fiber content of the ration is relatively low.

Certain minerals, particularly phosphorus, sodium, potassium, sulfur, and cobalt, are essential for maximum microfloral activity. The relative availability or solubility of these minerals in the rumen seems to be involved, as well as the content of the ration, at least with respect to phosphorus.

Certain feeds, such as high-quality alfalfa and certain commercial fermentation by-products, reportedly contain as yet unidentified factors that stimulate the activity of the rumen microflora. Ashing the alfalfa and feeding the ash as a supplement to low-quality roughage has been reported to increase the concentration of rumen bacteria and improve crude fiber digestion. This suggests that the mineral matter of alfalfa may be responsible for its unique qualities.

Bulk and density of a ration undoubtedly play a part in microfloral activity and therefore in digestibility of crude fiber. Bulk provides for distention of the rumen and enhances normal rumination and physiological function of the rumen itself. Indirectly, bulk is involved in rate of passage of material from the rumen, and it is logical to assume that material passing too rapidly from the rumen (finely ground or chopped roughages, for instance) is not subjected to the normal cellulytic bacterial action. Dense or high-concentrate rations that are low in fiber and therefore in bulk are not so dependent on rumen microflora as are the bulkier rations. The low-fiber rations are undoubtedly digested and utilized by cattle in a manner not much different from that of the monogastric animal. It is interesting that cattle on either an all-ground roughage ration or an all- or high-concentrate ration seldom ruminate or chew their cuds.

WASTAGES IN THE RUMEN

The products of fermentation and microbial action in the rumen are unfortunately not all used to advantage by the host. Ingested or ration protein and nonprotein nitrogen compounds are partially converted to ammonia and volatile fatty acids in the rumen. Although much of the ammonia is utilized by microorganisms and converted to bacterial protein, variable amounts of ammonia are absorbed from the rumen into the bloodstream and either used in the synthesis of amino acids or wastefully excreted in the urine. Some of the absorbed ammonia is reexcreted into the rumen by way of the saliva, where it may be further utilized by the microflora.

The extent of loss of protein or nitrogen accruing from excessive ammonia production in the rumen is not known, but undoubtedly it is affected by the level of protein in the ration, solubility of protein or nitrogenous material in the ration, type of carbohydrate in the ration, and concentration and character of microflora in the rumen. A constant intake of protein, and especially of urea, throughout the day, such as happens in self-feeding, as contrasted to once or twice daily feeding, improves the utilization of ammonia by the rumen microorganisms. Less soluble forms of nonprotein nitrogen—biuret, for example—are being investigated because of their promise in reducing the wastage from excessive ammonia production.

The microbial fermentation in the rumen results not only in the production of volatile fatty acids from carbohydrate material used as energy, but also in the formation of the gases, methane and carbon dioxide. Both gases are normally eliminated by way of the esophagus and serve no useful purpose, but rather represent a loss in energy from the ration.

Heat is also produced as a result of the fermentation in the rumen. Only in very cold weather, especially in the absence of shelter, do cattle derive any benefit from this heat production. On the contrary, elimination of this heat may impose a burden upon the animal and thus be responsible for wasting still more energy because of restlessness and rapid breathing.

A more efficient body heat elimination system in Brahman cattle, due to a larger skin surface, is said to explain this breed's superior adaptation to hot, humid climates.

NUTRIENT REQUIREMENTS OF BEEF CATTLE

The nutrient requirements of beef cattle closely parallel those of the microflora found in the rumen, at least from a qualitative standpoint. Because the requirements of both must be simultaneously supplied by the cattle ration, it is rather difficult to assess the requirements separately. For practical ration formulation it is unnecessary to do so. Quantitative nutrient requirements for rumen microorganisms have not been determined, and more information is needed before separate requirements can be established.

The nutrient requirements of beef cattle are discussed in the remainder of this chapter from a broad viewpoint under these headings: (1) feed capacity 'and bulk, (2) energy, (3) protein, (4) minerals, (5) vitamins, and (6) water. Specific requirements for different age groups and different feeding programs are discussed in the appropriate chapters.

FEED CAPACITY AND BULK

Cattle should be fed to capacity under all ordinary feeding conditions. Regardless of whether the aim is maximum daily gain in a steer finishing program or only maintenance of dry cows, performance is more favorable if the complete ration, or at least one item in the ration, is fed according to appetite. This procedure enables cattle to consume enough feed to satisfy their hunger and thus prevents the uneasiness and wasteful excess activity associated with a limited ration. Limited feeding of high-concentrate rations to growing cattle and dry cows is possible but usually not practical for economic reasons.

Condition and age both affect feed capacity, and there is much variation between animals of the same condition or age. Cattle on finishing rations voluntarily consume daily an amount of feed equal to 2.5 to 3 percent of their live weight (air-dry basis). Older cattle such as cows in good condition, and fleshy individuals such as mature bulls and fitted show cattle, consume less, even as low as 1.5 percent of their live weight. Thin, growthy yearling or older steers may consume up to 4 percent of their weight daily for short periods of time. In general, the lower the fiber content or bulk of a feed, the lower the intake.

Certain additions can be made to the ration to cause increased consumption, especially if a major portion of the ration is unpalatable. An example is the addition of molasses to a wintering ration consisting largely of ground corn cobs or similar low-quality roughage. Such an addition to an already palatable ration increases intake only temporarily and serves later merely as a replacement for an equal weight of ration being fed prior to the addition.

Evidence has been presented to suggest that acceptance or palatability of a ration, or the level of voluntary consumption, is related to the digestible nutrient content of the ration—that is, the more digestible the ration, the greater the daily consumption, and vice versa. This line of reasoning would relate daily feed intake to the quality of the ration as well as to the size, age, or condition of cattle. There is a growing body of thought that all animals, when given full access to a palatable ration, will eat only enough feed to meet their energy needs; that differences in energy needs, under hormonal control, are responsible for the genetic differences seen in rate of gain among animals.

Feeds high in water content, such as succulent spring pasture, winter small-grain pastures, and high-moisture silages, are apparently consumed at a lower level of dry matter or air-dry feed intake because of their high water content. The effect of moisture content would appear to be more physical than metabolic.

A certain minimum amount of bulk or roughage is required to maintain feed intake at a constantly high level in the ruminant; otherwise bloat and other digestive disturbances will be frequent. The practical minimum roughage level in finishing rations in most parts of the country is from 0.5 to 0.8 pound per 100 pounds live weight daily, or 20 to 30 percent when expressed as percentage of the ration. In areas where roughage nutrients are not cheaper than concentrate nutrients, rations containing less than 10 percent or even no roughage are fed. This practice is discussed in greater detail in Chapter 15. Apparently roughages can provide the necessary bulk, even though finely chopped, if fed in pelleted form, as discussed in Chapter 24 dealing with preparation of feeds.

ENERGY

The energy requirements of beef cattle have been studied by various methods with the result that requirements are expressed in a number of ways. Specific requirements as used in this text are given in terms of total digestible nutrients (TDN) and in terms of digestible energy. Daily requirements for energy and other nutrients are given, as is the required percentage content of the ration for all the various nutrients.

The total digestible nutrient content of a feed has been defined as the sum of all the digestible organic nutrients—protein, fiber, nitrogen-free extract, and fat (the latter being multiplied by 2.25 because its energy value is approximately 2.25 times that of protein and carbohydrate). The TDN as a measure of the energy value of feeds is subject to some criticism, mainly because feeds are not digested in the test tube exactly as they are in the cow. There is also the criticism that TDN assumes an equal caloric or energy value for digestible protein, nitrogen-free extract, and crude fiber, and a constant energy value for ether extract. It is now known that this assumption is invalid.

In the 1963 revision of beef cattle nutrient requirements prepared by the Sub-Committee on Beef Cattle Nutrition of the National Research Council, the energy requirement of beef cattle is expressed in terms of digestible energy (DE) or calories as well as in TDN. (In a later revision scheduled for publication in 1969, the energy requirement for finishing rations is also expressed in terms of net energy, which is discussed in the following paragraphs.) The difference in the gross energy content of feeds consumed and of the feces excreted in a digestion trial is a measure of the digestible energy content of a feed or ration. Gross energy of both feed and feces is determined by burning a sample of each in a bomb calorimeter and measuring the heat produced in the combustion process. The term "digestible energy" refers to a measure, expressed as therms, of the useful energy in a feed or ration that has been removed or absorbed from the feed by the process of digestion and absorption. The feed tables included in the Appendix contain values for digestible energy, which were computed from existing TDN values in feed tables as follows:

$$\text{therms digestible energy} = \frac{\text{lb TDN} \times 454 \times 4.41 \text{ kcal}}{1{,}000 \text{ kcal}}$$

Digestible-energy values and feed requirements computed in this fashion are only as reliable as the TDN values from which they were computed.

Metabolizable energy is a still more exacting method of describing the real worth of feeds, because it accounts for losses in digestion in the gastrointestinal tract as well as losses or wastages in metabolism in the cells or tissue. It is being used by the British nutritionists and, with more research, may in time replace TDN or DE in the United States as a means of describing feeds and nutrient requirements.

Net energy, yet another method of expressing energy content of the ration and long recognized as the ultimate or ideal method of describing or classifying feeds, is being used, especially in the West, in formulating rations for finishing cattle. Energy requirements are partitioned into an energy requirement for maintenance (NE_m) and a requirement for a given or desired level of gain or production (NE_p). The sum of the two requirements is obviously the total energy requirement for an animal. Knowing this requirement and the NE values for feeds for maintenance and gain, rations may be computed that will satisfy the requirement. One complicating factor is the question of voluntary feed intake, and, unfortunately, it is possible that in too many instances the intake level necessary to supply the computed NE requirement may either be more than the animal will be able to consume or less than a full feed. Most cattle feeders with practical experience may not accept the net energy method for this reason alone.

A deficiency of energy, due to a simple lack of sufficient total feed, is undoubtedly the most common deficiency in beef cattle rations. The results of low energy intake are slow growth or even loss in weight, stunting, failure to conceive, and increased disease and mortality. An energy deficiency is usually accompanied by deficiencies in all other nutrients but expecially in protein. Overstocking of pastures and ranges, especially in periods of prolonged drought, is the principal cause of energy deficiency. Cattle on finishing rations may receive rations adequate in total amount or weight to meet appetite needs but still not perform up to their capabilities in terms of gain, owing to a lack of sufficient energy in the ration.

PROTEIN

Protein requirements are usually expressed on the basis of percent of both the total and the digestible protein in the ration. In tables appearing in the appropriate chapters, requirements are also given in terms of daily requirements for total and digestible protein. The digestible protein values are approximately equal to 60 percent of the total protein for high roughage rations and to 75 percent for high-concentrate rations, owing to the differences in the digestibility of roughages and concentrates.

Ration nitrogen, from which bacterial protein can be synthesized in the rumen, can be supplied in nonprotein nitrogenous forms such as urea, as previously mentioned. Under ordinary circumstances up to one-third of the total protein requirement may safely be supplied in nonprotein nitrogenous forms. Some protein concentrates consisting almost entirely of nonprotein nitrogen are being successfully fed today, but in these situations the feeder must be alert to the danger of toxicity, as is further discussed in Chapter 17.

Symptoms of protein deficiency in beef cattle are poor growth, depressed appetite, reduced milk flow and light calf weaning weights, irregular estrus in cows, delayed onset of first estrus in heifers, delayed recycling of lactating cows, and loss of weight in extreme cases.

MINERALS

The requirements for calcium and phosphorus are quite well established, and numerous determinations for the content of these minerals in feeds have been made. The requirements for these minerals and their sources are given in appropriate tables in other chapters. Requirements for other minerals, except for salt, are not so well established and thus are not included.

Phosphorus. Phosphorus deficiency is the most common mineral deficiency in cattle and is most likely to occur in cattle being wintered on all-roughage rations or on mature, cured grass. Phosphorus deficiency is usually associated with use of feeds grown in phosphorus-deficient soils. Figure 27 shows that this mineral is deficient in the feeds raised in large areas of the country.

In the early stages of phosphorus deficiency or when a borderline deficiency exists, feed intake is decreased, gain is reduced, and milk production falls off, with a consequent reduction in suckling calf gains. Efficiency

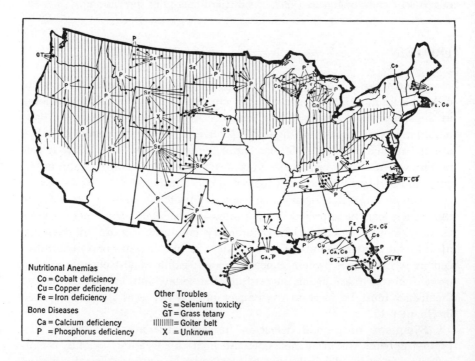

FIG. 27. Areas deficient in important minerals needed for the adequate nutrition of livestock. (U.S.D.A.)

of feed utilization is reduced in the feedlot, and if the deficiency is prolonged, blood phosphorus levels fall. Pica, or depraved appetite, results from severe prolonged deficiency and is usually accompanied by bone alterations resulting in lameness and stiff joints. In breeding cattle, phosphorus deficiency results in reduced percent calf crop weaned because of a combination of problems, discussed in Chapter 9.

As mentioned earlier, phosphorus may be one of the most critical nutrients for normal bacterial action in the rumen. Therefore, requirements of both the microorganisms and the host animal must be supplied in the ration. The relative availability or solubility of the various phosphorus supplements fed varies among the sources of phosphorus and even among batches of the same source. Dicalcium phosphate varies least in availability.

Calcium. A deficiency of calcium is most apt to occur when high-concentrate rations, such as finishing rations, are fed. This is especially true if the limited roughage portion of the ration is nonleguminous. If cows, especially those nursing calves, subsist principally on mature, weathered grasses or hay or cereal straw, they are almost certain to respond to calcium supplementation. Heavy corn silage rations are often borderline for calcium unless supplemented with legume hay. Calcium deficiency symptoms are not specific, except in extreme cases when fractures may occur because of depletion. Poor growth rate, inefficient feed conversion, and low ash content of the bones are usually the result of low calcium intake. Calcium deficiency is often associated with energy, protein, and phosphorus deficiencies.

Experiments conducted at the Indiana and Iowa stations showed practically no advantage in feeding calcium supplements to older steers being finished on well-balanced rations. However, the Kansas station concluded that calves fed a nonleguminous, low-calcium roughage, such as prairie hay or a combination of prairie hay and corn silage, should be supplied with finely ground limestone or some other mineral rich in calcium. As seen in Table 27, no benefit was derived from the addition of ground limestone to the ration of calves receiving 6 pounds of alfalfa hay. When only 2 pounds of alfalfa were fed, the feeding of limestone was profitable.

The calcium and phosphorus content of the principal supplements used to correct deficiencies of these elements is given in the Appendix tables. Note also the level of fluorine, a harmful compound in large amounts, in some of the supplements.

It will be noted from the nutrient requirement tables that the ratio of calcium required to phosphorus required is generally about 1:1. Recent research suggests that greatly altering this ratio is detrimental. This is especially true if considerably more calcium is added to the ration than phosphorus. For example, steamed bonemeal, usually considered a good, eco-

Table 27

Value of Adding Ground Limestone to a Low-Calcium Ration[a]
(375-Pound Calves, Fed 180 Days)

Hay Fed	Alfalfa Hay				Prairie Hay			
Other Roughage	None		Cane Silage		None		Cane Silage	
Mineral Fed	None	Ground Limestone	None	Ground Limestone	None	Ground Limestone	None	Ground Limestone
Daily gain (lb)	2.46	2.44	2.32	2.49	2.15	2.25	2.07	2.43
Average daily ration (lb)								
Shelled corn	9.74	9.60	9.77	9.51	9.83	10.09	9.65	9.85
Cottonseed meal	1.12	1.11	1.00	1.00	1.36	1.40	1.16	1.16
Hay	6.03	5.84	2.01	2.00	4.43	5.03	1.36	1.41
Cane silage	—	—	9.71	10.05	—	—	8.91	10.14
Ground limestone	—	0.11	—	0.10	—	0.12	—	0.10
Feed per cwt gain (lb)								
Shelled corn	396	394	422	382	457	448	466	406
Cottonseed meal	46	45	43	40	63	62	56	48
Hay	245	239	87	80	206	223	65	58
Cane silage	—	—	419	404	—	—	430	418
Ground limestone	—	5	—	4	—	5	—	4
Selling price per cwt	$13.25	$13.00	$12.75	$13.00	$12.50	$12.75	$12.50	$13.00

[a] Kansas Circular 151.

nomical phosphorus supplement, contains about 30 percent calcium but only 14 percent phosphorus. If this is the phosphorus supplement of choice, twice as much calcium is added as phosphorus, thus pushing the ratio out of balance. If the ration calcium content is already high in relation to phosphorus content, the problem may be especially acute. Phosphorus utilization is apparently impaired by a high calcium content in the ration. This problem needs, and is receiving, further research, but it currently appears that such phosphorus sources as diammonium phosphate and disodium or monosodium phosphate, which add no calcium, should be strongly considered in making up cattle mineral supplements.

Salt. The salt or sodium and chlorine requirements of full-fed beef cattle are met by including 0.5 percent salt in the total ration. Salt, of course, may be and often is satisfactorily fed free-choice rather than as part of a mixed ration.

Salt is essential to the growth and health of all kinds of livestock. Cattle exhibit a special eagerness for it and soon show signs of restlessness and malnutrition if it is long withheld. The form in which the salt is fed has an effect on the amount consumed. Cattle that have free access to granulated, flake, or loose salt will consume approximately twice as much as they will when salt is furnished in the form of compressed blocks. Whether enough salt is obtained from licking blocks kept in the feedbunk or in the pasture is a disputed subject among cattlemen. Weathering loss is a factor to be considered. At the Kansas station the flake form of salt weathered 24 percent per month, whereas the blocks weathered only 11 percent. In the states with a heavier rainfall, greater losses may be expected. Cattle accounted for the disappearance of 604 pounds and weathering for 141 pounds of block salt exposed in pastures from April 30 until October 21, or 175 days, in a study at the Illinois station. The total disappearance of salt per 2-year-old steer was approximately 3 pounds per month. Little difference was observed at the Illinois station between the salt consumption of yearling cattle wintered on good roughages in the drylot and that of the same cattle on good pasture the following summer. Evidence that salt consumption is greatly affected by the amount and character of the ration is shown in Table 28.

The use of salt to control free-choice protein supplement intake usually results in the consumption of considerably more salt than is required. Cattle may consume as much as 2 pounds of salt daily apparently without harm, provided they have free access to an abundant supply of drinking water to ensure elimination of the excess sodium and chlorine by way of the urine. Frozen waterers and stock tanks are to be avoided, as there is extreme danger of toxicity if water is unavailable for any length of time.

Other Major Minerals. Magnesium, potassium, and sulfur are required by cattle but the specific quantitative requirements have not been determined. Deficiencies of these minerals have been produced on purified diets only. Experimental work in this area is limited because there has been no apparent need for research as the content of these minerals in the feedstuffs customarily consumed by beef cattle is apparently sufficient to prevent deficiencies.

If urea supplies a considerable portion of the nitrogen in the ration, sulfur, either in organic or inorganic form, has been shown to be beneficial as a mineral additive. Sulfur, or the sulfur-bearing amino acid methionine, is known to facilitate bacterial synthesis of protein. Practical rations containing considerable roughage have not always been improved by the addition of sulfur, indicating that the sulfur content of such rations is probably already adequate.

Trace Minerals. Trace minerals are those found in only minute amounts in soils and plants. The content of such minerals in feedstuffs is closely associated with the amounts present in the soils in which the feedstuffs are grown. Figure 27 indicates areas where it has definitely been established that certain trace mineral deficiencies may occur. It will be seen that areas whose soils are most subject to leaching, either because of heavy rainfall or permeability, are apt to experience a greater incidence of trace mineral deficiencies in plants grown in such soils.

Iodine deficiency is usually evidenced by the production of goiterous calves that are either born dead or die soon after birth unless iodine is administered. The use of iodized salt containing 0.01 percent of stabilized potassium iodide in the pregnant cow's ration prevents deficiency symptoms.

Table 28
Salt Consumed by Beef Cattle Under Various Feeding Conditions

	Feeding Conditions	Daily Consumption per Head (lb)	
		Loose Salt	Block Salt
Calves			
Iowa	Full-fed in drylot	0.015	0.007
Kansas	Wintered on bluestem pasture	0.04	
	Wintered on sorgo silage	0.05	
	Wintered on sorgo silage and hay	0.06	
	Sorgo silage plus one-third feed of corn	0.05	
	Sorgo silage plus full feed of corn	0.013	
Montana	Wintered well on hay and straw	0.063	
	Limited amounts of hay and straw	0.183	
Yearlings			
Illinois	Wintered on hay and silage	0.18	0.096
Iowa	Full-fed in drylot		0.013[a]
	Full-fed in drylot		0.03[b]
Kansas	Full-fed in drylot	0.07	
Montana	Wintered well on hay and straw	0.08	
	Limited amounts of hay and straw	0.26	
Two-year-olds			
Illinois	Bluegrass-ladino pasture		0.09
Iowa	Full-fed in drylot	0.042	0.018
Iowa	Full feed of corn with corn silage		
	No protein concentrate		0.04
	1.5 lb protein concentrate		0.026
	3 lb protein concentrate		0.013
	Full feed of corn with legume hay		
	No protein concentrate		0.07
	1.5 lb protein concentrate		0.024
	3 lb protein concentrate		0.025

[a] Average of 2 lots.
[b] Average of 11 lots.

Iodized salt should be used as a general practice, if available, for the slight difference in cost over noniodized salt will usually be offset by the insurance provided.

Iron is undoubtedly required by beef cattle, but quantitative requirements are not established nor have consistent responses been obtained from its use in practical rations. It may therefore be assumed that feedstuffs commonly consumed by beef cattle contain adequate iron.

Copper requirements for beef cattle are very low, but the need for copper may become critical if excessive molybdenum and sulfates occur in the ration. Copper deficiency can be prevented by adding 0.25 to 0.5 percent of copper sulfate to the salt, fed free-choice. This insures a copper content of 4 to 8 parts per million in the total air-dry ration. Excess molybdenum apparently interferes with copper metabolism. Therefore, in areas where molybdenum toxicity is evident, additional copper should be fed. Although copper deficiency is almost exclusively a local problem, calves being fitted for show or sale by the use of nurse cows well beyond the normal time for weaning sometimes may develop copper deficiency. Generally poor performance along with intermittent to severe diarrhea and stunted growth are symptoms of copper deficiency. When copper deficiency is complicated by molybdenum toxicity, depigmentation of the haircoat is common.

Cobalt is required by rumen microflora to ensure adequate vitamin B_{12} synthesis, as this element is an integral part of the B_{12} molecule. The requirement has been set at 0.07 to 0.10 milligram per 100 pounds body weight. It is generally supplied in the ration, when needed, in the form of cobalt sulfate added to the salt or mineral mixture at the rate of 1 ounce per 100 pounds of salt.

Hay and pasture forage rations containing 0.01 to 0.07 part per million of cobalt have resulted in cobalt deficiency, whereas if the cobalt content is in the neighborhood of 0.10 part per million, performance is normal. Thus the requirement is apparently 0.10 part per million, expressed as content of the diet.

Symptoms of cobalt deficiency, like those of many other mineral deficiencies, are general rather than specific. Severe deficiency results in reduced feed intake, emaciation, weakness, and even death. Perhaps of more importance is the borderline deficiency, which may not be recognized and thus is not corrected. Such borderline deficiencies may result in generally reduced performance and inefficient feed conversion.

Manganese and *zinc* are undoubtedly required by beef cattle, but since forages contain 50 to 150 parts per million of manganese and 10 to 100 parts per million of zinc, the likelihood of a deficiency occurring under practical conditions is remote.

As indicated earlier, trace mineral deficiency is principally a geographical problem, and consultation with soils experts and nutritionists in specific areas is necessary to determine whether supplementation is needed. Trace-mineralized salt is available in most feed stores. The cost per unit of trace mineral may be quite high and its feeding may be unessential. This is especially likely if the rations being fed contain average to good quality roughages, grown in areas where the soils are adequate in trace mineral elements.

The elimination of most or all of the roughage in the newer high-concentrate rations has reawakened interest in the trace mineral question, and active research may yet show a greater need for trace minerals under such circumstances. Garrett of California[2] has suggested the following levels of the various inorganic minerals in the complete mixed rations being fed to feedlot cattle.

Element	Milligrams per Pound of Ration (90% Dry Matter)
Calcium	1,000 (calves, 1,400)
Chlorine	400.0
Cobalt	0.05
Copper	2.5
Fluorine	not more than 3.0
Iodine	0.08
Iron	10.0 (probably less)
Magnesium	300.0
Manganese	4.0
Molybdenum	0.05 (not over 9)
Phosphorus	1,000 (calves, 1,200)
Potassium	400.0
Selenium	0.05 (not over 2)
Sodium	200.0
Sulfur	450.0
Zinc	8.0

Toxic Minerals. Some minerals, although they may actually be required in minute amounts, may produce harmful effects if ingested in excess of these requirements.

Fluorine, contained in either undefluorinated or raw rock phosphate and mine washings or deposited upon the forage grown in areas subjected to fluoride-containing smoke, is toxic to beef cattle if consumed in excessive amounts. The usual fluorine content of the mineral supplements fed to beef cattle is shown in the Appendix tables.

Symptoms of fluorine toxicity are mottling and erosion of the tooth

[2] W. N. Garrett, *1964 Proceedings, California Animal Industry Conference.*

enamel and a softening and thickening of the bones with a resulting decrease in breaking strength. The maximum level for fluorine in the ration has been recommended at not more than 65 parts per million for feeder cattle and not more than 30 parts per million for breeding cattle kept on such rations for extended periods of time.

Selenium toxicity is a local problem occurring primarily in North Dakota, South Dakota, and the Rocky Mountain states. Cattle consuming feeds containing 8.5 parts per million of selenium for a considerable length of time show symptoms of chronic toxicity. Death occurs if feeds consumed contain 500 to 1,000 parts per million of selenium. Characteristic symptoms of selenium toxicity are loss of appetite, sloughing of hoofs, loss of hair from the tail, and eventually death. No effective treatment has been determined other than removal of animals from the affected areas. A minute amount of selenium in the ration of pregnant and lactating cows, in the form of sodium selenite, has reduced the incidence of "white muscle" disease.

Molybdenum is apparently an essential mineral, but the requirement is low, since more than 10 to 20 parts per million in forages results in toxic symptoms. Excess molybdenum interferes with copper metabolism as previously mentioned. Until more information is available, molybdenum should not be added to beef cattle mineral supplements.

VITAMINS

Vitamin deficiencies in general are less apt to occur in cattle feeding than are mineral deficiencies. This is because, first, as mentioned earlier, most of the vitamins required by cattle are synthesized by rumen microorganisms, and, second, the feeds usually consumed by cattle are fair to excellent sources of those vitamins that must be supplied in the ration.

Vitamin A and Carotene. Requirements for vitamin A and its precursor, carotene, are expressed as both carotene and vitamin A in the requirement tables in the appropriate chapters. Approximately 2 milligrams of carotene or 750 IU of vitamin A are required per pound of feed consumed for normal growth in young cattle, and for all finishing cattle. Dry pregnant cows, nursing cows, and bulls require 2.5 milligrams and 4 milligrams of carotene or 1,000 and 1,500 IU of vitamin A respectively, per pound of feed consumed.

Vitamin A, which is synthesized in the body from carotene obtained from the ration, may be stored in the body during periods of high carotene intake such as occur during the summer on lush pasture. Carotene reserves may be drawn upon during winter, and the amount stored of course affects the level of vitamin A or carotene necessary in the winter ration. Another factor affecting the overall practical problem of supplying vitamin A is the instability of carotene, evidenced by losses or destruction of the carotene in feeds through oxidation during storage. For example, hay may lose up to three fourths of its carotene content in one winter storage period.

Vitamin A deficiency in beef cattle is rather uncommon when good quality roughages are fed. However, when a combination of poor hay and concentrates low in carotene content, such as old corn, small grains, grain sorghums, or molasses, make up the ration, vitamin A deficiency may occur. A little-understood interrelationship between nitrate content of feeds and poor utilization of carotene apparently is causing considerable vitamin A deficiency today.

Common but not highly specific vitamin A deficiency symptoms in feedlot cattle are a marked reduction in feed intake, reduced gain, and poor feed conversion. These symptoms may become pronounced in cattle that are almost ready for market at the onset of hot weather.

Other vitamin A deficiency symptoms are night blindness (inability to adjust to sudden bright lights), incoordinated gait, and convulsive seizures in severe cases. Excessive lacrimation may also occur, but this symptom can easily be confused with watery eyes brought on by pink eye (see Chapter 27) or by irritation from cinders and dust during movement or shipment of the cattle. Diarrhea, from severe to intermittent, in both young and older cattle may result from vitamin A deficiency, but here, too, other causes are possible and may cause confusion. A characteristic deficiency symptom in feeder cattle is anasarca or generalized edema. The swellings are localized in the brisket area and in the knee and hock joints. Lameness usually accompanies the more severe cases.

Sexual activity declines in bulls suffering from vitamin A deficiency; spermatozoa decrease in numbers and motility and increase in number of abnormal forms. In breeding cows, estrus may continue, but cows conceive less readily. Vitamin A deficiency in the pregnant cow may result in abortion if it is sufficiently severe. Calves may be born dead or weak, and retained afterbirths are common in deficient cows. A suspected vitamin A deficiency can be verified by analysis of liver tissue for vitamin A content. A value that is below 4 micrograms of vitamin A per gram of fresh liver is considered critical, and from 4 to 10 micrograms are considered suboptimal.

Vitamin D. A deficiency of this vitamin is extremely unlikely under practical conditions, because beef cattle receive sufficient vitamin D from exposure to direct sunlight or from the consumption of sun-cured hay. In areas where days are extremely short or where cloudiness persists during most of the year, and especially during haying season, it is conceivable that a deficiency, especially in borderline form, may exist, but clear-cut cases of vitamin D deficiency are seldom reported.

The disease known as rickets, which results from poor calcification of bone, is the chief symptom of vitamin D deficiency produced under controlled experimental conditions. Posterior paralysis may result from fracture of vertebrae in severe cases. Poor performance may result from borderline deficiency. Adequate vitamin D is essential for efficient utilization of calcium and phosphorus and for bone formation.

Vitamin E. Muscular dystrophy or white muscle disease in calves between the ages of 2 to 12 weeks is the chief symptom of a deficiency of vitamin E. Certain limited geographical areas have reported this condition but it is not widespread. As mentioned in connection with mineral requirements, selenium level in the ration may be involved in this disease.

The quantitative requirement for vitamin E, expressed as tocopherol, is tentatively estimated to be less than 40 milligrams of tocopherol per 100 pounds of body weight daily. In affected areas, losses due to muscular dystrophy in calves may be reduced by feeding 2 to 3 pounds of grain during the last 60 days of pregnancy or by oral administration of tocopherol to both cow and calf shortly after parturition.

Vitamin K. This vitamin is synthesized by the rumen microflora in adequate amounts under normal feeding conditions. Moldy clover hay and clover pasture sometimes contain dicoumarol, a compound that prevents normal blood clotting. Therapy with vitamin K is usually effective in combating this condition.

The B Vitamins. Although requirements for most of the B vitamins such as riboflavin, thiamin, and biotin have been demonstrated for the young calf before it has developed a functioning rumen, attempts to improve the rations of cattle over 8 weeks of age with B-vitamin supplementation have not been generally successful. These vitamins, like vitamin K, are apparently synthesized by the microflora of the rumen at a rate sufficient to meet the needs of the animal. An example of how dietary nutrients may affect bacterial synthesis of B vitamins has been mentioned, namely the relationship between cobalt intake and B_{12} synthesis. Other such relationships may be discovered in the research being conducted in this field at the present time.

WATER

Water, because of its abundance and universal use, is seldom regarded as a feed, and yet it is one of the most essential nutrients for all animal life. The amount of water required by cattle, exclusive of the water contained in the ration, varies with the character of the feed, the amount of dry matter consumed, and the air temperature. Data on the water consumed on different rations and by cattle of different ages are extremely meager, but enough are available to permit a rough estimate of the daily water requirements of cattle full-fed in drylot. It can be seen from Table 29 that as the season advances from winter to summer, the amount of water consumed per 100 pounds live weight increases rather sharply. Similar data shown in Table 30 indicate that, in addition to temperature effect, increasing levels of feed intake result in the intake of more water.

Water should be kept slightly above freezing temperature during the winter by the use of insulated tanks and by tank heaters during extremely cold weather. Electric, oil, or gas heaters are all suitable, and the choice is usually based on cost of equipment, ease of operation, and freedom from mechanical difficulties. Heaters save labor in keeping the tanks ice-free, but cattle appear to thrive as well on ice-cold water as on water at moderate temperatures.

Table 29
Water Consumed by Cattle Full-Fed in Drylot

	Period	Daily Water Consumption (lb)	Consumption per 100-lb Live Weight (lb)
Calves			
Iowa	Feb. 4–14	26	4.8
	Mar. 5–15	27	4.4
	May 4–14	51	6.7
	July 3–13	65	7.1
Ohio	Apr. 15–Aug. 18	57	6.5
Yearlings			
Iowa	Feb. 4–14	29	3.4
	Mar. 5–15	25	2.7
Illinois	Aug. 5–19	92	8.8
	Aug. 22–Sept. 6	86	7.7
2-year-olds			
Iowa	Feb. 4–14	36	3.4
	Mar. 5–15	33	3.0

Table 30
Total Daily Water Intake[a] as Affected by Temperature and Level of Feed Intake (Dry Matter Basis)[b]

Temperature (Fahrenheit)			40°	50°	60°	70°	80°	90°
Gallons of Water per Pound of Dry Matter Intake			0.37	0.40	0.46	0.54	0.62	0.88
Body Weight (lb)	Expected Daily Gain (lb)	Dry Matter Daily (lb)	(gal)	(gal)	(gal)	(gal)	(gal)	(gal)
Cattle on Maintenance Rations[c]								
400	0.0	5.4	2.0	2.2	2.5	2.9	3.3	4.8
800	0.0	8.8	3.3	3.5	4.0	4.8	5.5	7.7
1,200	0.0	11.8	4.4	4.7	5.4	6.4	7.3	10.4
Wintering Weanling Calves								
400	1.0	9.9	3.7	4.0	4.5	5.3		
500	1.0	11.7	4.3	4.7	5.4	6.3		
600	1.0	13.5	5.0	5.4	6.2	7.3		
Wintering Yearling Cattle								
600	1.0	14.4	5.3	5.8	6.6	7.8		
800	0.7	16.2	6.0	6.5	7.4	8.7		

[a] Total water intake includes both the water drunk and that contained in the feed.
[b] *Journal of Animal Science*, 15:722–740.
[c] Animals that are neither gaining nor losing weight.

Part **III**

THE COMMERCIAL
COW-CALF PROGRAM

Chapter 7

The COMMERCIAL COW-CALF PROGRAM

The commercial cow-calf program, as the term is used in this discussion, refers to an operation consisting almost exclusively of grade or nonregistered mother cows and their suckling calves up to weaning time, the replacement heifers, and, of course, the bulls. Sometimes the steer calves and the surplus heifer calves are kept over to winter and summer before being sold as yearlings, in which case a stocker program has been added and the operation is no longer exclusively a cow-calf program. The weaner calves, sold at 6 to 8 months of age, are the principal production or source of income for the cow-calf man. Sale of cull cows and other surplus older cattle contribute income, of course, but this is incidental to the main objective, namely, the maximum return from the sale of heavy, high-quality calves.

FACTORS TO CONSIDER IN CHOICE OF PROGRAM

Before starting a commercial beef cow herd—or any beef cattle program, for that matter—one should be certain that the program is the one that will best utilize the feed production capabilities of the farm or ranch. Among the important considerations involved in choosing a program are:

1. Kind and amount of pasture to be utilized.
2. Homegrown supply of grain and roughages.
3. Season during which labor is least needed for other work.
4. Local market demands for feeder and/or slaughter cattle.
5. Proximity to market outlets and surplus feed supplies.
6. Climate, including temperature range, rain and snowfall, and relative humidity.

7. Available equipment and shelter.
8. Extent and availability of financial resources.
9. Training, skill, and experience of the operator.
10. Personal likes and dislikes.

No single program is best suited to all conditions, and each has its advantages and disadvantages. The cow-calf program is growing in popularity, as shown by the increasing numbers of new herds being established, especially east of the Mississippi River. Some believe that this is largely due to the periodic high prices that must be paid for stockers and feeders obtainable from the range states, but it is doubtful if this is the main explanation.

ADVANTAGES OF THE COW-CALF PROGRAM

Among the many reasons for the growing popularity of the cow-calf program over other forms of beef cattle production are these advantages:

1. Beef cows can produce more pounds of valuable product (calves) from poor to average pastures and low-grade harvested roughages.

FIG. 28. The commercial cow and calf program maximizes returns from pastures or range such as the Sandhills area of northwestern Nebraska, as shown by research at agricultural experiment stations and nearly a century of practical experience on the part of ranchers. (The Record Stockman.)

2. This program is less speculative; that is, there is less risk of losing large amounts of money because of rapidly declining prices.

3. A beef cow herd is a good stabilizer. The man with a successful cow herd is less likely to be an "inner and outer" trying to outguess the cattle market to his downfall. (There is probably reason to wonder whether stability, as used here, is cause or effect.)

4. The man with a combination of pasture, roughage, and grain to market through cattle can produce feeder cattle that best suit his needs by virtue of having under his control such management details as breeding and weaning dates, bred-in performance, and herd health.

5. The man with a cow herd can utilize labor and equipment that may already be on the farm and that would not otherwise be used—for example, a former dairy farm, well equipped with barns, silos, and fence and operated by family labor.

6. A cow herd can and usually does increase in value over the years by being graded up in quality through the use of good bulls and by growth in numbers.

7. This program is perhaps the most satisfying because it covers the gamut of experiences with cattle. (No doubt this is the main reason why the beef-breeding project is among the most popular of all projects for 4-H and FFA Club members.)

DISADVANTAGES OF THE COW-CALF PROGRAM

The cow-calf program has some real disadvantages and these must also be considered in choosing a cattle program:

1. The cow-calf program is a longtime program. Returns come slowly, and the program is not well adapted to tenant farming. Concerning this point someone has said, "Many tenants cannot wait that long and most landlords won't."

2. The program is inflexible; that is, it cannot readily be changed in size or method of management to adapt to unforeseen difficulties or to take advantage of unexpected higher cattle prices. The cowman seldom is able to capitalize on drastic price rises because, even though his cows may increase in value, he cannot sell them and stay in business.

3. A better grade of labor and management is required than is needed for certain other programs.

4. Losses due to disease, calving difficulties, and sterility are high compared with other cattle programs.

5. Average to excellent quality pastures, roughages, and concentrates

can be converted to more pounds of gain with some of the other cattle programs.

6. The volume of business or gross income is small for the large investment in cattle and land; thus the minimum-sized herd for profitability represents a comparatively large dollar investment.

When costs of land, fences, water developments, buildings if required, vehicles, and the cows and bulls are considered, an investment figure of $750 to $1000 per cow in the herd can be expected to prevail, and oddly enough, these figures have a way of being comparable throughout the country. Undoubtedly the strong demand for cow-calf land is responsible for the lack of "bargains" in range or pasture suitable for the production of beef calves. Table 31 shows the estimated costs of carrying a cow for one year in three selected states. Although the states have obviously different

FIG. 29. Income from a productive cow herd makes possible the clearing of land and application of seed and fertilizer, which are required to establish improved pastures such as this one in Georgia. (American Hereford Association.)

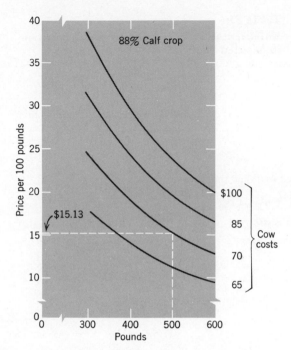

FIG. 30. Break-even prices in producing beef calves in Oklahoma. With an 88 percent calf crop, cow costs at $70 per year, and weaning weights of 500 pounds, the calf price needed to recover all costs is $15.13 per 100 pounds. (Oklahoma State University, Agriculture Extension Service.)

weather and feed production potentials, the annual costs of cow maintenance are quite comparable. The individual items vary, but the totals for pasture and feed are quite similar, indicating again that cow range and feed have about the same value throughout the land. Figure 30 also illustrates the narrow margin between costs and returns in the beef cow enterprise. The importance of calf weaning weights is also demonstrated. It is generally conceded that, if a beef cow herd is the major source of income, from 200 to 250 cows is a practical minimum-sized herd. Obviously, if the enterprise is a supplementary one, no set number is required, but fewer than 25 cows is probably not economically feasible.

THE CHOICE OF THE FOUNDATION BREEDING STOCK

Choosing the animals that are to be the foundation stock for the breeding herd is a matter of prime importance. Often insufficient attention is

Table 31
Estimated Annual Costs of Carrying a Beef Cow
in Selected States (1964)[a]

	States		
Item	Indiana	Minnesota	Texas
Feed costs			
Pasture[b]	$18.59	$13.50	$30.00
Harvested feeds and supplements	23.99	32.33	16.00
Total	$42.58	$45.83	$46.00
Nonfeed costs			
Interest and taxes[c]	$ 9.38	$10.00	$ 5.00
Labor	7.00	13.50	10.00
Bull costs	3.55	3.50	6.00
Buildings	4.00	5.25	—
Bedding	2.25	3.50	—
Miscellaneous	9.80	11.40	13.00
Grand total	$78.56	$92.98	$80.00

[a] Compiled from adaptations of extension economists' reports.
[b] Consists of interest and taxes on deeded land or of pasture leases.
[c] Consists of interest and taxes on cow and calf.

given to their selection. The young breeder, especially, in his eagerness to start operations, is likely to take too little time to consider properly just which animals will best suit his needs. He grows impatient at what seems to be a loss of time, and to get started he purchases the animals that are immediately available even though he may know they fall short of the kind he really wants. Such a procedure is to be avoided, for usually a herd that is established hastily in this way is found to be so unsatisfactory that it is soon replaced either by cattle of much higher merit or by some project entirely different from a beef-breeding herd.

Once a herd is established, it is usually best to raise one's own replacement heifers. This is because of the disease control problem present when outside animals are added, and because of the desire to make progress in improvement in performance and type. Decisions must be made almost every time a calf crop is weaned in order to choose the best heifers to keep, but since only a few cows are replaced each year, these decisions are not nearly so important as those involved in laying the foundation.

CHOOSING THE BREED

In some ways it is unfortunate that we have several breeds of beef cattle of the same general type, because many men waste considerable time and effort in attempting to decide which breed is best for their particular conditions.

The choice of breed, where straightbreeding is used, is, for practical purposes, unimportant to all but those cowmen operating in the Gulf Coast region. It is a well-established fact that the Brahman, the Santa Gertrudis, and the so-called exotic breeds that trace their foundation to an infusion of Brahman breeding, are better suited to this region, at least where performance is concerned. Otherwise all of the principal British breeds are well adapted to the climate and feed conditions of the remainder of the country. So much more variation exists between the individuals of any one breed than between the best, or even the average representatives, of any two of them that it is useless to argue over their respective merits. One thing is certain: If a man has a distinct preference or liking for a particular breed, that breed in all probability is the one for him to use because it is unlikely that he will ever be entirely satisfied with any other. In the absence of such a preference, however, he should choose the breed that is most popular in the community in which he lives. The very fact that it is prominent is a good indication that it is well suited to the prevailing environment. Moreover, such a choice will enable him to avail himself of the services of the outstanding bulls of the community and to use the selling agencies that have been perfected by the established breeders. But perhaps the most important advantages that come from such a choice are the opportunities to observe the methods employed by other breeders, to compare their cattle with his own, and to make purchases from herds about which he has a definite knowledge.

Although we have said that choice of breed is, in the final analysis, of minor importance in the commercial cow-calf program, the following statements by U.S. Department of Agriculture specialists effectively summarize the various aspects of the question:

1. Differences in preweaning and postweaning gain are relatively small among the three British breeds—Angus, Hereford, and Shorthorn—and the polled types of the last two. (See Table 32.)

2. The Charolais and the new breeds based on Brahman-European crossbred foundations grow faster, both before and after weaning, than the British breeds. (The same is true of crosses of these breeds with British breeds.)

Table 32

A Comparison of Calving and Weaning Records of Commercial Cow Herds of the Major Breeds of Beef Cattle[a]
(10-Year Average)

	Angus	Hereford	Shorthorn	Totals and Averages of All Breeds
Number of herds	165	171	38	374
Number of cows	5,011	4,910	934	10,855
Number of calves	4,779	4,679	904	10,362
Calf crop (%)	95.4	95.3	96.8	95.5
Average weaning age (days)	208	207	202	207
Average weaning weight (lb)	448	446	426	445
Average daily gain (lb)	1.76	1.77	1.71	1.76

[a] Charles R. Kyd, Missouri Extension Service. Information to the author.

3. If breed differences exist in efficiency of growth (feed consumed per unit of gain), they have not been established. Similarly, breed differences in fertility and longevity have not been clearly defined in most cases.

4. Among the British breeds, differences in meat palatability and tenderness are small.

5. The Charolais and its crosses, the Brahman and its crosses, and new breeds based on Brahman-European crosses produced carcasses with less external fat and higher yields of trimmed preferred retail cuts than British breeds. As compared with British types slaughtered at the same weights, these crosses ordinarily have less marbling and do not grade as high by U.S.D.A. quality grade standards.

6. Brahman cattle and breeds based on Brahman-European foundations have greater heat tolerance than European types and greater resistance to many insects and some diseases. As compared with British types, animals of these breeds are slower to reach sexual maturity, but brood cows are excellent mothers and have longer productive lives. The lean meat from the Brahman, and from breeds with part Brahman foundations, has been found in several experiments to be somewhat less tender than that of British breeds.

Breeders of each breed, whether producing purebreds or commercial cattle, should strive to improve the breed of their choice in the characteristics in which weaknesses are most apparent. In each breed, breeding stock can be found that is acceptable or even very strong in those points in which the breed needs improvement.

FIG. 31. F₁ cows resulting from the use of Brahman bulls on Hereford cows are proving to be excellent mother cows throughout the Gulf Coastal region. (American Brahman Breeders Association.)

CROSSBREEDING

The value of crossbreeding is discussed in some detail in Chapter 3. The Nebraska tests with three British breeds demonstrated the magnitude of the heterotic effect from both the first cross and the backcross. Other tests are under way throughout the country with three- and four-breed crossing. In the South, when the cross involves the Brahman or some of the newer breeds with Brahman ancestry such as the Brangus, Beefmaster, or Charbray, even greater heterotic response occurs than when British breeds are crossed. The ideal combination of breeds to use in a crossbreeding program depends to a large extent on the environment in the locality where the herd will be run and on the market demands in the area. Guidance in this matter can be obtained from beef cattle specialists in the state or area in question.

The crossing of breeds has been much less widely practiced with beef cattle in America than with either swine, sheep, or poultry. There is currently a more pronounced trend in this direction, however, and data are beginning to become available that will permit a sound appraisal of its value.

CHOICE OF ANIMALS FOR THE COMMERCIAL HERD

The following items should be carefully considered in choosing the animals that are to form the foundation of a new herd or are to serve as additions or replacements for a herd already in existence: (1) freedom from disease, (2) individuality, (3) performance records if available, (4) age, and (5) cost.

Freedom from Disease. The most important item in determining profits is percentage of calf crop, and the best insurance against poor calving percentage is a healthy herd. High selling prices due to extra quality and weight or low feed costs cannot offset low calving percentage and large death losses.

Important contagious reproductive diseases of breeding cattle that must and can be guarded against are brucellosis (contagious abortion), leptospirosis, and vibriosis in females and trichomoniasis in bulls. These diseases can be minimized by demanding, at the time of purchase, proof of negative results from tests conducted by qualified veterinarians on all breeding animals composing the foundation stock or additions to the herd. Replacements that are produced in one's own herd are likely to be free of these diseases if the herd in general is healthy; however, annual tests should be made on these additions as well. Compliance with federal and state regulations concerning these diseases is becoming mandatory in all parts of the country and, it is hoped, will continue until all herds are disease free.

Calfhood vaccination gives good protection against brucellosis, and approved vaccines are now in use for leptospirosis and vibriosis. Breeding animals should be bought subject to negative tests for these diseases in any case. Tests made more than 30 days before delivery of the animals to the buyer should not be accepted. All herd additions should be isolated from the rest of the herd for 60 to 90 days, and a retest should be made before they are released from quarantine.

Many other diseases affect breeding cattle—for example, tuberculosis, anthrax, and anaplasmosis—but these are comparatively uncommon. Nevertheless, the breeder should be aware of them and should consider the area of origin when buying foundation cattle. More details are available concerning these and other diseases in Chapter 27.

Individuality. As used here, individuality includes all those characteristics in an animal that may be noted from "eyeball" or visual inspection. Size, body type, quality, bone and set of legs, breed and sex character, and temperament all come under this heading. The extent to which visual appraisal of individuality is used to determine acceptance or rejection of breeding animals varies from using such an appraisal as the sole means of selection to completely ignoring the appearance of the animals. Some of the characteristics of breeding cattle are due to a combination of inheritance and environment, as mentioned in Chapter 3. This fact complicates selection. Often the first calves produced by the foundation stock or herd replacements bear little resemblance to their sire or dam and do not measure up to expectations. This is usually because environmental factors, such as degree of finish and previous treatment of introduced cattle, led the prospective breeder astray. Some of the characteristics mentioned, however, are highly heritable and can thus be expected to be passed on to offspring with considerable regularity.

Characteristics such as breed character or breed type contribute to the sales appeal of commercial cattle. These traits must be appraised visually, since they cannot be accurately measured with scales or calipers. Unfortunately, as experienced breeders know, these traits are no more reliably passed on to offspring than others previously mentioned. In addition, research indicates that these characteristics are not so closely associated with productivity as once believed. Observation of many near relatives is a good, though not perfect, guide when selecting young breeding cattle, insofar as the "hard to measure" characteristics are concerned.

Breeders and feeders alike often express a preference for a certain shade of red or a specific color pattern in their feeders or breeding cattle. Buyers have been known to pay as much as a dollar per hundredweight more for yellow-colored Hereford feeders, for instance, and shade of red may mean as much as several hundred dollars' difference in the selling price or cost of a bull. South Dakota workers have gathered data that indicate that shade of red in Herefords is not correlated with performance and that neither a premium nor a discount should be expected when selling or buying Herefords that are either extremely light or extremely dark colored.

Performance Records. With respect to grade females that are to serve as a foundation for a commercial herd, records of their performance or productive ability are seldom available. This is especially true if these females are bought in the range areas where herds are usually large and individual records are not kept on each cow or calf in the herd. This is not to imply that western ranch-bred females are not desirable foundation material. To the contrary, more good surplus females are apt to be found there than elsewhere because of the general awareness of economically important characteristics over a period of many years. In addition, cow numbers are relatively stable in these regions and proportionately fewer heifer calves or yearlings are needed for herd replacements or buildup.

More and more ranchers and operators of farm herds are participating in performance-testing programs. Weaning weight and weaning type score are the two principal kinds of information recorded for heifer calves. Breeders usually practice performance testing mainly to aid in selecting their own herd replacements, but the surplus females with good weights and weaning scores should make suitable foundation material for many prospective purchasers. If yearling heifer weight and type score can be obtained in addition to the information obtained at weaning time, so much the better. However, this information is unlikely to be available for most grade yearling heifers excepting those sold in dispersal sales.

The breeding ability of the sire or dam of prospective foundation females is another good guide in making selections. Usually a sizable number of a sire's progeny can be seen on one ranch or farm, and one should inspect as many of them as possible in order to evaluate the sire. An unselected sample of progeny should be seen—those from poor cows as well as those from the better cows. Calves that are creep-fed or otherwise pampered are not of much help in evaluating a bull's breeding worth.

The breeders of some localities have established reputations as good producers of foundation females. A new breeder would do well to ask for expert help, even when buying his females from such high-reputation areas, because not all herds in the area are outstanding, and prices are usually higher. Such an area is a good place to buy females, however, because more herds can be seen in less time and more good cattle will be offered for sale. The feedlot performance of large numbers of steers or heifers from a herd over a period of years is a good practical guide in seeking foundation females.

As for the bull or bulls to be used in a new commercial herd or as a replacement, nothing less than an excellent purebred bull on which performance data are available should be acceptable. Such data, to be complete, should include weaning weight corrected for the age of the calf and for the age of the dam, weaning type score, feed conversion data for at least a 120-day feeding period, and a 12- or 15-month weight and type score.

Records alone, of course, mean nothing. Obviously bulls with poor records should not be offered for sale, but some breeders, in their eagerness to make use of a "performance-tested" bull, mistakenly think that a bull with records, even though only average, should be better than a bull with no records.

Age. Age groups of females available for foundation material can be grouped according to numbers available, in descending order, as follows: (a) weanling heifer calves, (b) yearling heifers (usually bred), and (c) older cows with or without calves at side. From a quality standpoint the older cows, having withstood annual cullings, may be best. However, they

FIG. 32. Many ranchers wait until after the wintering period to decide which heifers to keep for replacements, in order to get more information concerning their growth rate and type. High condition in calves at weaning time, due to heavy-milking mothers, can mislead a breeder in selecting replacements. (American Angus Association.)

may be for sale because of advanced age and reasons related to age, such as bad disposition, eye problems, poor udder, and slow breeding tendency. Cows that are culled for such reasons are unlikely to be profitable additions to a herd. If mature cows are for sale owing to a forced reduction in cow numbers on a ranch or farm because of drought, settling of an estate, or the like, then such females are often very desirable foundation stock.

Western ranchers like to offer their surplus heifer calves and yearling heifers for sale in droves without undue culling or sorting. Therefore, if such females are bought to start a commercial herd, considerably more, even twice as many, should be bought than are needed to start the herd. The drove can be culled to size after the calves have been fed and handled in their new surroundings for 2 or 3 months, and the rejected females can be fed out for slaughter. Heifer calves offered for sale are usually of better quality than yearling heifers, especially if the above procedure is followed. Often the yearling heifer droves include late-dropped calves from the year before, or heifers that were intended as replacements in the original owner's herd but are now for sale because their subsequent development was disappointing. The effect of age on the future productive life of females is discussed in Chapter 4.

Age is also important in selecting bulls. Obviously a 4- or 5-year-old bull that has sired two or three crops of good-doing calves of the right type is a much safer bull to buy than a younger, unproven bull or a bull calf that has yet to demonstrate his breeding ability. However, not many such bulls are offered for sale, and often they are in great demand if of excellent quality. If the herd is already in production, a good plan is to buy a yearling bull to use on yearlings or 2-year-olds while the older bull is still in use. Thus he can be progeny-tested—that is, offspring sired by him can be compared with calves sired by the main herd bull. Young bulls should be used sparingly, as is discussed in Chapter 4. Great skill is required in selecting weanling bull calves as future herd sires, unless one has performance data on the calves themselves, as well as progeny data on the sire and dam of the calf, or unless one can see many close relatives in more mature stages of development.

Cost. Naturally cost is an important consideration in selecting commercial breeding cattle, but it should not crowd out all others. Often price becomes the all-important factor in deciding which animals to purchase. As a matter of fact, price should be the last consideration within reasonable limits. If closely culled grade females are bought, a 25 to 50 percent premium over feeder cattle market price is not unreasonable. If young heifers are bought, the added weight at disposal time as old cows will often offset the premium paid at purchase time. If a drove of females is bought in which no culling was permitted by the seller, naturally one would reasonably expect to pay a somewhat lower premium amounting to perhaps only 10 percent above market price, or no premium at all. Locally grown cattle of equal quality have an advantage with respect to price because of lower transportation and incidental costs. There is also less likelihood of losses due to shipping fever and similar diseases.

Good bulls can usually be bought for the equivalent of 2 to 3 choice grade finished steers. Producers of superior performance-tested bulls do and should expect somewhat higher prices because of the labor and expense involved in conducting such tests. Furthermore, not nearly all of the bulls tested will prove to be superior, hence the testing costs must be added to the bulls actually sold as superior or tested bulls.

SOURCES OF FOUNDATION FEMALES

Sources of good female breeding stock for the foundation of a commercial beef herd vary with the section of the country. Anyone wishing to start a herd in the range area has a relatively simple job with respect to this point because ranchers in the neighborhood are likely to be able to supply his wants. On the other hand, good females are harder to come by in areas where there are few herds or where most of the herds are small or are themselves expanding in numbers. Of course, even a buyer in these areas may secure surplus range area females through dealers or commission men or by direct purchase. Extra costs such as transportation and commission and the fact that little sorting can be done mean that this source may be more expensive. In the nonrange areas, then, more thought and time are required to obtain the right kind of females.

Some sources of females and methods of procurement are:

1. The occasional complete or intact herd available due to circumstances beyond the control of the owner or operator.
2. The normally surplus heifer calves and yearling heifers in the range areas.
3. Cow and calf pairs in the range areas, especially in periods of extended drought when forced herd reductions are necessary.
4. Dispersal sales held to settle estates, divide partnerships, and so forth.
5. Regularly scheduled auction sales, especially those held in states that have closely supervised health requirements.
6. Private treaty purchase from neighbors.

SOURCES OF BREEDING BULLS

Purebred herd bulls for use in establishing a commercial herd may be obtained from several sources, and the source is not so important as the bull himself. Satisfactory sources of breeding bulls, in descending order of choice, are:

1. A proven but still sound bull, purchased from another breeder.
2. A young performance-tested bull, purchased privately from a breeder or in a sale of performance-tested bulls.
3. A young bull from a reputable purebred breeder, after inspecting the sire, dam, and other near relatives.

4. Purebred bull sales where a screening committee has rigidly culled the sale offering.

METHODS OF ESTABLISHING A COMMERCIAL HERD

Because the average farmer or rancher seldom establishes more than one cow herd in a lifetime, it behooves him to make his foundation choices carefully and wisely. Seeking advice from an experienced neighbor, a reputable, prefessionally trained dealer or commission man, or a county agent or vocational agriculture teacher is especially advisable if the buyer is inexperienced. Some time-tested methods of starting a herd are:

1. Buy a complete herd with a good reputation, or the top half of such a herd, when a farm or ranch is changing hands. This method requires considerable capital, but returns come quickly, and ordinarily the quality is high unless some topping out has occurred. By all means the lower end of such a herd should not be purchased alone even if it seems to be a bargain.

2. Buy twice as many heifer calves as will ultimately make up the cow herd from a herd with a good reputation. The better half of the drove may be selected for foundation females on the basis of performance as shown by weight (or size if scales are unavailable), beef type or conformation, disposition, and uniformity after being fed a good growing ration for 3 to 4 months or until early spring. The remaining half of the heifers may be disposed of through any of several heifer feeding plans discussed later. Returns come more slowly from such foundation females, but the initial investment is comparatively low and considerable culling may be practiced.

3. Buy yearling heifers in the fall, again buying extras to permit culling before making final choices, after the cattle have spent several months on the purchaser's farm. As yearling heifers from the range area will probably already have been exposed unless guaranteed open, a pregnancy test should be made to reveal those not bred. The open heifers, with the culls, can then be fed out for the spring market. This method has the advantage over the purchase of heifer calves that the first calf crop is obtained one year earlier. Initial costs per head are higher and there is less opportunity for culling, however.

4. Buy cow and calf pairs, preferably from areas where numbers are being reduced for reasons other than simple culling—for example, where extended drought is forcing a drastic reduction in cow numbers. Such cows will ordinarily be bred when offered for sale in late summer or early fall. Steer calves and undesirable heifer calves at side can be sold or handled

in any one of the many programs discussed elsewhere. Naturally, such cows will not be the best cows on the farm or ranch from which they came, but the calves at side will tend toward the average of the herd of origin and will offer evidence of the quality of the herd. If such cattle are from a drought area, the buyer should take into consideration the lack of flesh or condition in the cows and the lack of weight and bloom in the calves resulting from the poor rations of their mothers.

5. Grading up the native beef females, or even dairy cows of the heavy breeds such as Holstein and Brown Swiss, available in the area. This method takes several years, but results are often quite good. Using at least three successive beef bulls of excellent quality and bred-in performance will result in good beef-type cows, especially if a performance-testing program is begun from the start and only the best heifers are retained. Especially good foundation females of this type are heifer calves produced in dairy herds where some of the cows have been artifically bred to beef bulls. This avoids the problem of too much milk being produced, as may be the case in the straightbred dairy females. The disease problem is greatly reduced by this method if the usual recommended procedures for disease control are followed. The initial outlay of capital is again comparatively low; consequently this method is well suited to young men just starting in the business. The selection of the herd bull is of the utmost importance in this method.

Chapter 8

BREEDING HERD
MANAGEMENT in WINTER

Herd management throughout the year should have as its aim the production of a 100 percent crop of uniformly high-quality, heavyweight calves. Items that contribute to the accomplishment of this aim are early sexual maturity of females, high conception rate, high calf livability, early rebreeding after calving, heavy and prolonged milk production, and longevity. Quality of management and plane of nutrition contribute to each of the items mentioned and are important determinants of profit or loss.

The cow-calf program is a low-gross-income enterprise at best; therefore, cash expenditures must be kept at a minimum to insure a net profit. As the largest item of cash expense is for winter feed, it is the item that requires the closest scrutiny. It has been correctly said that one cannot starve a profit out of cows, but it is equally true that supplying nutrients above the minimal requirement is uneconomical, especially when it applies to harvested or purchased feeds.

Sorting the herd into groups with similar feed requirements is the first step in planning an efficient feeding program for the winter. Usually the following groups of cattle are found in a typical well-managed commercial herd during the winter:

1. Dry cows.
2. Wet or nursing cows with calves, unless only spring calving is practiced.
3. Weaner calves, usually heifers.
4. Bred or open yearling and 2-year-old heifers.
5. Bulls.

In actual practice these five groups do not always need to be fed separately. If the herd is not large, the weaner calves, the bred yearlings if 2-year-old first calving is practiced, and the first-calf and oldest dry cows can all be fed together. The ration should be such that the weaners and yearling-bred heifers will gain enough to ensure maximum growth and development and such that the suckled-down first-calf heifers and old cows will mend to some extent before their next calving and lactation periods. The other main grouping would then contain the medium-aged dry cows and the open yearlings and bred 2-year-olds if heifers are bred to calve first as 3-year-olds.

If the herd has either a fall calving, a split calving, or a year-round calving program, the cows that are nursing calves should be handled separately, especially if creep-feeding is practiced, or they may be grouped with the first large group mentioned above. The bulls should be fed separately if only a spring-calving season is practiced, but of course they will be running with the wet cows if fall calving is used.

The nutritional plane and feeding and management plans depend on a number of factors:

1. Condition of cows at the end of the grazing season.
2. Age of cows.
3. Calving season—early spring versus late spring, for example.
4. Probable mineral and vitamin stores at the end of the summer grazing season.
5. Condition, amount, and species of forage if cows are to be wintered on cured range feed.
6. Probable weather conditions.
7. Labor available for feeding.
8. Availability of supplemental feeds.
9. Breed of cattle.

The four nutrients of greatest importance in wintering beef cattle are energy, protein, phosphorus, and vitamin A or its precursor carotene. It will be shown how the nine factors above relate to these nutrients and how they affect the choice of a feeding program for the herd in wintertime.

CONDITION OF COWS

This factor is extremely important, as it tells much about the type of pasture or range the cows have just experienced during the summer and fall grazing seasons. It is also a good indicator of the kind and amount of feed that will probably be available for winter, especially if it is a range operation and cows will be wintered on "set-aside" or ungrazed range. Obviously if cows have "summered hard" on drought-stricken or overgrazed range or pasture, the feeds reserved for winter use will also be poor. Unless

FIG. 33. Cows being wintered in northern New Mexico with only the shelter provided by the small mountains and draws. Often a high-energy mineral-vitamin-fortified cubed supplement is fed in these areas as insurance against deficiencies. (National Cottonseed Products Association, Inc.)

considerable culling or herd reduction has occurred to adjust cattle numbers to feed supply, the amount of winter feed will be short. On the other hand, if the cows are in fair or better flesh, they can lose up to 200 pounds as a result of deliberate or forced restriction of energy during the winter without serious harm. Such fleshy cows may be expected to have good liver vitamin A stores, but adequate protein and phosphorus must be supplied. Very thin cows require complete supplementation of their rations.

AGE OF COWS

First-calf heifers and especially those that calved first as 2-year-olds are the major concern. Inasmuch as nutrient requirements for lactation have priority over nutrients for growth, this age group will usually be suckled thin. The winter dry period will give them an opportunity to make

Table 33
Effect of Winter Gains of Cows on Weights of Calves[a]

High, Medium, and Low Winter Gains				
			Calves at Weaning	
	Number of Cows	Average Winter Gain (lb)	Average Weight (lb)	Average Age (days)
8 high winter-gaining lots	62	110.4	384.4	161.0
8 medium winter-gaining lots	50	66.9	370.2	162.3
8 low winter-gaining lots	61	21.2	362.4	160.1

Winter Gains versus Winter Losses of Cows					
	Cows			Calves		
	Average Winter Gain (lb)	Average Weight at End of Winter (lb)	Average Weight after Calving (lb)	Average Birth Weight (lb)	Average Weaning Weight (lb)	Average Age at Weaning (days)
11 highest winter-gaining cows	42.6	1,064.4	844.2	65.6	379.1	170.7
11 lowest winter-gaining cows	−62.7	981.3	781.3	64.6	326.9	177.1

[a] Montana Special Circular 7.

up for loss in growth; hence the ration should by all means contain sufficient energy, phosphorus, and protein. These first-calf heifers should not lose weight during winter, as is permissible with mature cows, but, rather, should gain up to 100 pounds before their next calving. Older cows, especially heavy milkers and cows with tooth problems, will also be suckled so thin that they will need to gain weight before calving again in the spring.

CALVING SEASON

The period from 30 days before through 90 days after calving is the most critical period of the year from a nutritional standpoint. Most of the fetal growth occurs during the last 30 days of gestation. Adequate protein, vitamin A, phosphorus, and energy will insure a strong calf with greatest chances of normal birth and survival. The early reestablishment of the estrus cycle and conception depend greatly on the plane of nutrition during this 4-month period, as discussed in Chapter 5. The relationship between the nutrient content of the cured forage available to the dry cows and the date when the calving season begins, in light of the above, is obvious. Thus the time to begin supplemental feeding, if required, is determined largely by the dates of the calving season. Fall-calving cows have winter feed requirements at least 50 percent higher than those of dry cows.

VITAMIN AND MINERAL STORES

This point is discussed separately from the subject of condition because differences in condition may not necessarily be associated with differences in vitamin and mineral reserves. The quality and species of forage available in late fall and winter have more to do with these reserves than quantity of feed during the entire grazing season. Thin cows may have had access to late-fall weeds, small-grain pastures, or browse, which ensured good liver vitamin A reserves. Judgment and close observation of range conditions may tell one more than condition of cows with respect to this point.

CONDITION, AMOUNT, AND SPECIES OF WINTER FORAGE

If cured range is to be the chief source of winter feed, the species of forage plants that are most prevalent will determine the supplemental feed needs. In programs where hay or silage is to be fed to cows in winter, the question is simplified because such feeds are usually either legume or nonlegume and the matter of balancing rations containing these feeds is simple by comparison. Some range plant species retain their feed value longer during the dormant season—for instance, black grama grass retains some green growth throughout the winter as do some of the browse plants—but others, notably the tallgrasses, may be devoid of carotene by February or March, as well as quite indigestible and thus low in all available nutrients.

WEATHER CONDITIONS

Either snow cover or extremely wet, muddy conditions may necessitate a complete dependence on supplemental feeding of the entire ration in some localities. The length of time during which such conditions prevail will of course greatly influence winter feed needs and the best time for the calving season. Temperature can affect the nutrient requirement itself, with energy and protein requirements for maintenance being most involved. Cows are adaptable to wide temperature ranges, however, and supplemental feeding adjustments for this reason are not so critical as some believe.

LABOR AVAILABLE FOR FEEDING

This factor probably has more to do with how and how often cows are fed than with what they are fed. Cows can be fed as infrequently as once weekly with success. Except possibly for checking the water supply and fences, little labor is required in managing cows during the winter until calving begins.

Table 34
Analysis of Prairie Grass at Different Seasons[a]
(Three-Year Average)

				Percent Composition of Dry Matter				
	Percent Dry Matter	Ash	Protein	Fat	Fiber	Nitrogen-Free Extract	Ca	P	Caro-tene[b]
Grass[c]									
November	82.39	5.01	2.53	1.74	40.02	50.40	0.253	0.046	14
January	94.85	5.92	2.57	1.57	40.86	48.74	0.309	0.039	trace
May	52.29	6.39	9.68	2.40	32.02	49.08	0.308	0.126	407
August	54.71	6.21	5.06	2.23	35.42	50.66	0.346	0.078	112
October	63.09	5.18	3.23	1.62	37.24	52.24	0.244	0.048	16

[a] Oklahoma Cattle Feeders' Day Report.
[b] Parts per million.
[c] Averages, by species, of the four predominant grasses: big bluestem, little bluestem, Indian, and switch.

AVAILABILITY OF SUPPLEMENTAL FEED

If distances between the source of supplemental feeds, when not home-grown, and location of the cow herd are great, concentrates are more apt to be used in supplementation programs than roughages, simply because of convenience and the comparative cost of transportation.

BREED OF CATTLE

The nutrient requirements of various breeds of cattle do not differ markedly, but breed may influence the condition of cows in the fall and thus have a bearing on supplemental feed needs. One breed of cow may summer better under extremes of weather, forage, and terrain than others, or one may travel or forage over a wider area in winter. Some breeds, especially some of the dairy $\times$ beef crosses, suckle down to a greater degree and thus may require a higher-energy winter ration.

In summary, it is evident that numerous factors prevent making hard and fast recommendations concerning the feeding of the cow herd during winter for each and every farm or ranch condition. Knowledge of the nutritional requirements of cows and replacement cattle and the composition of available forages is essential for the development of sound supplemental feeding programs.

LEVEL OF FEED INTAKE

The amount of feed a cow will consume is of interest because, if supplemental feeding is required, the total amount of feed required for an individual or herd must be determined, whether homegrown or purchased feeds are fed. Cows of approximately the same size will consume about the same amount of total air-dry feed regardless of condition, if fed free-choice or all they will clean up. However, because cows of the same body size will weigh less or more depending on whether the cows are thin or fleshy, it is necessary to use different factors for determining feed capacity for cows of varying condition. Cows that are fleshy, average, or thin with respect to condition will consume daily (on an air-dry basis) an amount of feed that is equivalent to approximately 1.75, 2, and 2.25 percent of their live weight, respectively. Thus cows that vary in weight because of differences in condition and not body size per se and, accordingly, weigh 1,100, 1,000, and 900 pounds, respectively, will each consume about 20 pounds of air-dry feed daily.

The nutrient requirements for breeding beef cattle, as recommended

Table 35

Daily Nutrient Requirements of Breeding Beef Cattle[a]
(Based on Air-Dry Feed Containing 90 Percent Dry Matter)

Body Weight (lb)	Average Daily Gain (lb)	Daily Feed per Animal (lb)	Total Protein (lb)	Digestible Protein (lb)	Digestible Energy For Maintenance (kcal)	Per Pound of Gain (kcal)	Total (kcal)	TDN[b] (lb)	Ca (gm)	P (gm)	Carotene[c] (mg)	Vitamin A (IU)
Wintering Pregnant Heifers												
700	1.5	20.0	1.5	0.9	—	—	20,000	10.0	15	14	50	20,000
900	0.8	18.0	1.4	0.8	—	—	18,000	9.0	13	12	45	18,000
1000	0.5	18.0	1.4	0.8	—	—	18,000	9.0	13	12	45	18,000
Wintering Mature Pregnant Cows												
800	1.5	22.0	1.7	1.0	—	—	22,000	11.0	16	15	55	22,000
1000	0.4	18.0	1.4	0.8	—	—	18,000	9.0	13	12	45	18,000
1200	0.0	18.0	1.4	0.8	—	—	18,000	9.0	13	12	45	18,000
1200	−0.5	17.6	1.3	0.8	—	—	15,000	7.5	13	12	44	17,600
Cows Nursing Calves, First 3 to 4 Months Postpartum												
900–1100	0.0	28.0	2.3	1.4			33,600	16.8	30	23	106	42,000
Bulls, Growth and Maintenance (Moderate Activity)												
600	2.3	16.2	2.0	1.2	—	—	20,200	10.1	21	15	62	24,300
1000	1.6	20.0	2.4	1.4	—	—	24,000	12.0	19	15	76	30,000
1400	1.0	24.7	2.4	1.4	—	—	28,400	14.2	17	16	94	37,100
1800	0.0	25.5	2.4	1.5	—	—	28,000	14.0	18	18	97	38,300

[a] *Nutrient Requirements of Beef Cattle*, Sub-Committee on Beef Cattle Nutrition, National Research Council, revised 1963.
[b] The TDN was calculated from digestible energy by assuming 2000 kcal of DE per pound of TDN.
[c] The carotene requirement was calculated from the vitamin A requirement assuming 400 IU of vitamin A per milligram of carotene.

Table 36
Nutrient Requirements of Breeding Beef Cattle Expressed as Percentage Composition of Air-Dry Rations[a]

Body Weight (lb)	Average Daily Gain (lb)	Daily Feed per Animal (lb)	Total Protein (%)	Percentage of Ration or Amount per Pound of Feed						
				Digestible Protein (%)	Digestible Energy (kcal/lb)	TDN[b] (%)	Ca (%)	P (%)	Carotene[c] (mg/lb)	Vitamin A (IU/lb)
Wintering Pregnant Heifers										
700	1.5	20.0	7.5	4.5	1000	50	0.16	0.15	2.5	1000
900	0.8	18.0	7.5	4.5	1000	50	0.16	0.15	2.5	1000
1000	0.5	18.0	7.5	4.5	1000	50	0.16	0.15	2.5	1000
Wintering Mature Pregnant Cows										
800	1.5	22.0	7.5	4.5	1000	50	0.16	0.15	2.5	1000
1000	0.4	18.0	7.5	4.5	1000	50	0.16	0.15	2.5	1000
1200	0.0	18.0	7.5	4.5	1000	50	0.16	0.15	2.5	1000
1200	-0.5	17.6	7.5	4.5	850	43	0.16	0.15	2.5	1000
Cows Nursing Calves, First 3 to 4 Months Postpartum										
900–1100	0.0	28.0	8.3	5.0	1200	60	0.24	0.18	3.8	1500
Bulls, Growth and Maintenance (Moderate Activity)										
600	2.3	16.2	12.5	7.5	1250	63	0.29	0.21	3.8	1500
1000	1.6	20.0	12.0	7.2	1200	60	0.21	0.17	3.8	1500
1400	1.0	24.7	10.0	6.0	1500	58	0.16	0.15	3.8	1500
1800	0.0	25.5	9.3	5.6	1100	55	0.15	0.15	3.8	1500

[a] *Nutrient Requirements of Beef Cattle*, Sub-Committee on Beef Cattle Nutrition, National Research Council, revised 1963.
[b] The TDN was calculated from digestible energy by assuming 2000 kcal of DE per pound of TDN.
[c] The carotene requirement was calculated from the vitamin A requirement assuming 400 IU of vitamin A per milligram of carotene.

by the National Research Council, expressed on the basis of daily needs and percentage in the ration, are shown in Tables 35 and 36. Morrison's feeding standards for breeding beef cattle are shown in Table 37. Rations that meet the needs of the various weight and age groups of breeding cattle, according to the requirements or standards indicated, can be computed exactly from information concerning the composition of feeds as found in the Appendix tables.

WINTER RATIONS FOR THE BREEDING HERD

The winter ration of beef cows should consist largely of the common farm or ranch roughages. Indeed it usually is possible to maintain dry cows satisfactorily during the winter on roughages alone in the farm regions, but in range areas, cured range may need supplementation. The roughage sources vary so widely between the farm and range regions as to warrant discussing them separately.

Table 37
Morrison's Feeding Standards for Breeding Beef Cattle[a]

	Requirements per Head Daily							
Class of Cattle and Weight (lb)	Dry Matter (lb)	Digestible Protein (lb)	Total Digestible Nutrients (lb)	Calcium (gm)	(lb)	Phosphorus (gm)	(lb)	Caro-tene (mg)
Pregnant cows								
900	13.1–18.4	0.65–0.70	6.9– 9.7	20	0.044	17	0.037	55
1000	14.2–20.2	0.70–0.80	7.5–10.5	20	0.044	17	0.037	55
1100	15.2–21.5	0.75–0.85	8.0–11.3	20	0.044	17	0.037	55
1200	16.3–22.8	0.80–0.90	8.6–12.0	20	0.044	17	0.037	55
Cows nursing calves								
900–1100	22.0–27.0	1.20–1.40	12.0–15.0	30	0.066	24	0.053	90
Growing cattle[b]								
700	14.2–16.5	0.87–0.98	8.9–10.2	17	0.037	15	0.033	40
800	15.9–18.3	0.90–1.00	9.5–10.9	16	0.035	15	0.033	45
900	17.3–19.7	0.93–1.03	10.1–11.5	16	0.035	15	0.033	50
Mature bulls[c]								
1400	17.2–19.0	1.19–1.31	11.0–12.2	14	0.031	14	0.031	84
1600	18.6–20.6	1.28–1.42	12.3–13.5	16	0.035	16	0.035	96
1800	20.4–22.6	1.40–1.54	13.5–14.9	18	0.041	18	0.041	108

[a] Taken by permission of The Morrison Publishing Company, Ithaca, N.Y., from *Feeds and Feeding*, 22nd edition, by Frank B. Morrison.
[b] Upper limits of range should be adequate for pregnant yearling or 2-year-old heifers.
[c] Assumed to be the same as for dairy bulls as shown in Morrison's tables.

WINTER FEEDING IN FARM REGIONS

In general, the harvested roughages found in the farming regions are higher in nutrient content, especially in energy or TDN, than the level required for feeding dry cows most economically, if full-fed. Thus either limited feeding of these roughages must be practiced or the ration must be a mixture of the better-quality roughage and a lower-grade roughage. For instance, a ration of legume hay can be balanced with access to a cornstalk field to achieve the most economical balance.

The roughages available in the farm regions can usually be grouped as follows:

LEGUME HAYS

No better feed than clover or alfalfa hay can be recommended for breeding cows, but an exclusive ration of legume hay for animals in average condition supplies more nutrients than are actually needed. If enough le-

FIG. 34. The stacks of mixed hay in the background, grown on irrigated meadows, will be used to winter the cows and their calves after weaning on this western Wyoming ranch. (The Record Stockman.)

gume hay is fed to furnish from 20 to 30 percent of the total air-dry feed requirement, the remainder being supplied by nonlegume roughages, the ration will be sufficiently well balanced for dry cows in ordinary condition. Soybean straw (obtained from threshed or combined beans) is high in crude fiber because many of the leaves are lost during combining. However, it is excellent for breeding cattle when fed rather liberally.

Lespedeza hay is an excellent feed for cows if cut at the proper stage of maturity. If it is overripe, the stems are very hard and the hay is much less palatable and digestible. The lespedeza straw obtained when the crop is cut and threshed for seed is practically worthless for feeding livestock.

Although it is true that good-quality legume hays, when fed as the sole ration, may supply more nutrients than needed, especially protein and carotene, it still may be the most economical feed to use if homegrown and if other roughages are unavailable. Mixed hay composed of about one-half legumes and one-half grass is excellent winter feed for cows, meeting their nutrient requirements almost perfectly.

GRASS HAYS

Grass hay composed principally of brome, orchard, or timothy and similar tall-growing bunch grasses is the principal grass hay fed to beef

FIG. 35. Prairie hay plus a protein supplement is a common feeding program for wintering stocker steers in the Nebraska Sandhills. (The Record Stockman.)

cows in the farm region. Improved wheatgrasses, found on the mountain meadows of the Northwest and sometimes on irrigated meadows, are especially nutritious if not overly mature when cut. Marsh hay, grown in the North Central states and elsewhere, may be of good quality but varies greatly with stage of cutting and state of the weather during curing. Coastal Bermuda, found in the southern region, has a composition similar to most grass hays. The principal deficiency of native hays is protein, but carotene and phosphorus may also be deficient if the hay is unduly mature when cut. Grass hays are usually cut when overly mature in order to get more tonnage, but actual pounds of digestible protein and energy may be reduced because of increased lignification and consequent lowered digestibility.

STRAW

Next to corn stover, the straw of small grain crops is the most abundant low-grade roughage material in the country. The value of straw for feeding purposes is determined largely by the stage at which the grain is cut and the amount of damage done by rains before baling or stacking. Good bright oat straw cut with a binder when a little short of being fully ripe and cured in well-constructed shocks has considerable feeding value, ranking only a little below timothy hay in the amount of total digestible nutrients. This method of harvesting is not widely practiced except in the North Central states and even there is fast passing from the picture. Oat straw obtained after a combine is much less valuable than threshed straw because it is cut when much riper and correspondingly higher in fiber. Moreover, most of the leaves and chaff, which are the best parts of the straw, are lost during combining and subsequent raking. Barley straw is somewhat less valuable as a feed than oat straw, and wheat straw has still less merit.

CORN AND SORGHUM FODDER AND STOVER

Corn stover—that is, the stalks left in the field behind a corn picker or combine—offers considerable possibility as a roughage for wintering beef cows. As an increasingly larger portion of the corn crop has been harvested with mechanical corn pickers, correspondingly less corn stover has been used for cow feed. This is because by the time corn is dry enough to pick for safe storage (15 to 18 percent moisture content) the stalk is dead ripe and many leaves have fallen, and the picker removes most of the remaining leaves. Stalk shredders are now available that pick up both stalk and leaf, delivering them into a wagon in shredded form, which can be either stored as stover or made into silage. The shredded stalks may mold if moisture content exceeds 16 percent when stored for dry roughage.

Field shelling or combining of corn earlier in the season when it still contains as much as 28 to 30 percent moisture, followed by drying of the shelled corn for safe storage or by ensiling the grain for feed, means that the corn stalk can be harvested as stover if allowed to field-cure, or as stover silage while the leaves still contain some green color. Such stover is less fibrous, hence more digestible, and because it contains more leaves it is more completely consumed. Some field shellers or combines cut and shred the stalk at the same time the grain is removed. Such chopped stover is best stored in a silo because it is likely to be too high in moisture to be stored as dry stover. Some water usually must be added if the stover is ensiled.

Sorghum stover is somewhat higher in feeding value than corn stover, since sorghums are sweeter because of their sugar content and thus are more completely consumed, and also because the proportion of leaf to stalk is higher, with the leaves being the more valuable portion. Furthermore, in a normal growing season the sorghum head ripens sufficiently for combining and safe storage before the stalk matures. In years when the harvesting season is delayed by weather or when an early frost kills the plant before it can be combined, sorghum stover deteriorates rather rapidly, especially in areas that receive considerable rainfall.

The increased use of grain dryers and the consequent earlier harvesting of corn and grain sorghums means that these two forages offer promise of increasing the supply of low-grade roughages for wintering cows in the farming regions. Cattle programs, except the cow-calf program, will not utilize these forages well, however.

As is the case with the grass hays and straws, corn and sorghum stover must be supplemented with at least the equivalent of 1 pound per day of a high-protein concentrate. Rations consisting principally of these forages should be supplemented with vitamin A or carotene, calcium, and phosphorus. Therefore, fortified protein supplements can be justified, unless some high-quality legume hay or winter small-grain pasture is available. The low palatability of shredded corn stover can be overcome with molasses, but the added cost may not be economically feasible.

CORN AND SORGHUM STALK FIELDS

It is estimated that approximately 90 percent of the corn in the Corn Belt is harvested for grain and the stalks left standing in the fields. These stalks can furnish a considerable amount of pasture for cattle during the

Table 38

Performance of Heifer Calves and Dry Beef Cows on
Cornstalk Fields[a]

Age	Heifer Calves (140 days)	Dry Cows (112 days)
Acres of stalk fields	24	24
Number of females	12	12
Fall weight (lb)	439(11/11)	1,041(11/30)
Spring weight (lb)	588(3/31)	1,139(3/22)
Winter gain (lb)	149	98
Supplemental feed per head (lb)		
Protein supplement	138	—
Oats	264	—
Hay	140	—
Salt and mineral	17	free choice
Feed cost per head	$15.12	$0.90

[a] *Results of Cattle Feeding Experiments*, Iowa State University AS-183, 1966.

late fall and early winter months, particularly for cows because they are capable of consuming and utilizing large quantities of coarse roughage. Ordinarily stalks may be relied upon to furnish the major sustenance of the breeding herd during the months of November, December, and January. However, the Iowa data summarized in Table 38 show that the full potential of cornstalk fields is probably not being realized on most Corn Belt farms. This experiment was conducted during only one winter and cow numbers were small, but the fact that the 12 cows in the Iowa study produced 12 calves with only a salt-mineral-vitamin supplement lends further support to the fact that the energy and protein requirement of bred cows is low so long as adequate trace-mineralized salt, phosphorus, and vitamin A are provided.

To use stalk fields for wintering cows, spring rather than fall plowing must be practiced, and some cash-grain farmers prefer not to postpone plowing, but this is a choice they will have to make. As a compromise, a stalk field may be reserved near the barns for cow feed, and the remainder of the corn ground may be plowed in the fall.

Field shelling of corn with a combine, as opposed to picking in the conventional manner, results in higher-quality forage if stalk fields are to be grazed, as is the case when stalk fields are to be harvested as stover or stover silage.

Sorghum stalk fields are well utilized by dry cows in late fall and early winter. A sorghum stalk field adjacent to a wheat pasture being grazed in winter makes an almost ideal combination because cattle on high-moisture wheat pasture crave a dry roughage such as the sorghum stalks left after combining the sorghums for grain. If cows are to be maintained solely on stalk fields for longer than 6 weeks, a protein supplement or some legume hay should be provided. Commercially prepared protein blocks, containing enough salt (about 20 percent) to limit protein intake to the desired level, are quite popular but expensive.

CORN AND SORGHUM STOVER SILAGE

Ensiling stover is preferable to shredding or grinding it if the equipment is available. It should be understood, however, that ensiling does not make the stover more nutritious; it merely renders it more palatable by making the hard, woody stalks soft and succulent. As a result, practically all of the stover silage is eaten, whereas approximately 30 percent of the weight of dry stover is refused, at least in the case of corn stover. Consequently a given quantity of stover will feed a herd of cows for a considerably longer time in the form of stover silage than it would as dry stover. Another advantage of stover silage over dry stover is that its high moisture content reduces the danger of impaction and other digestive disorders sometimes encountered among cows that are fed large quantities of dry corn stover.

In making stover silage, an amount of water approximately equal to the weight of air-dry stover must be added if a good quality of silage is to be obtained. Silage made from stover cut immediately after field shelling or combining will need somewhat less added water to bring the silage to about 30 percent dry matter content. Cows will keep in good thrifty condition on a daily ration of 40 to 50 pounds of stover silage and 4 to 6 pounds of legume hay, or on a full feed of green stover silage plus 1 pound of a protein concentrate daily and free access to oat straw.

CORN AND FORAGE SORGHUM SILAGES

In the Corn Belt and wherever corn and forage sorghums can be successfully grown, corn and sorghum silages are excellent, though not exactly economical, winter feeds for the breeding herd. Because they are moist and succulent, they tend to stimulate a good flow of milk in wet cows that calve in late winter or early spring, and such silages tend to keep the digestive system well regulated, resembling fresh grass in this respect. Their high yield per acre enables the farmer with a limited area to winter a considerable number of cattle. Their physical nature makes for ease and economy of storage, especially if stored in larger bunker or trench silos.

Corn and sorghum silages fed alone are not well balanced rations, as they are low in protein in comparison with their carbohydrate content. This can be remedied by using a legume roughage or a protein supplement along with silage. The oilseed meals, clover hay, and alfalfa are all rich in protein, and any of them can be used to supply the nitrogenous material or protein that the animals require. The addition of 10 pounds of urea per ton of silage at silo-filling time will provide the required supplemental nitrogen at a very economical figure.

Corn silage and, to a lesser extent, sorghum silage, because of their grain content, are usually considered higher-cost feeds than straw, stover, and other dry roughages, which have little cash sale value. Consequently these silages usually are fed to commercial cows in limited amounts, along with some cheap, dry roughage that is fed according to appetite. Sometimes no silage is fed until the cows start to calve, when feeding is begun at the rate of 20 to 30 pounds per head daily. Purebred cows, which are best kept in fairly good flesh to enhance the sale value of their offspring, will stay in about the desired condition on 5 to 8 pounds of legume hay and a full feed of corn or sorghum silage, which for an average-sized cow is about 40 pounds. When fed no hay or other dry roughage, cows will eat about 5 pounds of silage per 100 pounds body weight.

From the standpoint of the acreage required to produce the winter feed for beef cows, corn silage can be justified as a beef cow feed. For example, six cows consuming 50 pounds of silage daily for 4 months can be wintered on the forage produced on one acre if the corn silage yields 18 tons per acre, which is an average yield. Few other crops can do so well. Forage sorghums are a notable exception, and irrigated meadowland, producing either mixed hay or silage, might approach it.

HAY CROP SILAGE

Silage made from mixed grasses and legumes is even better than corn silage as a winter feed for dry cows. Its high protein, carotene, and mineral content makes it an ideal source of nutrients for cows during both gestation and lactation. It is sufficiently low in total digestible nutrients or digestible energy to permit it to be fed according to appetite. West Virginia station tests of corn silage and legume grass silage, summarized in Table 39, disclosed no significant difference between these feeds for pregnant and lactating beef cows when they were fed in quantities that furnished the same amount of dry matter. Making silage from grasses and legumes not needed

for pasture during the summer is an excellent way to extend the pasture
season, because good-quality hay-crop silage is similar in physical nature
and digestible nutrients to fresh pasture forage. Because of its high protein
content, no protein concentrate or legume hay is required; consequently,
additional roughage if fed may well be straw or stover.

MISCELLANEOUS ROUGHAGES

A variety of roughages that may be important in localized areas but
are not of widespread interest are also being utilized by beef cows.

Oat Hay and Oat-Vetch Hay. Oats, either seeded alone or with vetch,
makes excellent hay for cows, especially if cut in the soft dough stage.

Oat Silage. If the oat crop is cut for silage at the same stage as
for hay, the silage is equally good as feed for cows. In fact, because the
hazard of unfavorable hay-curing weather is eliminated and less oat grain
is shattered, an acre of oats cut for silage will winter more cows than
if cut for hay. Neither of these feeds requires supplementation for dry

Table 39
**Comparison of Corn Silage and Legume-Grass Silage for
Wintering Beef Cows**[a]

	Corn Silage	Legume-Grass Silage
Average initial weight (lb)	807	789
Average final weight (lb)	1,164	1,143
Average gain per head (lb)	357	354
Average winter ration (lb)		
First winter (yearling heifers)		
Silage	20	20
Alfalfa hay	6	6
Cracked corn	1	3
Soybean meal	0.75	—
Average next 4 winters (2- to 5-year-old cows)		
Silage	31.4[b]	27.7[b]
Legume-grass hay	6.5	6.5
Average winter gain (cows plus newborn calves) (lb)	94.2	79.8
Average birth weight of calves (lb)	66.2	67.3
Average weaning weight of calves (corrected to 180 days of age) (lb)	368.5	377.8

[a] West Virginia Agricultural Experiment Station, Mimeographed Report.
[b] Silages were fed in amounts that provided equal amounts of dry matter to both groups
of cows.

cows other than a salt-mineral mix. Access to some straw or a cured grass pasture will keep the cows on the silage ration more contented.

Sugarbeet-Top Silage. A full feed of this silage plus 2 to 4 pounds of legume hay makes a good cow wintering ration but this type of feed is naturally confined to areas that grow sugarbeets commercially. An all beet-top ration is apt to be too laxative, and limestone should be fed to counteract the diarrhea-producing agent, oxalic acid, which this silage contains.

Cannery Silage. In the vicinity of commercial canneries, pea vines and other vegetable residues can often be bought for ensiling purposes, or the silage can be bought directly from the company stack. Feeding some hay or even a pound or two of grain may be necessary if thin cows are fed this type of silage, because the silage is sometimes unpalatable and intake may not be adequate to meet all needs. Corn cannery waste is usually fed to other types of cattle, which utilize it better than do cows.

Turnips, Mangels, and Rutabagas. These root crops are more popular in Great Britain and northern Europe than in the United States. They usually contain only about 10 percent dry matter, and extremely large quantities must be consumed by cows if their energy requirements are to be met. In fact, because of the high water content of such feeds it is recommended that some form of dry roughage be fed in addition to the roots. If the dry roughage is nonleguminous, a half pound of high-protein concentrate should be added per day.

Prickly Pear. This plant of the cactus family is sometimes used as an emergency feed in the Southwest. The spines must first be removed by singeing, and a completely fortified supplement is required to balance the ration. It should always be considered as an emergency feed only. Nevertheless, thousands of cows have literally been saved from starvation by the prickly pear during extreme droughts such as those of 1953, 1957, and 1964.

Corn Cobs. Corn cobs have been used successfully as a source of energy for cows during the winter season. Because of low palatability and almost complete absence of protein, minerals, and vitamins, they must be fortified with complete supplements that furnish these essential nutrients. The cobs must be finely ground to make them more palatable and to facilitate thorough mixing with other feed materials. One part of ground ear corn, two parts of ground corn cobs, and one part of ground alfalfa hay fed free-choice through a self-feeder should maintain beef cows in satisfactory winter condition. If they increase noticeably in weight on this mixture, the amount of ground cobs should be increased. If they eat less than 2 pounds of the mixture per 100 pounds live weight, about 1 pound of 50 percent molasses feed per head should be added to the mixture daily.

Gin Trash. Although admittedly a poor-quality roughage, this by-product of the cotton ginning process may be used to help carry cows through a winter, especially in emergencies. If supplemented with a high-protein, high-energy supplement that is also fortified with minerals and vitamin A, a ration consisting largely of gin trash will support dry cows. From 1 to 2 pounds of fortified supplement will be needed for cows in average flesh. If extremely thin, cows require an additional 2 pounds of an energy concentrate such as molasses, sorghum, or corn. Feeding cows alfalfa hay one day and gin trash on the next will result in a balanced ration.

WINTER PASTURE

Fall-sown small grains, annual ryegrass, crimson clover, and fescue are examples of winter pastures that are utilized for wintering beef cows as well as other livestock. There is a limit to how far north such a program can be reliably used, but evidently a line drawn through Virginia, northern Kentucky, southern Illinois, northern Arkansas, Oklahoma, and the Texas Panhandle marks approximately the northern boundary where winter pastures can reasonably be depended upon. Fall and winter rainfall and some warm weather are essential for adequate growth in such pastures, but too much rain is a severe handicap because of the problem of grazing the wet fields. Fescue, a perennial grass, is grown in a large part of the wetter areas of the South because it forms a matted turf that helps support the weight of the cattle. In the western portion of the region outlined above, winter wheat and oats are the main crops used. The crop may or may not be intended for grain production in addition to whatever pasture it provides. The following conclusions from the Oklahoma station point out the value of and the problems involved in winter pasture utilization:

1. The pasture value of winter small grains is so high that livestock farmers may profitably use them entirely for pasture, without taking a grain crop.

2. The protein content of small-grain and annual ryegrass forage, when young, green, and succulent, is high—about 30 percent or more (dry-matter basis) as compared with around 42 percent in the usual high-protein supplement.

3. The carotene (provitamin A) content is exceedingly abundant. This is an important point, for winter rations in the Southwest are often seriously lacking in carotene. The forage is also high in minerals. Fiber is low—about the same as in alfalfa leaf meal.

4. Grain yield is not seriously affected by grazing until plants reach the jointing stage of growth.

5. The forage yield is about tripled if grains are completely pastured out instead of taking off the cattle when continued grazing begins to affect grain yield.

6. Forage production of the different varieties of the same crop tested differed enough to make it worthwhile to choose a variety specifically for pasture.

7. A good mixture for both early fall and late spring pasture would include either barley or rye, winter oats, and annual ryegrass.

8. On a low-phosphate soil, cows show a definite preference for pasture grown on plots where phosphate fertilizer has been applied.

9. It is more important to have plenty of succulent rapidly growing forage available to animals than it is to worry about possible differences in the palatability of pasture crops.

10. The protein, vitamin, and mineral content of small-grain and annual ryegrass forage cut at the stage of most rapid growth is so high that it apparently would be a valuable supplement for other feeds if dehydrated.

In the eastern and southern portion of the winter pasture belt, three major kinds of winter pasture are available: (*a*) permanent winter forages such as fescue, (*b*) self-reseeding annual forage crops, usually grown in connection with permanent summer pastures such as crimson clover in Bermuda pasture, and (*c*) annuals such as small grains or ryegrass planted on a prepared seedbed as in the winter wheat belt. An experiment concerned with evaluating various winter pastures was conducted by the Georgia station and is summarized in Table 40. On the basis of this and other tests conducted by the Georgia workers, the following conclusions may be made:

1. A seeding mixture of 2 bushels wheat, 25 pounds annual ryegrass, and 15 pounds crimson clover is a satisfactory pasture mixture for winter-calving beef cows in the Southeast.

2. A complete fertilizer, applied at the rate of 500 to 600 pounds per acre in a split application, is generally required for maximum production. Additional nitrogen, to the extent of 100 pounds per acre, in three topdressings is a good investment.

3. Fescue is a highly dependable permanent winter pasture for beef cows, but it is improved by the addition of crimson clover.

Although Bermuda grass is dormant during wintertime, it is being used as a year-round pasture in the Southeast and thus needs to be considered among the feeds for wintering beef cows in the nonrange area. The results of an Oklahoma test conducted at the Fort Reno station with spring-calving mature Hereford cows are shown in Table 41. Of interest is the fact that the cows in all three lots on improved Bermuda grass (Midland)

Table 40
Performance of Beef Cows on Winter Pasture[a]

Treatment	Drylot	Fescue Pasture	Crimson Clover Pasture	Winter Temporary Pasture
Number of cows	10	10	20	20
Total acres grazed	—	6	28	20
Date on test	1/18	1/18	2/9	1/18
Date off test	5/9	5/9	5/9	5/9
Days on test	112	112	84	112
Average cow weight on (lb)	958	955	920	957
Average cow weight off (lb)	896	916	1,015	1,099
Average daily gain, cows (lb)[b]	0.21	0.48	1.92	2.02
Number of calves born	6	6	15	20
Average birth weight, calves (lb)	65.8	68.2	68.0	68.1
Average calf weight off test (lb)	151.5	173.6	166.0	255.3
Average daily gain, calves (lb)	1.00	1.39	1.77	1.93
Daily milk production, 10th week of lactation (lb)	8.40	11.40	13.00	15.90
Average daily ration (lb)				
Cane silage	52.3	29.4	33.2	13.2
Oat straw	5.7	—	—	—
Cottonseed meal	1.5	—	—	—
Mineral	0.30	0.26	0.34	0.23
Gain per acre (lb) (cow and calf gains)	—	214	180	422
Cost				
Pasture per acre ($)	—	25.38	12.05	28.69
Pasture per head ($)	—	15.23	16.87	28.69
Supplemental feed per head ($)[c]	23.85	8.63	7.31	3.86
Total winter feed cost per head ($)	23.85	23.86	24.18	32.55
Cost per head daily (cents)	21.3	21.3	28.8	29.1

[a] Georgia Agricultural Experiment Station, Mimeographed Series N.S. 241, 1965.
[b] Corrected for an average loss of 120 pounds due to calving.
[c] Feed prices per ton: cane silage $5.24; oat straw, $12; cottonseed meal, $88.

produced heavier calves than were produced by cows on native pasture—which in this instance was an excellent stand of bluestem, buffalo, and grama grasses, usually considered to be superior cured forage for beef cows in wintertime, as well as good summer range. All cows received supplemental cottonseed meal and an injection of 1 million IU of vitamin A

Table 41

Supplemental Winter Feeding of Spring-Calving Cows on Bermuda Grass Pasture (December 3 to April 15)[a]

Pasture[b]	Bermuda		Bermuda		Bermuda		Native
Level of Cottonseed Meal (lb)	0		2		3		2
Phosphorus Supplement	−	+	−	+	−	+	+
Number of Cows	8	8	8	8	8	8	16
Cow Data							
Initial weight (lb)	1,102	1,073	1,084	1,096	1,093	1,117	1,066
Weight change (lb)							
To postcalving	−137	−123	−129	−107	−99	−101	−101
To weaning	−6	+43	+35	+25	+89	+86	+39
Average 24-hour milk (lb)	12.0	12.3	15.1	13.0	13.8	13.5	13.7
Plasma phosphorus (mg/100 ml),							
4 cows per treatment							
12/3/64	5.9	5.5	4.7	5.6	5.3	5.6	—
4/15/65	4.1	6.5	5.3	5.9	4.0	4.5	—
Calf Data							
Average birth weight (lb)	75.6	73.0	80.1	89.1	83.1	75.4	73.0
Average weaning weight (lb)[c]	471	481	483	490	500	478	444

[a] Oklahoma Miscellaneous Publication M.P. 79, U.S.D.A.-ARS cooperating.

[b] Each group of 16 cows grazed 140 acres year-round.

[c] Weaning weights adjusted to 205-day steer equivalents.

in early January. Half of the cows in each Bermuda grass lot received 36 grams of monosodium phosphate daily in the cottonseed meal as a supplemental source of phosphorus. The lot on native pasture received a 2:1 salt-bonemeal mixture free-choice. During the summer the cows on Bermuda grass pasture were rotated among four pastures at weekly intervals, and it should be noted that 200 pounds of nitrogen were applied per acre in three equal applications during the grazing season. The supplemental phosphorus proved beneficial in the lots fed less than 3 pounds of supplemental cottonseed meal, itself a good phosphorus source.

WINTER FEEDING IN RANGE REGIONS

Balancing the winter rations for beef cows in the range regions is quite different from that in the farming regions. The nutrient requirements of the cows do not differ if the calving season is similar, except possibly for some cows in areas at very high altitude, such as the North Park section in Colorado where cows are wintered without shelter and more energy may be required for maintenance. The main differences lie in the roughage sources available and in the time of year at which new spring growth begins.

The extensive nature of the cow-calf program in the range regions and the large size of ranches makes it possible or, more accurately, necessary to set aside a portion of the range for winter use, making a year's growth of grass and browse, in cured form, available for wintering dry cows and replacements. The variety of grasses, browse plants, and weeds found throughout the range area seems almost unlimited, and the nature of the available cured feed can vary tremendously. The types of range found in various parts of the range area may be classified and described as follows:

1. Tallgrass range. Located largely in the Plains States from Oklahoma northward; consists of big and little bluestem grasses, Indian grass, switch grasses, and others, but few browse plants. Intermediate rainfall and variable snow result in considerable wintertime deterioration of the forage. The average carrying capacity for year-round grazing is 10 to 15 acres per cow or animal unit.

2. Shortgrass range. Bounded on the east by the tallgrass range, on the west by the Rocky Mountains, and extending from Texas and New Mexico to Canada. The grama grasses, especially blue and side oats, dominate in the southern portion, with the wheatgrasses being introduced into the northern part. The low winter moisture conditions preserve the quality of the forage quite well. Only limited browse is available. From 20 to 30 acres are required per animal unit for year-round grazing.

3. Semidesert and sagebrush ranges. Intermingling types of range that dominate the southwestern area of the country, extending from New Mexico westward to California and northward into Wyoming, Utah, and Nevada. Black grama grass, numerous weeds, and browse species such as chamiza, winter fat, and Apache plume make this a good cattle-wintering area. But because of scant rainfall during the growing season, stocking rate is quite low if year-round grazing is necessary. In general, from 75 to 150 acres are required per cow annually for year-round grazing. Increased carrying capacity is possible when open-forest range (see below) is available for summer use.

4. Pacific bunchgrass range. Located from Washington to California, largely in the coastal area but extending eastward 200 to 300 miles in places. Once covered with permanent grasses and browse, this area is now characterized by annual grasses such as wild oat grass and cheat, and by filaree, burclover, and other forbs. Good fall and winter rainfall insure abundant feed for this season, but summer forage is dry and mature and

very low in quality. Acreage required per animal unit during the fall and winter growing season is 4 to 6 acres. Very little of the area is used for year-round grazing because of the extremely low nutrient content of the mature grass. Obviously, a fall-calving season is indicated for this unusual reversal of seasonal feed supply.

5. Open-forest range. Scattered throughout the West, principally at high altitudes and suited only for summer range. Most of the permanent grasses found in the West also occur in open-forest range, as do sedge and various annual grasses. The season is short and the available forage must be shared by domestic livestock and wild game, principally elk and deer. Because most of this range is federally owned and controlled, stocking rates and management practices are made a part of lease arrangements.

It is apparent that winter feed, in the form of cured range, varies greatly in the range regions, not only with respect to quantity but also nutritive quality and palatability. In sections that consist largely of grasses, especially the tallgrass and shortgrass types, supplemental protein and phosphorus are always required, as is evident from a comparison of Table 34 with the requirement tables, Tables 35 and 36.

The question of whether supplemental energy is required has been studied extensively by the Oklahoma station workers. The results of one longtime study are presented in Chapter 4. Another study, summarized in Table 42, shows the results of restricting the supplemental energy level of spring-calving cows considerably below the lowest level used in the first-mentioned Oklahoma study. One lot received supplemental energy well beyond what commercial cowmen customarily feed but at a level often fed by purebred breeders. The winter feeding program began each year in early November and ended about mid-April, by which time green feed was available and calves averaged about 1 month in age. Cottonseed cake or meal and ground milo were fed as a supplement to cured tallgrass range at rates that would accomplish the following winter weight change patterns:

Lot 1 (low). No gain the first winter as calves, with a loss of approximately 20 percent of fall weight during subsequent winters as bred females.

Lot 2 (moderate). Gain of 0.5 pound per head daily the first winter as calves, with a loss of 10 percent of fall weight during subsequent winters as bred females.

Lot 3 (high). Gain of 1 pound per head daily during the first winter as calves, then less than a 10 percent weight loss from fall weight during subsequent winters as bred females.

Lot 4 (very high). Self-fed a 50 percent concentrate mixture during the first winter as calves and during subsequent winters as bred females.

Table 42
Longtime Performance of Cows Wintered at Different Levels
(Seven Calf Crops)[a]

Wintering Level	Low	Moderate	High	Very High	Very High–Moderate[b]
Number of heifers started	30	30	30	15	15
Gain first winter, as calves (lb)	−9	97	140	280	268
Fall weight as 8-year-olds (lb)	1,146	1,182	1,262	1,372	1,172
Spring weight as 8½-year-olds (lb)[c]	926	1,011	1,139	1,248	971
Winter weight change (lb)	−220	−171	−123	−124	−201
Percent of cows still in herd	83.4	86.7	80.0	73.3	80.0
Average calving date	3/14	3/9	3/2	3/5	3/4
Average calf birth weight (lb)	75	77	79	75	77
Percent calf crop weaned	86.3	88.3	85.9	83.7	85.7
Average pounds calf weaned per cow[d]	442	465	481	452	451
Average daily milk production (lb)	11.2	12.7	12.1	10.6	10.2
Total supplement cost per cow	$19.53	$92.31	$207.28	$741.26	$308.23
Supplement cost per year	$ 2.79	$13.19	$ 29.61	$105.89	$ 44.03
Cow supplement and pasture cost per cwt calf[e]	$ 9.69	$11.44	$ 14.47	$ 32.29	$ 18.63

[a] Oklahoma Agricultural Experiment Station Miscellaneous Publication MP-76, 1965.
[b] Fed at the "very high" level during the first three winters, then at moderate rate.
[c] Weights of cows at 8½ years of age, weights taken about April 15, or about 6 weeks after calving.
[d] Corrected for sex and age differences of the calves.
[e] Includes $40 per cow for pasture and mineral cost.

Lot 5 (very high–moderate). Treated the same as Lot 4 for the first three winters, and as Lot 2 following.

The Lot 1 cows had to be fed wheat straw in drylot for about the first 2 months to ensure the desired weight loss. The summary table shows the performance of all lots for the first six calf crops, and it would appear that the level of energy fed did not affect calf birth weight, calving percentage, or calf weaning weight; thus each increment of increase in energy level fed unnecessarily increased the cost of the ration, at least of the last two levels. Other items of interest are the fact that the lower energy levels resulted in a later calving date and a lighter mature cow weight, and the very high level resulted in the loss of a greater number of cows from the test. Individual yearly weights of cows (not shown) were lower for the low-level cows for the first 5 years, indicating that maturity was delayed by the stress of lactation. The Oklahoma workers conclude their discussion of results as follows:

> The data presented for the individual yearly calf crops indicate that rather than select a level of wintering for the lifetime of the cow, consideration should be given to the life cycle feeding approach in which higher levels are used during growth and development of the female followed by lower levels after the cow has reached maturity since the major influence of the various levels on cow productivity occurs during the first three calf crops.

This conclusion fortifies the recommendation made earlier for sorting the cows in such a way that younger cows can be fed additional supplements if it is believed advisable after observing the condition of the cows.

It should be mentioned that drought is a common but unpredictable occurrence in the range areas. It is generally conceded by experienced cowmen and range management specialists that it is more economical to meet this situation by adjusting cattle numbers by culling cows heavily in the fall or by carrying over some steer calves from the previous fall to partially stock the range and sell as yearlings than it is to try to offset the feed shortage by heavier supplemental feeding. Drought-damaged range will be further damaged by overgrazing unless stocking rate is reduced.

ENERGY VERSUS PROTEIN SUPPLEMENTATION

A range must be carefully evaluated to determine what type of supplemental feeding program, if any, is called for. Energy content may be limited because of a shortage of forage, unpalatability of species in the pasture, low digestibility, poor range pasture layout, snow cover, and combinations of these factors. If it is unfeasible to adjust cattle numbers to meet the shortage of energy, a choice between several types of feed must be made, and economics or cost per unit of energy generally will determine the choice. In the range areas concentrates such as ground milo or corn or a 12 to 20 percent protein commercial range cube may be the preferred energy supplement, but if irrigated alfalfa meadows are accessible, this feed may be chosen because of its additional value as a protein, carotene, and phosphorus source.

An experiment of several years' duration conducted in the semidesert type of range in New Mexico is summarized in Table 43. This particular range provided considerable winter and early spring grazing during some years in the form of browse and annual weeds and grasses. The supplements were fed at the rate of 1 pound daily before calving and 2 pounds daily

Table 43
Response of Range Cows to Supplemental Feed[a]

Supplement Fed	Ground Milo	Cottonseed Meal Pellets
All years		
Weight loss of cow in calving season (lb)	46	24
Weaning weight of calves (lb)	387	389
Production per cow (lb)	321	311
Average-rainfall years		
Weaning weight of calves (lb)	424	430
Calves born following year (%)	91.7	90.6
Drought years		
Weaning weight of calves (lb)	366	366
Calves born following year (%)	87.4	84.1

[a] New Mexico Cattle Growers' Short Course, 1964.

from the beginning of the calving season until June or July when summer rains started the growth of the perennial grasses. In this experiment energy was apparently the limiting nutrient. It is interesting to note how limited energy, and probably multiple protein-mineral-vitamin deficiency, during a drought year can affect calf crop percentage the following year.

UREA IN WINTERING SUPPLEMENTS

Another chapter discusses how urea can be successfully substituted for a considerable portion of the supplemental plant proteins in cattle-finishing rations. While research on the subject of urea supplements for cattle being wintered on cured range forage is more limited, the results generally indicate poorer utilization of this economical nitrogen source. This is not surprising, since it is known that readily available sources of energy and minerals must be present in the rumen in order for the microflora to utilize urea nitrogen. Apparently the poor utilization of urea in this situation also affects the utilization of the forage itself. The resulting combined protein deficiency and poor forage utilization is responsible for such results as are shown in Table 44, which is compiled from an Oklahoma experiment on this subject.

The Oklahoma workers started with a set of weaner heifer calves and carried them through four successive winters and until they had weaned

Table 44
Effect of Level of Protein and Urea on Performance of Breeding Females Wintered on Tallgrass Cured Range[a]

		Supplements[b]		40% Protein[c] with Urea
		20% Protein	40% Protein	
Ingredients		Cottonseed Meal, Corn	Cottonseed Meal	Cottonseed Meal, Corn, Urea
Daily supplement fed (lb)		2	2	2
Winter weight change (lb)	*Days fed*			
Calves	162	− 9	6	− 13
Yearlings	171	− 15	− 8	2
2-year-olds	196	−141	−214	−146
3-year-olds	212	−311	−255	−321
Calf weaning weight (lb)				
2-year-old heifers		415	436	426
3-year-old heifers		347	405	381

[a] Oklahoma Agricultural Experiment Station Miscellaneous Publication MP-51, 1958.
[b] All supplements were equalized with respect to calcium and phosphorus.
[c] One-half of the protein equivalent was supplied by urea.

two calf crops. The similarity in performance between the lots on the 20 percent protein and the 40 percent protein-equivalent urea-containing supplement would indicate that little of the urea was utilized. The rations consumed by both these lots were too low in available protein, as evidenced by the fact that the heifers on the 2 pounds of 40 percent cottonseed meal outperformed the other two lots.

Most commercial range supplements today contain some urea in order to reduce the price per ton. When such supplements are adequately fortified with phosphorus, vitamin A, trace minerals, and a source of readily available energy such as molasses, they can be expected to perform satisfactorily if not more than 50 percent of the natural protein is replaced by nonprotein nitrogen. Such supplements may usually be obtained in block, cube, pellet, or liquid form, and some are prepared with sufficient salt content to permit self-feeding. Dehydrated alfalfa meal added to urea-containing range supplements at a rate that assures the daily consumption of at least 0.5 pound of the alfalfa meal, generally provides the necessary trace minerals, carotene (vitamin A precursor) and other as yet unidentified factors that apparently enhance urea utilization.

VITAMIN A IN WINTERING SUPPLEMENTS

The importance of vitamin A or its precursor, carotene, for growth and reproduction is discussed in Chapter 6. With reference to the winter feeding program for cows, the type and amount of available forages during the fall-winter period will have much to do with whether supplemental vitamin A is required. Fortunately cows and heifers can accumulate considerable vitamin A reserves in the liver, but if no green feed is available from frost in the fall until June or July or until after calving and the onset of the breeding season, the resulting vitamin A deficiency is likely to cause retained placentae, calf losses at or after calving, and poor conception rate for the succeeding calf crop. In certain range areas drought occurs frequently enough to warrant the use of vitamin A–fortified supplements. A daily intake of 50,000 IU of vitamin A should meet the needs of all cows. Field research, although leaving something to be desired in control, has shown that a single injection of an aqueous carrier containing 1 million units of vitamin A, either intramuscularly or intrarumenally, about 60 to 30 days before the calving season, is equally effective in preventing the symptoms of vitamin A deficiency.

Table 45 gives the results of an Oklahoma study that compares favorably with field trial results. Ranges that provide some weed or annual grass forage, due to intermittent spring showers, will provide adequate carotene under most circumstances. Stabilized, synthetic vitamin A is very inexpensive and should be included in virtually all protein supplements, which

Table 45
Effect of Vitamin A Injection of Cows Before Calving[a]

Treatment	No Vitamin A	Vitamin A[b]
Number of cows	77	78
Average weight upon injection (lb)	866	863
Average weight of cows, 5/7/63 (lb)	792	792
Weight change (lb)	−74	−71
Number of calves born	73	75
Number of calves at 112 days	66[c]	71
Weight of calves at 112 days (lb)	209	223
Date cows calved following year	3/10	3/11
Number of cows open following year	15	9

[a] Oklahoma Agricultural Experiment Station Miscellaneous Publication MP-76, 1965.
[b] One million IU vitamin A injected into rumen 1/31/63.
[c] Three calves were lost in this group to predators.

must be processed for the addition of other ingredients anyway. The question of whether to feed or to inject vitamin A appears to be a matter of relative cost and availability of labor. Heifers that are to be bred after wintering should be treated the same as bred cows with respect to supplemental vitamin A.

FREQUENCY OF FEEDING PROTEIN
SUPPLEMENT TO RANGE COWS

The experience of some ranchers and results of recent research show that beef cattle being wintered on cured range do not require their supplement daily. Table 46 summarizes an experiment with cows being wintered on shortgrass range in the Trans-Pecos area of West Texas. It made no difference whether the weekly allowance of supplement was fed at a given rate each day, three times weekly, or twice weekly. If anything, there may have been an advantage when some days were skipped. It has been observed that if cows do not expect to be fed daily, they will range out farther from the feedground or water source and thus will consume more and better-quality forage. Stocking rate in large pastures may be increased by this management practice because of the more uniform grazing pattern. Moving the feeding sites to various locations in large pastures can of course accomplish somewhat the same purpose if the terrain permits transporting the supplement throughout the pasture area.

When alfalfa hay is used as a winter protein supplement, the quantity fed must naturally be higher because of the lower protein content of the

Table 46
Effect of Interval of Feeding Cottonseed Cake to Beef Cows[a]

Frequency of Feeding	Daily	Three Times per Week	Two Times per Week
Winter gain (lb)			
1958–59	91	88	38
1959–60	−60	−62	−37
1960–61	−80	−121	−98
1961–62	−167	−176	−170
Percent calf crop weaned	81	89	86
Calf weight per cow (steer equivalent) (lb)	361	399	374

[a] Texas Agricultural Experiment Station Bulletin 1025, 1964.

Table 47

Effect of Interval of Feeding Alfalfa Hay to Stocker Heifer Calves[a]

Feeding Interval	Daily	Twice Weekly	Weekly
Winter gain (lb)	59	62	76
Summer gain (lb)	272	248	250
Total gain (lb)	331	310	326

[a] Nebraska Feeders' Day Progress Report, 1960.

hay. Nebraska workers fed heifer calves an allowance of 4 pounds of hay per day or an equivalent amount at twice-weekly or weekly intervals, with satisfactory results, as shown in Table 47. The summer compensatory gain of the heifers fed daily is evident.

A disadvantage of not feeding daily is that the herd will not be observed daily. If fences are in good condition and the water source does not need frequent checking, failure to see the herd daily is perhaps not serious, at least until beginning of the calving season. On very large spreads, much less labor is required if only the herds in one or two pastures are fed and checked daily.

SELF-FEEDING PROTEIN SUPPLEMENTS

A recent innovation that makes possible a saving in labor is the addition of sufficient salt to protein supplements to permit self-feeding. From 25 to 35 percent salt content in supplements fed as meal, cubes or pellets, blocks, or liquid will usually keep supplement consumption constant at about 2 pounds per head daily. Several factors affect the degree of success of this feeding method, including amount of forage available, location of feedground in relation to water supply, hardness of blocks if that is the form used, number of "boss" cattle in the herd, and added cost of preparation of the supplements. Salt-containing blocks seem to be gaining in favor in the area where large pastures are common. Regardless of the form of the high-salt supplements, plenty of unfrozen water must be available or salt toxicity will be a problem. Apparently the additional water is required for elimination of the excess sodium chloride by way of the urine and for prevention of dehydration.

CREEP-FEEDING FALL AND EARLY-WINTER CALVES

The question of whether creep-feeding is a profitable practice has been studied extensively. Most studies have been with spring-calving herds, in which case the pasture being utilized by both cows and calves would normally be ample in supply and good to excellent in quality. However, as has been mentioned, the roughage supply for cows in winter, especially in the range areas, is likely to be of lower quality and possibly in short supply. If cows calve in fall or winter, the added nutritional requirements for lactation, especially for the younger cows, are not likely to be met, resulting in reduced milk production and poor calf performance. Creep-feeding the calves to make up for the reduced milk supply provided by the mother cow would seem a practical solution to the problem.

Table 48
Value of Creep-Feeding Calves versus Supplemental Cow Feed for Young Fall-Calving Cows[a]

Lot Number	1	2	3
Cow Feeding Level	Low	Low	High
Creep-Feeding	Yes	No	No
Number of cows raising calves	27	25	29
Average weight per cow (lb)			
Initial (fall)	1,034	1,012	990
Winter change (203 days)	−283	−299	−194
Change to weaning	−42	−59	−6
Yearly change	24	32	60
Average weight per calf (lb)[b]			
Birth	73	72	73
Spring	222	160	207
Weaning	413	320	375
Average birth date of calves	10/21	10/26	11/3
Supplemental feed per head (lb)			
Cow			
Cottonseed meal	360	360	456
Milo	—	—	676
Calf (creep-fed)	944	—	—
Total feed cost per head ($)			
Cow	36.30	36.30	53.26
Calf	24.38	—	—
Total	60.68	36.30	53.26
Selling value per calf minus feed cost ($)	62.28	55.22	57.03

[a] Oklahoma Agricultural Experiment Station Bulletin, B-610, 1963.
[b] Corrected for sex and age.

The results of a 2-year experiment conducted by Oklahoma workers are summarized in Table 48, which shows that creep-feeding the calves is a better solution to the problem than extra feeding of the cow herself. Calves were dropped in late fall, beginning in late September. The cows were wintered on tallgrass range with one of two levels of supplementation. The creep-feeding, when used, extended from mid-January until weaning time in July. The creep ration consisted of 55 percent ground milo, 30 percent whole oats, 10 percent cottonseed meal, and 5 percent molasses. The results show that both creep-feeding the calves and feeding the cows additional feed were profitable practices, but feeding the calf itself was the more profitable of the two. It should be noted that the data are a summary of the first two calf crops of cows calving first as 2-year-olds. If only the data for the first calf crop had been shown, creep-feeding would have appeared even more favorable. The Oklahoma workers concluded that creep-feeding the calves from first-calf heifers calving in the fall should be profitable if the selling price from the resulting calves is not reduced by their extra weight and condition. This means that, in practice at the farm or ranch level, first-calf heifers would best be pastured or fed separately from the rest of the herd to take advantage of the potential profits that may accrue from creep-feeding.

The separate handling of first-calf heifers makes possible two other recommended management practices. To ensure quick rebreeding and high conception rate of cows nursing their first calves, added supplemental feed or the best pasture on the farm or ranch is recommended. Then too, because it is often desirable to breed these young cows to younger unrelated sires, pasturing these cows separately means they are unlikely to be exposed to their sires. Open heifers being bred for the first time may be handled together with these young cows.

WINTER MANAGEMENT OF BULLS

Winter is the proper time to condition bulls for the spring and summer breeding season of the following years. Bulls that have been running out on pasture with the breeding herd are likely to be thin and should have sufficient grain to put them in proper flesh before the arrival of spring. Young bulls should be fed more liberally than old mature ones because their growth requirements must be met before any improvement in condition can take place. A grain mixture composed largely of feeding stuffs that are growth-promoting rather than fattening is found best for bulls. Crushed oats make a good basal ingredient, to which corn, grain sorghum, or barley may be added only if a marked improvement in flesh is desired. Bulls that are in a run-down condition or that show signs of lack of thrift should have 2 to 4 pounds of protein concentrate per day, depending on age and weight. Mature bulls already carrying sufficient flesh may be wintered largely on choice roughages such as legume or mixed hay.

Bulls can be kept strong and vigorous during the winter only if they are given sufficient exercise as well as plenty of feed. To accomplish this end they should have the run of good-sized lots or, better yet, be turned with a few pregnant cows into an adjacent pasture or stalk field whenever the weather permits. When two or more bulls are kept, it is highly desirable to run them together because they will take more exercise when together than when they are alone. Although the turning together of old bulls that are strangers to each other is attended with some risk, a young bull may be put with an older one with comparative safety.

The importance of having bulls in strong condition (that does not mean fat, however) at the opening of the breeding season can scarcely be overemphasized. Valuable time is often lost by the use of a bull with impaired sex drive (libido) due to being either too fat or too thin. Examination of several semen samples for numbers and normalcy of sperm cells usually determines the fertility of a bull, and often a small or late calf crop can be prevented by such tests.

Western ranchers realize, perhaps better than other cattlemen, the importance of conditioning their bulls before the breeding season. They are well repaid for their pains, as shown more than 40 years ago by the data gathered on 15 ranches in northern Texas over a 3-year period.

On 10 of the 15 ranches the bulls were taken from the cow herds in the fall for conditioning and were returned about June 1 of the following year. On the other 5 ranches the bulls were kept with the cows during the entire year. The ranches on which the bulls were removed from the cow herds had a 77 percent average calf crop for the 3 years. On ranches where the bulls were not removed the average calf crop was 64 percent.

Whereas the better feed conditions prevalent in the central states make conditioning of bulls less urgent than on the range or in the coastal regions, more attention to this point on the part of all cattlemen will result in a material decrease in the number of cows that fail to raise calves through no fault of their own.

WINTERING BREEDING HEIFER CALVES

Rations for the heifer calves that are to be kept for herd replacements should be composed principally of the better-quality roughages available on the farm or ranch. The actual nutrient requirements of heifer calves are listed in Tables 65, 66, and 67 in Chapter 10. The level of gain desired depends on whether the heifers drop their first calves as 2- or 3-year-olds. If the heifers calve first as 2-year-olds, more energy is required during the winter in order to ensure adequate development prior to breeding the following summer. Meeting the requirements as listed under the subheading "Normal Growth for Heifers" in the previously mentioned tables will suffice. If the heifers are bred to calve as 3-year-olds, they can be wintered according to the requirements listed under the subheading "Wintering Weanling Calves" in the same tables. In either case the equivalent of 1.25 to 1.50 pounds of high-level protein concentrate should be fed daily.

BUILDINGS AND EQUIPMENT

One of the advantages claimed for the beef cow by her ardent supporters is that she requires a small outlay for shelter and equipment. Expensive barns are by no means necessary. In fact, in some instances they may be a positive disadvantage. An investment of much more than $100 per cow in buildings and equipment is an unrealistic financial outlay for a herd of beef cows kept solely to raise calves for the open market. The situation is of course quite different with respect to high-priced registered cattle. Even here the elaborate barns and lots or corrals so often seen in large purebred establishments are not indispensable, and sometimes large

FIG. 36. Dry pregnant beef cows can derive much of their early winter feed requirement, especially energy, from cornstalk fields. More snow than the light covering seen here would necessitate supplemental feeds, protein in particular. (Animal Science Department, Iowa State University.)

barns that were built at great cost have been discarded in favor of inexpensive open sheds in which the cattle seem to do better.

Dry cows prefer to live entirely in the open if the ground is well drained, affords a reasonably dry area where they may lie down, and is protected from high winds by hills or trees. Such a place is provided by rolling pastures that have some sheltered valleys protected by surrounding hills and thick stands of timber, where the cows may spend the entire winter in apparent comfort, even in near-zero weather.

A deep shed that opens to the south or southeast offers ideal shelter for dry cows when protection is needed in areas affected by wet, cold climates. Here the cows may run together in groups of 10 to 50 head or more and can be fed with a minimum of labor. Cows that are approaching parturition and those with very young calves should be handled apart from the rest of the herd if shed space is limited. However, after the calves are 10 days old, both cow and calf may be returned to the main herd. In a small herd the calves may be kept apart from the cows and allowed to nurse twice daily if a trap or pen is provided at one end of the shed where a deep, dry bed is available for the calves.

The lots occupied by cattle need not be large. If possible it is best to have two lots—a small paved lot adjoining the shed with about 50 or 60 square feet of space per cow and a much larger lot or trap into which the cattle may be turned for exercise when the ground is dry or frozen. If pasture land or stalk fields are available, the larger lot is unnecessary. The small paved lot, however, is almost indispensable in the nonrange areas during the spring when the ground is soft and muddy. Pavement is much more desirable in the lot just outside the shed than under the shed itself. This combination keeps the cattle more comfortable, reduces the labor of feeding, and repays its cost many times over in the extra amount of manure saved.

Chapter 9

BREEDING HERD
MANAGEMENT in SUMMER

In theory the feeding and management of the breeding herd during the summer is relatively simple, because the herd is then running on pasture and requires little attention. In practice, however, some of the cowman's most difficult problems may arise during this season of the year.

First in importance is the difficulty of estimating the forage supply, which is determined largely by weather conditions and can neither be definitely predicted nor controlled. Should the weather be unfavorable for the growth of grass, there may be a critical shortage of feed long before the usual end of the grazing season.

Second, the breeding season for most beef herds comes during the summer, making it necessary to inspect the herd regularly to determine whether the cows are being settled. Noting the dates on which a few familiar cows are observed to be in heat and then following up with an inspection 19 to 20 days later soon reveals whether the cows are being bred. Farmers and ranchers are invariably busy with fieldwork at this time of year, however, and are prone to limit inspections to once or twice weekly, especially if the herd is pastured some distance from the farmstead.

Third, early summer is the best time to castrate bull calves that were born on pasture or were too young to castrate while the herd was still in drylot. If this job is put off until the urgent fieldwork is finished, the calves are big and hard to handle when it is finally done. The same is true of such jobs as dehorning, vaccinating, and branding, which may not be done before cows go to pasture if calves were dropped late.

A fourth summertime problem is the need to attend to such parasites as hornflies, faceflies, and deerflies. In addition, in the range areas water shortage can be as serious a problem as grass shortage, and this also goes for farm areas that rely on farm ponds.

PASTURE

Pasture is relied upon almost entirely as feed for the breeding herd during the summer. On most farms and ranches no feed other than pasture is provided. As discussed elsewhere, pasture is considered to be so essential for the breeding herd that beef cows are not usually kept by operators who lack sufficient grass or other forage to carry them through the summer. Although it seems desirable to keep the breeding herd on pasture during the summer for as long as forage is plentiful enough to provide milk and nutritious feed for the calf, some successful operations are conducted completely in drylot. The year-round drylot system offers an opportunity for maximum use of equipment and labor and for actually harvesting more forage, either as hay, green chop, or silage, from high-yielding cultivated pasture seedings. The subject of drylot- or confinement-rearing of calves is treated in greater detail later in this chapter.

FIG. 37. An almost ideal pasture for a cow herd with spring-dropped calves. Addition of legumes to the predominantly grass pasture would improve both quality and quantity of forage. (American Angus Association.)

KINDS OF PASTURE

Pasture areas usually are occupied by perennial plants—that is, plants that do not die after ripening their seed but are dormant through the winter while continuing to grow year after year. Kentucky bluegrass, redtop, bromegrass, Bermuda grass, and nearly all of the native grasses of the Western Plains region such as the bluestems, the gramas, wheatgrasses, and buffalo, are perennials, and they together occupy by far the greater percentage of the total pasture area of the country.

Annuals—that is, plants that produce seed and die in one year—provide much forage for beef cattle and, when used to supplement perennial forages, they can increase both the performance of beef cattle and the carrying capacity of a farm or ranch. Such seeded annuals as straight Sudan, forage sorghum–Sudan hybrids, and small grains, when planted on productive soils, can carry as many as 3 to 5 cows per acre for the growing season, especially when fertilized and irrigated. Other annuals that reseed themselves, such as wild oats, lespedeza, filaree, and cheat, also are valuable pasture forages, but these usually are found in permanent pastures containing perennial species as well. The fact that these annuals grow most actively during seasons when the perennial grasses may be dormant makes them especially valuable.

Pastures can further be divided into temporary and permanent pastures. Temporary pastures may consist of only cultivated perennials, only annuals, or combinations of both. On the other hand, permanent pastures are more likely to consist of native perennials, and those in the nonhumid areas of the country are more apt to be spoken of as range than pasture.

ROTATION AND TEMPORARY PASTURE

The temporary pastures that consist of cultivated perennials such as alfalfa, clovers, and bromegrass as opposed to native perennials, are also referred to as rotation pastures and are found mainly in the Corn Belt where they are grown in a planned sequence with the cash crops of corn, wheat, and soybeans. They are usually preceded in the rotation by oats, used as a nurse crop. The acreages of rotation pastures and oats, are decreasing because it has become more economical to purchase fertilizers in the Midwest, especially nitrogen, than to depend on legumes, with their atmospheric nitrogen-fixing bacteria, to serve this function. Continuous cropping with corn is now an accepted practice in the more fertile areas of the country. The fact that the production from the rotation pastures and the oat crop returns considerably less cash income, if sold, than corn or soybeans, has also had much to do with the shifts in acreage mentioned. The claim that rotation pastures are required to maintain yields of cash crops is being disputed, and only time will tell whether the soil tilth, water-holding capacity, and micro- and macro-mineral content can be maintained without them.

Rotation pastures are usually allowed to stand for 1 to 3 years as part of the general crop rotation when used. These rotation seedings may be harvested for hay, haylage, or grass silage the first year and pastured for a year or two thereafter before they are plowed under for corn. Or they may be left standing for only one year, in which case the first crop is sometimes cut for hay or grass silage and the second crop is pastured. Or they may be grazed by beef cattle or other stock from early spring until late fall. Obviously the method of utilizing a particular field is determined largely by the need for additional pasture at the time it is ready for use.

The plant species most widely used in seeding rotation pastures consist largely of legumes, because the main reason for including a year of pasture in the crop rotation is to improve soil fertility. Consequently, alfalfa and red, alsike, and sweet clovers, either seeded alone or mixed with grasses, are commonly used. A mixture of red clover, sweet clover, and timothy is a satisfactory seeding for a pasture that is to be used for only one year, but alfalfa usually should be included in seedings that are to be allowed to stand 2 years or longer.

Advantages of Rotation Pastures

1. More feed is produced per acre, making possible a larger herd for a given pasture area.

2. A large percentage, if not all, of the forage consists of legumes.

3. Rotation pasture crops, as a rule, withstand the heat and drought of summer better than some perennial grasses.

4. The rotation pasture crop can be included in the regular farm crop sequence, making possible the utilization of the legumes grown primarily for soil improvement. At the same time those benefits accruing to the soil through pasturing and feeding on pasture are shared by different parts of the farm.

5. Any surplus forage can be utilized for hay or silage.

Of these advantages, the last two are most important. The others, important in some years, are open to question when a long period of time, involving all kinds of seasons and weather, is considered. With a good stand and ordinary weather conditions most legume-grass pasture combinations produce more and better forage than grasses alone on similar land.

However, experience shows that, under ordinary farming conditions, a perfect stand of legumes is secured in not more than 4 years out of 5, and in about 1 year out of 7 or 8 the crop is almost a total failure. At such times the common practice is to sow oats, rye, Sudan grass, or other non-legume crops for pasture.

As the advantages of rotation pastures have been mentioned, it is well to list the principal objections raised against them.

Disadvantages of Rotation Pastures

1. They require an annual outlay for seed, preparation of seedbed, and other expenses if the crop sequence is followed without interruption.

2. Stand of crop is often uncertain and sometimes a total failure.

3. Plants do not form a good turf; thus fields are often badly trampled in wet weather.

4. Adequate shade and water are provided with difficulty.

5. Rotation pasture crops sometimes cause bloat.

6. Some of the rotation pasture plants tend to ripen and dry up in late summer and early fall.

The last disadvantage may be dismissed with the statement that the clovers of the common rotation pasture crops are biennial legumes that ripen and dry up only in the fall of the second year. By that time the seedling pastures started in the spring of that year are large enough to graze and the stock can be shifted to them. Including alfalfa in the mixture will overcome this disadvantage.

Table 49
Performance of Calves on Different Pastures[a]

Kind of Pasture	Number of Cows	Number of Calves	Days on Pasture	Average Weaning Weight[b]	Average Total Gain of Calves	Average Daily Gain of Calves
Bluegrass only	1,878	1,799	176	378	262	1.49
Bluegrass and lespedeza mixture	2,235	2,120	176	414	294	1.67
Bluegrass in spring and lespedeza in summer (alone or after harvest of small grain)	5,847	5,610	176	470	343	1.95

[a] Charles R. Kyd, Missouri Extension Service, information to the author.
[b] All lots weaned at an average age of 200 days.

It is apparent that rotation forage crops have a real place in cattle raising. However, it would appear that their greatest use is to supplement rather than displace permanent pastures. The danger of not securing a stand, and thus being without any forage, is too great for the farmer to rely exclusively on rotation pastures for maintaining a breeding herd.

The best plan in all but the range areas is to depend on neither permanent nor rotation pasture exclusively for maintaining the breeding herd, but to provide enough of each to insure a continuous supply of nutritious, palatable forage throughout the summer. Some permanent pasture is especially advisable for use in early spring and late fall when rotation pastures often cannot be used without damage to the stand because of wet weather. Permanent pastures such as bluegrass and bromegrass, which form a thick, dense sod, are damaged much less seriously both by extremely wet weather and occasional periods of severe drought, when all pastures become little more than exercise lots. Figure 38 is just one example of how a long-season pasture program can be developed to fit local conditions.

Permanent pastures outside the range areas are usually ready for grazing earlier in the spring than rotation pastures, not only because they have a denser sod but because they are lower in moisture and thus less laxative. They are most productive during cool, moist weather and may be grazed heavily from the last of April to the middle of June. They are relatively dormant during July and August, but make considerable growth in the fall with the return of cool weather.

Most rotation pastures that are 1 or more years old are at their best in June, July, and August. Often they are so productive during the early summer that only a portion of the area is needed for grazing, making it possible to cut the remainder of the field. The second crop comes on quickly, and this fresh growth combines with the more mature forage of the grazed area to make an ideal pasture for beef cattle.

If rotation pastures are spring-seeded—as in the case of Sudan grass and the forage sorghum–Sudan hybrids, which are seeded alone, or sweet clover, Korean lespedeza, or alfalfa, which are seeded in small grain—they are usually ready to be grazed by late July or mid-August and furnish excellent grazing until the first killing frost.

Cattle should be removed from all legume pastures immediately after a killing frost, because frozen legume forage may cause serious bloat and scouring. However, such pastures may be safely grazed 2 or 3 weeks later after the stems and leaves are dead and brown. Legume pasture is much less palatable and nutritious after being frozen than while still green. Fortunately, at this season of the year permanent pastures usually are in excellent condition and are an ideal supplement to the cured-legume forage remaining in the fields.

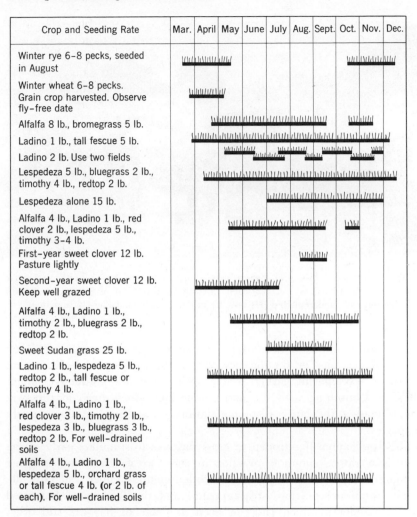

Crop and Seeding Rate	Mar.	April	May	June	July	Aug.	Sept.	Oct.	Nov.	Dec.
Winter rye 6–8 pecks, seeded in August										
Winter wheat 6–8 pecks. Grain crop harvested. Observe fly-free date										
Alfalfa 8 lb., bromegrass 5 lb.										
Ladino 1 lb., tall fescue 5 lb.										
Ladino 2 lb. Use two fields										
Lespedeza 5 lb., bluegrass 2 lb., timothy 4 lb., redtop 2 lb.										
Lespedeza alone 15 lb.										
Alfalfa 4 lb., Ladino 1 lb., red clover 2 lb., lespedeza 5 lb., timothy 3–4 lb.										
First-year sweet clover 12 lb. Pasture lightly										
Second-year sweet clover 12 lb. Keep well grazed										
Alfalfa 4 lb., Ladino 1 lb., timothy 2 lb., bluegrass 2 lb., redtop 2 lb.										
Sweet Sudan grass 25 lb.										
Ladino 1 lb., lespedeza 5 lb., redtop 2 lb., tall fescue or timothy 4 lb.										
Alfalfa 4 lb., Ladino 1 lb., red clover 3 lb., timothy 2 lb., lespedeza 3 lb., bluegrass 3 lb., redtop 2 lb. For well-drained soils										
Alfalfa 4 lb., Ladino 1 lb., lespedeza 5 lb., orchard grass or tall fescue 4 lb. (or 2 lb. of each). For well-drained soils										

FIG. 38. Grazing periods in the southern half of Illinois for various pasture seedings. Experiment station and extension agronomists of each state usually have prepared charts such as this for their respective areas. (Illinois Extension Circular 682.)

The length of the grazing season and the carrying capacity of temporary or rotation pastures vary with the latitude and with the species of plants making up the seeding. They also depend on weather conditions prevailing during the growing season and on soil fertility. In general, however, an average of 5, 7, and 9 months' grazing are obtained in the northern, central, and southern latitudes of the United States respectively. Pasture

FIG. 39. Beefmaster cows and heavy calves on high-carrying-capacity, irrigated
Coastal Bermuda grass pasture. Rotation grazing of this kind of pasture decreases
the amount of land required per cow by about one-third. (The Cattleman.)

specialists and plant breeders constantly strive to lengthen the grazing season
by developing strains or finding new species of grasses and legumes that
begin their development earlier in the spring and are productive longer
in the fall. Consulting with the pasture specialists in a given locality will
ensure obtaining the latest recommendations concerning suitable pasture
crop species.

The application of nitrogen to grass pastures to stimulate early spring
growth is a rather common practice in some areas. Irrigation is being
used more extensively as a pasture extender in midsummer and late fall.
Mowing, rotational grazing, strip grazing, and the feeding of harvested
pasture in drylot as green chop or even as silage or hay are still other
pasture extenders.

PERMANENT PASTURES

By far the majority of beef cows graze permanent pastures or ranges
in the summer. These pastures or ranges vary greatly in type and quality
depending on geographic location.

Some of the very best permanent or native grass range, such as that
of the Flint Hills region of Kansas where big and little bluestem are the
predominant grasses, may require only 4 or 5 acres per cow for summer

pasture. In contrast, in some parts of the West where annual rainfall is less than 8 inches, as many as 150 acres may be required per cow for the season. Unfortunately much of the native range area has such a low rainfall that improvement by introducing new species of grasses is unpromising because stands cannot be obtained with dependable regularity. In these situations, reducing the competition for moisture and fertility from brush and scrubby tree growth is the principal hope for improving carrying capacity. Figure 40 shows the extent of the acreage in the various states where brush removal can improve grazing and increase carrying capacity. Much research is being done in brush control and, although 2,4-D and 2,4,5-T sprays are producing good results, still more promising compounds are in the testing stage.

Overgrazing and burning, whether intentional or accidental, are perhaps the two most destructive practices with respect to permanent range. In both instances stands of desirable grasses are injured, and erosion and water runoff further reduce forage production. Weeds and less nutritious annual grasses and brush compete for moisture and fertility. Burning reduces forage production in the long run, even though growth may appear to start a few weeks earlier as a result of the burning. Kansas agronomists checked the effects of burning bluestem range over a long period of years and found the following differences in yield of forage in pounds per acre:

Unburned plot	2,502
Burned April 30	2,161
Burned April 10	1,934
Burned December 1	1,926
Burned March 20	1,845

Conversely, controlled burning is a recommended range-improvement practice in some instances where certain poisonous and noxious weeds, such as snakeweed, and undesirable brush can be controlled by such burning. Consultation with local range-management specialists is suggested for determining local recommendations in this regard.

Such pasture management practices as contouring, pitting of the sod, building of stock water ponds or tanks to help control grazing, use of drift fences, feeding supplements to prevent overgrazing in emergencies, rotation grazing to permit native grasses to set seed, rodent control, use of fireguards along highways, and other more localized improved management practices can enhance the productivity of permanent pastures. A discussion of all the peculiar problems of each range area is unfeasible in this text.

Permanent pastures for summer use in nonrange areas also vary widely in quality, carrying capacity, and length of grazing season, but at least

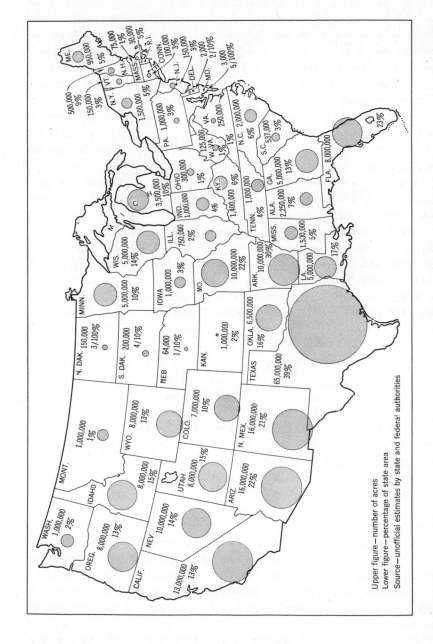

FIG. 40. Estimated acreage where brush removal can improve beef cattle gains in the United States. (Texas Livestock Journal.)

Upper figure—number of acres
Lower figure—percentage of state area
Source—unofficial estimates by state and federal authorities

in this situation lack of rainfall is ordinarily not the major concern. Soil infertility, overgrazing, erosion, and competition from weeds, underbrush, and scrub timber are the big problems here. Correcting soil mineral element deficiencies according to needs demonstrated by soil tests is the first step in pasture renovation. Seeding recommended mixtures of grasses and legumes according to proven procedures, followed by sound pasture management, makes possible carrying capacities as high as one cow and calf per acre for the summer pasture season in areas consisting largely of once-cultivated but now abandoned farmland. As mentioned earlier, brush control, by reducing competition for desirable grasses, can improve grazing in both range and nonrange areas.

GAINS MADE BY COWS ON PASTURE

The gains made by cows on pasture during the summer depend on a number of factors, including the abundance and nutritive value of the forage, the condition of the cows when turned onto pasture, the age of the suckling calves, and whether or not the calves are creep-fed. In most sections of the country the cows have calved before they are turned onto pasture and have lost considerable weight from parturition and lactation. In much of the range country, weight loss will have occurred throughout the winter. These losses from all causes often amount to 150 to 200 pounds per cow. Half or more of these losses usually is regained during the first 2 months on pasture while the grass is fresh and abundant, and the remainder is regained during the fall and early winter after the calves are weaned. Frequently the cows lose in weight during the middle of the summer when the pastures are scorched and dry and the demands of the nursing calves are greater. One advantage of creep-feeding calves on summer pasture is the reduction of stress on the cows.

CALVING ON PASTURE

There is no better place for a cow to deliver her calf than in a clean calving trap or pasture. Usually the danger from infection is much less than in the average cattle barn or drylot. However, if the cow requires assistance during parturition, it cannot be given so readily, especially if the pasture is a distance from the farmstead or headquarters. Consequently it is desirable to have a small grass calving pasture nearby in which cows well advanced in pregnancy may be kept. Under no circumstances should pregnant cows be allowed to remain in a pasture where there are hogs.

SUPPLEMENTARY FEED FOR COWS ON SUMMER PASTURE

Ranges or pasture properly stocked will insure the accumulation of enough reserve forage to prevent feed shortages during temporary droughts. However, in prolonged drought it is sometimes necessary to give the cows some supplemental feed if their milk flow and condition are to be maintained. Some cattlemen follow the practice of guarding their supply of grass against possible exhaustion by supplementing the pasture with other feeds from the very beginning, or soon after the beginning, of the grazing season. This is recommended on farms in the nonrange area where the pasture acreage is small and homegrown roughages are available, as it makes possible a greater carrying capacity. Table 50 shows how the pasture area was reduced 33 and 66 percent respectively at the Illinois station by using different amounts of supplementary feed.

One of the most satisfactory feeds for supplementing pastures is corn, sorghum, or grass silage. Silage is palatable, succulent, and usually economical. Seldom do cattle refuse to eat 15 to 20 pounds of silage per day, even when the supply of grass is abundant. On the other hand, dry hay or straw remains in the racks almost untouched as long as the supply of grass holds out.

Cowmen often make the mistake of delaying summer feeding too long. If it is to be a supplement to the grass and is to prolong the grazing period, feeding must be begun long before the grass is exhausted. The excuse so often heard—that the farmer hoped for a rain which would make feeding unnecessary—is invalid because a closely cropped pasture

Table 50
Effect of Supplementary Feed on the Pasture Area Required per Cow and Calf[a]

Pasture Acreage per Cow and Calf (Bluegrass)	Number of Days Pastured	Number of Days Supplement Was Fed	Average Daily Supplement per Cow for Period Fed (lb)		Average Daily Supplement per Cow for Entire Period (lb)	
			Cottonseed Meal	Corn Silage	Cottonseed Meal	Corn Silage
1.50	161	12	0.52	18.6	0.04	1.35
1.00	145	93	0.35	14.0	0.22	9.06
0.50	161	153	0.73	29.2	0.70	27.85

[a] Illinois Agricultural Experiment Station; average of three years' work.

seldom recovers as quickly after a rain as one that has had better treatment. It is much easier to discontinue feeding when it proves unnecessary than it is to repair damage to grass and cattle resulting from delaying supplemental feeding.

If silage, hay, or green chop is unavailable, it may be necessary to feed some grain to the cows during the summer. Six to eight pounds of grain a day per cow is usually sufficient, or specially designed emergency range pellets or blocks may be fed in amounts recommended by reliable manufacturers.

SUPPLEMENTARY PASTURE CROPS

Since pastures frequently become brown and bare toward the end of the grazing season, the farsighted cowman takes steps, if possible, to provide forage crops with which to supplement his regular pastures should a summer pasture shortage develop. In all but the range area and the winter wheat belt, clovers seeded in early spring in small grain are excellent for this purpose. Red, sweet, and alsike clovers, seeded alone or in simple mixtures, are, except in unusually dry weather, ready for light pasturing soon after the grain is havested. Usually good grazing is furnished by such "stubble pastures" from about the middle of August until late September. Close grazing of spring-sown clover is inadvisable if it is to be used two seasons, because close grazing tends to weaken the young plants and make them susceptible to winter killing. Clover pastures should be grazed with extreme caution after the first heavy frost, since frosted clover tends to cause both bloat and excessive scouring in cattle.

Sudan grass or forage sorghum–Sudan hybrids, seeded alone or with soybeans or cowpeas, is an excellent pasture crop for summer and early fall. These crops should preferably be sown in ploughed and carefully prepared ground. They should be grazed with care because of the danger from prussic acid poisoning. This poisoning is most likely to occur when abundant new growth follows a period of drought or when second growth is grazed following earlier grazing or cutting for hay. Use of one or two "tester" animals is often resorted to, but the forage can be analyzed to evaluate its safety.

SHADE AND WATER

Often too little attention is given to providing adequate shade and sufficient fresh water for cattle during hot weather. Many pastures are entirely without shade of any kind, and the cattle are exposed to the heat and glare of the summer sun. Tall-growing species of trees make the best shade for permanent pastures but in rotation or temporary pastures portable shades are preferable.

Table 51 summarizes an experiment conducted by California workers in the Imperial Valley under arid conditions. The effect of shades, both natural and artificial, under humid conditions is illustrated by the Louisiana experiment summarized in Table 52. The artificial shades were not as high as those recommended by the California workers, being only 7 feet high. Roofs of the Louisiana shades were made of 6-inch-thick layers of hay, straw, or pasture clippings supported by wire netting. It should be noted that the breeds used in this study were Herefords and Angus which are not so well adapted to the hot, humid weather of the coastal region as is the Brahman breed or its crosses. Thus the favorable response to shades reported here is unlikely to occur to the same extent if the latter breed or its crosses are used.

Water should be readily accessible to cows and calves in all pastures. Energy expended in walking great distances to water, or shade for that matter, is lost from a productive standpoint and thus should be kept to a minimum. Pasture areas that are more than 2 or 3 miles from water

Table 51

Effect of Type of Shade on Temperature Under Shade and Performance of Cattle[a]

Comparison	Galvanized Iron Check Shade	Air-cooled by Desert Evaporative Cooler	Burlap Roof Cooled by Sprinklers	Galvanized Iron Cooled by Sprinklers
Temperature Between 11 A.M. and 7 P.M., August 6				
Average outside air temperature (°F)	103.5	103.5	103.5	103.5
Air temperature under shade (°F)	104.0	96.5	101.5	101.5
Temperature of underside of roof (°F)	127.5	99.5	89.0	92.0
Ground temperature under shade (°F)	107.5	95.3	97.2	100.0
Cattle Performance (55 days)				
Number of Hereford steers	5	5	6	5
Average initial weight (lb)	430	428	429	420
Average daily gain (lb)[b]	0.69	1.05	0.80	0.89
Respiration rate on very hot, humid day	116	107	80	105

[a] Compiled from data in *Journal of Animal Science*, 184:10.
[b] Ration consisted of good alfalfa hay with 1 pound of barley per head added during the last 27 days.

Table 52

Effect of Type of Shade on Performance and Grazing Habits of Hereford and Angus Cows and Calves (Four-Year Summary)[a]

Treatment Class	Abundant Natural Shade		Scanty Natural Shade		Artificial Shade		Without Shade	
	Cows	Calves	Cows	Calves	Cows	Calves	Cows	Calves
Number	255	255	110	110	40	40	112	112
Average weight (lb)	1,013	336	1,037	349	990	354	960	322
Average daily gain (lb)	1.29[c]	1.85[c]	1.00[b]	1.64[b]	0.84	1.78[b]	−0.05	1.18
Grazing habits (6 A.M.– 7 P.M.)								
Grazing (hr and min)	4:16	3:05	4:13	2:45	3:58	3:06	3:47	2:42
Standing (hr and min)	4:16[c]	4:00[c]	4:30[c]	4:03[c]	4:55[b]	3:49[c]	6:04	5:06
Lying (hr and min)	4:28[c]	5:55[b]	4:17[c]	6:12[c]	4:07[b]	6:05[b]	3:09	5:12
Average maximum temperature (°F)	89.6	89.6	91.4	91.4	88.4	88.4	86.9	86.9
Average relative humidity (%)	63.1	63.1	62.4	62.4	62.8	62.8	64.0	64.0

[a] *Journal of Animal Science*, 15:59.

[b,c] Values significantly different, 5 percent and 1 percent level respectively, from those of the "without shade" treatment.

will not be well utilized and, still more serious, when water is not well distributed, overgrazing and permanent damage to the range occurs near the watering places.

Farm and ranch ponds should be fenced, and a tank with float control should be situated below the dam for best results. Water is cleaner and cattle do not damage the dam and pond itself by wading into the water. Some newly built ponds do not hold water at first, and purposely delaying fencing out the pond for a year or so will ensure trampling and sealing of the pond floor.

Metal or wooden tanks, partially covered, are most suitable for temporary pastures. Shades or covers over the tanks lower the temperatures of tank water with the result that somewhat less water is consumed during very warm weather.

In an Illinois experiment, conducted during June and July, water in tanks provided with shades 7 feet high averaged 11°F cooler during the day than water in unshaded tanks. Steers being full-fed corn on lush alfalfa pasture drank 9.4 percent more water in the lot in which the tank was unshaded. On such lush pastures, which are high in water content, excess consumption of drinking water conceivably could reduce intake of dry matter. It has not been proved experimentally that water consumed by cattle must be clean for normal performance, but average precautions are advisable.

PROTECTION FROM FLIES

Flies annually cause millions of dollars of loss to cattlemen. The hornfly is the principal species that annoys grazing cattle, and the stable fly is the more troublesome to animals confined in or watering at barns and feedlots during the summer months. Both species are bloodsuckers, and the blood loss suffered by animals of all ages during a long period of heavy fly infestation constitutes a serious tax upon the herd. The annoyance, disturbance, and reduced amount of grazing done by the cattle in the daytime are often capable of retarding milk production, growth, and fattening. Fortunately, cattle may be protected from both hornflies and stable flies by dipping or spraying the animals with an effective insecticide such as a solution of chlordane or DDT, as discussed in Chapter 28. The first application should be made in the spring when the cattle are turned onto pasture, and additional sprayings should be made during the summer at intervals of 4 to 6 weeks. Effective rubbers of various kinds are also available. Sheds, barns, and feeding equipment used by the cattle during the summer should also be sprayed with an effective residual spray to destroy house and stable flies, which breed in moist straw and manure around the farmstead or ranch headquarters.

SALT

Cattle on grass usually consume a somewhat larger quantity of salt than they do in the drylot. Keeping block salt constantly before the cattle is strongly recommended. A mature cow requires about 0.1 pound of block salt per day, or 18 pounds during a 6 months' grazing season. About 20 pounds a head should be supplied, since a small amount is dissolved by the rain.

Location of the salt in a pasture can be effectively used to ensure more efficient range utilization. Successful ranchers often use salt to entice cattle into relatively inaccessible or isolated areas in mountain ranges that the cattle would otherwise not utilize. In areas such as most of the Southeastern and Gulf Coastal regions of the country, where year-round mineral supplementation is absolutely recommended for cattle on pasture, loose salt is usually offered free-choice or in combination with minerals in weatherproof salt and mineral feeders.

MINERAL SUPPLEMENTATION ON SUMMER PASTURE

With the exception of salt, phosphorus is the mineral most apt to be deficient in summer pasture or range. This deficiency is more likely to occur in areas where soils are deficient in phosphorus, as indicated in Fig. 27.

If cows are wintered on harvested forages of good quality, and especially if protein supplements such as soybean or cottonseed oil meals are fed, phosphorus-deficient or low phosphorus-content summer pasture or range is not so serious insofar as reproduction is concerned. On the other hand, if cows graze such mineral-deficient forage in summer and then are wintered on dry, cured forage of the same type, serious reductions in percent calf crop weaned and in weights of all ages of cattle are bound to follow.

New Mexico Experiment Station workers compared the production of two herds handled under similar conditions, except that one herd had year-round access to a mineral mixture composed of half salt and half steamed bonemeal (a common source of supplemental phosphorus in cattle rations). The pertinent data from their 7-year study are shown in Table 53. The annual consumption of mineral mix, including salt, was 51.7, 30.2, 34.5, and 73 pounds, respectively, for cows, calves over 7 months, yearlings, and bulls. Since half the mixture was salt, which was needed in any case, the consumption of 26 pounds steamed bonemeal, valued at less than $1 per cow annually, resulted in a 9 percent larger calf crop weaned, and

Table 53
Effect of Mineral Supplements on Weight and Production of Range Cattle (Results of 7 Years)[a]

Measure of Production	Cows Not Fed Minerals	Cows Fed Minerals	Gain from Mineral Feeding
Cows calving (%)	90.4	92.2	1.8
Calves died (%)	10.8	2.7	8.1
Cows weaning calves (%)	80.7	89.7	9.0
Average weight of calves (lb)[b]	408	442	34
Average production per cow (lb)	336	389	53
Gain of yearling steers (lb)	321	353	32
Gain of 2-year-old heifers (lb)	202	278	76

[a] New Mexico Agricultural Experiment Station Bulletin 359.
[b] Average weight of calves multiplied by percent of cows weaning calves.

those calves weaned averaged 53 more pounds in weight. Improved performance for the other classes of stock was equally striking. Texas workers got similar responses from adding phosphorus to the drinking water in soluble form. Supplementation with calcium alone (the other important constituent of bonemeal) on summer pasture has not produced similar results.

Not all cattle respond similarly to phosphorus supplementation on pasture or range. Consultation with pasture specialists or animal nutritionists in the local area is recommended in all situations.

BREEDING ON PASTURE

Cows that are to calve in the springtime must be bred during the early summer months. At this time they are, of course, on pasture, where their heat periods are likely to pass unobserved. It is usually advisable to pasture-breed during the grazing season. It is, of course, understood

FIG. 41. A well-located creep-feeder in use in a purebred operation. (American Angus Association.)

that a bull used in pasture mating should be sound and mature. Also, the herd should be divided, where practical, so that there are not more than 25 or 30 cows with each bull. Rotation of bulls every 3 weeks is practiced by some breeders using single-sire herds. Purebred breeders do not use this method for obvious reasons.

Many breeders of purebred cattle seldom resort to pasture breeding. The importance of knowing the breeding dates and of being absolutely certain which bull has made the service makes hand mating preferable. There is always risk that a bull running in a pasture may jump a fence and get with cows that are being kept for mating with another bull, or get with heifers that are too young to be bred. The man with valuable purebred cattle cannot afford to take such risks.

If artificial insemination is used on summer-pastured cows, the problems of heat detection and inconvenience are compounded. Use of small pastures through at least two heat cycles for each cow reduces the disadvantages of artificial insemination of spring-calving cows. A cleanup bull is recommended for breeding the unsettled cows.

CREEP-FEEDING CALVES ON PASTURE

In the commercial cow-calf program where spring calving is used, the decision of whether to creep-feed depends on the method of disposal or marketing of the calves. Experimental results do not all agree on this subject, but in most instances creep-feeding does not pay if the weaned calves are sold as feeder calves. If, on the other hand, the calves are of insufficient quality to sell well as feeders but rather are sold as slaughter calves, creep-feeding results in a larger return. Table 54 illustrates how the ultimate objective of a particular cow-calf program determines whether creep-feeding is advisable. Note that, although the feeding of concentrates increased gains materially, the added value of the extra gain was not sufficient to pay for the feed, to say nothing of some extra labor and equipment costs.

After weaning, the steer calves in this test were placed on a drylot baby-beef finishing program and fed to approximately 650 pounds final weight. Admittedly this is a light selling weight, and the differences shown in Table 55 would undoubtedly have narrowed with a longer feeding period. However, the inadvisability of creep-feeding for anything but calves that are to be sold as slaughter calves is again illustrated. Time in the feedlot was reduced, of course, because the creep-fed calves reached market weight sooner. Consequently some labor was also saved, but if the breeder also fed out his own calves, the labor saved during the finishing period would be offset by the extra labor required for creep feeding in summer.

Table 54
Creep-Feeding Beef Calves (166 Days)[a]

	Not Creep-Fed	Creep-Fed	Creep-Fed (Molasses)
Average birth date, March	4	5	5
Number of calves	15	15	15
Steers	7	7	7
Heifers	8	8	8
Average feed per head (lb)			
Corn		188	330
Oats		94	198
Cottonseed meal		31	66
Molasses			66
Average weight per calf (lb)			
Initial	163	166	164
Final	454	483	521
Total gain (lb)	291	317	357
Average daily gain (lb)	1.75	1.91	2.15
Feed cost per head ($)		9.50	21.43
Financial ($)			
Appraised price per cwt			
Slaughter	11.50	13.50	16.50
Feeder	16.00	16.50	16.50
Value per calf[b]			
Slaughter	50.60	63.18	83.32
Feeder	70.40	77.22	83.32
Value per calf minus feed cost[c]	70.40	67.72	61.89

[a] Oklahoma Agricultural Experiment Station Miscellaneous Publication MP-34.
[b] Weights were shrunk 3 percent.
[c] Highest value per calf (as feeders) minus feed cost.

For programs in which creep feeding is recommended—the fat-calf and the purebred programs—recommendations as to which concentrate mixtures to use and where and what type of creep feeder to build, are made in Chapters 22 and 23.

Creep-fed calves may in many instances actually sell for less per hundredweight as feeders than the same quality of calves not creep-fed. Cattle feeders who want to utilize fall pasture and stalk fields or who simply winter their calves on a stocker program feel that it does not pay to buy calves with extra condition and bloom because the calves usually lose it when handled as stockers. The added weight the feeder has to buy may only result in lower and more costly gains in the feedlot.

Table 55
Finishing Steer Calves in Drylot After Creep-Feeding[a]

	Not Creep-Fed	Creep-Fed	Creep-Fed (Molasses)
Number of days fed	117	89	75
Average weight per calf (lb)			
Initial[b]	440	464	531
Final[c]	687	635	665
Daily gain to selling	2.11	1.92	1.79
Average daily ration (lb)			
Ground yellow corn	10.25	9.09	10.45
Cottonseed meal	1.50	1.50	1.50
Alfalfa hay	0.98	0.99	0.99
Prairie hay	2.50	2.72	2.17
Financial ($)			
Feed cost per cwt gain	18.19	18.36	21.61
Initial cost[d]	77.00	83.52	95.58
Feed cost per steer	44.93	31.41	28.96
Selling price per cwt	21.00	21.13	22.00
Total value per steer	144.27	134.18	146.30
Return per steer	22.35	19.25	21.76
Carcass grade			
Low prime	1		
High choice	3		4
Average choice	3	4	3
Low choice		3	
Average dressing percent	60.1	5.89	59.0

Profit Summary, Both Phases ($)

	Not Creep-Fed	Creep-Fed	Creep-Fed (Molasses)
Value per steer when sold	144.27	134.18	146.30
Feed cost	44.93	40.91	50.39
Return (value of steer minus feed cost)	99.34	93.27	95.71

[a] Oklahoma Agricultural Experiment Station Miscellaneous Publication MP-34.
[b] Weighed after an overnight shrink.
[c] Weights were shrunk 3 percent.
[d] Based on value as feeder steers at weaning time.

DRY COWS AND TWO-YEAR-OLD HEIFERS

Dry cows that will calve in the fall and yearling or 2-year-old heifers that will not calve until the next spring require comparatively little attention during the summer. Ordinarily the cows with calves are given the best pasture, whereas the dry she-stock is required to get along on somewhat less abundant forage. It is also a common practice to take the dry cows to the pasture farthest removed from the farmstead or headquarters, as they do not require the daily attention needed by the unbred cows and those with calves. Only in very dry weather or pronounced overstocking is feed other than grass necessary.

YEAR-ROUND DRYLOT MANAGEMENT OF BEEF COWS

Pasture is not absolutely necessary for the well-being of cattle in summer, as was proved rather conclusively at the Illinois station 50 years ago. Ten heifers were kept in a small paved lot from the time they were weaned until they were 4½ years old. This period covered four summers, during the last two of which the cows were suckling spring-born calves. During the entire 4 years the cows received nothing except corn silage supplemented with cottonseed meal at the rate of 2½ pounds of cottonseed meal per 100 pounds of silage fed. The daily feed of silage for the mature cows was 40 pounds. These cows, kept constantly in the drylot, maintained their weight nearly as well as other cows that were on pasture each summer, and they produced calves that were in every way normal. The only trouble experienced during the summer was a few cases of foot rot among the calves.

In an Iowa experiment, cows nursing calves consumed daily 114 pounds of an alfalfa-grass mixture, fed as green chop. These green-chop-fed cows gained an average of 7 pounds for the summer period as compared with a loss of 70 pounds for comparable wet cows on a permanent pasture. Weaned calf weights were 370 and 405 pounds respectively for the calves nursing the permanent-pasture and green-chop cow groups. However, because of the added charge made for the labor of harvesting and feeding the green chop, the heavier calves returned a smaller margin of profit per calf.

A more extensive test, still under way, is being conducted by the Texas Agricultural Experiment Station workers at the Rolling Plains Livestock Research Station at Spur, Texas. After six calf crops they have concluded that the advantages and disadvantages of a continuous drylot system for cows are the following:

Advantages

1. Investment in land is lower. A small operator can enlarge his herd without buying more land.

2. Closer observation of cattle facilitates selection, physical attention, and record keeping.

3. Artificial insemination is easier to use and a more uniform calf crop is possible.

4. There is less hazard from drought through storage of reserve feed.

5. Drylot calves are easy to precondition, wean, and start on finishing rations.

Disadvantages

1. More labor and equipment are required.

2. There is greater confinement of the operator; labor is required for daily feeding, or there must be a pasture where cattle can graze while the operator is away.

3. Disease problems may be more acute, especially for the calves.

The Texas experiment was started with two groups of 36 Hereford weaner heifer calves, with one group being continuously drylotted and the other run on pasture. Each group was further subdivided into three groups during the winter for comparisons of supplemental energy levels. Table 56 shows the supplemental levels of feed fed and the performance of the cows and calves through six calf crops.

The drylot cows were fed 1 pound of cottonseed meal, 2 pounds of sorghum grain, and 50 to 55 pounds of sorghum silage or green chop after May 1, while pasture cows received no feed other than grass. Iodized salt was fed both groups year-round and a 50-50 bonemeal-salt mix was provided for 6 weeks in the spring. A creep-feed was provided for the

Table 56
Effect of Method of Management and Energy Level on Performance of Beef Cows (1960–1965)[a]

| | Level of Energy | | | Average, All Levels of Energy | |
	Low	Medium	High		
Winter daily feed (lb)					
Cottonseed meal	1.25	0.75	—		
Sorghum grain	—	2.00	5.0		
Sorghum silage[b]	30.00	35.00	40.00		

Calf production	%	lb	%	lb	%	lb	%	lb
Pasture	90	468	85	454	91	469	89	464
Drylot	85	483	85	488	92	466	88	478
Average 345 calves	87	476	85	471	92	468	88	471

[a] Texas Agricultural Experiment Station, Spur Technical Report Number 2.
[b] Silage was fed to the drylot cows only. Pasture cows grazed cured range plus the supplements indicated.

drylot calves to compensate for the grass consumed by the calves on pasture. The drylot cows, in general, have been heavier than the pasture cows, indicating that a still lower level of feed might be used. The different energy levels tested have resulted in comparable results in both drylot and pasture cows. After six calf crops, the greatest difference evident is in the number of cows lost from the project. Only four cows have been lost from the drylot while ten have been lost from the pasture group. The reasons for removal are so variable that no concrete explanation can be offered as yet. Percent calf crop weaned and weaning weights are so highly comparable that one must conclude that, with respect to performance, drylotting cows can be as successful as running cows on pasture. With respect to cost, the pasture cows have, over the six years, had the advantage, as the summer feed for the drylot cows has been expensive. Using the low energy levels only, the pasture and drylot cows produced net returns per cow of $22.05 and $21.20 respectively for the last calf crop weaned. The Texas workers suggest that these data indicate that cows in drylot can produce calves as efficiently as cows on pasture. They add that if drylot cows could be grazed on more economical temporary pastures for 90 to 120 days during the summer, an additional $10 to $15 could be added to the net return per cow.

These experiments, together with the knowledge available concerning the wide use of soiling crops—that is, harvesting and feeding forage crops in a fresh state—in Europe, prove conclusively that pastures are not indispensable in raising beef cattle. However, the great amount of labor involved in caring for cattle under such conditions precludes any wide use of such methods in this country at the present time, unless automatic feeding systems, silo unloaders and auger bunks for feeding silage, or field choppers and self-feeding bunk wagons for soiling or feeding green chop, are installed and used year-round in relatively large operations. Drylot summertime rations for cows nursing calves should be compounded according to the requirements shown in Chapter 10 for nursing cows.

Because most calves are "worked"—that is, dehorned, castrated, marked, and vaccinated during summer and weaned in the fall—brief discussions of these jobs are appropriate in connection with summer management of the cow herd. This does not imply, however, that such chores can only be done then.

DEHORNING

Although a nicely shaped pair of horns undoubtedly adds to an animal's appearance in the show ring or purebred herd, the presence of horns on commercial cattle is considered objectionable by most cowmen. Horns are the cause of so much loss to the packer in damaged hides and bruised carcasses that polled or dehorned cattle normally sell from $1 to $2 per hundredweight higher than horned cattle of equal merit in other respects. This is particularly true when the animals are shipped some distance and are kept in crowded cars or trucks for several hours. Horns are also objectionable on the farm and in the feedlot. Cattle with horns require more shed room per animal as well as more space at the feed bunk or hay rack and in the truck or car when shipped. Among horned cattle there is always a tendency for some to be "bossy" and to keep the timid ones away from their share of shelter and feed. Dehorning tends to curb the aggressiveness of such animals, thus lessening feedlot disturbances. Even hornless cattle, however, establish a "peck order," with some being more aggressive or shy than others.

Dehorning by Breeding. Using a polled bull results in a majority of hornless calves. If such a bull is a "pure" polled, carrying in his germ plasm no factor for producing horns, all his calves will be polled, even

FIG. 42. A purebred Polled Hereford bull being used to dehorn calves through heredity. If the bull is homozygous for the polled character, the horned cows will produce only polled calves. (American Hereford Association.)

though all their dams have horns. If, however, the bull is an impure polled (the product, let us say, of a pure polled bull and a horned cow), only half his calves from horned cows will in the long run be polled, and the rest will have horns. This explains why Angus bulls (which are almost invariably "pure" polled) seldom sire any but polled calves, whereas some polled bulls of the Hereford and Shorthorn breeds often sire a number of horned offspring.

Dehorning by Use of Chemicals. A satisfactory way to prevent the growth of horns in newborn calves is to treat the horn button with a strong chemical. This is best done when the calf is 1 to 5 days old but can be delayed for as long as 2 weeks. Caustic sticks and pastes composed of such alkalies as potassium, sodium, and calcium hydroxides have been used for years but because they tend to burn the calves' heads, and their dams during the act of nursing, they have been largely replaced by improved preparations. One such newer product, marketed as "Pol," contains antimony trichloride, salicylic acid, and flexible collodion. This material dries faster, adheres more firmly, and is water-resistant, making it much safer than the older product.

In removing horns with caustic one should proceed somewhat as follows. With a pair of hand clippers or ordinary scissors, clip away the hair as closely as possible to expose the budding horn. If paste or liquid caustic is used, a thin coating should be applied with the applicator to each horn area, avoiding an excessive amount because it may spread to the skin surrounding the horn, causing serious burns. Inexperienced operators may avoid such an accident by applying a thin coating of vaseline or lard to the surrounding skin, leaving only the horn button exposed.

The effect of the caustic is to deaden the matrix or root of the horn. In a few days a scab appears over each horn button; this soon drops off, leaving a smooth spot of skin devoid of hair, no larger than a dime. These spots are soon hidden by growth of adjacent hair and no trace of the operation is apparent.

Because caustic, to be successful, should be used while the calf is very young, this dehorning method requires considerable labor when calves are born on pasture. Consequently it is used mainly on farms where the calves can easily be caught and treated. In a performance-tested herd, calves will usually be caught for tattooing or tagging anyway, and dehorning can be done at that time.

THE BELL DEHORNER

A heated dehorning iron is a popular tool for dehorning young calves on many ranches and farms. After the bell-shaped iron has been heated

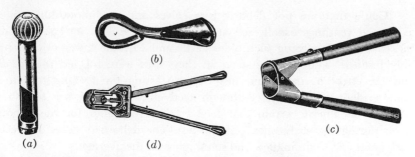

FIG. 43. Dehorning tools, listed in the order of age of calves for which they are best suited (a) dehorning tube, 1 to 3 months; (b) dehorning spoon, 3 to 5 months; (c) Barnes dehorners (available in two sizes), 5 to 12 months; (d) Leavitt clippers, 12 to 24 months; (e) dehorning saw (not shown), over 24 months. (O. M. Franklin Serum Company, Denver, Colorado.)

to the proper temperature it is fitted over the horn button and held firmly against the head until the horn matrix has been destroyed. Care is necessary to make the burn deep enough to destroy the horn tissue but not so deep as to produce a bad sore. The operation is more painful and requires more skill than chemical dehorning but it may be used to dehorn calves up to 2 months of age. Fire irons and electrically heated irons are available from most commercial sources.

DEHORNING OLDER CALVES

In large herds of commercial cattle, especially in the range area where cows and calves are dispersed over a wide area, dehorning is usually postponed until the calving season is practically over, in order that all calves may be dehorned, castrated, vaccinated, and branded with only one working of the herd. At this time the horn buttons of most calves are well developed but still soft and loosely attached to the skin and can easily be removed with a knifelike dehorning tool that separates the horn button from the adjoining skin with little blood loss. Such tools are called "gougers," "spoons," or "tubes," depending on their shape and the way in which they are used (see Fig. 43, a and b). Either a dehorning saw or some type of clippers is used in dehorning cattle older than 4 months (see Fig. 43, c and d).

The Barnes dehorner shown in Fig. 43 literally lifts the horn out by the roots and crushes the blood vessels so that only a trickle of blood is produced. However, it is suitable only for dehorning calves and short yearlings.

Cattle that are not dehorned as calves and are allowed to go over as horned yearlings usually sell at a discount. The data in Table 57 show the effects of dehorning such older cattle and why the lower price is probably justified. The yearling steers in these tests were full-fed for 132 days and the effect of dehorning just before starting the feeding trial is still quite evident at 40 days. Although feed efficiency data are not shown, one can be almost certain that feed costs were higher for the dehorned steers during the early phases of the test. The dehorned steers did recover and, as fat cattle, shrank less and sold higher on the average.

Experimental data such as those shown in Table 58 support the idea that the clipper produces less shock and fewer aftereffects when used on yearling cattle, as compared with the saw. Pulling the arteries to reduce bleeding also appears to be a worthwhile practice for yearling cattle.

In calves, special precautions are unnecessary to prevent excessive bleeding. Although there may be considerable bleeding immediately after the horn is removed, especially if calves are excited or heated up, it usually diminishes rapidly as a heavy clot is formed and the capillaries contract. If heavy bleeding persists to the point where the calf becomes weak, the cut artery should be grasped with an artery forceps or tweezers and twisted so that it breaks off deep within the skull.

Table 57
Effect of Dehorning on Feedlot Performance of Yearling Steers[a]

| Trial Number | 1 | | 2 | |
Treatment	Horned	Dehorned	Horned	Dehorned
Number of steers	11	10	9	10
Initial weight (lb)	676	680	693	686
Average gain or loss after dehorning (lb)				
1 day	2	−33	4	−27
5 days	34	−9	15	−20
11 days	45	−15	21	−14
28 days	74	39	57	46
40 days	100	50	80	62
132 days	227	237	276	241
Average daily gain (lb)	1.87	1.85	2.09	1.82
Number of steers bruised	2	0	3	2
Shrinkage (%)[b]	2.25	1.86	3.87	3.43
Selling price per cwt	$20.73	$22.57	$21.44	$21.20

[a] South Dakota Agricultural Experiment Station Bulletin 489, 1960.
[b] Shrinkage includes transit loss during a 60-mile haul.

FIG. 44. The entire family often helps at calf-working time on the range. Corrals with headgates, and sometimes tilting tables, are used on many ranches where less labor is available. (American Hereford Association.)

EQUIPMENT FOR DEHORNING

Little equipment is necessary for dehorning calves. Often they are simply thrown to the ground and held firmly, or sometimes they are snubbed to a fencepost. However, if a number of calves are to be dehorned it is advisable to construct or buy a regular dehorning chute. If herds are large and calves will be worked in readily accessible areas, special tilting tables are justified.

CASTRATION

Bull calves are castrated for economic reasons. At the present time, steers sell higher per hundredweight than bulls whether sold as stockers, feeders, or fat cattle. Castration results in improved texture, tenderness, and flavor of the beef, and produces a quieter disposition, which is an asset in the feedlot. It is noted elsewhere that if market discrimination against bulls were reduced, the feeding of bull calves for slaughter might become a common practice because of their superior performance in the feedlot and higher yield of retail cuts of meat.

Table 58
Effect of Method of Dehorning on Performance of Yearling Heifers[a]

Method of Dehorning	Clippers	Saw	Clippers, Artery Pulled
Number of heifers	8	7	8
Average weight at dehorning (lb)	613	612	616
Average final recovery weight (40 days) (lb)	674	659	682
Average daily gain (lb)	1.52	1.19	1.66

[a] South Dakota Agricultural Experiment Station Bulletin 489, 1960.

AGE AND SEASON FOR CASTRATION

Castration is best done when calves are 4 to 10 weeks old. Like dehorning, castration should be performed when weather conditions are favorable. The spring and fall months are considered the best, although winter castration is not objectionable if it is done on a mild, bright day and if adequate protection is furnished during the following night. As a rule, calves born during the winter and early spring are castrated just before being turned to pasture, whereas those born during summer and fall are allowed to go until they are brought in from the pasture in the fall. On the range, castration is performed at the summer or fall roundup or "work" as it is referred to by ranchers.

METHODS OF CASTRATION

Young calves are usually thrown to be castrated, but animals older than 4 months may be operated on better while standing. In throwing, the calf is placed on either side and the feet are held or "hogtied." This position exposes the scrotum to the operator, who stands or kneels alongside the calf's rump. If not absolutely clean, the scrotum should first be cleansed with a mild antiseptic to remove any dirt or filth that might otherwise contaminate the wound. The operator's hands and the knife should always be clean.

Several satisfactory methods are used for surgical removal of the testicles, the more common methods being the following:

1. Removal of the lower third of the scrotum, exposing both testicles. In calves 3 months of age or less, the testicles can then easily be removed by working each one loose and simply pulling it from the scrotum. In

older calves where bleeding may be more severe, the testicle should be worked loose, then the cord should be severed as high into the scrotum or as near the body as possible. This may be done either by scraping with a knife blade until the cord comes apart or by crushing the cord with an emasculatome or clamp.

2. A sharp-pointed knife may be inserted into the side of the scrotum all the way through to the opposite side. Then with one downward stroke the scrotum can be split to the bottom, exposing the testicles for removal as described above.

FIG. 45. A sorting chute is useful for dividing calves from their dams for such summer jobs as dehorning, castrating, and branding, or for weaning in the fall. The work is made easier and there is less stress on both cowboys and cattle. (Bell Ranch, New Mexico.)

3. Incisions may be made on the front side of each half of the scrotum from the middle to the lower end, permitting a ready grasp of the testicle which can then be removed as above. Show steers are usually castrated in this manner to ensure a full, well-shaped cod.

Larger calves and short yearlings can be castrated standing if well secured. Pulling the tail sharply upwards and holding it firmly prevents the animal from kicking the operator. Any of the three methods described above can then be used for removing the testicles. All three methods ensure good drainage, which is an important factor in trouble-free castration. Removing as much cord as possible with the testicle prevents stagginess.

TREATMENT AFTER CASTRATION

After the testicles are removed the scrotum should be examined to make sure that the incisions are sufficiently large and low to afford proper drainage. The calves should be put in a clean box stall or a small, grassy trap or lot with their mothers for a few hours until bleeding stops and all mothers have claimed their calves. The calves should have a clean, dry place to lie and should not have access to lots covered with mud or manure or to sheds that have not recently been cleaned and rebedded. Daily observations should be made for a week or 10 days to make sure that any swelling due to faulty drainage is promptly relieved by reopening the incision and treating the wound with an antiseptic solution. If castration is done in fly season, repellents should by all means be used.

CASTRATION PINCERS

A unique castration pincers called the Burdizzo or emasculator has been in use for some time. These pincers or "clamps" and others of similar design have blunt jaws that close with enormous force when sufficient pressure is exerted on the handles to lock them into the closed position. In using the clamps the object is to crush or sever the spermatic cord and the blood vessels that supply the testicle so that the testicle degenerates for want of circulation.

The advantage of the Burdizzo over the knife is that there is no blood loss and the skin is unbroken, eliminating the problem of blowflies. Nevertheless, the jaws of the clamps cut deeply into the tissue, and there is some danger that the scrotum itself may slough off for want of blood supply if it is clamped all the way across. This, however, may be prevented by severing each cord separately, one a little higher than the other, so that some of the blood vessels leading to the lower part of the scrotum will escape injury. An unusually attractive, well-shaped cod results when animals are emasculated in this manner.

At first it appeared that the Burdizzo method of castration would eventually supersede the knife, especially in Texas and adjoining states where the presence of blowflies and screwworms made castration with a knife unsafe after March 1. However, it was soon evident that use of the clamps under practical conditions did not always sever the cord and blood vessels sufficiently well to arrest the development of the testicle. The result was a calf called a "slip," which began to show signs of stagginess when about a year old. When the Burdizzo was used by inexperienced operators, the percentage of slips sometimes amounted to 10 or 15 percent. Slips are less frequent when calves are castrated by an improved type of Burdizzo that holds the cord in place when the jaws are closed (Fig. 46a).

A rather new castration instrument called the elastrator has been used with some success in castrating calves. It is a forceps-like instrument that slips a strong elastic band around the scrotum close to its attachment to the groin. The pressure exerted by the rubber band shuts off the blood supply to the scrotum and testicles, causing them to slough off. Some cattle feeders object to steers that have been thus castrated because, with no cods, they appear light and shallow in the twist.

The virtual elimination of the screwworm fly has of course lessened the dangers from open-wound castration methods.

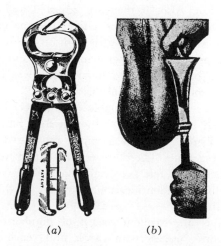

(a)   (b)

FIG. 46. (a) Burdizzo castration pincers; (b) correct method of using to ensure crushing of spermatic cord. (O. M. Franklin Serum Company, Denver, Colorado.)

MARKING

It is highly desirable that all animals in the herd bear some mark or tag whereby each can be positively identified. In the range areas, marking or branding with a registered brand is required by law to establish ownership. In all herds it is desirable to establish ancestry or pedigree, and it is a necessity if performance testing is practiced.

The method used for marking depends on the object for which it is done. When the purpose is to establish ownership, as it is on the range and in poorly fenced pastures, permanency and ease of recognition are of paramount importance. Under such conditions branding with a hot iron is probably the best method. Although there have been objections to branding because of the pain inflicted and the damage to the hide, no perfect substitute for the iron has yet been devised. However, a brand should be no larger than necessary to permit easy identification at a distance of 30 or 40 feet, and no deeper than needed to destroy the hair follicles.

Other ownership marks sometimes used are ear notching or slitting and bobbing. Another mark of this sort is the slitting of the dewlap in a way that causes one or two "wattles" of skin to hang from the neck. The principal objection to such marks is that they detract from the animal's appearance.

Methods used to establish the identity of each animal in the herd include numbered neck straps and chains, ear tags, horn brands, numbered fire brands, and tattoos. The principal objection to neck chains is that they are frequently lost. Also, if they are adjusted to fit the cattle when turned onto pasture in the spring, the chains become either too tight or too loose later in the summer as the cattle gain or lose flesh. It is especially difficult to keep the neck chains of growing calves and yearlings properly adjusted. Cattle have been known to hang themselves on neck chains in brushy country or on fences and feedbunks.

Many kinds of ear tags are available at relatively low cost. They are usually easy to attach but are also easily lost and remain in place only a short time. Unnumbered plastic tags can be purchased that are attached by means of a piercing plug. These are convenient in that any numbering system can be applied with felt pens and renewed if necessary. One side may be used for individual identification, leaving the other for the number of dam or sire or both.

Horn branding is an excellent means of numbering mature horned cattle but is, of course, useless with calves and yearlings, and obviously the polled breeds cannot be horn branded. Horn brands must be reburned every few years. This is the common method of numbering cattle at sales

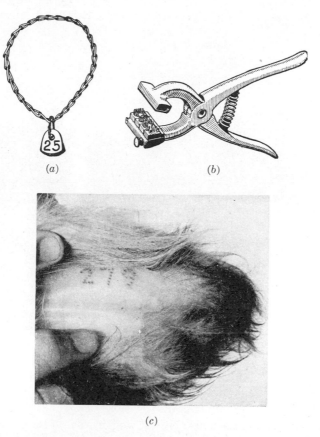

(a)   (b)

(c)

FIG. 47. Devices for marking cattle: (a) adjustable neck chain; (b) tattooing
instrument; (c) a correctly tattooed ear. (O. M. Franklin Serum Company, Denver,
Colorado.)

of purebred Shorthorn, Hereford, and other horned breeds, where each
animal's horn brand is made to correspond with the catalog number. Half-
inch copper brands are the proper size for this purpose.

The best method yet devised for identifying breeding cattle is tattooing
the ear with indelible ink. The advantages of this method are that the
mark is permanent and in no way disfigures the animal. Its only serious
objection is that the animal must be caught and the ear examined at
close range before the mark can be read. Tattooing outfits, including an
initial letter and a set of figures, may be purchased from any stockman's
supply house. All purebred record associations now require that calves be
properly tattooed before they are accepted for registration. Black ink is

FIG. 48. A 6-inch easily read, well-located freeze brand. Although still somewhat experimental, this method of marking is painless and does not damage the hide. (North Dakota Agriculture Extension Service.)

preferred for all but the black breeds, and in their case green ink seems more satisfactory.

A promising new method of marking cattle for individual identity is freeze-branding. The method is painless, eliminates danger of infection, does no damage to the hide, and results in an easy-to-read brand when well done. Essentially the method consists of destroying the hair-pigmentation cells in the skin by use of a copper or steel branding iron that has been held in an alcohol or gasoline bath chilled with dry ice to —158°F. The hair must be closely clipped and the chilled iron held in place for 40 to 60 seconds. If the technique is successful, the hair and surface layer of skin will slough off in 4 to 6 weeks, to be replaced by permanently white hairs. At present the success rate is about 50 percent, but the method is new and improvements in the technique are certain to occur. This brand is not a substitute for fire brands in legally establishing ownership of cattle at present, but may become so in time if the method is perfected. Figure 48 illustrates the type of brand that may result from freeze-branding.

VACCINATION

Calves born on farms in areas that are infested with blackleg and malignant edema should be vaccinated against these diseases before they are 8 months old (see Chapter 27). Vaccination at a much younger age is advisable if an outbreak has occurred in the vicinity of the farm within the preceding 2 years. Vaccinating at the regular summer work is the practice on most ranches to reduce the number of times the cattle must be gathered.

Heifer calves that are to be kept or sold for breeding should be vaccinated for contagious abortion at 4 to 8 months of age unless the herd is "accredited" (see Chapter 27) as a result of annual clean tests. Although the value of calfhood vaccination is questioned by some breeders and veterinarians, cattle that are so vaccinated and later pass the blood-test are regarded by most breeders as being much less likely to contract the disease than unvaccinated animals. While calfhood vaccination is more urgently needed in purebred than in grade herds because of the greater risk of

FIG. 49. Pickups, jeeps, and even helicopters have replaced some of the cow horses and ranch hands as a laborsaving practice on many ranches. Much of the western range is so rough in terrain, however, that horses will never entirely disappear from use. (American Hereford Association.)

exposure, the vaccination of grade heifers that are to be added to the breeding herd is recommended, since they too may be exposed to the disease through contacts with cattle on adjoining farms. Vaccination for abortion must be done by a veterinarian, who tattoos a "V" and the year of injection in the right ear. Other vaccinations such as those for leptospirosis and vibriosis should be done according to a veterinarian's instructions. State laws vary with respect to health certificates for vaccinated females that are to be shipped to another ranch or state.

WEANING

Beef calves ordinarily are weaned at 6 to 8 months of age. Usually the date is determined largely by necessary changes in the management of the herd. Spring-born calves are usually weaned about the first of October and fall calves are weaned in the spring when the cow herd is taken to an outlying pasture. Fall-born calves are often allowed to nurse until mid-summer, since green pasture causes cows to increase their milk production if they are still being suckled when they are changed from drylot to pasture.

Calves that have been running with their dams should be removed from them once and for all. If they can be placed beyond earshot of one another, so much the better. The practice of turning them back on the second or third day to suck out the cows cannot be recommended, because it tends to prolong the period during which the calves pine for their mothers, and may give rise to digestive disorders. The question of whether to move the cows or the calves from a pasture is a much argued one. Each farmer or rancher must, by experience, work out a system that seems best. Small weaning pastures with tight fences for either the cows or calves will save much labor in keeping the cows and calves apart and under control. Selling and delivering calves at weaning time of course eliminates many of these problems.

Part IV

THE STOCKER AND FINISHING PROGRAMS

Chapter 10

OPERATION of
the STOCKER PROGRAM.

A discussion of the development of young cattle logically follows the chapters dealing with feeding and managing the cow herd. The question of finishing cattle for slaughter is treated in a separate section, leaving this chapter to discuss the subject of calves that are not to be finished immediately but are to be handled in a way that achieves maximum growth at the lowest possible feed cost.

This stocker period, as it is called, consists of the period following weaning up to the time when the calves, if steers, are sold or put into the feedlot for finishing. The stocker program also applies to the replacement heifers in a cow-calf program that are handled so as to insure normal growth and development. It is customary to refer to the animals on such a program as *stockers,* and they are considered as being on a growing, rather than a finishing, ration. *Feeders,* on the other hand, are older calves and yearlings carrying more finish and bloom, which therefore are placed on higher-energy finishing rations in order to take advantage of their extra condition.

The stocker program is sometimes used as the sole program on a particular farm or ranch, but as often as not it is either conducted in connection with a cow-calf operation or it precedes the finishing program. In the latter circumstance, the program is often referred to as a "warm-up" or "conditioner" program. When the stocker program is the only beef cattle program on a farm it is usually managed in one of two ways. In one plan, calves or light yearlings are bought in the fall to be wintered on roughage rations in drylot and sold in the spring to buyers who need "grass cattle" for the summer, or who need feeders to put on a summer finishing program. In the second plan the operator has summer pasture to utilize and lighter-weight calves are bought in the fall to be wintered as in the first plan, but usually at a slower rate of gain. Instead of being sold in the spring, they are grazed all summer, or until pastures mature and deteriorate in the fall, or until winter sets in.

A third plan, limited to areas where winter small grains are grown, makes use of these winter pastures and, while still a stocker program in every sense of the term, has distinct features of its own and is more fully discussed later in this chapter.

There is a tendency for more and more of the calves to be handled and fed according to the first plan, and even this program tends to be shortened in order to reduce the time from weaning to market by whatever segments of the industry are involved. This trend will undoubtedly accelerate because of rising nonfeed costs associated with the industry.

If the farm or ranch also has a cow-calf program, the calf crop from the herd can be handled in the stocker program in the same manner as purchased calves. Ordinarily, however, a farm or ranch suited to the cow-calf program may as well be stocked to capacity with cows to increase the size of herd and thus reduce the fixed costs per calf raised. Keeping one's own calves for further grazing as yearlings means that, unless the operation is a large one, both the cow-calf and the stocker program will be reduced in size. If calves from one's own herd are fed out in drylot as baby beeves, that is another matter, as it would not be likely to reduce the maximum size of cow herd. Under certain semiarid range conditions, keeping one's steer calves over to run as yearlings provides a flexibility factor that allows for adjustment in cattle numbers to fit the almost certain fluctuation in grass or feed supply resulting from periodic droughts.

Perhaps the most common type of stocker program is the combination stocker-feeder program, which is especially adapted to the Corn Belt and the irrigated sections of the Southwest where high-yielding corn and sorghum silage crops can be produced. The cattle feeder purchases steer calves or light yearlings of good to choice quality in the fall, or as early as August 1 if available. Sometimes the cattle are wanted early in order to utilize fall growth in small-grain stubble or stalk fields and similar aftermath. After most of these feeds are salvaged and gains have begun to slow down, these cattle should be brought to drylot. They are then fed a heavy feed of roughage, preferably corn or sorghum silage plus supplement or legume hay, until the cattle reach a stage of condition where gains are reduced below a level that permits economical conversion of roughages to gains. This point is usually reached after about 2 to 3 months with yearlings and 4 to 5 months with calves. Having reached feeder condition, the steers are finished out on a full feed of grain, either in drylot or on pasture, to be sold in late summer or fall.

Lower grades of steers, heavy yearlings, and most heifers are less suited to the stocker program, because such cattle usually should be placed on high-energy finishing rations soon after they are obtained. A warm-up on a rather high-energy stocker ration for about 50 to 60 days is an exception to this rule that is being practiced by southwestern and western feeders using "Okie" stockers. Another exception would be choice or prime grade heifer calves that have sufficient quality to justify carrying them to high choice or low prime condition.

ADVANTAGES AND DISADVANTAGES OF THE STOCKER PROGRAM

This program has the following advantages over other programs:

1. The program is adapted to an intensive type of farming—that is, a large volume of business can be done on either small or large farms that can produce large tonnages of roughages.

2. Returns come quickly, as early as 4 to 6 months, if not preceded or followed by other programs. In some instances this quick turnover permits feeding two to three droves of cattle per year.

3. If used in winter only, this program is completed by the time labor is needed for spring and summer farm work.

4. Feedlots that are poorly located with respect to a fat cattle market or a source of grain can grow cattle for feeders on a contract basis.

5. Stockers can utilize large quantities of harvested roughages and aftermath, thus cheapening the price of feeders if the same owner is also to finish them.

6. The stocker program is quite flexible, adjustments in size being easily made.

7. Death losses are lower than in the cow-calf program.

8. Little equipment is required for a farm-sized operation, but the use of mechanized equipment for handling silage or other roughage reduces labor requirements in the large commercial or custom feedlot.

9. Little capital is required if a grower contract can be arranged with a feedlot, thus not requiring the purchase of cattle.

Some inherent weaknesses of the stocker program are:

1. Much capital or available credit is required as compared with the cow-calf program, unless the program is a contract arrangement and someone else supplies at least the major share of the capital.

2. Buying and selling skills are extremely important in this program because shrink and other losses, on both ends, and mistakes in judging

quality or in judging the health of the stockers can quickly offset the economical gains that may be made.

3. The stocker program may have conflicting labor loads; for example, roughage-harvesting time and the winter feeding period.

4. The program carries above-average risk for the owner-feeder. Total gains made are not large in proportion to the weight purchased; hence profits must come both from a favorable price spread between buying and selling prices and from added weight made at feed and labor costs that are lower than the selling price. The latter is not difficult to achieve with stockers but, as mentioned, the total weight gain per head to which this factor may be applied is comparatively small. Thus the profits made because of economical gains can be lost when the cattle are sold if the selling price is much below the purchase price.

Large commercial feedlot operators who have a more or less continuous type of operation would do well to make most of their stocker purchases during the favorable fall season, carrying at least some of the cattle along on stocker rations until they are needed to replace finished cattle in the finishing pens. Sometimes early spring is a favorable time to buy stockers in the winter small-grain belt, but the cattle feeder should not depend on this situation without alternatives.

MARKET CLASSES AND GRADES
OF STOCKERS AND FEEDERS

Throughout the year, many thousands of cattle are marketed daily in the large central markets and auction markets, and by direct sale or purchase on farms and ranches. These cattle are of every kind, displaying a wide range of combinations of the various characteristics such as sex, age, weight, size, conformation, breeding, and condition. There is fortunately a market for each kind of cattle and the variation in prices received reflects both the supply and the demand for each kind and the variation in suitability to the purpose for which each kind is purchased.

The need for standardized terms to describe the various kinds of cattle has long been realized. As early as 1902, H. W. Mumford of the Illinois station published the results of a lengthy study of the subject. In 1918 when the Bureau of Markets, now the Bureau of Agricultural Economics, of the U.S. Department of Agriculture inaugurated its market-reporting service on livestock at Chicago, the market classes and grades of cattle suggested by Mumford were used to establish the terminology and classification of cattle for market-reporting purposes. It is a tribute to the pioneer workers in this field that the classes and grades of cattle have since been changed only slightly. The use of uniform descriptions of all classes and grades of cattle by producers, selling and buying agencies, and packers throughout the country has contributed much toward the orderly marketing of stocker, feeder, and slaughter cattle.

In recent years there has been some criticism of the grading system for stockers and feeders, because performance in the feedlot is not always related to grade—that is, the higher-grading stockers are not always the fastest and most efficient gainers. This justifiable criticism does not alter the positive features of the present system, which should continue to be useful until one that evaluates more than just the subjective features of feeder cattle can be developed. A number of experiment stations are pursuing this problem, but results to date are not consistent enough to warrant discarding the long-used system.

Table 59
Market Classes and Grades of Stocker and Feeder Cattle[a]

Class (Sex Condition)	Market Grade
Steer calves and yearlings	Prime, choice, good, medium, common, inferior
Heifer calves and yearlings	Prime, choice, good, medium, common, inferior
Cows	Prime, choice, good, medium, common, inferior
Bulls	Ungraded
Stags	Ungraded

[a] Based on U.S.D.A. Circular 505.

The official standards for live cattle as developed by the Department of Agriculture and as generally used today provide for segregation according to (1) use, such as stocker, feeder, slaughter, (2) class, determined by sex condition, and (3) grade, determined by the conformation and apparent desirability of the animal for its particular use. Table 59 shows the grouping of feeder cattle according to sex condition and grade.

The grade assigned to a stocker or feeder is based on a correct appraisal of the extent to which the animal meets certain established standards of conformation, quality, breeding, constitution and capacity, and condition or finish. Figure 50 shows the standards for stocker and feeder calves with respect to the various grades. In practice, the upper three grades are further divided into three subgrades. For example, in the choice grade, market men speak of low, average, and high choice, although these subgrades are not a part of the standard government grading system.

A brief description of the characteristics that typify each grade of stocker and feeder follows.

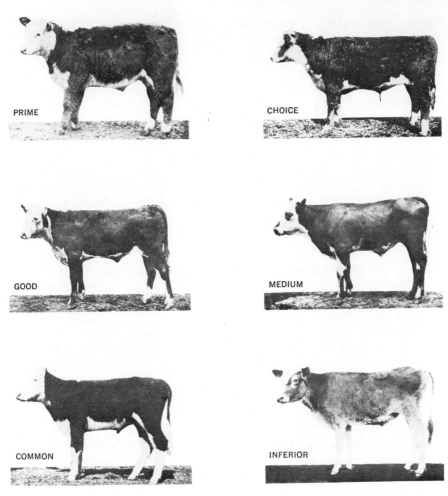

FIG. 50. U.S. grades of stocker and feeder steers. In planning a feeding program the choice of grade depends on many factors, but mainly on the amount and quality of the feed supply. (U.S.D.A.)

Prime stocker and feeder cattle approach the ideal in beef type or conformation, possessing all the characteristics, except high degree of finish, that will result in prime slaughter cattle if they are fed long enough. Thick natural fleshing, symmetry and balance, and quality are developed to the highest degree. A drove of such cattle is uniform in type and color and, because only a small proportion of the stocker and feeder cattle supply is good enough in conformation to make this grade, they command a premium price when either sold or purchased. In the trade such cattle are usually spoken of as "reputation brands," and they are ordinarily not found in central markets, but rather are either sold at feeder cattle auctions or are contracted for on the farm or ranch.

Prime calves usually carry considerable bloom and finish, often as a result of creep-feeding, and should be handled in a program that makes the fullest use of these characteristics. It seldom pays to place prime grade calves on a stocker program; they should rather be put on full feed as soon as practicable. Prime grade home-raised calves make ideal feeders for the baby-beef program, in which case they are probably finished as

FIG. 51. A drove of high-choice and prime calves. Dehorning would have increased the selling price of these calves by 5 to 10 percent. Note uniformity, good heads, width of body, and straightness of lines. (American Hereford Association.)

rapidly as possible. Prime grade feeders should be fed to prime slaughter grade in order to command the selling price needed to offset the premium purchase price. It is worth repeating that the correlation between grade and performance in the feedlot is low, or even negative in some instances. Therefore, by all means a man who feeds prime grade cattle should be almost certain of a premium market for his cattle, as such cattle can be very risky because of high costs and the long feeding period required.

Prime grade heifers are seldom available for feeding purposes because heifers of such quality are usually retained as herd replacements or are sold at higher prices than they would bring as feeders, to add to or start other breeding herds.

Choice stocker and feeder cattle, especially those from the middle to lower end of the grade, are the most numerous kind of cattle in American feedlots. Choice grade cattle resemble those of prime grade but lack some of the eye appeal. Whereas they are moderately wide, muscular, and low

FIG. 52. A drove of choice heifer calves, suitable either as herd replacements or as feeders. Large droves of uniform Angus feeders are less numerous at present than Herefords and consequently they usually command a premium price. (American Angus Association.)

set, they may have more scale and frame and some signs of coarseness and thus may lack the refinement and quality of prime grade feeders. Choice grade feeders usually show evidence of being "good-doing" cattle, with considerable thickness and depth, but they may be somewhat lacking in straightness of lines and in balance. They are more uneven as to type, weight, and breed characteristics. Cattle feeders who choose this grade are often able to top out part of a drove for further feeding to prime grade. Cattle of choice grade are usually fed for at least 5 months if put on feed as yearlings and at least 7 or 8 months if started on full feed as calves.

Good grade stockers and feeders have the appearance of thrifty cattle with moderate thickness, but they are usually quite rangy, upstanding, and lacking in balance. Good grade cattle usually show the color markings of the principal beef breeds. Many of the yearling heifers sold as feeders fall in this grade. Many cattle of this grade are thinner than those in the upper two grades. An alert feeder buyer often spots a potential low choice slaughter cattle end on a drove of good grade feeder cattle, which could well be fed longer to upgrade them into choice.

Many steers originating in the winter small-grain pasture and bluestem areas are cattle of good grade, although cattle of the better grades come from these areas as well. "Okie" cattle, a term used to describe many of the stockers and feeders originating in the Southeast and Coastal regions, fit into the good grade. If they appear to have a cross or two of Brahman breeding they may be called "No. 2 Okies" and probably belong in the medium grade.

Medium feeders are lacking in beef conformation. They are long, shallow, and narrow and often show evidence of mixed dairy or Brahman breeding. Frequently they are the result of crossing British beef bulls on cows of either dairy or native breeding. Such cattle may not have made a profit for the breeder, but they may be quite profitable for the feeder. Because of their growthiness and size and because they are often older for their weight as a result of poor nutrition, they often gain quite well. Gains must be very economical and the purchase price must be low, because such cattle usually do not grade better than standard or good on the rail and thus will never sell very high. The feeding period is short, because these cattle soon reach the point where the gains cost more than the selling price warrants. Expert judgment is required in feeding medium grade cattle, because speculative risks are high. Heifers of this grade are rarely available in large numbers, because they will have been more profitably marketed as slaughter calves.

As slaughter cattle, medium grade feeders usually grade standard if young, or commercial if mature. It is not uncommon, though, to find these

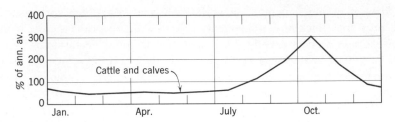

FIG. 53. Seasonally heavy shipments of stockers and feeders occur in the fall months
of September and October. (U.S.D.A.)

feeders being graded good on the rail after finishing. They produce carcasses
of high cutability but their dressing percent is low.

Common and inferior feeders, at the bottom of the quality scale, can
best be described by saying that they lack beef characteristics. They are
nondescript in breeding and color and are often unthrifty. Although some
cattle of this grade may make good gains for short periods, the low selling
price and the rather numerous stunted or "dogied" individuals that perform poorly make this grade of feeder a very risky proposition. Carcasses
from cattle of this grade generally grade utility.

It is seldom advisable to feed anything but the good and choice grades
of cattle in strictly stocker feeding programs. Because the lower grades
seldom sell well enough to someone else as feeders for further feeding,
a grower program usually does not pay. The man who finishes the lower
grade of cattle usually does best not to delay unduly long in getting his
cattle fed out and marketed.

BUYING STOCKERS AND FEEDERS

Stocker programs should be planned to take advantage of the seasonal
fluctuations in supply, and therefore price, of such cattle. Figure 53 shows
that the fall months are the heaviest in movement of stocker and feeder
cattle, whereas the spring and early summer months are the lightest. Figure
54 shows that prices react to the supply situation and that stockers and
feeders can be bought much more favorably in the fall months.

SOURCES OF STOCKER AND FEEDER CATTLE

Principal sources of stocker and feeder cattle and the relative importance of each source vary specifically from area to area and even from

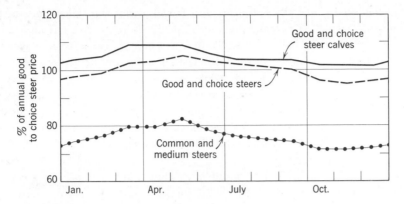

FIG. 54. Seasonality in prices of stocker and feeder cattle at Kansas City, one of the more important central markets for such cattle. (U.S.D.A.)

community to community. The principal sources in general are (1) public (terminal) markets, either from dealers or commission firms, (2) local dealers, (3) direct from ranchers or farmers, (4) auctions, and (5) contract arrangements.

A survey of 8 counties in Illinois, which included 123 cattle feeders and 270 different lots of cattle, supplied some interesting information concerning the source and area of origin of feeders. The counties surveyed are typical of much of the Corn Belt area. The survey showed that the

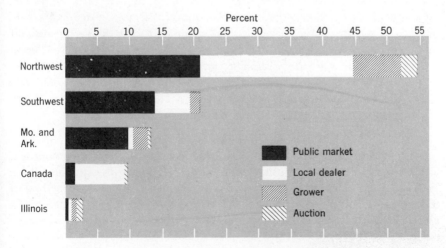

FIG. 55. Percentage of cattle moving to 123 feeders from each area of origin and through each source of purchase. (Illinois Agricultural Experiment Station.)

cattle feeders bought their stockers and feeders from the following sources: public stockyards, 46 percent; local dealers, 36 percent; direct from grower, 12 percent; auctions, 4 percent; other, 2 percent. Data with respect to area of origin are shown in Fig. 55. Ranchers in the Northwest sell more of their feeders to local dealers who, in turn, sell them to cattle feeders. Southwestern ranchers and farmer-breeders with smaller herds in Missouri and Arkansas sell mainly through public stockyards. Even here, dealers play an important role, as they will have purchased the feeders from the rancher at the public stockyards and immediately offered them for sale to prospective feeder-buyers.

Outside the Corn Belt the public market has less relative importance as a feeder cattle source. Buying in auctions and direct from neighbors is more common where the number of feeders fed per farm is smaller and cattle feeding is relatively less important. Since completion of the study referred to above, the trend toward less selling through public stockyards has continued and, in fact, accelerated. There is greater use of auction sales and of direct dealing.

METHODS OF BUYING

There is an old saying, "Well bought is half sold." Certainly this adage applies to the buying of stocker and feeder cattle. An inexperienced feeder should not attempt to buy his own cattle, but should place an order with a dealer, commission firm, or order buyer who knows his business and, preferably, also knows the feed situation of the particular farm or feedlot, the capabilities of the feeder, and how much risk the feeder can afford to take. The commission charge is small compared with the savings usually made. This does not mean that ranchers or dealers who have stocker or feeder cattle for sale are dishonest, but naturally they place their own or the seller's interests uppermost, and the prospective buyer should also follow the good business practice of having a knowledgeable person represent him on the other side of the transaction.

The methods of buying stockers and feeders are as follows:

1. *Buying direct from rancher or neighbor* by oral or written contract. Cattle for fall delivery are usually contracted for during the summer. Price, pencil shrink, weighing conditions, sort, if any, date of delivery, method of transportation, and arrangements for paying for the cattle should all be agreed upon in advance and preferably verified with a written, signed contract. Beginners should by all means avoid driving out to the range areas, shopping for bargains from ranch to ranch. Much time and money can be wasted and cattle are not often bought satisfactorily in this manner.

Many experienced feeders believe that the increasing use of this practice by cattle feeders has upset normal channels of buying and selling feeder cattle, sometimes to the detriment of both buyer and seller.

2. *Placing an order with a commission firm or dealer.* In this arrangement the buyer makes his wants known to a dealer, preferably 3 or 4 months in advance of desired delivery. The commission firm or dealer locates the cattle, negotiates for them, and makes all necessary arrangements for shipping. The cattle are paid for upon delivery, with the commission firm or dealer acting as go-between. Ranch or delivery-point weight, less the customary 2 to 4 percent pencil shrink, is the weight actually paid for by the buyer. Freight, insurance, feed costs in transit, commission, and miscellaneous costs are added to the original negotiated price. Sometimes the commission firm or dealer pays for and assumes ownership of the cattle in order to compute more simply the incidental costs and values, in case shipments from a particular ranch are to be divided and sold to more than one buyer. Some commission firms also pay for cattle that were not specifically ordered and assume ownership themselves, hoping to sell the feeders again at a profit. Cattle bought through a commission firm are usually not seen by the buyer until delivery. The buyer thus depends on the reliability of the firm that he deals with.

3. *Buying direct from a dealer.* Some feeders prefer buying direct from a dealer at his yards, because they can see what they are buying and the "asking price" usually covers all costs. Naturally the asking price is as high as the demand allows and it may seem unduly high compared with prices published in many livestock reports because freight and other incidental costs are included. Sometimes ranch weight and sometimes dealer's yard weight is used to determine price. This buying method is convenient because dealers are often located in the midst of the feeding territory, although every terminal market also has numerous dealers.

4. *Buying in person at feeder cattle auctions.* The number of feeder cattle auctions is increasing, especially in areas that are developing rapidly in the cow-calf business, such as the Southeast, the Corn Belt, and the South. A special once-a-year kind of auction is sponsored by cattlemen's associations, with large numbers of stocker and feeder cattle being brought together and sorted or graded into lots that are uniform in grade and weight. They are then auctioned in truckload or smaller lots to the highest bidder.

When a good-sized crowd is present, bidders may become overenthusiastic and actually pay more than the cattle are worth. At auction sales cattle may sell with more fill than cattle bought elsewhere, and cattle that appear uniform when purchased, as a result of the sorting, may soon grow uneven

FIG. 56. Well-managed auction sales with convenient truck and rail loading facilities are gaining in prestige and popularity as a source of stockers and feeders. (Schnell Livestock Auction Market, Dickinson, North Dakota.)

because they came from several different farms or ranches. On the other hand, because buying costs, freight costs, and profits to middlemen may be lower or absent, delivered or on-the-farm costs are also often lower than when cattle are bought by other methods.

Buying in "community sale" auctions is done by many smaller feeders. Buying at auctions requires great skill, especially if cattle must be bought "by the head" rather than by weight. It is difficult to put together a uniform lot of cattle and the chance of buying diseased, stale, or overexposed cattle is great. It should be said that most auctions of this kind operate under the highest standards, and in some localities they are the only source of stocker and feeder cattle. If the yards cover more than 100,000 square yards in area, they are "posted"—that is, they come under the supervision of the U.S. Department of Agriculture Packers and Stockyards Act, making them subject to enforcement of prescribed rules of sanitation, scales inspection, and management in general.

TRANSPORTATION OF STOCKERS AND FEEDERS

Railroads haul most of the stocker and feeder cattle from the range states to destinations east of the Mississippi River, because rail shipment

is cheaper than other methods for such distances. The cattle are delivered
to terminal markets, dealers' yards, and sometimes to holding pens on railroad sidings near the feeders' farms or feedlots. Trucks are used to move
cattle shorter distances, such as from producer to loading point and from
terminal markets, dealers' yards, and auctions to the farm.

Many stocker and feeder cattle, of course, are also trucked direct from
producer or auction sale to feedlot if distances are not great. Double-decked
trucks, with an extra driver, have become highly competitive with railroads
for long hauls, because they can carry heavier loads than formerly and
have appreciably reduced the time spent on the road. Most cattle that
move from the Southeast to the Arizona-California area now move by
truck.

FIG. 57. Mixed yearlings being loaded at a country point in Idaho. Fast, high-
priority trains on the main lines shorten the time en route to the Corn Belt destination by many hours, reducing shrink and degree of stress due to long delays.
(The Record Stockman.)

Railroads are required by law to unload, feed, water, and rest cattle every 28 hours, except when the owner gives written permission for a "36-hour release," which is permissible if the haul takes less than 36 hours. This helps insure delivery of the cattle in better condition with less shrink, because they will not be delayed for the additional minimum 5 hours otherwise required for a rest stop.

SHRINK FROM SOURCE TO FEEDLOT

Shrink may be of two kinds in stocker and feeder cattle. *Excretory shrink* is the loss in weight from excretion of manure and urine, and this shrink can be regained in a short time. *Tissue shrink* is actual loss of flesh and body water, and this type of shrink is regained much more slowly. Length of the trip, previous feeding, and temperature affect both amount and type of shrink. Most of the shrink on a short trip on a hot day with a heavy fill at loading time is excretory shrink. "Sappy," milk-fat calves, shipped on a week-long trip, may incur heavy tissue shrink, especially if they do not eat or drink adequately at rest stops, as they seldom do.

It is difficult to generalize as to what is normal shrink for stocker and feeder cattle. Calves may shrink up to 10 percent or more while heavy cattle may not shrink over 4 or 5 percent. The time required to return to pay weight is perhaps as important as pounds of shrink in transit, because all feed fed during the shrink recovery period is added cost. This period may vary from less than a week for older, heavy cattle to as much as a month for calves or stale cattle. An outbreak of shipping fever during this shrink-recovery period of course drastically alters the time required to return to pay weight.

HOMEGROWN STOCKERS AND FEEDERS

Even in the heavy cattle feeding areas in the Corn Belt there are large numbers of brood cows in comparatively small farm herds, and this is also true throughout the Southeast and in the Great Lakes region where cow numbers have increased greatly during the last 20 years. Calves or yearlings produced in these herds are often fed out on the farm where they are produced, but still others are offered for sale locally. There are both advantages and disadvantages to buying local or native feeder and stocker cattle.

Advantages of native or homegrown stockers and feeders:

1. Native stockers and feeders are more likely to escape shipping fever and similar diseases, especially if bought direct on the farm where produced.

2. Delivered price may be lower because of lower freight and buying costs or because of less demand for local calves.

3. Shrink, especially tissue shrink, may be lower. Of course, there is no shrink at all, except that resulting from weaning, in homegrown cattle that are fed on the farm where they are produced.

4. Native stockers and feeders do not have to adjust to sudden changes in weather, altitude, and the like.

5. Native cattle are somewhat accustomed to the feeds produced in the area.

Disadvantages of native or homegrown stockers and feeders:

1. They are likely to be less uniform in quality, condition, and age.

2. Native or homegrown stockers and feeders may be of lower quality than those purchased in the traditional range areas, but this is not always true.

3. Such cattle are likely to be fleshier and thus do not make such large compensatory gains on roughages as shipped-in cattle are apt to do.

4. Several purchases must be combined to meet the needs of cattle feeders who want to feed more than a car or truckload of cattle.

5. Native stockers and feeders may not be available in large enough numbers when wanted.

Pay weight is the weight at the producer's farm or at the loading point, less whatever allowance for shrink, if any, the buyer can negotiate for. The method for determining pay weight should be agreed upon in advance in order to prevent misunderstanding. The most common practice in the range area is to allow the buyer a 3 percent shrink—that is, the calves or yearlings are weighed at the producer's ranch or at the loading point, then 3 percent of the weight is subtracted. Such shrink is generally referred to as "pencil shrink." Final settlement is based on the calculated shrunk weight. If cattle must be driven some distance to weigh and load, no pencil shrink is expected by the buyer. If cattle are lotted overnight without feed or water, only 1 or 2 percent or possibly no shrink is deducted. The delivered weight is subtracted from pay weight to determine how much shrink occurred to the weight of feeder cattle actually paid for.

DISEASE PROBLEMS AND DEATH LOSS IN STOCKER AND FEEDER CATTLE

Losses due to disease and accident are highest in stocker and feeder calves and lowest in older, heavy cattle. A survey of reports on 113 droves of cattle fed by Farm Bureau Farm Management Service cooperators in

Illinois showed that death losses of calves were 2.43 percent; yearlings, 0.85 percent; 2-year-olds and over, 0.43 percent. Most of these losses are caused by exposure to certain viruses as yet not well described, and to stress, which weakens the calves' resistance. Exposure occurs somewhere between the producer's farm or ranch and the feedlot. Shipping fever or "stress fever" is the principal disease that causes losses. This and other diseases are more fully discussed in Chapter 27.

Results of one of many similar experiments recently conducted on the effect of high-level feeding of antibiotics and sulfa drugs are shown in Table 60. Low levels of 75 milligrams daily of any of several broad-spectrum antibiotics have been fed to stocker calves, but usually this level does not prevent shipping or stress fever. It improves gains over a period of 120 to 150 days by about 5 to 10 percent. A higher level of 350 to 500 milligrams daily over the first 3- or 4-week period is apparently required to reduce the incidence of shipping fever and death loss. It should be noted that the gains reported in Table 60 are unusually high. This is because "off-truck" empty weights were initially used in the study in order to reduce

Table 60
Effect of High Levels of Antibiotic and Sulfa Drugs in Conditioner Supplements (28 days)[a]

	Control	350 mg Aureomycin	350 mg Sulfamethazine	Aureomycin plus Sulfamethazine
Number of animals	20	20	20	20
Average initial weight (lb)	441	416	427	421
Average final weight (lb)	500	493	489	494
Gain per animal (lb)	59	77	62	73
Average daily gain (lb)	2.11	2.75	2.21	2.61
Daily feed (lb)				
Corn	2.0	1.9	1.9	1.8
Soybean meal	2.0	1.9	1.9	1.8
Corn silage	17.8	18.9	17.7	18.3
Corn cobs[b]	0.9	0.6	0.8	0.6
Pounds of feed per pound of gain				
Corn	0.9	0.7	0.9	0.7
Soybean meal	0.9	0.7	0.9	0.7
Corn silage	8.4	6.9	8.0	7.0
Corn cobs	0.4	0.2	0.4	0.2

[a] Indiana Research Progress Report 250, 1966.

[b] Three pounds of ground corn cobs per head daily were fed only the first 11 days.

the variability in results that would be attributable to differences in initial fill.

Responses to sulfa drugs in the ration are usually negative, as in this trial, but some trials have been reported where responses were obtained from sulfa drugs dissolved in the drinking water. Many veterinarians prescribe the use of "electrolytes," sold under various trade names, in the drinking water to insure rehydration to offset the loss of fluids during weaning, shipment, and adjustment to new feeds and surroundings. It is not completely understood how these chemicals function, but evidently they are involved in the maintenance of the proper acid-base balance in the cellular fluids. Stocker calves that are extremely tired and stressed often drink but do not eat for some time after arrival, and perhaps medication by way of the drinking water is indicated in this situation.

Vaccines that are administered once in the summer as calves are last worked, and again when they are shipped at weaning time, appear helpful as a means of reducing the incidence of the shipping fever problem. Another promising method of pretreatment is early weaning of the calves and getting them accustomed to grain and hay. The injection of tranquilizers, vitamins, or antibiotics and a number of other practices are not especially promising in most situations. More research on the subject of pretreatment, conditioning, or "backgrounding," though it is expensive and difficult to conduct, is badly needed.

"LAID-IN" COST OF STOCKERS AND FEEDERS

The buyer of stocker and feeder cattle is of course interested in buying his cattle as cheaply as possible without sacrificing quality and bred-in performance. The rancher or breeder is just as interested in selling as high as possible. However, the buyer always pays more than the grower receives. There are numerous items of expense from ranch to feedlot, and these are all added to the purchase price by the time the cattle are "laid in" or safely unloaded at their new home. Not all of the following items are included in every case, although most of them are. Charges naturally vary somewhat from area to area.

Items of expense in procuring stocker and feeder cattle:

1. *Freight.* Freight varies with distance and method of hauling, but it ranges from 15 cents per hundredweight for locally bought cattle to as much as $2 per hundred for cattle requiring a long haul.

2. *Commission and buying expense.* This charge applies only when a buyer has placed an order with a commission firm or order buyer. It

may range from 10 to 20 cents per hundredweight depending on services rendered, but generally averages out at about $1 per head.

3. *Transit insurance.* Insurance is usually figured by the head, but it can be broken down to a hundredweight basis and varies from 3 to 6 cents per hundredweight depending on distance hauled.

4. *Feed and labor of feeding en route.* Feeding cost varies greatly depending on how many rest stops must be made, but it may amount to 10 or 15 cents per hundredweight. Trucked cattle are seldom fed on the road.

5. *Trucking from railroad to feedlot or farm.* The size of this item depends on distance hauled, and it may vary from a low of 15 cents to 50 cents per hundredweight if the haul is as much as 200 miles.

6. *Personal expenses on the buying trip.* Some buyers who go to the West to buy cattle may mark the trip off as vacation expense, but usually one or more special buying trips are necessary, even if only to the local feeder yard. In any case, this expense must be borne by the newly acquired stocker or feeder cattle. A reasonable estimate of this item is from a few to 20 cents per hundredweight.

The items listed above usually add a total of 75 cents to $3 per hundredweight to the purchase price of stockers and feeders and must be added in calculating the laid-in cost. Locally purchased cattle, hauled in the buyer's truck, naturally have little of this type of cost added, but the feeding enterprise must carry its share of the depreciation and maintenance costs of the owner's truck.

Shrink and death loss occurring on the farm or in the feedlot have already been mentioned, but actually these items can well be added to the laid-in price of the feeders before final calculations of profit or loss are made. The same may be said for financing costs, or interest and carrying charges on borrowed money used to buy the feeder cattle. For example, a 6 percent loan for 1 year on a long-fed calf costing $100 and weighing 400 pounds at time of purchase adds $1.50 per hundredweight to the cost of the calf.

FEEDING AND MANAGEMENT OF STOCKER CATTLE

In planning the management of stocker cattle it should be remembered that such animals are kept for three reasons: (1) to insure a supply of the right kind of cattle for the finishing lot at the proper time; (2) to utilize farm roughages, otherwise unmarketable; or (3) to "cheapen down" the feeder cattle. No matter which of these objects is uppermost in the owner's mind, economy in feeding and management is of utmost importance.

GAINS OF STOCKER CATTLE

Stocker cattle should be fed with the minimum outlay for feed consistent with normal growth and efficient feed conversion. Increase in condition much beyond that represented by normal growth may not be desirable. In fact, a noticeable increase in condition is usually undesirable if the stockers are to make the most efficient use of the finishing ration that subsequently will be fed.

The amount of gain necessary to account for normal growth depends, of course, on the age of the animal. Table 61 shows the results of an early experiment that demonstrates the levels of daily gain that constitute normal growth without fattening for annual periods up to 1, 2, and 3 years of age. Growth is most rapid in young animals and gradually decreases as the animals approach maturity. Thus calves will make greater gains than yearlings and yearlings greater gains than 2-year-olds when all are on the plane of nutrition that permits normal growth but little or no improvement in condition. In fact, even yearling steers grow so slowly that it is seldom advisable to carry them on nonfattening rations. Because they require a large amount of feed for maintenance, gains that represent only slow growth in older cattle are inefficient and costly.

As far as practicable, the stocker ration should prepare the cattle for making the maximum use of the finishing ration, or of grass if they are to be grazed without grain. The amount of gain made on grass or in the feedlot varies inversely with the amount of gain made during the stocker period. This inverse relationship—that is, the tendency for lightly wintered

Table 61
Yearly Gains Made by Cattle on Different Planes of Nutrition[a]

Age of Cattle		Group 1 (Maximum Growth Without Fattening)	Group 2 (Growth Distinctly Retarded)
		(lb)	(lb)
30 to 360 days	Weight gained	409.7	241.6
	Daily gain	1.24	0.73
360 to 720 days	Weight gained	320.8	235.6
	Daily gain	0.89	0.65
720 to 1,080 days	Weight gained	126.9	121.2
	Daily gain	0.35	0.34

[a] Missouri Research Bulletin 43.

Table 62

Effect of Winter Gains on Gain Made on Grass During the Following Summer[a]

	Winter Gain (lb)	Gain Following Summer and Fall (lb)	Total for Year (lb)
Average of 4 lots, making less than 50 lb winter gain	20	257	277
Average of 7 lots, making 50 to 99 lb winter gain	77	235	312
Average of 6 lots, making 100 to 149 lb winter gain	121	227	353
Average of 4 lots, making 150 to 199 lb winter gain	167	212	379
Average of 5 lots, making 200 or more lb winter gain	222	165	387

[a] Oregon Bulletin 182.

cattle to gain faster and for more heavily wintered cattle to gain slower in a subsequent grazing or finishing program—is the result of a physiological phenomenon called compensatory gain. The data in Table 62 aptly illustrate this type of relationship in yearlings on grass. If cattle are being wintered for the purpose of converting grass or pasture into gain the following summer, the winter gains should not be so great that the cattle's ability to utilize pasture will be impaired.

California workers have reported the results of a comprehensive study that explains, in part at least, the phenomenon of compensatory gain. In order to correct for the large differences in fill that occur when cattle are fed on rations of widely differing roughage content, they slaughtered some of their animals at the beginning of the test and others at the end of each of the three periods in a three-phase study. They then determined the empty-body composition of the slaughtered animals and calculated the energy gain made in each period from known amounts of feed. The report is too long and involved to review in detail in this text, but it is suggested for further study. In essence, the California group found that, even after correcting for differences in fill and body composition, compensatory gain did in fact occur. They concluded that this might be explained by an improvement in feed utilization and an increase in feed intake by those animals previously fed on the lower-energy rations.

The amount of gain desired for weaner calves during the winter, or on a grower program at any time for that matter, depends largely on the way the cattle are to be handled the following summer or subsequent feeding period. If they are to be fed concentrates on pasture, they should be wintered better than if they are to be only grazed. Also, if they are

Table 63
Effect of Level of Wintering Gain on Summer and Total Gain[a]

| | | | Daily Summer Gain and Total Winter plus Summer Gain (Summer Programs = 100 Days) | | | | | | | |
| | Winter Gain (160 Days) | | Pasture Only | | Full-Fed on Pasture | | Full-Fed in Drylot | | All Summer Programs | |
	Average (lb)	Total (lb)	Average Daily Summer Gain (lb)	Total Summer + Winter Gain (lb)	Average Daily Summer Gain (lb)	Total Summer + Winter Gain (lb)	Average Daily Summer Gain (lb)	Total Summer + Winter Gain (lb)	Average Daily Gain— 100 days (lb)	Total Gain— 260 days (lb)
Number of Steers	60		20		20		20			
Lot										
1	0.92	146	1.42	288	2.70	416	2.35	381	2.17	363
2	1.25	200	1.35	335	2.00	400	2.12	412	1.82	382
3	1.51	231	1.10	341	1.55	386	2.00	431	1.55	386
4	1.64	262	1.05	367	1.60	422	1.90	452	1.51	413
5	1.65	263	0.92	355	1.82	445	2.12	475	1.62	425

[a] Illinois Cattle Feeders' Day Report.

to be grazed only until midsummer and then full-fed for the late-fall market, they should be wintered at a higher level than would be advisable if they were to be grazed the entire summer and fall.

Table 63 shows the effect of level of winter gain on the gains made the following summer when cattle are handled under three different summer management programs. Total gains for winter and summer are also shown. In fertile areas where the yields of roughages such as corn and sorghum silage are high, wintering gains can often be made more economically than summer gains resulting from finishing rations fed on pasture. Therefore, it seldom pays to skimp on wintering or grower rations for cattle that are to be finished immediately afterward.

Although it is true that the smaller the gain made by young cattle during the winter, the greater the gain on pasture or in drylot the following summer, the winter and summer gains are not exactly inversely proportional, as is often believed. For example, if one lot of steers gains twice as much during the winter as a second lot, its summer gains are not limited to half those of the second lot, but will probably be 70 to 90 percent as much. Consequently, it usually happens that the cattle that make the largest stocker gains also make the most economical and the largest total gain for the entire period. Although the effect of the stocker gain on the finishing gain varies widely from year to year and between different droves of cattle, a good rule for the practical cattleman is that, for every additional pound

Table 64
Recommended Gains for Stocker Cattle (120–150 Days)

Method of Feeding and Management the Following Summer	Calves		Yearlings	
	Total Gain (lb)	Average Daily Gain (lb)	Total Gain (lb)	Average Daily Gain (lb)
1. Grazed entire summer; sold as feeders or started on feed in late fall (November)	115	0.75	100	0.66
2. Grazed until August 1; full-fed until late fall or early winter (November or December)	150	1.00	115	0.75
3. Pastured only until about June 1; then full-fed on pasture and marketed late fall (November)	185	1.25	150	1.00
4. Full-fed during summer in drylot; marketed in early fall (September)	225	1.50	185	1.25
5. Full-fed during entire summer on pasture; marketed in fall (October)	225	1.50	185	1.25

that stocker calves gain during the growing period they will gain 0.5 pound less in drylot or on grass.

Recommendations for level of stocker gain for stockers that are to be handled subsequently according to the more common steer management plans are shown in Table 64. Recommendations for feeding replacement heifers are found in the two preceding chapters.

NUTRIENT REQUIREMENTS FOR STOCKER CATTLE

National Research Council recommendations for the daily nutrient needs for stocker steer calves and yearlings and for normal growth in replacement heifers are shown in Table 65, and the recommended percentage nutrient content of rations for the same classes of cattle is shown in Table 66. If more or less gain is desired than that shown in the NRC recommendations, the total digestible nutrient (TDN) and digestible energy (DE) values need to be adjusted upward or downward slightly, but this does not apply to the remaining values. Morrison's feeding standards for the same classes of cattle are shown in Table 67.

As already stated, the feeding of stocker cattle is not likely to prove profitable unless the ration consists largely of farm-grown roughages. The presence of much grain in such rations is unjustified. As a matter of fact, it is seldom necessary to use any grain at all for yearling stocker cattle. Calves usually require 3 to 5 pounds of grain daily to ensure satisfactory gains unless they are fed almost a full feed of corn or sorghum silage. In the absence of some kind of legume roughage, it is highly advisable to feed a high-level protein supplement to supply the protein requirements of the growing animals. Usually 1 pound of such feeds per head daily is sufficient. If 4 or more pounds of good legume hay or its equivalent in silage is fed, protein concentrates may be dispensed with entirely. Urea may be added at the rate of 10 to 20 pounds per ton at silo-filling time to corn and sorghum silage with reasonable success in meeting the protein requirements.

The importance of supplying adequate protein in the stocker ration is shown by results of the feeding experiments summarized in Table 68. The addition of 0.50 pound of cottonseed cake to a full feed of prairie hay increased the daily gain to 0.58 pound, whereas the addition of another 0.50 pound of cake further increased the gain by 0.37 pound. Little further increase in daily gain was obtained when the supplement feeding rate was increased to 1.5 pounds. This test shows that calves should have from 0.75 to 1 pound daily of high-level protein supplement when they are consuming nonlegume roughages if they are to make satisfactory gains.

Table 65
Daily Nutrient Requirements of Stocker Cattle (Based on Air-Dry Feed Containing 90 Percent Dry Matter)[a]

Body Weight (lb)	Average Daily Gain (lb)	Daily Feed per Animal (lb)	Total Protein (lb)	Digestible Protein (lb)	Digestible Energy			TDN[b] (lb)	Ca (gm)	P (gm)	Carotene[c] (mg)	Vitamin A (IU)
					For Maintenance (kcal)	Per Pound of Gain (kcal)	Total (kcal)					
Normal Growth, Heifers and Steers												
400	1.6	12.2	1.4	0.9	6600	3900	12800	6.4	16	11	23	9200
600	1.4	16.4	1.5	0.9	9000	5300	16400	8.2	16	12	31	12300
800	1.2	19.1	1.5	0.9	11200	6600	19100	9.6	16	13	36	14300
1000	1.0	21.1	1.6	1.0	13300	7800	21100	10.6	14	14	40	15800
Wintering Weanling Calves												
400	1.0	10.5	1.1	0.7	6600	3900	10500	5.3	13	10	20	7900
500	1.0	12.6	1.3	0.8	7900	4700	12600	6.3	13	10	24	9500
600	1.0	14.3	1.3	0.8	9000	5300	14300	7.2	13	10	27	10700
Wintering Yearling Cattle												
600	1.0	14.3	1.2	0.7	9000	5300	14300	7.2	13	11	27	10700
800	0.7	15.8	1.2	0.7	11200	6600	15800	7.9	13	12	30	11900
900	0.5	15.8	1.2	0.7	12200	7200	15800	7.9	13	12	30	11900

[a] _Nutrient Requirements of Beef Cattle_, Sub-Committee on Beef Cattle Nutrition, National Research Council, revised 1963.
[b] The TDN was calculated from digestible energy by assuming 2000 kcal of DE per pound of TDN.
[c] The carotene requirement was calculated from the vitamin A requirement assuming 400 IU of vitamin A per milligram of carotene.

Table 66
Nutrient Requirements of Stocker Cattle Expressed as Percentage Composition of Air-Dry Rations[a]

Body Weight (lb)	Average Daily Gain (lb)	Daily Feed per Animal (lb)	Total Protein (%)	Digestible Protein (%)	Digestible Energy (kcal/lb)	TDN[b] (%)	Ca (%)	P (%)	Carotene[c] (mg/lb)	Vitamin A (IU/lb)
					Percentage of Ration or Amount per Pound of Feed					
Normal Growth, Heifers and Steers										
400	1.6	12.2	11.7	7.0	1050	53	0.29	0.21	1.9	750
600	1.4	16.4	9.3	5.6	1000	50	0.21	0.16	1.9	750
800	1.2	19.1	7.8	4.7	1000	50	0.18	0.15	1.9	750
1000	1.0	21.1	7.8	4.7	1000	50	0.15	0.15	1.9	750
Wintering Weanling Calves										
400	1.0	10.5	10.3	6.2	1000	50	0.27	0.21	1.9	750
500	1.0	12.6	10.3	6.2	1000	50	0.23	0.18	1.9	750
600	1.0	14.3	9.1	5.5	1000	50	0.20	0.16	1.9	750
Wintering Yearling Cattle										
600	1.0	14.3	8.3	5.0	1000	50	0.20	0.17	1.9	750
800	0.7	15.8	7.5	4.5	1000	50	0.18	0.17	1.9	750
900	0.5	15.8	7.5	4.5	1000	50	0.18	0.17	1.9	750

[a] *Nutrient Requirements of Beef Cattle*, Sub-Committee on Beef Cattle Nutrition, National Research Council, revised 1963.
[b] The TDN was calculated from digestible energy by assuming 2000 kcal of DE per pound of TDN.
[c] The carotene requirement was calculated from the vitamin A requirement assuming 400 IU of vitamin A per milligram of carotene.

277

Table 67
Morrison's Feeding Standards for Stocker Cattle[a]

			Requirements per Head Daily					
Body Weight (lb)	Dry Matter (lb)	Digestible Protein (lb)	Total Digestible Nutrients (lb)	Calcium (gm)	(lb)	Phosphorus (gm)	(lb)	Carotene (mg)
Growing Cattle, Fed for Rapid Growth								
300	7.2– 9.0	0.67–0.77	5.1– 6.2	18	0.040	13	0.029	20
400	9.1–11.4	0.76–0.87	6.2– 7.2	20	0.044	15	0.033	25
500	10.7–13.0	0.81–0.92	7.2– 8.4	19	0.042	15	0.033	30
600	12.4–14.7	0.84–0.95	8.1– 9.3	18	0.040	15	0.033	35
700	14.2–16.5	0.87–0.98	8.9–10.2	17	0.037	15	0.033	40
800	15.9–18.3	0.90–1.00	9.5–10.9	16	0.035	15	0.033	45
900	17.3–19.7	0.93–1.03	10.1–11.5	16	0.035	15	0.033	50
1,000	18.6–21.0	0.95–1.05	10.6–12.0	15	0.033	15	0.033	55
Wintering Calves, to Gain 0.75 to 1 lb per Head Daily								
300	7.0– 8.3	0.52–0.58	3.9– 4.6	16	0.035	12	0.026	17
400	8.7–10.3	0.63–0.70	4.8– 5.7	16	0.035	12	0.026	25
500	10.3–12.1	0.71–0.78	5.7– 6.7	16	0.035	12	0.026	30
600	11.7–13.9	0.79–0.88	6.5– 7.7	16	0.035	12	0.026	35
Wintering Yearling Cattle, to Gain 0.50 to 0.75 lb per Head Daily								
600	11.6–13.3	0.67–0.75	6.3– 7.2	16	0.035	12	0.026	35
700	12.9–14.8	0.76–0.83	7.0– 8.0	16	0.035	12	0.026	40
800	14.2–16.3	0.83–0.90	7.7– 8.8	16	0.035	12	0.026	45

[a] Taken by permission of The Morrison Publishing Company, Ithaca, N.Y., from *Feeds and Feeding*, 22nd edition, by Frank B. Morrison.

On the other hand, results secured at the Indiana station, reported in Table 69, in which 2 to 2.5 pounds of protein-mineral-vitamin concentrate were fed to calves wintered on oat straw, corn cobs, and soybean straw, indicate that relatively high levels of nitrogenous supplements, as well as minerals and vitamins, must be fed with low-grade roughages if the feed nutrients in such roughages are to be efficiently utilized.

Complex protein-mineral-vitamin supplements are beneficial when added to low-quality roughages as just shown. However, such supplements are not required when such good-quality roughages as legume-grass or corn silage are fed, as shown by results of Illinois studies reported in Table 70. Had the complex supplements been fed at levels that would supply

Table 68
Importance of Adequate Protein in the Rations of Stocker Cattle[a]

Concentrate Fed	None	0.5 lb Cottonseed Cake	0.75 lb Cottonseed Cake	1 lb Cottonseed Cake	1.5 lb Cottonseed Cake
Number of trials averaged	5	2	3	3	2
Average initial weight (lb)	408	418	433	385	363
Average final weight (lb)	433	543	596	564	548
Average winter gain (lb)	25	125	163	179	185
Average daily gain (lb)	0.15	0.73	0.92	1.10	1.16
Average prairie hay eaten daily (lb)	10.6	12.7	13.7	13.2	11.6
Average summer gain (lb)	259	231	222	212	190
Average total gain (lb)	284	356	385	391	375

[a] Nebraska Bulletin 357.

Table 69
Value of Low-Grade Roughages for Wintering Stocker Cattle When Fed with a Complete Supplement[a]

Roughage Fed	First Trial 112 Days			Second Trial 147 Days		
	Oat Straw	Corn Cobs	Corn Silage	Soybean Straw	Corn Cobs	Corn Silage
Average initial weight (lb)	485	479	478	480	478	481
Average final weight (lb)	589	671	721	595	698	806
Average total gain (lb)	104	192	243	115	220	325
Average daily gain (lb)	0.93	1.72	2.18	0.78	1.50	2.21
Average daily feed (lb)						
Roughage	12.4	12.8	31.0	13.3	13.4	37.0
Supplement A[b]	3.5	3.5	3.5	3.5	3.5	3.5
Minerals[c]	Free choice	Free choice	Free choice	0.06	0.05	0.04
Feed per hundredweight gain (lb)						
Roughage	1,337	742	1,423	1,707	890	1,671
Supplement A	377	202	160	449	233	158

[a] Indiana Mimeographs AH 47 and AH 59.
[b] Supplement A consisted of 2.25 lb soybean meal, 1 lb molasses feed, 0.18 lb bonemeal, 0.06 lb salt, and 0.01 lb vitamin A concentrate.
[c] Mineral mixture fed free-choice during both tests: 2 parts steamed bonemeal, 1 part iodized salt, plus 1 oz. cobalt sulfate per 100 lb of salt.

Table 70

Comparison of Simple and Complex Supplements for Good-Quality Roughages Fed to Stocker Calves[a]

Type of Silage	Corn Silage		Legume-Grass Silage[b]	
Supplement Fed	Purdue Supplement A	Soybean Meal, Ground Shelled Corn	Purdue Supplement A	Ground Shelled Corn
Days fed	125	125	160	160
Average initial weight (lb)	333	320	400	400
Average daily gain (lb)	2.19	2.17	1.51	1.65
Average daily ration (lb)				
Silage	27.5	26.6	22.5	22.5
Protein supplement	3.5	1.25	3.5	—
Ground shelled corn	—	2.00	—	3.5
Minerals	0.08	0.11	0.06	0.07
Feed cost per cwt gain ($)	15.31	12.74	21.58	14.05
Winter feed bill per head ($)	42.09	34.61	49.85	36.95

[a] Illinois Cattle Feeders' Day Reports.
[b] Contained 150 lb of ground shelled corn preservative per ton of silage.

only the usually recommended level of protein concentrate, the results would not have been so overwhelmingly in favor of simple high-energy concentrates from a cost standpoint.

ROUGHAGES USED IN STOCKER PROGRAMS

The character of the roughage fed should depend somewhat on the age of the cattle. Yearling steers can make good use of corn or sorghum stover, cobs, and straw. Calves, on the other hand, should be fed a limited amount of such materials. If possible, corn, sorghum, or grass silage and legume hay should furnish at least 75 percent of the dry matter of the roughage ration for calves and 50 percent for yearlings. The remainder may well consist of straw or other low-quality roughage.

It is important that the cattle be given all they will eat of some component of the ration; otherwise their hunger is unsatisfied and they are restless and waste some of their energy in moving about. As a matter of economy, the better practice is to limit the quantities of legume hay and silage to the amounts actually required to produce the desired gains, and to keep a supply of cheaper feed such as cottonseed hulls or straw before the cattle at all times. Stocker cattle consume the equivalent of about 2.5 percent of their live weight daily, in total air-dry ration, if they are in average flesh. Thin stockers consume slightly more, up to 3 percent, and, conversely, fleshier stockers consume less.

Corn and Sorghum Silage. Silage made from high-yielding corn and sorghum varieties is the most satisfactory feed for stocker cattle. The palatability of such silage ensures a sufficient consumption of feed to produce rapid and efficient gains. Its succulent nature makes it a good substitute for grass, and it keeps the digestive system in normal condition. Silage is usually economical enough to permit liberal feeding without unduly increasing feed costs. In fact, when yield per acre, ease of feeding, and the amount required to produce a given gain are considered, corn and sorghum silages are usually the cheapest feeds that can be fed to young stocker cattle. It is not uncommon to find reports of experiments where 1 ton of stocker gains are made per acre of corn or sorghums harvested and fed as silage.

Weanling calves should be fed as much silage as they will consume, which averages approximately 30 pounds or 5 to 6 percent of body weight daily if they are fed only a little hay in addition. Corn and sorghum silages are low in both protein and minerals. Therefore, approximately 1 pound of protein supplement and 0.10 pound of finely ground limestone or dicalcium phosphate should be fed daily per head unless 4 or more pounds of leafy legume hay are present in the ration. If limestone is used, a phosphorus source should also be included.

Legume-Grass Silage. Silage made from mixed grasses and legumes cut at the proper stage of maturity and carefully ensiled is an excellent feed for stocker cattle. Such silage has a high protein and mineral content and does not need to be supplemented with a protein concentrate. However, it is much lower in digestible or net energy than corn or sorghum silage, and smaller gains result from its use, unless grain is added as a preservative, at the rate of 150 to 200 pounds a ton, or unless it is supplemented at feeding time with some grain or other energy source. If corn or sorghum silage and legume-grass silage are both to be fed to a drove of stocker cattle, the legume-grass silage should be fed out first, since it usually is less palatable than corn or sorghum silage.

The addition of a low level of a dry roughage to a heavy feed of silage being fed to stockers is recommended. Although the reason for the beneficial responses obtained from such additions is not well established, the physiological well-being of both the rumen microorganisms and the host animal are apparently involved. Table 71 shows the results of one of several such trials. In this trial, dry matter intake was constant and

Table 71

Effect of a Small Hay Addition to a Silage Stocker Ration Fed to Calves[a]

	Steer Calves (133 Days)	
Roughages Fed	Legume-Grass Silage	Legume-Grass Silage + Hay
Number of calves	10	10
Initial weight (lb)	435	435
Average daily gain (lb)	1.68	1.80
Average daily ration (lb)		
Silage	28.7	22.8
Hay	—	2.2
Ground shelled corn	3.5	3.5
Mineral mixture	0.07	0.07
Feed cost per cwt gain ($)	13.26	12.45

[a] Illinois Cattle Feeders' Day Report.

the small amount of hay fed was comparable to the silage replaced with respect to species and stage of maturity when cut. Large commercial feedlots, which utilize fenceline bunks and automatic unloading wagons or trucks, may find it economically unfeasible to feed a dry roughage in the form of long hay.

Table 72

A Comparison of Different Silages for Yearling Stocker Steers[a]

	Corn Silage	Legume Silage[b]	Sorgo Silage	Barley Silage
Percentage of moisture when fed	63.7	74.1	72.9	77.5
Percentage of protein	3.0	4.4	2.1	2.7
Yield per acre, tons	8.5–9.0	—	15.0	—
Preservative added (lb)	—	78	—	45
Initial weight of cattle (lb)	738.8	739.7	738.8	739.9
Average daily gain, 126 days (lb)	1.85	1.34	1.21	1.00
Average daily ration (lb)				
Silage	32.1	38.1	35.1	40.2
Alfalfa hay	6.6	6.7	6.6	6.8
Feed per cwt gain (lb)				
Silage	1,734	2,834	2,890	4,012
Alfalfa hay	357	497	547	679

[a] Missouri Mimeographed Report.
[b] The legume silage was made from a mixed seeding of alfalfa, sweet clover, and red clover, which was ensiled the last week of May.

It must be recognized that quality of silage can vary greatly, and differences in the feeding value between two samples of legume-grass silage, for instance, may well be greater than between the silages made from two different crop species. Table 72 and Fig. 58 show results from well-controlled tests, but varietal changes, variation in methods of making and feeding silage, fertilizer treatments, and even class and breeding of stocker

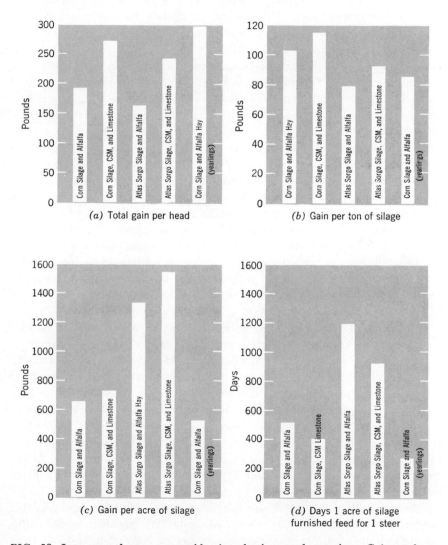

FIG. 58. Important factors to consider in selecting stocker rations. Gain made per head and per acre during the following summer are additional important factors. (Nebraska Mimeograph Cattle Circular 146.)

cattle to which the silage is fed can be responsible for results that may differ widely from those reported in these and other tables.

Legume-Grass Hay. Mixed hays containing up to 50 percent legumes such as alfalfa, red clover, or lespedeza are excellent feed for stocker cattle, especially if the hays are cut before they are unduly mature and high in crude fiber, and if they are properly cured. In all but the western sections of the country, such hays are often lower in quality than the same crops would be if preserved as silage. This is because good hay-curing days are few in most parts of the country, especially for the first cutting.

Energy supplementation of mixed-hay stocker rations is required, as it is with legume-grass silages, with 2 to 4 pounds of grain daily being the level usually fed. Protein supplementation may or may not be required, depending on the quality of the roughage. In general, if the hay consumed daily contains the equivalent of 4 pounds of good-quality legume hay, supplementation with a protein concentrate increases gains, but no more than if a similar amount of a high-energy concentrate such as corn or ground grain sorghum is added. Carotene or vitamin A, calcium, and phosphorus supplements may be needed in very poor quality mixed-hay rations.

Haylage. Haylage is legume-grass or straight-legume forage that has been allowed to wilt in the windrow until it has been reduced to about 50 percent moisture. The forage is then usually chopped and stored in a conventional silo or, more commonly, in an oxygen-free silo such as the Harvestore. The forage undergoes normal silage fermentation and, except for the difference in moisture content, has a feeding value comparable to that of silage or hay that was properly made from the same crop. Supplementation of this feed thus presents no problems that were not discussed in connection with silage and hay.

The use of haylage is increasing despite the added cost in storage and harvesting equipment. This must be because of the convenience of harvesting and feeding, the reduction in field-harvesting losses, and the reduction in losses traceable to poor haymaking weather on the one hand or to high fermentation losses of unwilted silage on the other.

Grass Hay. Straight-grass hays, such as prairie hay, Coastal Bermuda, timothy, Sudan or Johnson grass, may satisfactorily make up a major portion of the ration of stocker cattle, but fairly complete supplementation with protein, minerals, and vitamins is essential for normal growth. Added energy supplementation is needed if any improvement in condition of the stockers is desired.

Stalk Fields. Stalk fields furnish much cheap feed for stocker cattle during the late fall and early winter. Many feeders buy their feeder cattle, especially if they handle yearlings, in October or November and run them on stubble and stalk fields until about the first of January, when they are put into the feedlot and started on finishing rations. Yearlings are better than calves for this method of handling, as they are better able to make use of the coarse roughage that stalk fields furnish. Stocker cattle, particularly calves, should never be maintained wholly on stalks because stalks are quite low in net energy and seriously deficient in protein. Unless there is access to a good bluegrass or clover pasture or winter small grains, cattle on stalk fields should be fed 4 to 6 pounds of legume hay or 1 to 1.5 pounds of protein concentrate per head daily. Protein blocks or liquid supplements that can be self-fed fit this situation nicely.

WINTER SMALL-GRAIN PASTURES

Winter wheat pastures are widely used in Oklahoma, Texas, and Kansas for wintering all classes of cattle, but calves make especially good use of this rather uncertain feed supply. Extensive snows, wet weather, or prolonged dry weather all make this source of feed rather unreliable, but a stocker program remains the most profitable one for a man with such pasture to sell through cattle.

Good wheat pasture alone will produce acceptable stocker gains, as shown by Table 73. Although the subject of hormones is discussed in detail elsewhere, note the response obtained from stilbestrol on both the silage and pasture programs. Wheat pasture is also excellent feed for cows, especially if they are fall-calving cows, but the main disadvantage of this program is that when the wheat pasture fails to materialize or hold up because of adverse weather conditions, cows must either be sold or moved, or fed on roughages that will have a very high cost if purchased or a high sale value if homegrown. Usually a large area is affected by the same weather pattern, which accounts for the general high cost of feed when purchases are necessary.

Wheat pasture poisoning, also referred to as grass tetany when it occurs on certain grass pastures in some parts of the world, can present real problems to the man who runs cattle on rapidly growing wheat pasture. The condition follows a rapid course, with usually a lapse of only 6 to 10 hours between onset and coma which, unless the condition is treated, is nearly always followed by death. Initially, affected animals appear nervous and excited. The third eyelid flickers, and there is twitching in the muscles of the extremities, followed by tetany, contractions, and convulsions. Rapid breathing and a pounding heart immediately precede coma and death. Cows are most often affected, especially those that are nursing calves, but other classes of cattle can be involved.

The exact cause of this condition is incompletely understood, but sup-

Table 73

Response of Stilbestrol-Implanted Stocker Heifer Calves
on Sorghum Silage or Wheat Pasture[a]

	Sorghum Silage + Protein Supplement[b]	Sorghum Silage + Supplemental Ration[c]	Wheat Pasture	Wheat Pasture + Supplemental Ration[d]
Number of heifers	10	10	10	10
Initial weight (lb)	472	474	471	477
Final weight (lb)	553	667	667	701
Total gain (88 days) (lb)	81	193	196	224
Daily gain (all calves) (lb)	0.92	2.19	2.22	2.54
Daily gain (lb)				
Controls	0.87	1.87	2.03	2.58
12 mg stilbestrol implant	0.95	2.50	2.42	2.51
Daily supplemental feeds[c,d]				
Mixed feed	—	9.64	—	9.34
Cottonseed meal	1.50	1.00	—	—
Mineral and salt	Free choice	Free choice	Free choice	Free choice

[a] Oklahoma Miscellaneous Publication 79, 1967.
[b] 1.5 lb cottonseed meal daily.
[c] Supplemental ration consisting of 77 percent ground milo, 8 percent molasses, and 15 percent chopped alfalfa hay fed at a level consumed by cattle on wheat pasture when provided free-choice plus 1 lb cottonseed meal daily as supplemental protein.
[d] Supplemental ration described in footnote (c) provided free-choice. No additional protein supplement was fed.

plementation with magnesium will prevent it from developing. Approximately 6 grams of magnesium per day, usually provided in the form of magnesium sulfate added to the salt and mineral supplement, is adequate. Commercial supplements are obtainable in the wheat pasture grazing area, and these should be fed according to directions. Affected animals can be treated with intravenous or intraperitoneal injections of calcium gluconate solutions fortified with magnesium and phosphorus. A second treatment may be required in advanced cases.

In the southeastern states there is extensive use of winter oats and fescue, a perennial grass that remains green throughout much of the winter, in wintering cattle. Wheat and rye are also used to a lesser extent. The Louisiana feeding and grazing trials reported in Table 74 illustrate what may be expected of calves on various combinations of rations in this important cow-calf area. This area is turning more and more to stocker or grower programs, to utilize winter pasture, to increase the output or volume of

Table 74

A Comparison of Six Methods of Wintering Stocker Calves and Effect on Spring Gains (Three Years, 1961–1963)[a]

Group Number	1	2	3	4	5	6
Treatment	Coastal Bermuda Hay	Hay + Grain	Wheat Pasture + Hay	Kentucky 31 Fescue Pasture	Fescue + Grain	Fescue and Wheat Pasture
Winter Period						
Number of acres	Drylot	Drylot	15	5	5	10
Number of calves	10	10	10	10	10	11.3
Initial weight, fall (lb)	454	456	456	456	456	456
Weight, spring (lb)	484	570	579	506	587	575
Winter gain, 105 days (lb)	30	114	123	50	131	119
Average daily gain (lb)	0.28	1.08	1.17	0.47	1.24	1.13
Grain fed per calf per day (lb)	—	5	—	—	5	—
Hay fed per calf per day (lb)	11.4	9.1	8.0	9.8	9.8	9.8
Number of days hay was fed	105	105	19	2[b]	2[b]	2[b]
Number of days grazing:						
fescue	—	—	—	105	105	67
wheat	—	—	86	—	—	38
Cost per cwt of gain[c]	$50.13	$22.04	$28.59	$24.72	$19.50	$17.22
Spring Period						
Weight, summer (lb)	626	677	699	648	692	702
Spring gain, 87 days (lb)	142	107	120	142	105	127
Average daily gain (lb)	1.63	1.23	1.37	1.63	1.20	1.45
Both Periods						
Total gain, 192 days (lb)	172	221	243	192	236	246
Average daily gain (lb)	0.89	1.15	1.26	1.00	1.22	1.28

[a] Red River Valley Station, Louisiana Agricultural Experiment Station, 1966.

[b] Hay fed when snow occurred.

[c] Feed and pasture cost: Coastal hay, $25 per ton; grain ration, $50 per ton; fescue pasture, $24 per acre; wheat pasture, $21.50 per acre.

business, and to spread the marketing period for their calves. Note that compensatory gain in the summer is related to rate of winter gain, as previously discussed.

CONCENTRATES FOR STOCKER CATTLE

Whether stocker cattle should be fed concentrates during the stocker program depends on the amount and character of roughage fed and on the subsequent feeding plan. Calves that are not fed silage should be fed 2 to 4 pounds of grain daily because they are unable to consume enough dry roughage to gain more than a pound a day. However, calves that are fed a full feed of high grain-content corn or sorghum silage and 1 pound of a protein concentrate or 4 or 5 pounds of legume hay need not be fed grain unless they are to be marketed following only a short feed on finishing rations. Then the feeding of some grain is advisable because the cattle will have a higher finish and sell for a better price when marketed. The effect of such low-level grain additions is shown in Table 75. Newly arrived calves fed on "warm-up" or "backgrounding" rations for only a short period respond especially well to 3 to 4 pounds of grain. Because their total intake of feed is low, owing to prior stress and the likely presence of a low-grade fever, the high energy content of the grain meets their energy requirements where roughage alone would not.

Except when poor to very poor quality roughages are fed, there is apparently little difference between the various protein concentrates for

Table 75
Effect of Feeding During the Winter a Half Ration of Grain to Calves That Are to Be Grazed the Following Summer

	Missouri[a] Average of 3 trials		Kansas[a] Average of 3 trials	
	Grain and Roughage	Roughage Only	Grain and Roughage	Roughage Only
Average initial weight (lb)	342	348	350	349
Average winter ration (lb)				
Shelled corn	3.8	—	4.6	—
Protein concentrate	.5	—	1.0	1.0
Corn or cane silage	11.2	14.3	18.4	24.0
Legume hay	4.2	4.5	2.0	2.0
Average daily gain (lb)	1.70	.88	1.89	1.34
Grazed without grain	56 days	56 days	90 days	90 days
Total gain on pasture (lb)	22	57	98	123
Average daily gain on pasture (lb)	.34	1.02	1.09	1.38
Full fed grain on pasture	112 days	112 days	100 days	100 days
Total gain (lb)	269	276	256	263
Average daily gain (lb)	2.41	2.47	2.56	2.63
Final weight (lb)	912	823	961	917
Market value per cwt	$14.17	$13.75	$14.92	$14.58
Shelled corn fed (bu)	36.5	24.3	36.9	26.2

[a] Mimeographed reports of cattle-feeding experiments.

Table 76
Recommended Rations for Stocker Cattle

Rations		Calves (lb)	Yearlings (lb)
I.	Alfalfa or clover hay	5–6	4–5
	Corn or sorghum silage	25	35
II.	Clover hay	8–10	6–8
	Corn or oats	4–5	5–6
	Straw or hulls[a]	2–4	12–15
III.	Soybean, cottonseed, or linseed meal	1–1.5	1.0
	Corn or sorghum silage	20–25	20.0
	Straw or hulls	2–3	10–12
IV.	Grass hay	10–12	16–20
	Corn, sorghum, or oats	4–5	5–6
	Soybean, cottonseed, or linseed meal	1–1.5	1–1.5
V.	Legume-grass or oat silage	20–25	35–45
	Corn, sorghum, or oats	4–5	5–6
	Straw or mixed hay	2.0	2.0
Expected daily gain		1.4–1.8	1.5–2.0

[a] All straw and hulls should be fed according to appetite.

feeding stocker cattle, provided they are fed in sufficient amounts to furnish as much protein or its equivalent in nonprotein nitrogen as is present in 1 pound of the oilseed meals. It has been demonstrated at the Fort Hays branch of the Kansas station that 2 pounds of ground barley or wheat, 3 pounds of wheat bran, or 4 pounds of alfalfa hay are as good supplements to a full feed of silage as 1 pound of linseed, cottonseed, or soybean meal.

RECOMMENDED RATIONS FOR STOCKER CATTLE

Stocker cattle are being successfully fed on a wide variety of rations, and it speaks well for this program that it can be adapted to the efficient utilization of so many different roughages. Some recommended rations, based on the more commonly used roughages and concentrates, are found in Table 76.

SUMMER MANAGEMENT OF YOUNG CATTLE

The problems of managing stockers on pasture during the summer are relatively simple in comparison with those arising in drylot. With the availability of spring pasture, feeding problems are largely solved because good pasture is an ideal ration for young growing animals. It is nutritious and rich in growth-promoting and protective vitamins.

Pastures to be used by young breeding animals should not be heavily stocked unless there are adequate facilities for feeding harvested feeds if the supply of grass becomes short. Grass alone ordinarily provides a satisfactory ration for growing cattle during the lush growing season. With stocker steers or grade yearling heifers intended for a commercial breeding herd, economy of maintenance usually requires that pasture alone, if it is available, should form the ration for the entire summer, even though the cattle lose some flesh and may not grow quite so rapidly. If the grass fails because of dry weather, pastures should be supplemented by feeding silage, hay, chopped green corn, or hybrid Sudan.

SUMMER GAINS OF YOUNG CATTLE ON PASTURE

The gains made during the summer by young cattle on pasture do, of course, vary greatly with the condition of the animals at the beginning of the grazing season, with the kind and amount of forage available, and with whether they were implanted with hormones. Also, gains vary from

FIG. 59. Yearling stocker steers can advantageously convert high-quality legume-grass rotation pastures to gains without the use of grain. Monthly gains of 75 to 100 pounds per acre may be expected during the peak pasture months. (University of Illinois.)

FIG. 60. Many early-day cattle ranchers believed that sheep would "poison" or ruin a cattle pasture. Contrary to legend, these steers and sheep are effectively demonstrating on use of range to increase the production and income on a New Mexico ranch. Prolonged overstocking with either species can permanently damage a pasture, of course. (Western Livestock Journal.)

year to year because of weather conditions, which greatly affect the amount, palatability, and nutritive value of the forage. Occasionally a severe drought and the discomfort caused by flies and oppressive heat may cause a midsummer loss of much of the gain put on during the spring. As a consequence, the gain for the entire season is disappointingly small. However, such years are offset by unusually favorable seasons, when the gains made are almost double those obtained during an average year.

Yearling cattle weighing approximately 550 to 650 pounds when turned onto pasture gain at the rate of 1.25 to 1.75 pounds a day—or 200 to 300 pounds for the season—on good permanent pasture such as bluegrass, bromegrass, or the bluestem pastures of Kansas. Rotation pastures that consist largely of legumes, thus retaining their nutritive value and palatability throughout the summer, may usually be counted on for about 50 percent more gain per head than is obtained from permanent pastures during the same season.

STOCKER OR GROWER CONTRACTS

This subject was briefly mentioned in connection with the question of risk reduction for the operator of the stocker program. There is every indication that this method of financing the stocker program will increase in use. It has much to offer both the small crop farmer who has neither the capital nor the risk capability to buy calves for himself and the feedlot operator who specializes in finishing cattle on high-energy rations. It permits both operators to specialize and to feed only one type of ration. Often roughages can be grown very efficiently in one area and grain in another. The grower contract thus permits both crops to be efficiently fed where they are grown, vastly reducing the cost of transportation of feeds. Large commercial feedlots, of course, can and often do carry out both steps in the grower-finishing operation.

The program is not new in that, for some years, much of the Kansas bluestem pasture was used for growing yearlings owned by feeders in Iowa and elsewhere. Much of the new interest, though, is in drylot feeding of stockers, and not always in winter as has been customary. Many smaller feeders in the Greeley, Colorado, area, for example, now feed the corn silage produced on their fertile, irrigated fields to stockers owned by one of several large feedlots in the vicinity. Formerly they may have owned and finished the cattle themselves or may even have sold the corn crop itself, as green corn, to the feedlot when this seemed to present the better chance for profit.

There are several kinds of contracts. The two most common are based either on a fixed cost for the gain, or on a feed cost plus an extra charge for labor and lot rental. Such contracts should be in written form, and the basis for adjusting for death loss and for computing amount of gain should be well understood in advance by both parties.

Chapter 11

The CATTLE FINISHING PROGRAM

The finishing of cattle for the slaughter markets is one of the most important enterprises of the American agricultural economy. Cattle finishing once was conducted almost exclusively in the corn-growing regions, but important cattle-finishing areas are now located throughout the United States, including Hawaii. The rapid development of this enterprise is illustrated by the tenfold increase in feeding in the Texas-Oklahoma Panhandle sorghum belt, even while the country as a whole was doubling the number of cattle fed during the past 20 years. As the 1960s were drawing to a close, this same Panhandle and adjoining areas in New Mexico and Kansas continued to be the most rapidly growing cattle-feeding center in the nation, supplanting California and Arizona, which underwent a similar development at the close of World War II. The Platte Valley areas in Colorado and Nebraska, always important feeding areas, were also experiencing large numerical increases in total numbers fed.

Figure 61 and Table 77 illustrate the increases seen in the various sections of the country during 1959–1967. Figure 62 shows that the number of cattle fed, per farm or feedlot feeding cattle, remains smallest in the Corn Belt. Also, the larger lots are mainly located in the 16 western states. In fact, 53 percent of all cattle on feed January 1, 1967, were located in those 16 states. Two-thirds of the cattle on feed in those states were in lots of 1,000-head or greater capacity. The number of cattle feeders in the Corn Belt, although still large, is decreasing, and the size of lot is increasing. The majority are still farmer-feeders and not custom or commercial feedlot operators such as are found in the 16 western states. Figure 63 shows the fed cattle marketings by states for 1967 and illustrates that the Corn Belt remains among the leaders in the cattle-finishing business.

Perhaps equally as important as the increases in total numbers fed

293

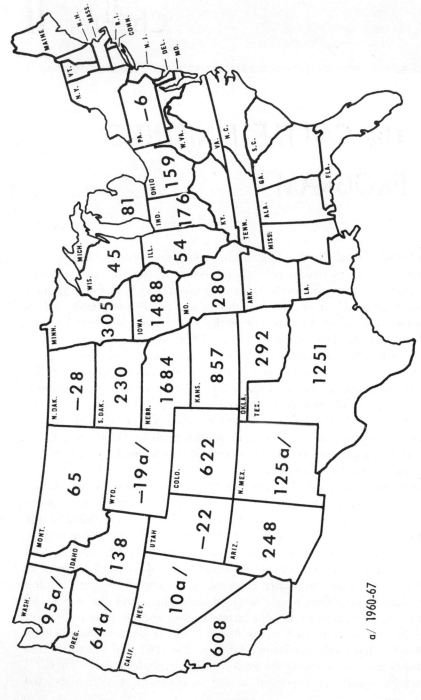

a/ 1960-67

FIG. 61. Increases in fed cattle marketings, by states (in thousands), during 1959–1967 show that the western Corn Belt, the Panhandle area, and California have continued to strengthen their position as cattle-feeding centers. (U.S.D.A.)

Table 77

**Number of Fed Cattle Marketed by States and Regions
with Comparisons Between 1959 and 1967**[a]

State or Region	Number Marketed (1,000 head)		United States Rank		Percentage of United States Total, 1967	Percent Increase Between 1959 and 1967
	1959	1967	1959	1967		
Iowa	2,569	4,057	1	1	18.7	58
Nebraska	1,373	3,057	3	2	14.1	123
Kansas	455	1,312	7	6	6.1	188
Minnesota	564	869	6	8	4.0	55
Missouri	408	688	9	9	3.2	69
South Dakota	388	618	11	11	2.9	59
West North Central	**5,757**	**10,601**			**48.9**	**84**
Illinois	1,227	1,281	4	7	5.9	4
Indiana	320	496	12	12	2.3	55
Ohio	283	442	13	13	2.0	56
Michigan	159	240	17	17	1.1	51
East North Central	**1,989**	**2,459**			**11.3**	**24**
California	1,441	2,049	2	3	9.5	42
Texas	403	1,651	10	4	7.6	310
Colorado	708	1,330	5	5	6.1	88
Arizona	410	658	8	10	3.0	60
Oklahoma	132	414	16	14	1.9	221
Idaho	227	365	14	15	1.7	61
Washington	223	318	15	16	1.5	43
New Mexico	113	238	18	18	1.1	110
Western	**3,657**	**7,026**			**32.4**	**92**
Total	**11,403**	**20,086**			**92.6**	

[a] Prepared from data published by U.S.D.A.

are the changes in the nature of the business. Although Corn Belt farmer-feeders continue to feed many cattle, commercial feedlots are feeding even more, and for them it is usually their only enterprise, as contrasted to the farmer-feeder who continues to look upon cattle finishing as a supplementary enterprise to crop production. Changes that have occurred and some that are still taking place, in no particular order of importance, may be listed as follows:

1. The number of cattle feeders is decreasing, especially in the Corn Belt states.

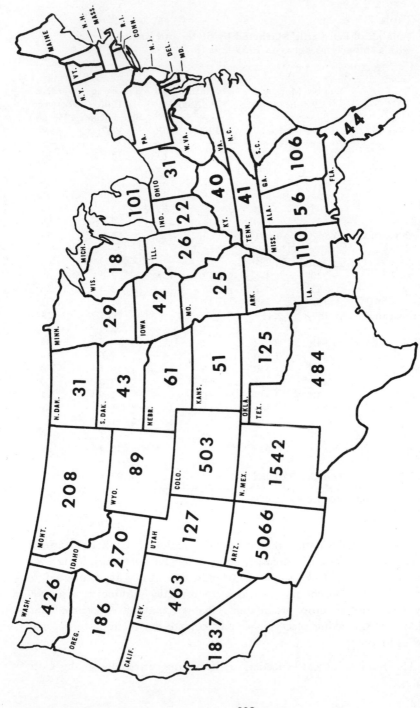

FIG. 62. The average number of cattle on feed, per feedlot, on January 1, 1968, illustrates that cattle feeders in the Corn Belt still feed cattle as a supplementary enterprise, but that cattle feeding in the Southwest is big business. (U.S.D.A.)

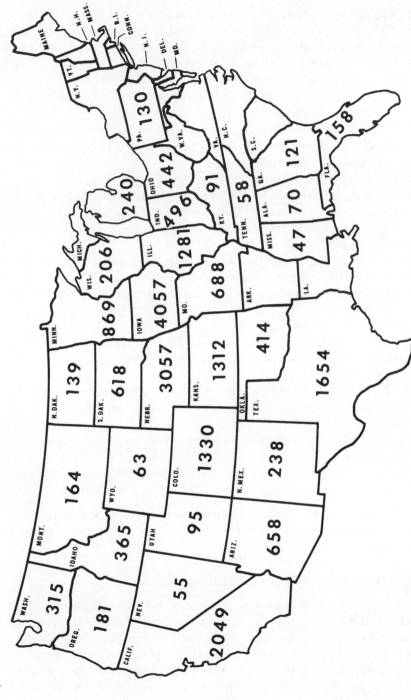

FIG. 63. Fed-cattle marketings for the year 1967, by states (in thousands), show generally that cattle are being fed near the centers of feed production. California and Arizona are exceptions. (U.S.D.A.)

2. The average number of cattle fed per feeder is increasing.

3. The number of cattle fed on some kind of contractual arrangement is greater, with many nonfarm people in the business.

4. The energy content of finishing rations is increasing, with a corresponding reduction in roughage content.

5. A greater variety of additives is used, particularly hormones, antibiotics, vitamins, and minerals.

6. The length of the finishing period is being reduced, not because cattle are sold at lighter weights but largely because initial weights are heavier and gains are more rapid.

7. Fewer cattle are bought and sold in central markets.

8. More of the plainer grades of cattle are being fed, whereas formerly most such cattle were slaughtered off grass.

9. Grain processing is more sophisticated, with much grain now being steam-processed and rolled.

10. Less labor is required, because of automation, with a corresponding increase in capital investment in machinery.

11. Rations are selected in many instances through least-cost formulations by computers.

12. Many new by-product feeds are in use.

13. More nonprotein nitrogen sources, such as urea, are being used as protein-concentrate replacements because of the lower costs made possible by this substitution.

14. The problems of disease control are intensified because of greater stress and concentration of animals in the feedlots, with less individual attention.

15. Air and water pollution problems occur near urban areas as a result of manure accumulation.

16. More and more feeders contract for cattle procurement and sale, nutritional advice, and veterinary service.

In summary, cattle feeding has become big business, and in no sense of the word can it be called a way of life, a term often used to describe general farming.

REASONS FOR FEEDING CATTLE

Cattle are usually fed for one or more of the following reasons: (1) to obtain better than current prices for farm-grown grains; (2) to market roughages and pasture at a profit; (3) to maintain and improve the fertility of the farmland; (4) to convert purchased resources, namely feed, into a more valuable finished product, namely beef. These reasons may be combined by saying that cattle are fed to obtain the highest net return for feeds, whether farm-grown or purchased. This statement implies that cattle are finished principally on feeds grown on the farm or in the region where the cattle are fed. This is largely true, although there are large-scale feeders and feedlot operators who buy most or nearly all of their feeds and transport them for as many as 1,000 miles. Such men feed cattle for the same reason that other men operate factories, namely, to make a profit by combining raw materials, which are of lower value in their natural form, into a product for which there is a strong demand and which, therefore, can be sold for a much higher price than the cost of the raw materials.

The high cost of freight on both feed and cattle is tending to push cattle feeding back into the major feed-producing areas. Strong demand for beef in an area cannot alone long hold a large cattle-feeding business, because this demand can be more economically supplied with beef in the carcass, by truck or rail, especially if freight rates favor such a trend.

ADVANTAGES AND DISADVANTAGES OF THE CATTLE-FINISHING PROGRAM

The cattle-finishing program offers much opportunity for profit, but at the same time results in as many failures, financially speaking, as all the other beef cattle programs combined. This is mainly because (1) feeds used in finishing rations are relatively expensive, (2) feed conversion is less efficient for finishing purposes than for growth, (3) price fluctuations may erase "price spread," one of the sources of profit, and (4) nonfeed costs are not related to cattle prices and are continually increasing.

Very economical gains in the stocker program or very cheap maintenance rations in the cow-calf program can often offset mistakes in judgment at buying and selling time, but no such opportunity presents itself in the finishing program. Sometimes when both stocker and finishing programs are conducted successively on the same farm, all the money made in the stocker program and more is lost in the finishing program because of poor planning or a disastrous decline in prices. For this and many other reasons, a thorough understanding of this program, the riskiest of all commercial cattle programs, is essential.

Advantages of the Finishing Program. The large number of participants is evidence of belief in the following advantages usually ascribed to the finishing program:

1. Because of its intensive nature it affords an opportunity to market at a profit large quantities of both roughages and grains.

2. Large profits are occasionally made as a result of favorable price rises during the period of ownership of a drove of feeder cattle.

3. A large volume of manure high in fertility is produced.

4. It is a relatively short-time program, making it possible to turn more than one drove of cattle per year in some types of finishing programs or to finish off a drove of cattle between peak labor requirements in farming operations.

5. There is flexibility with respect to number, weight, and grade of cattle, as well as to length of feeding period and type of ration fed.

6. Death losses are relatively low as compared with some of the other programs such as the cow-calf program.

Disadvantages of the Finishing Program. Even when the cattle finishing program is the obvious choice for best utilization of the pastures, harvested roughages, and grains found on a farm or in the area, certain disadvantages are inherent in this program and should be recognized:

1. The program is speculative, with the result that large amounts of money are sometimes lost. As discussed elsewhere, some finishing programs carry less risk than others.

2. Skilled buying and selling are required. This accounts for such common sayings as "well bought is half sold" with reference to feeder cattle.

3. Relatively large amounts of capital or credit must be readily available, to make the necessary periodic purchases of feeder cattle and, in some instances, feed as well.

4. Specialized feedlot and feed preparation equipment is required.

5. Skill and knowledge of feeding principles is very necessary in the finishing program.

6. Serious labor conflicts may develop in the farmer-feeder programs, especially with summer finishing programs.

7. The program is rather confining, requiring constant attention, although pushbutton feeding systems minimize such confinement. The program may interfere with weekend absences and long vacations.

8. Problems arise with respect to manure disposal and increased fly and bird populations.

9. Government support prices on feed grains assure a fairly strong price on feeds; thus feeders cannot offset lower cattle prices with correspondingly lower feed costs.

SOURCES OF PROFIT

A discussion of profits that may be expected from a cattle-feeding enterprise must include a breakdown of the following two sources of potential profit:

1. Price spread—the difference between buying and selling price per 100 pounds. This is also sometimes simply referred to as "margin" or

"spread" and is not to be confused with "feeding margin." It is applied to the purchased weight only.

2. Feeding margin—the difference between the feed cost of producing 100 pounds of gain and the selling price per hundredweight. Feeding margin applies only to the weight added during feeding, and only feed costs are used in its calculation, excluding nonfeed costs.

The following example illustrates how these two factors function:

1. Price spread
   700 lb yearling purchased @ $24 cwt delivered.
   1,100 lb choice steer sold @ $26 cwt net.
   $26 − $24 (7 cwt) = $14 potential profit due to price spread.
2. Feeding margin
   1,100 lb − 700 lb = lb gain @ cost of $21 cwt gain.
   $26 − $21 (4 cwt) = $20 potential profit due to feeding margin.
   $14 + $20 = $34 total returns above cost of feed and yearling feeder.

After paying for the feeder steer, feed, and buying and marketing costs, a total of $34 potential profit remains, from which must be deducted the following additional costs: labor; interest on investment in steers, feed, equipment, and buildings; depreciation on machinery, equipment, and buildings; death loss; veterinary costs; insurance. If, after paying all of these "out of pocket" costs, any of the $34 remains, this is net profit. These incidental costs are referred to as nonfeed costs and range from 5 to 15 cents per head per day in the feedlot. Many cattle feeders lump all of the incidental costs together and hope the price spread takes care of these items.

When portions of the drylot rations or the pasture are so low in quality that they have no sale value, many farmer-feeders simply do not charge the cattle with these feeds because they would not otherwise be used. Examples are cornstalk fields, meadow aftermath, or permanent pasture that cannot be harvested for hay. The same may be said for labor that is available and would otherwise not be utilized.

Increasingly, feeder cattle, especially calves, are fed on a minus price spread—that is, the cattle sell for less per hundredweight than they cost. This places the entire burden for profit on feeding margin. The relative importance of the two sources of potential profit depends on age, weight, sex, and grade of the feeder cattle, the price of feed, and the relationship between the selling price of cattle and the cost or value of feed and other items that contribute to the total cost of gains. These relationships are discussed further in Chapters 12 and 13.

Table 78 gives some comparisons that illustrate that net returns due to favorable price spreads are generally smaller and proportionately less of the total return for the calf programs than for those programs using

Table 78
Effect of Price Spread and Feeding Margin on Profits in Various Steer-Finishing Programs[a]

Feeding Year	Dollars per 100 Pounds of Steer					Return Above Feed and Steer Cost per Head[d]
	Price Paid	Selling Price	Feed Cost	Price Spread[b]	Feeding Margin[c]	
Long-Fed Good-to-Choice Steer Calves						
1962–63	$29.53	$23.41	$17.51	$ −6.12	$ 5.90	$ 7.17
1961–62	26.56	26.46	15.97	−0.10	10.49	59.76
1960–61	26.48	23.11	16.28	−3.37	6.83	22.80
1960–63 average	**27.52**	**24.33**	**16.59**	**−3.19**	**7.74**	**29.91**
Long-Fed Good-to-Choice Yearling Steers on Pasture						
1962–63	$27.29	$23.11	$19.42	$ −4.18	$ 3.69	$ −6.55
1961–62	25.43	26.98	17.10	1.55	9.88	59.30
1960–61	24.89	22.93	16.46	−1.96	6.47	22.29
1960–63 average	**25.87**	**24.34**	**17.66**	**−1.53**	**6.68**	**25.01**
Long-Fed Good-to-Choice Yearling Steers in Drylot						
1962–63	$27.48	$23.23	$20.09	$ −4.25	$ 3.14	$ −12.88
1961–62	25.61	26.15	17.26	0.54	8.89	45.64
1960–61	25.22	22.45	18.64	−2.77	3.81	3.10
1960–63 average	**26.10**	**23.94**	**18.66**	**−2.16**	**5.28**	**11.95**
Short-Fed Good-to-Choice Yearling Steers						
1962–63	$25.02	$22.08	$19.29	$ −2.94	$ 2.79	$ −9.45
1961–62	24.16	24.45	17.52	0.29	6.93	26.80
1960–61	23.72	22.90	16.81	−0.82	6.09	18.70
1960–63 average	**24.30**	**23.14**	**17.87**	**−1.16**	**5.27**	**12.02**

[a] 1964 Annual Report, Farm Bureau Farm Management Service, prepared by Illinois Agricultural Experiment Station.
[b] Price paid − selling price = price spread.
[c] Selling price − feed cost = feeding margin.
[d] Does not take nonfeed costs into account.

older or heavier cattle. The table also shows that older and heavier feeder cattle produce less net return from the feeding margin. This merely reflects the principle that older cattle convert feeds to gain less efficiently than do calves. The 3-year averages for all programs show a negative price spread but in a few individual years a positive price spread prevailed. It is doubtful if price spread will be consistently reliable as a source of profit in the future when feeding good to choice or better cattle. Programs that must rely on favorable price spreads are much more speculative and carry correspondingly greater risk. This is undoubtedly the chief reason why feeders the country over are feeding more calves, fewer yearlings, and almost no 2-year-olds.

Note in Table 78 that in the feeding year 1962–63 the highest-priced feeder cattle fed, namely the long-fed good to choice steer calves, were the only ones that showed a return over steer and feed cost. They did this in spite of the greatest negative price spread, and because of a higher feeding margin resulting from a combination of lower cost of gain and high selling price.

METHODS OF FINISHING CATTLE

A wide variety of programs are used in finishing cattle for market. Although there is no best method, even for a particular feeding situation, some methods are likely to prove more satisfactory than others when certain conditions prevail.

Successful feeders choose the one program, or combination of programs, that best fits their needs, and they change only when necessary to adjust to changing conditions. Under some conditions a combination of two or more programs offers some special advantages, especially on larger farms. Among the factors to be considered in the choice of feeding program are the following:

1. *Length of feeding period.* Usually cattle that are fed grain or concentrates for fewer than 100 days are called "short-fed" cattle. Ordinarily, less than 25 bushels of corn or similar concentrate will be fed per head to such cattle. On the other hand, if cattle are fed finishing rations for 8 to 10 months, they are spoken of as "long-fed" cattle. These cattle (usually choice or prime steer calves or light yearlings) ordinarily grade high choice or prime after finishing, provided they were of sufficiently good conformation. They will have consumed 75 bushels or more of corn or similar feeds per head. Anything falling about midway between these limits is regarded as a medium feeding period.

2. *Relative amount of grain or concentrates in the ration.* This may vary all the way from a full feed of concentrates (1.5 to 2 pounds for each 100 pounds of body weight daily) to a limited feed (0.5 to 1 pound per 100 pounds of body weight per day). Cattle are full-fed in order to finish them quickly. It may not always be the most economical way to produce a pound of gain but feeders follow this practice so that there will be sufficient finish on the cattle to insure satisfactory grade in the carcass and, of course, to market their grain and shorten the feeding period. Full

FIG. 64. Long-fed cattle such as these near-prime steers assume considerable importance in a few Corn Belt areas but, in general, choice is the grade of preference of most cattle feeders. Remodeling so as to mechanize even the small-sized feedlots in this area is under way in order to reduce labor costs. (Corn Belt Farm Dailies.)

feeding usually follows a period of feeding on growing rations or a period of grazing in farmer-feeder situations.

Limited feeding is usually practiced either during a wintering or grazing period in which principally growth, rather than fattening, is desired, or when a heavy feed of roughages plus some concentrates is fed just prior to a period of full feeding. The latter is done in order to cheapen the gains made during the entire feeding program. Roughages and pastures, when adequately supplemented, make cheap gains, but, when unsupplemented, often scarcely do more than maintain the weight of cattle. This fact accounts for the practice of limited feeding of supplemental high-energy concentrates. When all feeds are purchased, roughage nutrients cost more than concentrate nutrients as well as require more processing and handling labor; hence the very strong trend toward full feeding of high-concentrate rations by commercial feedlots.

3. *Systems of feeding.* Cattle may be fed finishing rations in drylot or on pasture, or both. For example, many cattle, especially calves, may first be wintered on a grower or stocker ration consisting almost entirely of roughage plus, at most, a limited grain feed. They are then fed either a limited feed or a full feed of grain on pasture during spring and summer, after which they are finally finished in drylot on a full feed for about 60 days. This combination of programs furnishes a market for harvested roughages and grains as well as pasture, which in the final analysis is one of the primary reasons for feeding cattle, on farms at least.

4. *Time of year when fed.* For the most part, cattle finished principally in drylot in the farming regions are fed during the winter months, whereas those finished on pasture are naturally fed in the spring and summer or during the pasture-growing season. Most specialized feedyard operators feed cattle in drylot on a year-round basis, and this trend is developing even in the farm areas, to keep expensive equipment in constant use.

THE KIND OF CATTLE TO FEED

Most feeding programs require a certain kind of cattle if they are to be most successful. Conversely, each type of cattle should be fed according to a rather well recognized plan if the cattle are to return the greatest profit. During a period when prices are rising and profits are large, a particular drove of cattle may return a profit almost regardless of the way they are fed, but when the reverse is true, the kind of cattle purchased must be well suited to the plan of feeding if a profit is to be made.

Feeder cattle vary widely in those characteristics that affect their suitability for different feeding conditions and that are therefore important to prospective buyers. Their chief differences are as follows:

1. *Age.* Calves, yearlings, and 2-year-olds.
2. *Condition.* Thin, medium, and fleshy.
3. *Weight.* Light, medium, and heavy. Weight, of course is the result of both age and condition. For example, a calf 9 months old in fleshy condition may weigh as much as or more than a very thin yearling.
4. *Sex.* Steers, bulls, heifers, and cows. Steers may be any age from 4 to 40 months, but feeder heifers are seldom older than 24 months. If older, they probably have produced calves and are called heiferettes.
5. *Breeding.* Beef, mixed, and dairy. Cattle are said to show beef breeding when they show the characteristics of the Angus, Hereford, Shorthorn, or Brahman breeds. A drove of cattle of mixed breeding usually contains at least some animals that have both beef and dairy ancestry. Such breeds as Charolais, Charbray, Braford, Santa Gertrudis, and Beefmaster are nearer

the beef category, but they possess distinctive qualities that perhaps someday will result in a classification of their own.

6. *Grade*. Fancy, choice, good, medium, and common. As discussed in Chapter 13, the grade of an animal indicates its all-round desirability for the use to which it will be put—in the case of a feeder, its desirability for feeding. The grade of a feeder steer or heifer is determined chiefly by its conformation, quality, and breeding.

7. *Origin*. Some cattle feeders attach some importance to the region from which the cattle have been shipped. Cattle from certain areas in Texas and New Mexico are very popular with some Corn Belt feeders. Others maintain that cattle from the mountainous Northwest are more hardy and growthy. It seems doubtful that any preference should be shown for a drove of cattle solely because they come from a certain area. Every important cattle-breeding area of the country produces some excellent cattle and some that are only mediocre. Each drove should be judged on its own merit and not on that of another shipment from the same locality.

CONTRACT FEEDING

Occasionally a large rancher, packer, or cattle speculator enters into a contract with a cattle feeder to finish cattle without a change of ownership. The chief reasons why larger feeders and feedlot operators are increasingly resorting to feeding cattle on contract are (1) the risk is almost entirely eliminated for the feedlot operator, in that the owner of the cattle must take the loss if prices should fall or if, for any of several other reasons, the cattle perform poorly, and (2) less capital is required by the feeder, permitting him to use his own often limited capital to enlarge his operation or even lay in a feed supply in advance as a hedge against seasonal price increases. By becoming involved in a contract, the contract feeder does forego the occasional opportunity to make large profits when prices rise spectacularly, as they do perhaps one year out of five or six; he does this so as to assure a guaranteed income or return from his operation. Some feeders fill only half or a portion of their lots with contract-fed cattle and thus, as it were, play the market both ways.

KINDS OF FEEDER CONTRACTS

Contracts are essentially of two kinds and the choice of which to use depends on factors too numerous to discuss fully here. The two types are:

1. *Gain-in-weight contract*. The feeder agrees to feed a certain ration for a given period of time, or until cattle attain a certain weight. The

owner of the cattle agrees to pay a certain amount per pound of gain. The owner also pays for all medicines and veterinary costs, but the feeder may or may not be paid for the weight gains of any cattle that die. The weighing arrangements should be specifically spelled out for the best interests of both parties.

2. *Feed and yardage contract.* The feeder agrees to provide all feed, labor, and equipment required, and he is paid an agreed-upon price for feed, per ton, as well as a yardage charge, which may either be built into the feed charge or may be a separate item—for example, 7 or 8 cents per head per day. The owner of the cattle specifies the ration, and a minimum length of feeding period is stipulated by the feedlot owner. The owner of the cattle again provides all veterinary service and drugs. Because a feedlot may be feeding several owners' cattle simultaneously, each owner expects an accurate accounting of all feed fed his particular cattle.

Contract forms are obtainable from various sources for both types of contracts, but none will completely cover every situation. Therefore, an experienced lawyer should be consulted. Most owners like to add some type of incentive clause that will provide a motive for better than average care or husbandry of his cattle on the part of the feedlot operator.

Contract feeding has never been popular with farmer-feeders because seldom have both parties been satisfied with results of the arrangement. If prices are rising and the cattle return the owner a handsome profit, the feeder is unhappy that he does not own the cattle himself and thereby receive the owner's profits as well as his own. And if prices are falling and the cattle are sold for less than expected, the owner is dissatisfied when he finds himself taking home less money than he was offered for his feeder cattle the previous fall. Because it takes two parties to make a contract and usually only one is interested in such an arrangement, not much contract feeding is done by the average farmer-feeder.

INVESTMENTS IN FEEDLOTS

The capital investments in a feedlot for acreage, buildings, corrals, feed storage, feed-handling and feed-processing equipment, manure-handling equipment, transportation, and utilities can be enormous. Expressed on a per head capacity basis, the investment may range from as little as $25 in the semiarid Southwest for a 20,000-head lot to as high as $400 in the Corn Belt for a completely automated lot with environmental control for 200 head.

The costs of depreciation, maintenance, taxes, and insurance on the feedlot facility generally make up the major portion of the nonfeed costs

FIG. 65. Commercial feedyards such as this one in the Texas Panhandle sorghum area are responsible for much of the phenomenal increase in numbers of cattle fed in the Southwest. Low investment in feedlot and shelter, combined with commercial-type feed storage and processing, makes low nonfeed costs possible. Choice cattle are common in the feedlots of this area, no doubt a reflection of the proximity of ranches that produce mainly such cattle. (Western Livestock Journal.)

in operating a feedlot. A number of factors determine the amount of these nonfeed costs. The environment, including temperature, relative humidity, rainfall, and drainage, determines to a large extent the type of buildings, if any, and feedlot floor that are required. The type of ration fed is a large factor. For example, silage and high-moisture grain storage can be very expensive, in terms of investment capital at least. Steam-processing of grain requires high-cost equipment, especially if small numbers are fed. Concrete feeding floors and concrete bunks can also be expensive, but nevertheless are an excellent investment in the Corn Belt where seasonally heavy rain and snowfall, coupled with flat terrain and heavy soils that prevent drainage, make them almost essential.

Example data, shown in Tables 79 and 80, for Ohio and California respectively, show something of the magnitude of these costs and, more important, how size of feedlot affects the per head nonfeed costs. In the Ohio report the data are shown on the basis of cost per hundredweight gain. It is evident that on the smaller farms the nonfeed costs amount to $10 per hundredweight gain, or about $37 per head fed. On the larger farms the nonfeed costs amount to $5.70 per hundredweight gain, or about $26 per head. For contrast, the California data in Tables 80 and 81 show both the effect of differences in necessary investments in feedlot facilities

Table 79
Effect of Number of Cattle on Feed on Nonfeed Costs in Ohio Feedlots, 1958–1964[a]

	Average Number of Cattle Fed and Number of Farms			
Cattle	53	123	238	640
Farms	30	72	32	34
Nonfeed costs per cwt gain				
Labor	$ 3.18	$ 1.80	$ 1.23	$ 1.08
Tractor power	0.72	0.38	0.39	0.30
Veterinary and drugs	0.23	0.17	0.13	0.25
Truck, auto, telephone, electricity	0.25	0.29	0.21	0.30
Buildings and equipment	3.51	2.91	2.43	2.18
Taxes and interest	2.11	1.81	1.74	1.59
Total	$10.00	$ 7.36	$ 6.13	$ 5.70
Miscellaneous performance and cost data				
Average total gain (lb)	369	452	469	460
Average number days fed	225	259	261	258
Average daily gain (lb)	1.63	1.72	1.82	1.83
Purchase weight (lb)	581	574	570	551
Sale weight (lb)	950	1,026	1,039	1,011
Building investment per cwt	$24.65	$20.22	$16.07	$14.71
Death loss (%)	1.18	1.33	.86	1.65

[a] Ohio Agricultural Experiment Station Report, 1966.

Table 80
Average Replacement Value of Investments per Head Capacity in California Feedlots, 1964[a]

	Type of Operation			Capacity				
Investment	Farmer-Feeder	Commercial Feeder	Custom Feeder	Up to 4,000	4,000–9,000	9,000–16,000	Over 16,000	State Average
Land	$ 7.84	$ 8.61	$12.05	$15.35	$ 9.81	$17.90	$ 5.82	$10.29
Feedlot	17.84	17.19	18.55	24.22	20.06	16.37	18.06	18.12
Feed mill	20.77	20.14	17.49	27.27	22.00	17.42	18.10	18.87
Feed trucks	3.73	3.35	2.75	5.20	3.38	3.19	2.81	3.11
Other	2.77	3.27	3.47	5.58	5.21	3.77	2.16	3.21
Total except land	$45.11	$43.95	$42.26	$62.27	$50.65	$40.75	$41.13	$43.31
Total with land	$52.95	$52.56	$54.31	$77.62	$60.46	$58.65	$46.95	$53.60

[a] *Cattle Feeding in California*, Bank of America, Economic Research Department, 1965.

Table 81

Average Daily Nonfeed Costs per Head Fed in Eighty-One California Feedlots (Cents)[a]

Cost Items	Type of Operation			Numbers Fed				
	Farmer-Feeder	Com-mercial Feeder	Custom Feeder	Under 4,000	4,000–10,000	10,000–26,000	Over 26,000	State Average
Salaries and wages[b]	3.67	3.10	3.21	4.68	3.52	2.97	3.29	3.27
Taxes, interest, insurance[c]	1.28	1.29	0.82	2.32	1.42	0.93	0.89	1.00
Utilities	0.52	0.45	0.37	0.59	0.51	0.41	0.40	0.42
Gasoline, oil, grease	0.37	0.27	0.24	0.53	0.29	0.26	0.26	0.27
Depreciation	1.41	1.00	0.88	2.59	1.23	0.88	0.93	1.00
Repairs	0.79	0.88	0.81	1.11	0.62	0.73	0.87	0.82
Veterinary fees, medical	0.66	0.29	0.50	0.85	0.41	0.64	0.41	0.49
Nutrition services	0.06	0.02	0.12	0.01	0.05	0.15	0.07	0.09
Legal and accounting	0.07	0.07	0.17	0.05	0.15	0.07	0.16	0.13
Trucking and freight	0.04	0.12	0.08	0.15	0.28	0.02	0.08	0.08
Promotion	0.02	0.05	0.06	0.01	0.04	0.03	0.06	0.05
Other costs[d]	0.22	0.33	0.22	0.28	0.29	0.19	0.26	0.24
Total gross costs[e]	9.11	7.87	7.48	13.17	8.81	7.28	7.68	7.86
Manure credit	−0.43	−0.38	−0.42	−0.43	−0.13	−0.27	−0.54	−0.41
Total net cost per day	8.68	7.49	7.06	12.74	8.68	7.01	7.14	7.45

[a] *Cattle Feeding in California*, Bank of America, Economic Research Department, 1965.
[b] All salaries and wages and payroll taxes.
[c] Property taxes, interest on investment (computed at 6 percent of book value of total investment in feedlot, mill, storage, office, trucks, etc.) and insurance costs.
[d] Other costs included market reports, odor control, rental fees, and livestock commissions.
[e] Death losses are not included. If death losses average 1 percent and loss occurs halfway through feeding period and animal lost weighs 800 lb, is worth 22.5 cents per pound, and feeding period is 150 days, daily cost per head fed for death loss is 1.2 cents. Death loss is a cost that should not be omitted. Because California cattle feeders use lot accounting and charge deaths to the entire lot costs, and because the loss is higher for calves than for yearlings, an average death loss was not included in the analysis.

and the effect of still larger size of lot. The average custom feeder, who bought nearly all his feed and fed few or no cattle of his own, had nonfeed costs of only 7.06 cents per day. The California farmer-feeder's costs were higher in general, but still below all the Ohio categories. It should be mentioned that since California feeders normally feed for shorter periods of time and have a faster turnover rate than those in Ohio, or the entire Corn Belt for that matter, they can reduce their daily nonfeed charges because they are more likely to have the lot full the year around. Their turnover rate was 1.59 for the lots sampled—that is, 1.59 head were fed per space per year. It appears from the California study that the nonfeed cost reduction, due to increasing numbers fed, levels off at about 26,000-head capacity.

The fact that many farmer-feeders use existing buildings already on the farm, either dairy, horse, or beef barns, means that some nonfeed cost items need not be included for them when calculating relative costs of feeding cattle in different areas or in different situations. Remodeling old structures for mechanization may actually cost more than newer, more efficiently designed buildings. The more costly grain storage facilities, such as oxygen-free storage for high-moisture corn or sorghum, in those areas where these crops are grown in surplus, can be offset by lower purchase and transportation costs of grain. No single area of the country has all the advantages or disadvantages when it comes to costs of feeding cattle.

EXPANSION OF COMMERCIAL FEEDYARDS

Most of the increase in numbers of cattle finished in California, Arizona, Colorado, and in the grain sorghum belt is accounted for by the increase in commercial feedyard operations. Operators of these feedyards carry on primarily a custom or contract feeding operation, seldom owning all of the cattle being fed. As a rule the operators of such feedyards feed their own cattle only when the outlook for a substantially favorable price spread is indicated or when vacancies occur because of insufficient patronage. Low per unit cost and low risk are essential for profit in this program. Therefore, the operator must try to keep his feedyard filled to capacity the year around with customers' cattle.

The finishing rations fed in commercial feedyards vary widely, mainly because the crops grown specifically for cattle feed vary, depending on the area, but also because the by-products of commercially grown crops vary still more. Generally speaking, feedyard operators feed rations of two main types, depending mainly on the source and type of roughage fed. The first category of feeders feed primarily drylot rations consisting of small grains, sorghum, or corn, and hay (often dehydrated) or straw, and some dried crop by-products such as sugarbeet pulp, citrus pulp, culled dried fruit, or dried surplus truck crops. The second category includes those who rely heavily on green or wet feeds—either fresh chopped alfalfa, corn, or sorghum, and, in some cases, silages made from these crops. Naturally the green feed or silage is supplemented with varying amounts of concentrates, depending on the level of gain desired. The feeding program of those in the latter category is similar in principle to the stocker program discussed in Chapter 10, at least in the early stages before the cattle are placed on finishing rations containing a higher level of high-energy concentrates. Actually a feedlot may carry on both programs, but the trend seems to be for a lot to specialize in one or the other.

Because most cattle fed in commercial feedyards are fed on a contract or custom basis, it may be interesting to see what the daily per head ration

Table 82

Typical Drylot and Green Feed Rations, with
Costs per Steer Finished[a]

Feed	Price per Ton ($)	Drylot Ration (120 days)		Green Feed Ration (180 days)	
		Pounds Fed	Total Cost ($)	Pounds Fed	Total Cost ($)
Alfalfa hay	27.00	625	8.43	360	4.86
Barley straw	18.00	625	5.63	270	2.43
Barley	45.00	1,155	25.99	540	12.15
Sugarbeet pulp	56.00	336	9.41	180	5.04
Cottonseed meal	66.00	291	9.60	180	5.94
Molasses	21.00	372	3.91	360	3.78
Fortified supplement	120.00	18	1.08	22	1.32
Green feed or silage	9.00	—	—	4,500	20.25
Total feed costs			64.05		55.77
Average feed costs per day			0.53		0.31

[a] *Cattle Feeding in California*, Bank of America, 1956.

cost was in an earlier California survey conducted by the Bank of America. Daily ration costs, such as those summarized in Table 82, must be added to the daily per head nonfeed costs to arrive at a minimum daily charge which an operator of a commercial feedyard must make in order to meet costs when feeds are priced as indicated and when the feedyard is filled to capacity.

Successful commercial feedyards generally have the following characteristics:

1. They are large enough to permit labor specialization; that is, each man does one particular job during an entire day.

2. Precision feed-processing and mixing equipment and mechanical feed-handling equipment are used, with large numbers on feed making such equipment economical and practical.

3. Specialists such as veterinarians, nutritionists, and marketing men are employed, with large numbers on feed again making this practical.

4. Feeds and incidentals are bought in wholesale quantities, often with competitive bids.

5. Buyers for customers' cattle buy many cattle direct from the feedyards, partly because of the large numbers available and, in some cases,

because of the reputation for "killing well" that cattle from a particular yard may have.

It is doubtful if the commercial feedyard will assume as important a role in the traditional cattle-feeding areas of the Corn Belt as it has in the West and Southwest. Most feedyards presently found in the Corn Belt are not operating on a custom basis. Furthermore, they are built around the utilization of some special by-product such as the corn cobs and off-shaped kernels produced by a seed corn producer or the cannery waste resulting from the processing of sweet corn or peas. Individual farmer-feeders are increasing the size of their operations, but they will be limited by the roughage-producing capacity of their cropping systems. Naturally such farmer-feeders are as desirous of reducing their fixed, per head, non-feed costs as are the commercial feedyard operators.

Chapter 12

The IMPORTANCE of
AGE and SEX in
GROWTH and FINISHING

Changes in consumer demand and in production techniques practiced by cattle breeders and feeders alike have been responsible for a gradual shift to younger and lighter slaughter cattle. How far this shift will go is hard to determine with certainty, because there are factors that tend to counter this trend.

At present the average weight of slaughter steers is near 1,000 pounds for the country as a whole, but in certain areas steers weighing 875 to 925 pounds are in greatest demand. Heifers generally go to market at least 100 pounds lighter than steers. Naturally, as the carcass weights of slaughter cattle change, the age at which cattle are placed on feed and finally slaughtered changes too. Whereas as recently as 1940 the majority of the cattle on finishing rations were yearlings or older, today well over half are calves. This is especially true in areas such as California, Arizona, and West Texas where expansion in cattle feeding has been greatest. Even in the traditional cattle-finishing state of Iowa a recent survey showed that 43 percent of cattle feeders fed calves.

More heifers are being fed than formerly, also, as may be seen in Table 83, which shows that heifer beef production more than doubled during the 20 years following 1946. This reflects the fact that a smaller portion of the heifer calf crop is kept back for replacements or to build new herds, and that those not kept for breeding purposes are being grain-fed rather than sold as slaughter calves or grass-fat heifers. These changes add to the need for knowledge concerning the effect of age and sex of feeder cattle in finishing programs.

Table 83
Estimated Composition of Beef Production, 1946–1966[a]

Year	Total Production[b] (mil. lb)	Beef Production by Class Steer (%)	Heifer (%)	Cow[c] (%)	Fed Beef Quantity (mil. lb)	Percentage of Total
1946	9,373	52	10	38	3,427	37
1950	9,534	57	9	34	4,446	47
1954	12,963	55	12	33	5,319	41
1958	13,330	60	15	25	6,760	51
1962	15,298	61	20	19	9,896	65
1966	19,694	56	24	21	13,207	67

[a] U.S.D.A., *Livestock and Meat Situation*, 1967.
[b] Includes production for farm consumption.
[c] Includes bull and stag beef, which averaged about 2 percent of total annual supply.

EFFECT OF AGE ON RATE OF GAIN

A discussion of the effect of age on the daily gains made by cattle must be treated from two standpoints: (1) its effect on cattle that have been sufficiently well fed from birth to permit growth and fattening at the same time, and (2) its effect on cattle that are placed on a finishing ration after having been grazed or fed for some time on a nutritional plane that produces only growth. Inasmuch as cattle make their most rapid growth during the first year and grow more slowly as they approach maturity, it is evident that in cattle fed a full feed from weaning onward the rate of gain varies inversely with age. Young purebred cattle that are being developed for show purposes and are fed liberally from the time they are old enough to eat, gain about 70 percent as much the second year as they do the first and 50 percent as much the third year as they do the second.

An entirely different situation exists when thin cattle of different ages are placed in the finishing pen and fed liberally for perhaps the first time in their lives. In this case the larger capacity of the older steers gives them a decided advantage over the younger cattle, with the result that they gain much more rapidly. Here the daily gains vary directly with the age of the animals, instead of inversely, as shown in Table 84.

To illustrate the benefit of the application of new research findings, notably the use of vitamin A, antibiotics, and diethylstilbestrol, the data

Table 84

The Effect of Age on Rate of Gain Made by Feeder Steers on Full Feed[a]

	Calves		Yearlings		Two-Year-Olds	
	Days Fed	Average Daily Gain (lb)	Days Fed	Average Daily Gain (lb)	Days Fed	Average Daily Gain (lb)
Indiana	270	1.93	180	2.38	180	2.57
Indiana	270	1.82	210	2.06	180	2.65
Indiana	270	1.93	210	2.24	180	2.27
Ohio	175	2.19	147	2.42	119	2.81
Ohio	182	2.23	154	2.32	126	2.65
Iowa	240	2.22	180	2.64	120	2.76
Iowa	240	2.32	160	2.47	150	2.48
Average of 7 trials	235	2.09	177	2.36	151	2.60

[a] All trials were conducted before advent of the use of stilbestrol; consequently, gains are approximately 10 to 20 percent below presently accepted standards.

from a recent Indiana study are shown in Table 85. The differences between the rates of gain of calves and yearlings remain the same in magnitude as those shown in the older studies in Table 84—namely, about 0.2 pound per day. However, for the full-fed lots of the recent study, both calves and yearlings, the rates of gain are about 0.4 pound greater than those observed earlier. This more rapid gain resulted in both a shorter feeding period and more efficient gains, as can be seen by comparing the feed requirements in the recent Indiana trial and the summary of the older trials shown in Table 86. It is of course possible that the genetic growth potential of the cattle was sufficiently different to explain all of the differences, but this is unlikely. The shorter feeding period itself would explain part of the more rapid gain.

EFFECT OF AGE ON ECONOMY OF GAINS

Numerous experiments have shown that young animals are more efficient in feed conversion than are older cattle. The principal explanation is found in the fact that increases in the body weight of young animals

Table 85

Effect of Age and Energy Level on Performance of Beef Steers[a]

	Calves		Yearlings	
Energy Level	Low	High	Low	High
Number of calves	19	18	19	19
Average initial weight (lb)	448	450	745	746
Average final weight (lb)	1,072	1,084	1,145	1,164
Days fed	305	250	182	154
Average daily gain (lb)	2.04	2.53	2.19	2.70
Daily feed (lb)				
Ground corn	2.0	11.8	2.0	15.1
Supplement A	2.0	2.0	2.0	2.0
Corn silage	38.8	15.9	51.7	19.4
Feed per cwt gain				
Air-dry feed (lb)	849	654	1,001	748
Total TDN (lb)	484	539	557	626
Feed cost ($)	15.10	16.10	16.50	18.00
Carcass data				
Hot weight (lb)	649	654	679	698
Loin-eye area (sq in.)	11.00	10.39	12.25	11.71
Backfat thickness (in.)	0.55	0.78	0.63	0.54
Number of choice carcasses	9	1	4	5

[a] Indiana Agricultural Experiment Station Research Progress Report 113, 1964.

Table 86

Effect of Age of Cattle on Amounts of Feed Required to Attain Choice Grade[a]

	Calves		Yearlings		Two-Year-Olds	
Feed	Pounds	Bushels or Tons	Pounds	Bushels or Tons	Pounds	Bushels or Tons
Shelled corn	2,735	48.8	2,696	48.1	2,811	50.2
Protein concentrate	452	0.23	405	0.20	372	0.19
Dry roughage	1,139	0.57	1,183	0.59	1,203	0.60

[a] Average of the seven trials appearing in Table 84

Table 87
Influence of Age on Variation in Economy of Gains
Throughout the Feeding Period[a]

	Feed per Hundredweight Gain (lb)			
Period	Three-Year-Old Steers	Two-Year-Old Steers	Yearlings	Calves
First 100 days				
Corn	586	597	534	431
Alfalfa	363	346	304	221
Second 100 days				
Corn	1,282	1,085	916	623
Alfalfa	320	272	219	148
Total 200 days				
Corn	835	798	702	529
Alfalfa	350	314	266	186

[a] Nebraska Bulletin 229.

are due partly to the growth of muscles, bones, and vital organs, whereas the body increases of older cattle consist largely of fat deposits. Fat contains much less water and a great deal more energy than an equal weight of any other kind of animal tissue. Thus more feed is required for its formation. Other factors contributing to the more economical gains made by young animals are (1) a slightly larger consumption of feed in proportion to body weight and (2) the much greater maintenance requirement of the older cattle, especially when expressed on a daily basis. As a result of their relatively large feed consumption, younger cattle use a smaller percentage of the ration to satisfy maintenance requirements than is used for this purpose by older animals, thus making available a greater percentage of the total feed consumed for the production of growth and fat.

The more efficient gains made by young cattle become more and more evident as the feeding period progresses. This fact is well illustrated in Table 87, which shows the results obtained at the Nebraska station with steers of various ages fed for a period of 200 days. Whereas in calves the total feed required per 100 pounds of gain during the last half of the period was only 18 percent above that required during the first, in the yearling, 2-year-old, and 3-year-old steers it was 35, 44, and 69 percent greater, respectively. These figures bear out the fact that calves may be fed profitably over a longer period than older cattle. If necessary, they can be held for a considerable length of time, awaiting a favorable market,

Table 88

Effect of Length of Feeding Period and Finish on Cost of Gain of Choice Steers Fed in Drylot[a]

Gain	400-lb Calf	640-lb Yearling	840-lb Two-Year-Old
First 100 lb	$ 13.05[b]	$ 15.75	$ 16.05
Second 100 lb	14.55	18.45	19.65
Third 100 lb	16.35	21.45	25.20
Fourth 100 lb	18.60	26.25	35.10
Fifth 100 lb	23.40	33.60	$24.45 for 50 lb
Sixth 100 lb	26.10	—	—
Seventh 100 lb	32.85	—	—
Total gain (lb)	700	500	450
Total cost	$144.90	$115.50	$120.45
Average cost per cwt gain	$ 20.70	$ 23.10	$ 26.77

[a] Data from U.S.D.A. Technical Bulletin 900.

[b] Based on corn at $1.50 per bushel, with other feeds comparably priced.

and all the while they will be making fair gains at not too great a cost. Mature steers, on the other hand, are held at great expense. After being on full feed for 150 or 160 days, they gain very slowly and inefficiently.

The fact that feeding cattle to excessive finish is uneconomical is further demonstrated by the data in Table 88. It will be noted that, although the total investment in feed required to finish a calf is higher than for the older steers, the cost of gain is less, making it more likely to realize a profit from the feeding margin, as indicated earlier.

A practical question arising in the minds of cattle feeders relative to age of feeder cattle, is whether to buy the earlier and thus heavier calves from a rancher or whether to buy the later and therefore lighter and younger calves. An Oklahoma study sheds some light on this question, and the pertinent results are shown in Table 89. The calves were all half-brothers, but there was a 15-day average difference in ages. Despite the fact that the heavier calves outweighed the lighter ones by 70 pounds at the start of the test, they gained equally as well as the lighter ones. The phenomenon of compensatory gain usually seen in lighter cattle did not occur, possibly because all calves had been creep-fed and thus were probably in the same condition or degree of fatness. The advantage held by the lighter calves with respect to feed efficiency would have justified paying an extra dollar per hundredweight for them as weaners. Before a feeder pays a premium for light calves, he should know that they are light because they are young or undernourished.

Table 89

**Effect of Age and Weight of Calves on Subsequent
Performance in the Feedlot When Full-Fed[a]**

	Weight Group		Difference
	Heavy	Light	Heavy — Light
Number of calves	104	107	
Initial age (days)	219	207	12
Initial weight (lb)	514	444	70
Final age (days)	398	386	12
Final weight (lb)	942	866	76
Gain on test (lb)	428	422	6
Average daily gain (lb)	2.39	2.36	0.03
Dressing percent	63.1	62.4	0.7
Carcass grade[b]	10.7	10.9	−0.2
Feed per pound of gain (lb)	9.80	9.34	0.46

[a] Oklahoma Agricultural Experiment Station Miscellaneous Publication 67, 1962.

[b] Low choice = 10; average choice = 11; high choice = 12.

EFFECT OF AGE ON LENGTH OF FEEDING PERIOD

It has been mentioned that young cattle can be fed longer with profit than old cattle. If feeder steers of different ages are started on feed in approximately the same state of flesh or condition, the time required to finish each age of feeder to the same grade varies inversely with the age of the cattle. When cattle are finished for their grade, they should be marketed as soon as possible. To hold them longer will require both feed and labor for which little will be realized, either from increased weight or improved carcass merit.

On the basis of experiments such as that reported in Table 90, as well as the experiences of practical feeders, it may be stated that 4 to 5 months of heavy feeding are required for yearling feeder steers to be put in choice condition, and from 8 to 9 months for calves. It should, of course, be understood that large numbers of cattle are marketed after a relatively short feed before they are really finished. The increasing demand for leanness in beef cuts may make short feeding more practical, but it must be recognized that with short-fed cattle profits must come from favorable price spread. Short-fed cattle do not provide a market for much feed, and a drastic shift to this program by large numbers of feeders would tend to increase the demand for the lower grades of feeders commonly

Table 90
Feedlot Performance and Carcass Merit of Steers Fed to
Different Market Weights[a]

Age of Steers	Long-Fed Yearlings	Medium-Fed Yearlings	Long-Fed Calves	Medium-Fed Calves
Number of steers	10	10	10	10
Days on feed	308	140	392	266
Average weight (lb)				
Initial	688	688	463	460
Final	1,442	1,114	1,246	1,057
Average daily gain (lb)	2.45	3.04	2.00	2.25
Average daily feed (lb)				
Corn silage	16.84	19.10	17.45	18.58
Corn, high-moisture	17.92	16.52	12.52	10.89
Soybean meal (50% protein)	1.25	1.25	1.25	1.25
Vitamin A supplement	0.50	0.50	0.50	0.50
Feed per pound of gain (lb)				
(85% dry-matter basis)	9.67	7.67	9.63	8.13
Feed cost per cwt gain[b]	$17.89	$13.93	$17.35	$14.36
Dressing percent	66.5	63.4	65.8	61.5
Fat thickness over 12th rib (in.)	1.43	0.60	1.01	0.51
Loin-eye area (sq in.)	11.75	10.38	12.14	10.90
Carcass grade[c]				
Conformation	20.6	19.2	21.4	20.3
Quality	20.6	19.1	20.8	19.8
Overall	20.4	18.9	20.8	19.7
Total retailable product, percent				
of carcass weight	55.2	67.0	56.4	61.5

[a] Illinois Cattle Feeder's Day Report, 1966.
[b] Feed cost per ton: corn, $40; corn silage, $8; haylage, $20; (85 percent dry-matter basis); soybean meal, $80; vitamin A supplement, $50.
[c] Low choice = 19; average choice = 20; high choice = 21.

used. Thus feeder cattle costs would be too high to permit the necessary price spreads.

EFFECT OF AGE ON TOTAL GAIN REQUIRED TO FINISH

There is not a great deal of difference in the increase in weight that must be realized by feeder steers of different ages to attain the same degree of finish. If the cattle are in equal condition when put on feed, the total gain necessary to finish decreases slightly with advanced age. In general this gain is approximately as shown:

Age	Total Gain Necessary to Finish (lb)
Two-year-olds	300–400
Yearlings	400–500
Calves	450–550

When expressed in terms of the ratio of initial weight as feeders to final weight, however, the differences in total gains are significant. Calves practically double their weight while in the feedlot, yearlings increase in weight approximately 70 percent, and 2-year-olds from 30 to 40 percent. Obviously these percentages are greatly affected by the condition of the steers when started on feed. The older the feeder, the smaller the fixed nonfeed costs per head. However, if a feedlot is kept constantly filled with cattle the year around, nonfeed costs will still be lower for calves because some items of cost, such as interest, are related to size of animal or investment per animal. If profits are being made from feeding margin rather than price spread, then the age of cattle that requires the greatest amount of gain to reach desired grade is favored.

EFFECT OF AGE ON TOTAL FEED CONSUMED

The age of the cattle has comparatively little effect on the total amount of feed required to attain a given degree of finish, provided all ages are given a full feed of grain and a reasonable amount of high-quality roughages throughout the feeding period. The longer feeding period of the younger animals tends to make up for their smaller daily consumption of feeds, so that by the time they reach the desired finish they have eaten about as much grain, protein concentrate, hay, and other feeds as older cattle would have eaten in a shorter time. Consequently the age of the cattle has relatively little effect on the number of animals that should be purchased to utilize a given amount of feed, assuming, of course, that the feed supply is equally well suited for finishing calves, yearlings, and 2-year-old steers. (See Table 86.)

EFFECT OF AGE ON QUALITY OF FEEDS USED

Although age of cattle has little effect on the amounts of grain, protein concentrate, and good-quality hay required to attain a good or choice finish, it greatly influences the possibility of limiting the grain ration to something less than a full feed and supplying more roughage to take its place. Such a practice may be fairly satisfactory with long yearlings and 2-year-olds, but not with calves, which lack the capacity for large amounts of bulky feeds. Two-year-old steers may be fed large amounts of silage or given free access to good-quality hay without depressing their consumption of grain below the amount required to produce a satisfactory finish, whereas such a practice with calves would be almost certain to reduce the grade of the finished calves.

With the advent of acceptable high-energy rations—that is, rations with less roughage and a higher caloric density than formerly—feeding programs for calves are being used in which they consume as much energy as older cattle, expressed on the basis of energy or calories per hundred pounds of body weight. The result is that calves can be fed even more efficiently, as compared with older cattle, than formerly. This entire subject needs more research, but if the feeding of high-energy rations to calves does not result in excessive fat deposition and therefore does not reduce cutability or yield of lean retail cuts, then we shall surely see the trend toward feeding more calves than yearlings accelerated still further.

EFFECT OF AGE ON PASTURE UTILIZATION

Because of their limited capacity for feed, yearlings finish less rapidly on pasture than do older cattle. (There are relatively few calves available for grazing in the spring because most of the calves born the previous year are yearlings by then.) Fresh grass is a bulky feed with a high moisture and low energy content. Therefore yearlings scarcely get more digestible nutrients from pasture than they need for maximum growth, leaving little for the improvement of condition. Two-year-old steers, on the other hand, are almost full-grown and nearly all the nutrients consumed above maintenance requirements are available for the production of fat. For this reason older cattle are more popular than yearlings in the better grazing areas of the United States and in Argentina and New Zealand where many cattle are sold for slaughter directly off grass without the feeding of any grain. Even when grain is fed on pasture, yearlings finish less satisfactorily than 2-year-olds because they tend to eat too much of the green pasture forage and less of the high-energy grain. This does not mean that yearlings make less gain per acre on pasture than 2-year-olds, but their gain is in the form of growth rather than fat.

EFFECT OF AGE ON CAPITAL INVESTMENT

Although feeder calves usually cost $2 to $4 a hundred more than yearling feeders of the same grade and $3 to $5 a hundred more than

2-year-olds, they cost much less per head because of their lighter weight. This is a factor of great importance when feeder cattle are high in price. For example, a 400-pound calf at 30 cents a pound will cost $120, whereas a 650-pound yearling at 26 cents will cost $170 and a 900-pound 2-year-old at 24 cents will cost $215. Thus $21,500 would be required to buy 100 of the 2-year-old cattle, but only $17,000 and $12,000 would be needed for the same number of yearlings and calves respectively.

Calves, on the other hand, tie up capital in cattle and feed for a longer period, as they must be fed for a longer time. They also require a slightly higher capital investment in buildings and equipment, because they require somewhat better shelter than do older cattle.

EFFECT OF AGE ON FLEXIBILITY OF FEEDING AND MANAGEMENT

Young cattle have a marked advantage over older cattle with respect to suitability for a greater variety of feeding programs. Calves may be put on feed immediately or carried on roughage and pasture for 5 to 12 months before being started on a finishing ration. Yearlings, too, may be roughed through the winter but are less satisfactory than calves for such a feeding plan because their gains are relatively slow and expensive. Two-year-old steers are usually started on feed soon after their arrival at the farm or feedlot, as they will make little gain on feeds that provide only enough digestible nutrients for maintenance and growth. Because calves have a long growing period, they may be held beyond the marketing date originally selected more safely than older cattle, if such a change in feeding plans seems advisable.

EFFECT OF AGE ON CARCASS COMPOSITION

From an industry standpoint, the age at which to slaughter beef cattle is when maximum muscling has occurred but before excessive fattening renders carcasses wasteful and feed efficiency poor. Unfortunately not all parts of the animal body mature or reach maximum development at the same age, and the problem is made more difficult because there are wide variations between breeds of cattle, and even within breeds, in this respect. In other words, the chronological age at which cattle reach physiological maturity is not constant or fixed for the species. Some of the British breeds reach the point where yield of lean cuts begins to decline, on a percentage basis, considerably sooner than do the dual-purpose or dairy breeds. Thus there is a genetic basis for this variation, which means that, through selection, age of maturity can be reduced within a breed—or, if the opposite condition is desired, crossing with a larger, slower maturing breed can be used.

It would seem that the trend among breeders would be toward slower maturity and thus larger cattle. Certain items of nonfeed costs that are determined by the number of cattle fed rather than weight or size—such as labor, taxes, and grazing fees on leased land—discourage the use of smaller, rapidly maturing cattle. One problem associated with deciding when to stop feeding a drove of cattle in order to ensure maximum cutability or meatiness and yet not stop too soon is knowing body composition of the live animal. Kansas workers have conducted an interesting study with 64 half-brother Angus steer calves that sheds some light on this question. The pertinent data are shown in Table 91. One group of 8 steers was slaughtered at the start of the test when the calves weighed 351 pounds. Then, after feeding on a finishing ration for 56 days and each 28 days thereafter, further groups of 8 steers were slaughtered. Skeletal, organ, and muscle weights and muscle areas were obtained as measures of growth. Ether extract (fat) within the muscle was chemically determined to measure degree of finish or quality, the principal determinant of carcass grade.

Although this study represents only one breed, and evidently an early-

Table 91
Effect of Age and Length of Feeding Period on Performance and Body Composition of Beef Steers Fed a Finishing Ration (0–224 Days)[a]

Group	Age (days)	Days Fed	Daily Gain (lb)	Slaughter Weight (lb)	Carcass Weight (lb)	Carcass Grade	Yield Grade[b]	Percent Trimmed Cuts[c]	Total Bone (lb)	*Longissimus dorsi* Weight (lb)	Percent Ether Extract
1	240	0	—	351	188	Good	2.25	51.7	19.6	3.0	5.6
2	296	56	2.35	447	255	Good	2.28	51.6	21.3	4.0	6.8
3	324	84	2.42	493	298	Good	2.35	51.4	23.3	4.5	7.9
4	352	112	2.28	525	328	Good	3.02	50.0	25.5	5.0	11.5
5	380	140	2.37	631	391	Good	3.24	49.5	29.5	5.5	13.3
6	408	168	2.38	682	431	Good	3.12	49.8	30.7	6.0	13.8
7	436	196	2.50	785	488	Choice	3.70	48.4	32.7	6.4	20.1
8	464	224	2.32	835	522	Choice	3.84	48.1	31.9	6.8	23.9

[a] Kansas Agricultural Experiment Station Bulletin 507, 1967.
[b] Yield grade or cutability score, determined by using a U.S.D.A. equation that accounts for variations in fat thickness, percent kidney, pelvic, and heart fat, carcass weight, and area of rib eye or *Longissimus dorsi*. Larger values mean lower yield of edible product.
[c] Estimated percent of carcass weight in boneless, closely trimmed retail cuts from the round, loin, rib, and chuck.

maturing one at that, the data show that growth occurred rather steadily to about 14.5 months of age. Other data not shown indicated that the muscles of the round region followed this pattern, but note in Table 91 that the loin or rib-eye muscle was still increasing in weight when the test ended. As skeletal and muscle growth tended to decelerate, fat deposition within the muscle increased rapidly as reflected in the improvement in grade. The reduction in yield grade is largely a result of fat deposition outside the muscle—that is, on the outside of the carcass (bark)—and in depot fat regions such as the kidney area. The percent edible portion of the carcass decreases steadily with fat deposition.

The Kansas workers concluded that the optimum slaughter age for their test steers was 14.5 months at a weight of 785 pounds, and after a feeding period of 196 days. Feed efficiency was not reported but, based on other data, this age also corresponds closely with the age at which cost of gains increases rapidly. On the side of feeding for a slightly longer period it should be mentioned that a certain amount of apparent excess fat cover is required to prevent excessive shrink in the packer's cooler and to ensure longer shelf-life for the retailer.

DECIDING ON THE AGE OF CATTLE TO FEED

It should be apparent after the foregoing discussion that age is often a very important factor in the purchase of feeder cattle. For example, calves are unsatisfactory for a short feed and should not be bought by the feeder who plans to market his cattle after feeding them only 4 or 5 months. Similarly, 2-year-old steers should not be bought in the fall by the man who expects to carry his cattle into the late summer or fall of the next year, because their gains will be very expensive if they are fed that long.

Two-year-old steers are much better than calves or yearlings for utilizing large quantities of roughage such as silage and sorghum fodder because they have greater digestive capacity. On the other hand, yearlings are better than either calves or 2-year-olds for utilizing stalk fields and late fall pastures, because they are old enough to use such feeds to advantage yet not too old to make gains in the form of growth while on rations that are insufficiently abundant or nutritious to promote the formation of fat.

Some men buy cattle of the age and weight that can most easily be made ready for market by the time they expect the highest prices for finished cattle the following year. This is ideal in theory, but often too many cattlemen make the same "guess" as to when the market will be high, and the increased number of cattle then going to market at the same time causes unsatisfactory prices after all. As a rule, the better plan is to purchase cattle of the age that is able to utilize the available feeds to best advantage and that will be in condition to sell advantageously when the feed supply is gone or when selling is necessary to avoid a labor shortage in the fields. Table 92 contains data collected under average farm conditions in a Corn Belt state, which can be used to estimate potential profit or loss from a feeder program using steers of various ages as well as heifer calves. Actual feeder and feed costs existing at a certain location or time may be substituted for those used in the table to calculate necessary selling or break-even price to pay all but nonfeed costs.

Table 92
Economic Data on Various Ages and Sexes of Feeder Cattle Fed to Good to Choice Grade[a]

	Steers			Heifer
Item	Calves	Yearlings	2-Year-Olds	Calves
Average purchase weight (lb)	420	600	860	400
Average daily gain (lb)[b]	1.6	1.7	1.9	1.5
Average total gain (lb)	580	500	310	440
Average sale weight (lb)	1,000	1,100	1,170	840
Feed per cwt gain				
Corn (bu)	7.7	11.0	13.0	8.0
Supplement (lb)	40.0	59.0	61.0	47.0
Silage (lb)	193.0	261.0	462.0	173.0
Hay (lb)	294.0	307.0	251.0	410.0
Cost[c]	$ 16.85	$ 21.68	$ 24.46	$ 18.25
Financial statement				
Purchase cost	$ 92.00	$126.00	$163.00	$ 72.00
Saleable feed cost, total	73.00	88.00	66.00	57.00
Roughage cost	25.00	20.00	10.00	23.00
Death loss cost	2.00	1.00	—	2.00
Total cost	$192.00	$235.00	$239.00	$154.00
Break-even price	$ 19.20	$ 21.36	$ 20.42	$ 18.33

[a] These data were constructed from several hundred Farm Bureau Farm Management Service Cooperator records in Illinois in a recent year and are representative of feeding and management programs used on many Corn Belt farms.
[b] The gains shown are calculated on the basis of purchase and sale weights and thus include shrink on both ends.
[c] Based on the following feed costs: corn, $1.25 per bushel; supplement, $90 per ton; silage, $10 per ton; hay, $20 per ton.

IMPORTANCE OF SEX IN CATTLE FEEDING

On the basis of sex differences, feeder cattle may be either steers, heifers, cows, or bulls. From the standpoint of numbers, steers are more important than the other three classes combined and are the only kind of cattle to be found in fairly large numbers throughout the year. Table 83 shows the composition of the annual United States cattle slaughter by class or sex condition. It will be seen that steers make up over half of the total and that two-thirds of the cattle now slaughtered are classed as fed cattle. It is a safe assumption that a considerably larger percentage of the steers are fed before slaughter than are the other classes. The trade in feeder heifers assumes considerable volume only during the fall and early winter months. Breeding-age bulls are usually as valuable right off grass as after feeding awhile. Consequently these bulls are usually sold at the end of the breeding season to be used as ground beef or processed meats.

Yearling heifers are better suited for a short rather than long feed. Many such heifers have probably been bred before being sold and are likely to show unmistakable signs of pregnancy if fed longer than 90 to 120 days. Even if evidence of pregnancy is not a question, it is seldom profitable to feed heifers for as long as steers of the same condition and quality because the market does not require female beef to be in such high condition as steer beef.

The most attractive heifer carcasses are produced by animals weighing 700 to 900 pounds, showing good condition and finish, but not so fat as to be wasteful. Heifers of this description sometimes sell for nearly the same price as steers, and loads of mixed yearling steers and heifers often sell without sorting. Animals in choice condition at this weight, if started on feed as calves, are 12 to 15 months old, when there is little likelihood of the females' being pregnant.

Heifer calves finish a little earlier than steer calves. This, together with the fact that heifers need not be so well finished as steers, means that heifers are ready for the market 6 to 10 weeks before steers started on feed at the same time.

The gains made by heifers while on feed are somewhat smaller and more costly than those made by steers, as shown in Table 93, because of the slower rate of growth of heifers. As previously discussed, the largest and cheapest gains are secured during the period when the growth rate is most rapid. From this one would expect that, as between two animals, the one having the higher growth rate would show both more rapid and more economical gains. But this point has little practical importance because steers are usually fed several weeks longer than heifers. Thus any advantage

Table 93
**Comparative Feeding Qualities of Steer and Heifer Calves
When Both Are Fed the Same Length of Time**

| | Minnesota | | Missouri | | | |
| | | | Full-Fed in Drylot | | One-half Grain Ration in Winter; Full-Fed on Grass in Summer | |
	Steer Calves	Heifer Calves	Steer Calves	Heifer Calves	Steer Calves	Heifer Calves
Days fed	217	217	182	182	322	322
Initial weight (lb)	451	449	359	358	358	354
Average daily gain (lb)	2.35	2.27	2.16	1.94	1.84	1.64
Average daily ration (lb)						
Shelled corn	12.1	12.1	8.0	8.0	8.3	8.0
Protein concentrate	2.0	2.0	1.1	1.1	1.2	1.1
Alfalfa hay	1.6	2.1	3.2	3.2	1.8	1.9
Corn silage	7.6	8.9	8.7	8.9	5.5	5.5
					(Pasture)	(Pasture)
Feed per cwt gain (lb)						
Shelled corn	517	535	369	410	452	483
Protein concentrate	85	88	53	58	65	69
Alfalfa hay	69	93	115	162	97	118
Corn silage	322	391	400	450	297	336
Return above cost of feed	$9.43	$2.60	$8.30	$8.16	$38.43	$21.14
Dressing percent	—	—	57.3	59.1	59.5	60.3

they may have over heifers in rate and economy of gains at the time the heifers are marketed is likely to disappear during the remaining weeks they are continued on feed, when their gains are both slow and expensive relative to those made earlier. Indeed it frequently happens that heifers make larger average daily gains and also cheaper gains during a period of 5 or 6 months than do steers during a feeding period that is 2 or 3 months longer, as shown in data from the Illinois station in Table 94.

A recent study by Tennessee workers, using Angus calves purchased in auction sales and all approximately the same in weight, shows that grade of calves changes some of the relationships mentioned earlier. In Chapter 13 the effect of grade on performance is discussed in detail, but note in Table 95 that heifers of varying feeder grade tend to grade more nearly alike as slaughter cattle than do steers. The result of this difference is that lower grades of heifers are often more profitable than lower-grade

Table 94

Comparison of Steers and Heifers When Marketed as Required Finish Is Attained[a]

	Iowa (Average of 3 Tests)		Illinois	
	Steer Calves	Heifer Calves	Steer Calves	Heifer Calves
Number of days fed	240	170	200	140
Average initial weight (lb)	415	406	379	379
Average daily gain (lb)	2.18	2.18	2.35	2.56
Average daily ration (lb)				
Shelled corn	10.7	9.1	10.1	9.3
Protein concentrate	1.5	1.5	1.5	1.4
Alfalfa hay	4.9	5.4	2.0	2.0
Corn silage	—	—	8.1	8.2
Feed per cwt gain (lb)				
Shelled corn	492	433	428	363
Protein concentrate	65	65	63	54
Alfalfa hay	226	247	85	78
Corn silage	—	—	343	319
Date marketed	July 18	May 6	July 13	May 14
Return above feed cost per head	$11.70[b]	$0.78[b]	$23.98	$10.94

[a] Mimeographed reports of calf-feeding experiments.

[b] Average of last 2 years; first year discarded because of abnormal market conditions.

steers, because the spread between grades of feeder steers is narrower than that for feeder heifers. Note in Table 95 that rib-eye area, expressed as square inches per hundredweight of carcass, favored the heifers in all cases. This measurement is often used as an indicator of overall meatiness and cutability. The apparent muscularity of the heifer carcass is offset by a greater thickness of outside fat.

A more important factor than the relative rate and economy of the gains made by steers and heifers is their relative price when purchased and marketed. In the Tennessee study the heifers sold for a price much nearer the purchase price than did the steers. Thus they had much less negative price spread to overcome in order to show a profit. The medium heifers showed a slight positive price spread, whereas the choice steers were faced with a $5.03 negative price spread.

Steers usually are preferred to heifers for roughing through the winter and grazing on pasture the following summer, because they grow more and therefore make larger gains on nonfattening feeds. Moreover, yearling heifers, after a summer on pasture and a short period of grain feeding in the drylot, command lower prices than steers because of the competition from the larger number of grass-fat heifers that are marketed in the fall. Short-fed steers do not experience serious competition from grass-fed steers, as most of the grass-fed steers are sold for further feeding rather than for immediate slaughter.

Table 95
Performance of Feeder Cattle of Different Grades and Sexes[a]

	Steers			Heifers		
	Choice	Good	Medium	Choice	Good	Medium
Number of animals	30	30	30	30	30	29
Average days on feed	203	203	204	179	179	179
Average weight and gain per head (lb)						
Initial weight	479	488	492	485	481	488
Final weight	892	917	916	807	830	821
Total gain	413	428	425	321	349	333
Daily gain	2.04	2.12	2.10	1.79	1.95	1.86
Financial statement						
Feed cost per cwt gain ($)	16.10	15.70	16.37	16.50	15.74	16.26
Initial value per cwt ($)	28.58	27.78	25.18	27.45	25.24	23.05
Final value per cwt ($)	23.55	23.38	22.53	23.75	23.67	23.23
Return per head over feed cost ($)	4.65	9.76	11.63	4.27	19.29	22.96
Final slaughter grade[b]	12.5	11.7	10.8	12.4	12.1	11.4
Carcass data						
U.S.D.A. grade[b]	12.3	12.1	10.9	11.7	11.7	10.9
Hot carcass weight (lb)	544	553	545	495	502	495
Dressing percent	60.2	59.5	58.8	60.8	60.0	59.9
Rib-eye area (sq in.)	11.22	11.23	10.97	10.88	11.06	11.41
Rib-eye area per cwt carcass (sq in.)	2.06	2.03	2.01	2.20	2.20	2.31
Marbling score[c]	6.1	5.7	5.2	5.6	5.5	5.1
Fat thickness (in.)	0.50	0.42	0.37	0.53	0.48	0.46
Shear force, 1-in. core (lb)	15.51	15.34	15.17	15.24	14.59	15.49

[a] Tennessee Agricultural Experiment Station Bulletin 381, 1964.
[b] Average good = 10; high good = 11; low choice = 12; average choice = 13.
[c] Traces = 3; slight = 4; small = 5; modest = 6; moderate = 7.

EFFECT OF PREGNANCY

The once rather prevalent practice of breeding yearling feeder heifers, either accidentally or by intent, has become relatively uncommon. Ranchers, by using more fencing and better range management practices, can more easily graze their yearling heifers apart from the cow herd and thus prevent breeding. Farmers are not so apt to keep heifers over as yearlings, making the problem of bred feeder heifers from that source much less common than formerly. This trend has undoubtedly had much to do with the reduction in the spread between feeder steers and heifers, and this follows through to the slaughter stage as well, where packers no longer discriminate so generally against heifers. Selling heifers on grade and yield would of course completely eliminate any justification on the part of packers for buying slaughter heifers at a lower price to protect against possible advanced pregnancies and a reduced dressing percent. Chapter 21 includes a discussion of MGA, a synthetic hormone approved for feeding to heifers, which reduces the disturbing effects of the normal estrus cycle seen in feedlot heifers. The use of this hormone may well eliminate the practice of breeding feedlot heifers in order to quiet them.

An injectible form of diethylstilbestrol, marketed under the trade name of Repositol, is being used by veterinarians, with approximately 70 percent effectiveness, in aborting pregnant heifers prior to the fourth or fifth month of pregnancy. There are some temporary physiological side effects, such as uterine prolapse in extreme cases, but in general the practice is successful and feedlot performance continues to be normal.

SPAYED VERSUS OPEN HEIFERS

One of the principal objections to feeding heifers is the disturbance caused by their coming in heat. When a carload or more of heifers are fed together, hardly a day passes without one or more animals being in this condition, so that the herd is frequently in a state of excitement and unrest. Obviously such conditions are not conducive to rapid and economical gains.

Spaying is sometimes used to avoid the disturbances caused by in-heat heifers. More heifers are not spayed mainly because of the extra cost and the failure of the market to pay sufficient premium for them to cover the cost. The principal advantages claimed for spaying are a more tranquil disposition in the feedlot and a somewhat higher price for the heifers when they are marketed. However, spayed heifers seldom have made as large or as economical gains or have attained as high a finish as open heifers in feeding experiments where they have been compared. Apparently, the removal of the ovaries retards growth and development of young heifers in much the same way that castration retards growth and development of male calves.

Because spayed heifers gain less rapidly than open heifers, spaying cannot be recommended on the ground that it will lessen activity in the feedlot and thus result in faster gains. As a matter of fact, disturbances caused by in-heat heifers usually are most noticeable during the first few weeks of the feeding period and tend to become less frequent as the heifers approach market finish. Consequently, any advantage of spayed heifers must lie in the certainty that they are not pregnant and therefore may be purchased for a long feeding period without danger of advanced pregnancy by the time they are ready for market.

FEEDLOT PERFORMANCE OF BULLS

Undoubtedly the research with hormone-like compounds during the past decade in the feeding of steers and heifers has been responsible for revived interest in investigating the performance of slaughter bulls. It has long been observed that bulls outgain steers and do it more efficiently, but for several reasons the feeding of young bulls has not been an accepted practice in America as it has in Europe. Price discrimination by packer buyers, in part justified by a lower dressing percent, lower carcass grade, and a mandatory label of "bull" by government graders, has undoubtedly been the principal deterrent. Feeders themselves have been convinced that the disposition or temperament of bulls was a big disadvantage in the feedlot. The results of a number of recent experiments have caused breeders and feeders alike to reexamine the validity of the practice of castration. The three studies to be reviewed here indicate something of the potential, in terms of increased supply of acceptable lean beef produced at high efficiency, that may follow if the practice of castration is abandoned.

Wyoming workers, in cooperation with Safeway Stores, Inc., and Armour and Company, compared Angus bull and steer calves of similar genetic background. The experiment included a consumer acceptance study, which sets this study apart from most others bearing on this subject. The pertinent data are shown in Table 96. In brief, the Wyoming workers concluded that bulls, when compared with steers, weaned heavier and gained faster in the feedlot, and produced meatier carcasses with less trim-fat, but graded a full grade lower because of less marbling, coarser texture, and darker color. Consumers preferred bull chucks, but steer steaks.

Table 96

Effect of Castration on Production Traits and Selected Carcass Measurement[a]

	Bulls	Steers	Difference Bull-Steer
Number	19	19	
Weaning weight (lb)	353	330	23
Initial feedlot weight (lb)	422	391	31
Slaughter weight (lb)	963	877	86
Gain per day on feed (lb)	2.21	1.97	0.24
Dressing percent	63.7	62.9	0.80
Carcass gain per day of age (lb)	1.22	1.13	0.09
Area of rib eye (sq in.)	13.4	10.5	2.9
Area of rib eye per cwt carcass (sq in.)	2.32	2.07	0.25
Fat thickness at 12th rib (in.)	0.34	0.67	−0.33
Estimated yield of retail cuts (%)	50.3	46.3	4.0
Percent cooler shrink	3.19	3.71	−0.52
Carcass grade	Good	Choice	

[a] Wyoming Agricultural Experiment Station Bulletin 417, 1964.

The findings in a Nebraska study with Angus cattle closely parallel those of the Wyoming study, but heifers were included in this test, and the design permitted the evaluation of feed efficiency differences as well as the effect of length of feeding period because the cattle were fed for two different lengths of time. The Nebraska workers found that bulls outgained both steers and heifers and did so more efficiently. The bull carcasses contained more lean and less fat, but were tougher and graded lower. When comparable retail prices, adjusted for quality grade, were applied to the retail-trimmed cuts of all carcasses, the value of carcasses less feed costs favored bulls over steers and steers over heifers.

It should be realized that at the retail level bull beef probably will not be obviously labeled as such. However, if a packer requests that a U.S. Department of Agriculture grader grade the bull carcass, the grader is required to label it "bull," which means that the packer will probably have to sell the carcass at a substantial discount. Nongovernment-graded young bull carcasses might sell at the same price as steer or heifer carcasses. There is strong likelihood that government standards for grading young bulls will be altered to lower the penalty of leaving bulls intact or uncastrated.

Implantation with estrogen-like synthetic hormones has been used to reduce the activity of bulls in the feedlot. Texas workers have conducted

Table 97
Effect of Sex on Production and Carcass Traits[a]

	Bulls		Steers		Heifers	
Days Fed	211	255	211	255	219	259
Number	25	24	27	24	18	8
Average initial weight (lb)	435	425	432	409	413	403
Average daily gain (lb)	2.22	1.95	1.78	1.64	1.62	1.45
TDN per pound of gain (lb)	5.14	6.05	6.42	6.85	6.58	7.08
Carcass grade	G+	G+	C−	C−	C−	C−
Tenderness[b]	13.2	17.0	9.8	10.2	10.5	12.8
Rib-eye area (sq in.)	11.4	11.7	9.9	9.9	10.0	9.4
Trimmed and boned cuts (%)	56.1	54.7	53.4	51.4	50.3	49.8
Retail value of carcass less feed cost	$238	$232	$191	$179	$168	$153

[a] Animal Science Department, University of Nebraska, 1964.
[b] Warner-Bratzler Shear Values—Larger numbers indicate more pressure required to cut a 1-inch core of the loin-eye muscle.

Table 98
Comparison of Diethylstilbestrol-Implanted Bulls and Steers (182 Days)[a]

	Bulls	Steers	Difference Bull-Steer
Stilbestrol implantation level (mg)	30	30	
Number of animals	8	10	
Average initial weight (lb)	403	397	6
Average final weight (lb)	917	808	109
Average daily gain (lb)	2.82	2.26	0.56
Shrink (%)	5.0	4.9	0.1
Dressing percent (chilled)	61.1	60.8	0.3
Carcass grades			
Choice	—	3	—
Good	7	7	—
Standard	1	—	—
Feed per cwt gain (lb)	814	851	−37
Feed cost per cwt gain	$15.74	$16.32	−$0.58
Daily ration (lb)	22.9	19.2	3.7
Rib-eye area (sq in.)	11.43	9.80	1.63
Fat thickness (in.)	0.38	0.55	−0.17
Estimated lean cuts (%)	51.2	50.4	0.8

[a] Texas Agricultural Experiment Station, Substation #1, Beeville, 1963.

numerous studies on the subject, and the one summarized in Table 98 demonstrates typical results. As in the studies mentioned above, bulls outperformed steers and, if anything, implantation with hormones accentuated the differences between bulls and steers in this respect.

FEEDING COWS

Large numbers of cows are marketed each fall by farmers and ranchers. Many of these cows have been suckling calves all summer and are very thin; consequently, they frequently are considered satisfactory for feeding, particularly if they can be bought for only about half the cost per pound of choice feeder calves and yearlings.

Although feeder cows often appear to be worth the money when they are bought in the fall, they frequently prove to be a rather sorry bargain by the time they are marketed. Many misfortunes may happen to a drove of cows during the feeding period, which will reduce the profits materially. In the first place, they are mature animals and consequently all gains are in the form of fat and are very expensive. Gains made by cows are often

Table 99

Gains and Feed Consumption of Feeder Cows While Being Fattened for Market

	Nebraska 66 days		Missouri 111 days	Illinois 154 days	
	Shelled Corn	Ground Ear Corn	Shelled Corn	1st 42 Days	Last 112 Days
Average initial weight (lb)	1,006	987	1,004	713.5	
Average daily gain (lb)	1.34	2.09	2.24	1.32	2.44
Average daily ration (lb)					
Corn or grain mix	18.5	26.1	9.9	—	14.4
Protein concentrate	—	—	1.2	1.0	1.0
Corn silage	—	—	41.3	40.0	21.7
Legume hay	8.9	8.1	6.7	2.0	2.0
Feed per cwt gain (lb)					
Corn	1,384	1,246	441	490	
Protein concentrate	—	—	55	47	
Corn silage	—	—	1,843	1,244	
Legume hay	666	389	302	93	

50 to 100 percent more costly than those made by calves or yearlings. Such costly gains require a relatively large price spread to prevent a financial loss. However, the price of finished cows is seldom very high, so that large price spreads are the exception rather than the rule. In fact, with ordinary price conditions, very skillful buying and feeding are necessary to realize a profit on a cow-feeding venture. These points are well illustrated in Table 99.

Another objection to buying cows for feeding is that most of them have been bred some 3 to 5 months before being marketed. Thus some of them will calve before they are finished, and a considerable percentage of those that do not will be sufficiently advanced in pregnancy to be penalized in price when they are sold. Men who buy cows to feed frequently change their plans upon discovering that most of the cows appear to be pregnant and decide to keep them until fall and finish them after the calves are weaned. Then, however, the grain already fed yields little return, because the finish put on during the winter is largely lost during the suckling period. Seldom do all the cows prove to be in calf, and the barren ones may be sold directly off pasture the following fall, but if sold at this time of year they may not bring much more per pound than they cost a year earlier.

Cows fed during a period of rising prices occasionally return very satisfactory profits and yield a high return on the amount of money invested. Also, the return realized per bushel of corn fed may be high, because cows usually are fed a limited grain ration or a full feed for only a short time.

In summary, the feeding of cows must be considered among the most speculative of all cattle-finishing programs and one that only the most experienced feeder should attempt.

Chapter 13

The IMPORTANCE of GRADE and BREED in FEEDER CATTLE

Which age and sex of cattle to buy are only two of the many questions that confront the man needing feeder cattle. In addition, he must decide between animals of one breed and those of another, and between higher-grade feeder steers that might cost upward of 30 cents per pound, and nondescript cattle that can be purchased at a considerably lower figure.

At most seasons of the year feeder cattle of all kinds are available in fairly large numbers at any of the larger central markets and auction sales. Naturally certain markets or areas have a reputation as sources of unusually good selections of cattle of a certain breed or grade. For example, Omaha and Denver are recognized as being among the best markets at which to buy high-grade Hereford or Angus calves; Oklahoma City and Kansas City as good places to buy heavy grass steers in late summer or early fall, suitable for a short feed; East St. Louis and St. Paul as places where lower-grade feeders, especially with some Holstein breeding, are found in large numbers. If a large number of cattle of a given kind are to be bought, it is usually best to go to that market or area where the supply is likely to be largest. However, if only a load or two are wanted, the market nearest at hand is usually able to furnish them at the lowest cost, freight and other expenses considered.

KIND OF CATTLE TO BUY

There is no hard and fast rule as to which kind of cattle is best to feed. Many factors must be taken into account in deciding this question.

338

Without doubt the strength of the demand in relation to the supply, both for the feeder animals and for the finished beeves, is likely to be an important factor in determining the profit made from the feeding operation. Lower-grade cattle may prove just as profitable, or more so, as choice or prime feeders if they are purchased when the supply of such animals is greater than the demand and if sold at a market that has a strong demand for good-grade fat cattle. On the other hand, the supply of prime, well-finished steers is often so small that they enjoy a market all their own. The profit made on such cattle may then be relatively large, even though the price paid for them as feeders seemed at the time unreasonably high compared with quotations for plainer cattle.

The conditions under which the cattle are to be fed and the way they are to be handled should receive first consideration in deciding which kind of cattle to buy. For a long feed in which a liberal allowance of grain will be furnished, only the better grades of feeders should be purchased. Cattle of good breeding and of the proper type will, under such conditions, be choice or better when marketed and will command the highest market price. Medium or common feeders, on the other hand, can never be made into "market toppers," no matter how long they are fed nor how excellent the ration. To feed such cattle large quantities of expensive feed is likely to prove unprofitable. As a rule, they should be fed on high-energy rations for a comparatively short time.

KIND OF CATTLE BEING FED

As mentioned in Chapter 10, grade in a stocker or feeder steer is determined by his predicted grade as a slaughter animal. Feeding finishing rations to a choice or higher-grade feeder for a longer time than is usually required to finish him to his corresponding slaughter grade will generally not be profitable, because only a small percentage of this grade of feeder can be upgraded. Also, a few animals may not live up to expectations, thus not making the predicted slaughter grade after the usual length of feeding period.

On the other hand, lower-grading feeders can usually be upgraded one or two grades. There has been a substantial increase in the feeding of the lower grades of feeders, especially in western and southwestern feedlots where it has always been popular, but the practice is not confined to these areas, as many cattle feeders in Colorado, the Corn Belt, and elsewhere are eager to take advantage of the possibilities for upgrading.

Table 100 shows the percentages of the various grades of carcasses in the total United States beef production over a recent period of 20 years.

Table 100

Percentage Composition of the United States Beef Supply, By Grades (1946–1966)[a]

Years	Prime	Choice	Good	Standard	Commer-cial[c]	Utility	Canner, Cutter
					Percentage of Total Beef by Grade[b]		
1946–1948	6	28	19		17	17	14
1949–1951	7	34	19		15	15	11
1952–1954	5	34	18		16	13	13
1955–1957	4	33	21		16	13	13
1958–1960	4	36	27		14	10	9
1961–1963	3	47	18	11	4	9	8
1964	3	49	17	11	4	8	9
1965	4	47	17	9	5	8	12
1966	4	49	18	7	4	8	10

[a] U.S.D.A., *Livestock and Meat Situation*, May 1967.
[b] Data are 3-year averages from 1946 through 1963; included are graded beef and estimate for ungraded beef.
[c] Includes standard and commercial grades from 1946 through 1960.

It is interesting to see that prime grade makes up a very small portion of the total and that it continues to decrease. Choice grade has increased, good has remained level, and standard grade, added only a few years ago, is about equal to the total decrease in commercial and utility grades. These changes largely reflect a growing demand for grain-fed beef.

MARKET CLASSES AND GRADES OF SLAUGHTER CATTLE

Official U.S. Department of Agriculture standards have been established for slaughter cattle and beef carcasses just as for stockers and feeders. The slaughter grades, sometimes called "on foot" or "live" grades, are based on the predicted carcass grade of an animal after slaughter. Color of fat and lean, interspersion of fat and lean (marbling), and age as determined by bone condition are important factors in determining carcass grade that cannot be accurately assessed by examination of the live animal. For this reason, slaughter and carcass grades do not always correspond, but experienced buyers seldom miss their evaluations to any great degree.

Figure 66 shows rear and side views of the more common grades of slaughter steers. Slaughter heifers should have corresponding conformation and degree of finish. Table 101 lists the various grades of slaughter cattle.

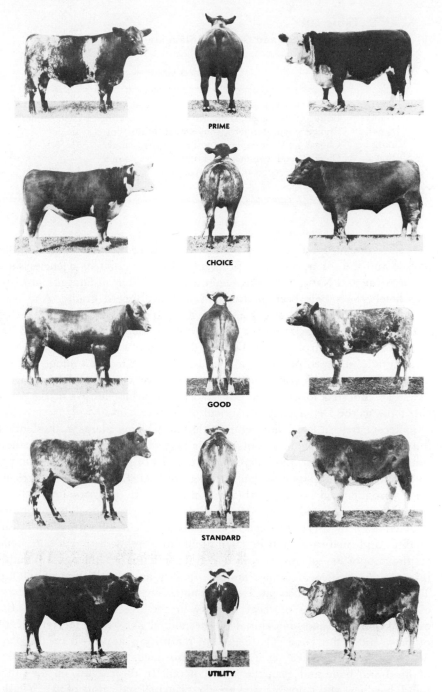

PRIME

CHOICE

GOOD

STANDARD

UTILITY

FIG. 66. Steers illustrating the U.S.D.A. grades of slaughter steers. It is important to be able to judge when a steer is finished for his grade, as economy of gain and length of feeding period are closely related (U.S.D.A.)

Table 101

Market Classes and Grades of Slaughter Cattle[a]

Market Class	Market Grade
Steers	Prime, choice, good, standard, commercial, utility, cutter
Heifers	Prime, choice, good, standard, commercial, utility, cutter, canner
Cows	Choice, good, standard, commercial, utility, cutter, canner
Bulls	Choice, good, commercial, utility, cutter, canner
Stags	Choice, good, commercial, utility, cutter, canner
Calves	Prime, choice, good, commercial, utility, cull

[a] Based on U.S.D.A., *Livestock and Meat Situation.*

Government grading of beef carcasses is done at the request of the packer and thus is a voluntary program. Most government agencies such as the Army and Navy, and many chain stores as well as restaurants, airlines, and hotels, buy their beef on the basis of government grades. Any beef entering interstate commerce must be government-inspected and labeled but not necessarily government-graded. Government-graded beef can be identified by the purple shield, bearing the letters "USDA" and the name of the grade, stamped on practically all retail cuts. Almost all major packers have their own system of labeling the respective grades of beef as well. These are referred to as "in-house" grades and do not always fit the same categories as the Department of Agriculture grades.

Recent federal legislation requires that all packing plants in the United States, whether selling in or out of their respective states, meet inspection standards equivalent to those heretofore required of plants that were subject to federal inspection for interstate shipment of beef. Department of Agriculture personnel will be responsible for inspection in this situation.

The specific grade assigned to a live slaughter animal by a grader is determined by its relative excellence with respect to conformation, finish, quality, and maturity or age. Conformation is related to cut-out value or the proportion of higher-priced cuts in relation to the whole. Finish refers to the degree of fatness and the quality and distribution of the fat. The latter factor is associated with palatability, tenderness, and quality of the individual cuts of meat. Quality in the live animal refers to the overall symmetry and smoothness of the animal as well as the refinement of head, hide, and bone. The degree of maturity is appraised on the basis of the animal's physical characteristics associated with age, such as size of head and bone and even the length of tail. In the carcass, on the other hand, hardness and color of bone are the principal indicators of age. Youthfulness and finish are both believed to be associated with palatability. Preliminary research data indicate that tenderness and palatability are heritable. It is conceivable that, in the future, selection for these traits may result in cuts that are tender and palatable without the necessity of feeding to the degree of finish found in high choice or prime cattle.

Since more and more cattle are being sold and bought on a grade and yield basis, it would be helpful if feeders could evaluate the degree of marbling, the principal determinant of carcass grade, in the live animal. This is difficult to do in every instance, but there are some indicators that can be used. Breed has an effect, and it is generally conceded that Angus cattle marble earliest, with the other British breeds intermediate and dairy and dual-purpose breeds last. Crosses between these breeds or types will usually be intermediate between the parent breeds. Age is another factor; the older the animal, the greater the marbling. Length of feeding period on high-concentrate rations is also highly correlated with degree of marbling. Within a breed or type, exterior finish, as seen in the cod or udder regions, brisket, about the hooks and pins, and along the back and loin edge, can also be used with some degree of reliability.

FIG. 67. Colorado-bred good to choice Angus-Charolais crossbred calves, which are acquiring a reputation as rapid, efficient gainers. Combining heavier weights with faster feedlot gains reduces nonfeed costs through more rapid turnover. (Western Livestock Journal.)

Feeders usually attempt to feed cattle just long enough to bring the majority of a drove of cattle barely into the desired grade. Cattle are graded by full grades on the rail, and, if they are bought by grade, there is usually no premium for carcasses in the middle or upper third of the grade. Progeny-testing programs help to identify sires whose offspring tend to marble earlier than the average of the breed.

IMPORTANCE OF TYPE

Cattle feeders differ in their opinions as to which type of steer within a breed or class of cattle is most profitable to feed. Some prefer small, compact, low-set feeders, believing that they finish more readily, have a

Table 102

Effect of Type of Feeder Steers on Rate and Economy of Gain and on Carcass Merit

	Calves Average of 3 Trials[a]		Yearlings Average of 9 Trials[b]		
	Small Type	Conven- tional Type	Compact Type	Medium Type	Rangy Type
Average initial weight (lb)	338	412	722	752	785
Average total gain (lb)	351	438	350	360	373
Average daily gain (lb)	1.79	2.14	2.08	2.14	2.22
Percentage gain (gain ÷ initial weight)	104	106	48.5	47.9	47.5
Total feed consumed daily (lb)	12.5	15.1	—	—	—
Daily feed consumption per cwt body weight (lb)	2.43	2.42	—	—	—
Total feed consumed per cwt gain (lb)	702	707	—	—	—
Digestible nutrients per cwt gain (lb)	476	479	—	—	—
Feeder grade[c]	5.1	4.4	—	—	—
Fat steer grade[c]	4.4	4.3	5.3	5.5	5.6
Carcass grade[c]	4.5	4.4	5.1	5.1	5.1
Dressing percent	57.7	58.9	57.4	57.7	58.4

[a] Colorado Agricultural Experiment Station, Mimeographed Report.
[b] New Mexico Agricultural Experiment Station, *Journal of Animal Science*, 5:4, 331–337.
[c] Grading systems: (1) Colorado: 6 = prime, 5 = choice, 4 = good; (2) New Mexico: 4 = choice, 5 = low choice, 6 = high good.

higher dressing percentage, and produce more shapely carcasses than the conventional type of steer. Others favor large, growthy cattle with plenty of bone and substance, believing that such feeders make faster and more economical gains and return more profit because of their greater weight. Numerous tests have been conducted to obtain information on these points, but few have disclosed consistent differences between the two types. Although the large-type steers usually have gained at a somewhat faster rate, seldom has there been a significant difference in feed efficiency, as seen in Table 102. It should be mentioned that, within the small or larger types, the faster gainers will gain more efficiently.

Although the small and large types of cattle have not, as a rule, shown marked differences when both were full-fed for the same length of time, it is possible that one type may be better suited than the other for a deferred system of feeding in which young animals are carried on roughage and pasture for 6 to 12 months before they are given a finishing ration. Such an experiment has been conducted at the Kansas, Ohio, and Oklahoma stations under the auspices of the American Hereford Association. For this method of feeding, the medium type was slightly more profitable than either

FIG. 68. Prime steers such as these, bred and fed in the Corn Belt, provide a market for at least 75 bushels of corn per head, but steers fed to this grade find a limited market today. (American Shorthorn Association.)

Table 103

Relative Value of Small, Medium, and Large Hereford Calves for a Deferred System of Feeding[a]

	Small	Medium	Large
Average initial weight (lb)	401	419	442
Wintering period 150 days			
Average gain (lb)	169	172	197
Summer grazing period 98 days			
Average gain (lb)	108	115	114
Late summer full-feeding period 101 days			
Average gain (lb)	228	233	231
Total gain 349 days	505	520	542
Final weight (lb)	906	939	984
Feed cost per head			
Winter period	$28.54	$28.40	$29.51
Summer grazing period	11.00	11.00	11.00
Full-fed period	47.20	48.02	49.43
Total feed cost per head	$86.74	$87.42	$89.94
Feed cost per cwt gain	$17.18	$16.81	$16.59
Selling price per cwt	$25.53	$25.78	$24.87
Dressing percent	57.8	58.1	57.7
Average carcass grade	Low choice[b]	High good[b]	High good[b]

[a] *American Hereford Journal*, 14:22, 21.
[b] Old grading standards.

the small or large type, if all calves were purchased at the same price per pound (see Table 103). It should be added, before leaving this topic, that all three types or sizes of cattle mentioned can be found within a particular breed.

IMPORTANCE OF GRADE

Apart from overall type or size within a breed, as just discussed, is the matter of feeder grade. In Chapter 10, which deals with the stocker program, it is noted that there is a wide range in grade, reflecting differences ranging from the ultimate in beef type, as seen in the British breeds, to common and inferior at the other end of the scale. Some old and much new research casts considerable doubt on the suitability of the present stocker and feeder grading system as a means of sorting cattle into groups with predictable performance.

Performance-testing programs generally have brought to light the fact that conformation grade and performance on feed, or even on the rail, are not highly associated. Some researchers go so far as to say that they are negatively correlated—that is, the higher the grade, the poorer the performance in the feedlot and the poorer the cutability or yield of trimmed cuts. Perhaps a more middle-of-the-road approach is to say that within each grade there are excellent and poor performers, and that performance records on the sires of stockers and feeders or on their half-siblings are a better guide to their ability than the grade assigned to them.

An Iowa study summarized in Table 104 illustrates that the lower or plainer grades of cattle can be profitable. The reason is that there is usually a wider spread in the purchase price between grades of feeders than there is the selling price of the same cattle when sold for slaughter. In the Iowa study the choice and good grade steers were Herefords, the medium steers were of mixed breeding comparable to the "Okies" that are so popular in the West, and the common steers were Holsteins. Unfortunately, all lots were fed the same length of time. If the medium and common steers had been sold 50 days earlier, the advantage they showed would have been even greater. In practice, plainer cattle should not be fed longer than 120 to 150 days. The question that must be asked, though, is whether the upgrading of the plainer cattle that resulted would have occurred on a shorter feed. Figure 69, which diagrams this part of the Iowa study, shows that the choice grade steers actually slipped away from average choice to low choice, but the lower grades were upgraded by two grades. It should be mentioned that the entire reduction in selling price necessary to break even that is shown by the plainer cattle can be accounted for by their lower purchase price.

The results obtained in the Iowa study may not hold for many years in the future. As more and more cattle feeders change over to lower grades, the relative cost of these grades of feeders will surely increase because of the increased demand. It can also be said with certainty that the supply will be reduced, because the breeder of these plainer cattle will be unsatisfied with the lower price he has been receiving and will upgrade his cattle. The straight Holsteins, which of course come from dairies, will continue to be available, although even here changes are occurring, as many dairymen are now breeding only their very best cows to dairy bulls, preferring to use beef bulls on the average and lower producers. The resulting crossbred offspring will probably grade good as feeders and thus will cost the feeder more than straight Holsteins.

Table 104

Feedlot Performance, Economic Returns, and Carcass Evaluation of Four Feeder Grades of Yearling Steers[a]

Cattle Fed 198 Days Market to Market (18 Head per Group)	Feeder Grade			
	Choice	Good	Medium	Common
Feedlot Performance				
Purchase weight (lb)	762	763	760	824
Slaughter weight (lb)	1,255	1,292	1,258	1,359
Average daily gain market to market (lb)	2.49	2.67	2.52	2.70
Feed consumed per steer per day (lb)[b]	21.9	23.0	22.2	24.6
Feed required per cwt gain (lb)	879	861	881	911
Carcass Measurements				
Average carcass weight, shrunk (lb)	776	793	757	800
Dressing percent	61.8	61.4	60.2	58.9
Federal carcass grade	C−	C−	C−	G
Rib-eye area per cwt carcass (sq in.)	1.5	1.5	1.5	1.5
Fat cover per cwt carcass (in.)	0.09	0.09	0.06	0.02
Predicted retail yield of carcass (%)	67.1	66.6	68.7	71.4
Financial Values				
Cost per steer				
Original cost ($)	190.50	179.30	168.72	158.21
Feedlot costs (feed + 3 cents per pound of gain) ($)	97.61	102.16	98.60	109.68
Shipping costs to packing plant, $	2.51	2.58	2.52	2.72
Total cost to packing plant ($)	290.62	285.04	269.84	270.61
Purchase price per pound (cents)	25.0	23.5	22.2	19.2
Feed cost + 3 cents nonfeed cost per pound of gain (cents)	19.8	19.5	19.8	20.5
Shipping cost per pound to packing plant (cents)	0.2	0.2	0.2	0.2
Break-even price per pound at packing plant (cents)	23.2	22.1	21.4	19.9

[a] *Iowa Farm Science*, 20:2,223.

[b] Feed consumed included 5 lb mixed hay and 1 lb supplement per steer per day plus balance of ration in rolled shelled corn. Feed prices: corn, $1.12 per bushel; hay, $20 per ton; supplement, $100 per ton.

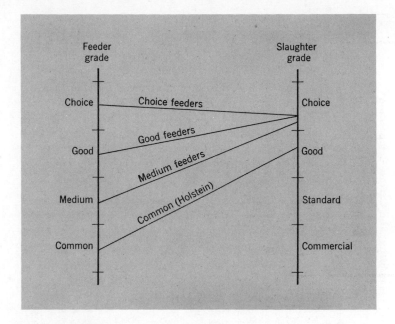

FIG. 69. Lower grades of feeder steers can usually be upgraded in the feedlot, whereas higher-grade feeders seldom can be and, in fact, often are reduced in grade during the finishing process. (Iowa State University.)

EFFECT OF BREED

There have been numerous tests with the objective of comparing breeds with respect to feedlot performance and carcass merit, and, as mentioned in Chapter 7 in connection with choice of breed for cow-calf programs, drawing conclusions from a single test is unwise to say the least. A study of the many tests does, however, establish a pattern, and the exhaustive study conducted by Tennessee researchers and summarized in Table 105 can serve as a reliable evaluation of breeds with respect to the traits investigated.

Calves were obtained from various sources at an average age of about 5 months and were full-fed a high-concentrate ration until they either weighed 900 pounds or reached 20 months of age, whichever occurred first. British and Zebu types did not differ significantly in any production traits. The dairy type, when both dairy breeds were combined, was inferior in all traits except gain. Angus steers dressed higher than all other breeds and graded highest as well. However, the Angus along with the straight Brahman and Jersey cattle were less efficient in feed conversion. With re-

Table 105

Effect of Type and Breed of Feeder Cattle on Performance, Palatability, and Carcass Composition[a]

Breed and Type	Hereford	Angus	Brahman	Brahman X[b]	Santa Gertrudis	Holstein	Jersey
Number fed	32	29	22	10	12	24	25
Performance data							
Average initial weight (lb)	334	338	305	340	368	291	219
Average final weight (lb)	885	865	835	911	893	909	791
Feeding period (days)	310	315	362	303	298	294	368
Average daily gain (lb)	1.84	1.76	1.50	1.90	1.91	2.16	1.56
24-hour live shrink (%)	5.6	6.2	5.2	5.1	6.1	6.9	6.1
Dressing percent	62.2	63.4	62.4	62.4	62.8	59.7	57.5
Feed per cwt gain (lb)	869	909	936	858	892	776	959
Carcass data							
Carcass grade	G+	C−	St+	G−	G−	St	St
Fat thickness (mm)	18.5	19.3	11.5	14.3	16.5	9.1	10.8
Kidney fat (%)	4.1	5.0	3.7	4.3	4.1	4.5	6.5
Rib-eye area (sq in.)	9.4	9.5	9.2	9.9	9.7	8.8	8.0
Loin steak evaluation							
Shear values[c]	4.9	5.6	6.3	5.9	5.4	5.2	5.0
Flavor[d]	7.4	7.3	6.8	7.3	7.3	7.2	7.4
Cooking loss (%)	25.4	25.9	26.6	28.5	27.1	24.2	22.7

[a] Tennessee Agricultural Experiment Station, *Journal of Animal Science*, 22:702, 1001–1008, 1963.
[b] Brahman sires on either Hereford or Angus cows.
[c] Warner-Bratzler value on 0.5-inch core.
[d] Extremely poor = 4; excellent = 9.

spect to carcass traits, Brahman carcasses scored lowest on tenderness and flavor, with all other breeds scoring fairly even with one another. The carcasses of the British breeds were fatter and graded higher, but the higher grade did not result in greater tenderness and flavor.

Before leaving the discussion of the effect of conformation and quality on feedlot performance, some of the problems encountered in feeding the plainer grades of cattle should be mentioned. California station workers at Davis conducted a trial comparing high-grade Herefords, produced in two herds cooperating in the Agricultural Extension Service herd improvement program, with "No. 2 Okies," purchased through an order buyer and shipped from Fort Worth, Texas. The term "Okies" is a colloquialism used largely in the South and West to describe feeder cattle that show a preponderance of British breeding but also about one-fourth Zebu breeding. Most of these cattle originate in the Gulf Coast region and the Southeast generally, and they currently make up about 50 percent of the cattle fed in Arizona and California.

In the Davis experiment, half of each type of calf was fed out immediately following purchase in the fall. The other half was wintered on foothill range and then fed out the following fall. Each year half of the Okies were fed as long as the Herefords and the other half to the same weight achieved by the Herefords. Eighty percent of the Okies had to be castrated and 70 percent dehorned after arrival at the feedlot. A quotation from the California report is of interest: "The Okies arrived during cold, foggy weather. One was dead on arrival. Four more calves died and many were sick during the first weeks; another died following castration."

Results of the test are summarized in Table 106. The Herefords outgained the Okie calves in the feedlot and on the range, and they were more efficient. As yearlings in the feedlot, all groups made comparable gains, but the Okies were more efficient. All of the calves were more efficient than the yearlings but, surprisingly, the Hereford calves outgained the Hereford yearlings. Undoubtedly the initial sickness of the Okies and the combined stress of shipping, castration, and dehorning had an adverse effect, but the circumstances concerning these calves do not differ greatly from those occurring in many calves of this grade shipped a comparable distance.

As noted in reviews of other trials comparing grades, in the California study the higher-grade feeders did not improve in grade upon finishing, but the Okies did. Furthermore, the Okies had greater rib-eye area, thinner outside fat covering or bark, and a resultant higher yield-grade, denoting better cutability. Based on the delivered cost per hundredweight as listed, the Hereford calves had a $1 per hundredweight lower break-even cost, but the Okie yearlings showed a $2 to $3 per hundredweight lower break-even cost than the Herefords. Death losses are not calculated into the

Table 106
Feedlot Performance and Carcass Merit of Hereford and Okie Calves and Yearlings[a]

Feeder Age	Calves				Yearlings			
	Siskiyou Herefords	Humbolt Herefords	Time-Fed Okies	Weight-Fed Okies	Siskiyou Herefords	Humbolt Herefords	Time-Fed Okies	Weight-Fed Okies
Number fed	30	30	30	20	30	30	10	18
Feedlot data								
Days on feed	191	191	191	244	153	153	148	189
Initial weight (lb)	425	435	383	376	636	654	532	531
Final weight (lb)	988	997	879	988	1,025	1,051	922	1,010
Average daily gain (lb)	2.95	2.94	2.62	2.51	2.55	2.60	2.64	2.53
Feed per cwt gain (lb)	641	661	743	788	937	895	868	902
Carcass data								
Feeder grade	C	C	G	G	C	C	G−	G
Live slaughter grade	C	C	G+	C−	C−	C	C−	C−
Carcass grade	C	C	C−	C	C	C	C−	C
Yield grade[b]	4.0	3.9	3.2	3.5	3.7	3.4	2.9	3.4
Fat cover (in.)	0.60	0.56	0.41	0.52	0.52	0.50	0.36	0.49
Rib-eye area (sq in.)	10.5	10.8	10.7	11.5	10.9	11.9	11.3	11.4
Financial data								
Delivered price per cwt	$ 30.00	$ 29.00	$ 26.40	$ 26.40	$ 24.00	$ 23.00	$ 20.40	$ 20.40
Cost per head	127.59	126.23	107.24	106.14	152.54	150.46	108.45	108.31
Feed cost per cwt gain	17.35	17.91	20.12	20.98	25.76	24.62	23.86	23.49
Break-even price	22.80	22.75	23.63	23.74	24.67	23.61	21.87	21.86

[a] California Cattle Feeders' Day Report, 1966.

[b] U.S.D.A. yield grades estimate cutability or yield of trimmed lean cuts. Lower numbers indicate higher yield.

352

Table 107
Time and Feed Required to Finish Beef Cattle to Various Slaughter Grades[a]

Average Initial Weight (lb)	Average Days Fed	Days ± from Middle Grade	Average Final Weight (lb)	Average Total Gain (lb)	Average Daily Gain (lb)	Total Corn Fed (bu)	Corn per cwt Gain (lb)	Cost per cwt Gain[b] ($)	Average Carcass Grade
			Choice Steer Calves						
	145	−90	771	291	2.01	24.4	405	17.37	Good
480	235	0	938	458	1.95	42.8	476	19.29	Choice
	358	+123	1,163	683	1.91	67.4	553	21.48	Prime
			Choice Heifer Calves						
	154	−83	714	284	1.84	23.6	398	17.22	Good
430	237	0	871	441	1.86	41.8	484	19.54	Good
	268	+31	932	502	1.89	49.2	507	20.14	Choice
			Medium Yearling Steers						
	69	−52	934	154	2.23	11.0	400	22.28	Comm.
780	121	0	1,073	293	2.43	23.2	444	22.26	Good
	164	+43	1,153	373	2.48	34.2	514	24.82	Choice
			Good Yearling Steers						
	95	−52	891	231	2.43	20.3	377	17.65	Comm.
660	147	0	1,015	355	2.42	34.8	453	19.50	Good
	208	+61	1,135	475	2.28	53.2	538	21.78	Choice
			Choice Yearling Steers						
	111	−85	962	252	2.27	23.3	385	19.36	Good
710	196	0	1,126	416	2.12	43.5	465	21.54	Choice
	346	+150	1,383	673	1.94	83.7	618	25.17	Prime
			Good Yearling Heifers (Bred When Purchased)						
	73	−73	798	183	2.49	15.7	476	20.36	Comm.
615	146	0	908	293	2.01	25.8	492	21.83	Good
	203	+57	1,126	411	2.03	39.7	541	22.97	Choice
			Good Yearling Heifers (Open)						
	80	−65	687	162	2.01	9.7	337	16.30	Comm.
525	155	0	704	279	1.79	23.2	466	20.39	Low good
	204	+49	883	358	1.76	32.5	492	21.10	Good
			Good Two-Year-Old Steers						
	28	−68	826	81	2.85	5.4	337	12.11	Comm.
745	96	0	976	231	2.41	23.5	568	19.48	Good
	200	+104	1,192	447	2.23	56.9	713	24.10	Choice

[a] Illinois Agricultural Experiment Station, unpublished data.
[b] Feed prices: corn, $1.50 per bushel; protein concentrate, $90 per ton; clover hay, $20 per ton; corn silage, $12 per ton.

break-even prices as given, but of course in a practical situation they would be included.

In summary, in comparing the merits of the plainer grades of feeders with feeders of choice or better grades, the following statements seem appropriate:

1. Lower-grade feeder cattle are usually the offspring of larger parents of the Brahman, Holstein, Brown Swiss, or Charolais breeds crossed on British breeds and thus can be expected to outgain straight British-bred feeders.

2. Lower-grade feeders are usually thinner in condition, owing to their tendency to grow larger and mature later and to the fact that their area of origin is likely to provide a lower plane of nutrition. Thinner cattle should, again, gain more rapidly than fleshier cattle.

3. The lower grades of feeders usually are from 2 to 3 months older than straight British-bred feeders as found for sale.

4. In practice, lower-grade feeders are usually fed for shorter lengths of time, again contributing to higher average daily gains from purchase to sale weight.

5. The lower grades of cattle are much more likely to make a profit for the feeder than for the breeder.

6. As British-bred cattle acquire more bred-in performance and the price spread between grades of feeders becomes narrower, the lower grades will lose much of the profit advantage they now enjoy in many instances.

EFFECT OF GRADE ON TIME AND FEED REQUIRED TO ATTAIN A GIVEN FINISH

The results of an extensive experiment conducted at the Illinois station with the more common classes and grades of feeder cattle are summarized in Table 107. The plan was to feed similar rations to three lots of a certain class and grade of feeder, with the first lot being slaughtered when it was felt that the cattle would yield carcasses grading one grade below the feeder grade, the second when it reached the slaughter grade corresponding to the feeder grade, and the third when it reached the slaughter grade above the feeder grade. It will be seen that a poorer job of estimating the carcass grade was done in the heifer lots than in the steer lots.

Chapter 14

ENERGY in
the FINISHING RATION

The principal difference between finishing rations, or those fed to cattle being readied for slaughter, and rations that are fed to cows or to stockers is in their energy content. Improvement in condition or finish is the principal aim in finishing programs, and this is brought about by feeding a ration that supplies energy in excess of the amount needed for maintenance and growth—because, of course, growth occurs simultaneously with fattening in younger cattle, especially in the calf-finishing programs.

Corn, sorghum, and barley are the principal high-energy feeds used in finishing rations but in some localities other sources of energy, such as molasses, are of some importance. In this discussion the term "grain" is generally used in speaking of "high-energy" feeds, because its meaning is more fully understood. High-energy feeds are also quite generally referred to as carbonaceous concentrates because of their high carbohydrate content, in contrast to protein concentrates that are so termed because of their above-average protein or nitrogen content. It should be mentioned that, generally speaking, protein concentrates, especially the oilseed meals, are also quite high in energy content.

During the 1960s beef cattle nutritionists have probably devoted more time and research funds to study of the energy portion of the ration than to that of any other nutrients. The relative increase in cost of metabolizable or net energy supplied by roughages, combined with increasing labor and machinery costs for processing roughage, has been responsible for much of the interest in high-concentrate finishing rations.

Because the ruminant's digestive system is uniquely adapted to the utilization of high-bulk, high-fiber feeds, it was formerly believed that rations containing mainly grain should be fed only near the end of a finishing period and after feeding a stocker or growing ration for some time. Few digestive disturbances occurred under this system because the shift from low or medium to high energy in the ration was made gradually, perhaps over a 3- or 4-week period.

Recent research in steam processing and rolling of grains has resulted in high-concentrate rations with the "bulky" characteristic of rations that contain some roughage. These rations are readily eaten and, although fewer pounds are consumed, their greater caloric density results in a higher energy intake. The rate of gain is not consistently improved, but fewer pounds of feed are required per pound of gain.

Year-round feeding of finishing rations has become commonplace, meaning of course that many cattle are on finishing rations during the hot summertime. It has rather recently been discovered that the fiber content of rations consumed during hot weather affects maintenance requirements materially. Apparently digestion and metabolism of the energy in crude fiber increases body heat production more than does the utilization of concentrates. This is wasted energy in the sense that its end result produces neither growth nor gain. And the dissipation of this extra heat itself, because of the increased respiration rate and restlessness, requires further energy. Hence the increased maintenance requirement and reduced efficiency when higher-roughage rations are fed. For these reasons it is quite common for the rations fed in southwestern feedlots in the summer to contain as little as 5 percent roughage as contrasted to the more typical 20 percent.

Apart from their feeding qualities, high-concentrate rations lend themselves to mechanical processing and handling. In addition, the manure disposal problem is lessened, as fewer pounds of manure are produced because of reduced feed intake and lower bulk in the ration.

High-concentrate rations seem best suited to yearlings rather than calves, and to the lower grades of feeders. Apparently there is a tendency for younger cattle of the British breeds to fatten too early and at too light a weight on high-concentrate rations. Thus there still seems to be a place for the combination stocker-feeder programs in areas where both roughages and grain are produced. This may well mean that the higher grades of feeders will be fed where corn and sorghum silages are economically produced, and the "Okie" type of cattle will be fed mainly in the parts of the country where high-concentrate rations are most feasible.

FULL FEED DEFINED

Most of the cattle finished for market are given a full feed of grain during part of the feeding peroid. On some farms and in most commercial feedlots, cattle are placed on a full feed as quickly as possible, perhaps within 2 or 3 weeks after they are put into the feedlot; on others a full feed of grain is still fed during only the last month before the cattle are sold.

Opinions differ as to what constitutes a full feed of grain. Most feeders consider cattle to be on a full feed of grain when they are fed all the grain they will clean up in an hour or so if fed twice daily. Yet, if they are also fed all the good-quality legume hay and corn silage they want, they can be induced to eat still more grain by limiting the roughage to what they will clean up readily.

It is therefore highly desirable that the term "full feed" be defined more accurately than by saying simply that it is the amount of grain that cattle will eat. The term "full feed" may be applied to the roughage part of the ration as well as the grain. If the grain and hay were limited but the cattle were fed all the corn silage they would eat, we would then say they were given a *full feed of silage*. However, unless otherwise indicated, the term refers to a full feed of grain. The amount of grain that cattle will eat varies much too widely for the term to have much value to the inexperienced feeder who is attempting to feed his cattle in accordance with approved practices.

A much better way to define the term is to speak of the amount of grain they *should eat* if they are to be regarded as full-fed cattle. With this thought in mind we shall define a *minimum full feed of grain* as being 1.8 pounds daily per 100 pounds live weight, including the grain in any corn silage that is fed. Anything over this amount is, of course, a full feed without qualification.

A very heavy consumption of concentrates during the early part of the feeding period is likely to "burn out" the cattle and result in sluggish appetites and low gains toward the end. This is especially true for long-fed cattle. The Iowa calves in Table 108 are a good example of cattle that were fed too much corn during the first half of the feeding period. Whereas they averaged 10.2 pounds of corn per day during the third month and 13 pounds during the sixth, their average consumption was only 10.2 pounds during the eighth month and 11.2 pounds during the ninth.

After cattle have been worked up to a full feed, they should be kept there until they are marketed. Consequently, it is highly important that they be fed so that their consumption of grain does not fall below 1.8 pounds per 100 pounds live weight even during the last month they are on feed.

HOW MUCH FEED WILL CATTLE EAT?

The previous discussion as to what constitutes a full feed of grain has brought out the fact that the amount of grain or roughage eaten can be controlled by increasing or decreasing the other components of the ration.

Table 108

Feed Eaten per 100 Pounds Live Weight by Cattle of Different Ages

	First Month	Second Month	Third Month	Fourth Month	Fifth Month	Sixth Month	Seventh Month	Eighth Month	Ninth Month	Average Total Period
Shelled corn (lb)										
Calves										
Iowa[a]	1.6	1.8	1.8	1.8	1.8	1.6	1.4	1.2	1.2	1.6
Indiana[b]	1.3	1.5	1.5	1.6	1.8	1.7	—	—	—	1.6
Nebraska[c]	*	*	*	*	*	*	*	—	—	1.9
Morrison[d]	*	*	*	*	*	*	*	—	—	1.7
Yearlings										
Iowa	1.4	1.8	1.7	1.6	1.5	1.4	1.3	—	—	1.6
Indiana	1.3	1.7	1.7	1.7	1.6	1.6	*	—	—	1.6
Nebraska	*	*	*	*	*	*	*	—	—	1.8
Morrison	*	*	*	*	*	*	—	—	—	1.7
2-year-olds										
Iowa	1.3	1.9	1.7	1.7	—	—	—	—	—	1.7
Indiana	1.1	1.5	1.5	1.7	1.6	1.5	*	—	—	1.5
Nebraska	*	*	*	*	*	*	*	—	—	1.6
Morrison	*	*	*	*	*	—	—	—	—	1.6

Table 108 (Continued)

	First Month	Second Month	Third Month	Fourth Month	Fifth Month	Sixth Month	Seventh Month	Eighth Month	Ninth Month	Average Total Period
Total air-dry feed (lb)										
Calves										
Iowa	2.6	2.6	2.6	2.5	2.4	2.4	2.1	1.9	1.8	2.3
Indiana	2.7	2.7	2.6	2.5	2.4	2.3	—	—	—	2.5
Nebraska	*	*	*	*	*	*	*	—	—	2.6
Morrison	*	*	*	*	*	*	*	—	—	2.5
Yearlings										
Iowa	2.7	2.7	2.6	2.4	2.2	2.1	1.9	—	—	2.4
Indiana	2.9	2.7	2.6	2.5	2.2	2.1	—	—	—	2.4
Nebraska	*	*	*	*	*	*	*	—	—	2.4
Morrison	*	*	*	*	*	*	—	—	—	2.5
2-year-olds										
Iowa	2.4	2.6	2.3	2.1	—	—	—	—	—	2.4
Indiana	2.6	2.5	2.4	2.4	2.2	1.9	—	—	—	2.3
Nebraska	*	*	*	*	*	*	*	—	—	2.2
Morrison	*	*	*	*	*	*	—	—	—	2.4

* Monthly amounts not reported.
[a] Iowa Bulletin 271.
[b] Indiana Bulletin 136.
[c] Nebraska Bulletin 229.
[d] Morrison, *Feeds and Feeding*, 21st edition, p. 799.

If cattle are being fed principally to utilize roughage, the grain is fed in limited amounts in order to induce the cattle to consume large quantities of roughage. But if the purpose of feeding is to market as much grain through cattle as possible, it is roughage that is limited. This leads to such questions as how much total feed cattle will eat and what the replacement value of roughage is in terms of grain.

The lower half of Table 108 shows that the total air-dry feed eaten per 100 pounds live weight decreases gradually during the grain-feeding period. During the first quarter it is about 2.7 pounds; during the last quarter it is often little more than 2 pounds per 100 pounds live weight. However, this does not mean that the daily consumption of feeds has declined. Rather it has constantly increased because the cattle have increased in weight faster than their feed intake per unit of weight has diminished. Furthermore, if grain content of the ration is also increased as the feeding period progresses, the increasing caloric density may offset reduced intake of ration.

Between cattle of different ages the differences in feed consumption per 100 pounds live weight are too small to be of practical importance. Table 108 shows that the Iowa calves were grain-fed for 9 months, which is about 1 month longer than most calves are fed. Had they been marketed after 8 months, their average daily consumption of feed for the total period would have been 2.4 pounds. On the other hand, the Indiana calves were fed for only 6 months. Had they been fed 2 months longer, their daily consumption of both corn and hay probably would have been about 2.4 pounds per 100 pounds live weight.

In the case of the 2-year-old steers, both the Indiana and Nebraska steers were fed somewhat longer than most cattle of this age. Had they been fed only 5 months, it seems reasonable to believe that their average total feed consumption per 100 pounds live weight would also have been close to 2.4 pounds. Consequently, 2.4 pounds per 100 pounds live weight appears to be a good figure to use in estimating the average daily total feed consumption of cattle that are fed a full feed of grain after about the first month or 6 weeks of the feeding period.

The approximate consumption during different parts of the feeding period may be obtained by assuming the total air-dry feed to be 2.7, 2.5, 2.3, and 2.1 percent of the live weight for the first, second, third, and last quarters of the feeding period, respectively. This is on the assumption that calves are fed grain for approximately 250 days, yearlings 200 days, and 2-year-old steers 150 days. For cattle fed a shorter time the estimates for the last quarter should be disregarded; and for a feeding period of only one-half the usual length, only 2.7 and 2.5 percent will apply. To make month-by-month estimates of the grain or total feed consumed daily, the monthly averages given in Table 108 may be used.

In discussing a full feed of grain and the total feed consumed daily, no mention was made of the amount of protein concentrate fed. This item of the ration has been intentionally omitted from the discussion, as the feeding of 1 to 2 pounds of protein supplement has no significant effect on the amount of grain and roughage consumed. Consequently, it may be ignored in estimating the amount of grain required for a full feed and the total air-dry feed required to satisfy the appetites of full-fed cattle.

REPLACEMENT VALUE OF ROUGHAGE

It appears from the data in Table 109 that grain and air-dry roughage are interchangeable on a pound-for-pound basis for cattle that are fed at least half of a minimum full feed of grain. The replacement value of silage in terms of grain is more difficult to determine because it varies greatly in moisture content. However, assuming that the air-dry roughage equivalent of silage is 33.3 percent of the weight fed, grain and silage will replace each other on an air-dry basis within the limits of practical error.

A SIMPLE METHOD FOR ESTIMATING FEED REQUIRED TO FINISH CATTLE

The factors developed in the preceding paragraphs in regard to the feed consumption of finishing cattle are valuable in estimating the amount of grain and roughage required to finish a given drove of cattle. These factors are also useful in checking the ration at different stages of the feeding period to see whether grain and roughage are being fed in about the required amounts. The following example shows how these factors are used.

Given: 30 steers, weighing 700 pounds when purchased, which are to be fed to choice finish. How much shelled corn or sorghum and hay will be needed?

Solution:

Estimated time to be fed, 150 days

Estimated average daily gain, 2.6 lb

Calculated final weights $(700 + 150 \times 2.6)$ lb = 1,090 lb

Calculated average weight $(700 + 1,090)/2$ = 895 lb

Total feed required daily per 100 lb live weight = 2.4 lb

Total feed required daily per head, 2.4 lb $\times$ 8.95 = 21.5 lb

Total feed required daily by 30 steers, 30 $\times$ 21.5 = 645 lb

Average grain eaten daily per 100 lb live weight by cattle on full feed, 1.8 lb

Average grain eaten daily by 895-lb steer, $8.95 \times 1.8 = 16.2$ lb
Average grain eaten daily by 30 steers, $30 \times 16.2 = 486$ lb
Average hay eaten daily by 30 steers $= 645 - 486 = 159$ lb
Average hay eaten daily per steer $= 159/30 = 5.3$ lb
Estimated total grain needed for 150 days $= 150 \times 486 =$ 72,900 lb $= 1,302$ bu corn or 1,565 bu sorghum
Estimated hay needed for 150 days $= 150 \times 159 = 23,850$ lb $= 12$ tons.

Table 109
Replacement Value of Grain and Roughage in the Daily Ration of Feeder Cattle

Replacement Made	Average Weight During Experiment (lb)	Average Daily Ration			Average Total Air-Dry Feed per Day (lb)[a]	Time Fed (days)	Air-Dry Feed per 100 Pounds Live Weight (lb)
		Grain (lb)	Hay (lb)	Silage (lb)			
Grain replaced by hay							
Nebraska							
Heavy feed of grain	1,200	19.9	8.8	—	28.7	84	2.50
Medium feed of grain	1,156	17.2	9.7	—	26.8	84	2.33
Light feed of grain	1,124	12.0	14.0	—	26.0	84	2.31
Nebraska							
Heavy feed of grain	1,028	16.4	9.7	—	26.1	140	2.54
Medium feed of grain	1,039	14.0	12.6	—	26.6	140	2.56
Light feed of grain	1,042	12.1	15.5	—	27.6	140	2.65
Ohio							
Full feed of corn	923	13.3	3.2	13.7	22.0	240	2.38
¾ feed of corn	914	10.1	6.2	14.1	21.7	240	2.38
½ feed of corn	889	6.7	8.7	14.2	21.1	240	2.37
Grain replaced by silage							
Kansas							
Full feed of barley	936	14.0	—	17.8	21.2	180	2.27
⅔ feed of barley	916	9.4	—	33.6	22.8	180	2.49
⅓ feed of barley	909	4.7	—	41.8	21.4	180	2.36
Iowa							
Full feed of corn	900	13.2	1.1	17.7	21.4	175	2.38
½ feed of corn	862	6.4	1.1	32.3	20.4	175	2.37

[a] Air-dry weight of silage assumed to be 40 percent of weight of silage fed.

Similarly, if it is desired to know how much grain and hay will be needed during the last month of the feeding period, a close estimate may be made by assuming that approximately 2 pounds of grain and hay combined will be eaten daily per 100 pounds live weight. About 1.7 pounds of this amount will be grain and 0.3 pound will be hay (see Table 110). Multiplying these quantities by the estimated weight of the cattle and dividing the results by 100 will give the approximate average daily consumption of grain and hay during the last month of the feeding period.

Table 110
Average Amounts of Feed Consumed by Full-Fed Cattle During Different Parts of the Feeding Period[a]

| Quarter of Feeding Period | Feed per 100 Pounds Live Weight | | | Concentrate: Roughage (approximate ratio) |
	Total Air-Dry Feed (lb)	Grain (lb)	Air-Dry Roughage (lb)	
First	2.7	1.7	1.0	65:35
Second	2.5	1.9	0.6	75:25
Third	2.3	1.8	0.5	80:20
Last	2.1	1.7	0.4	85:15
Av. entire period	2.4	1.8	0.6	75:25

[a] Average length of grain feeding period: 2-year-old steers, 150 days; yearlings, 200 days; calves, 250 days.

The amount of silage required in this example can be determined by increasing the daily hay consumption by 3 or by dividing it by the factor 0.33. Because of the palatability of silage, the grain in the silage will be consumed without reducing the regular grain consumption to any great extent.

NUTRIENT REQUIREMENTS OF FEEDER CATTLE

The simple rules of thumb just discussed for estimating total feed and energy requirements, and those for protein requirements discussed in Chapter 16, simply serve as guides in the formulation of rations and in determining the total feed requirements of a drove of feeder cattle. For those wishing to formulate rations more exactly, or for those wishing to feed a complete mixed ration, Tables 111 and 112 give the National Research Council requirements for feeder calves, yearlings, and 2-year-old

Table 111

Daily Nutrient Requirements of Feeder Cattle[a]
(Based on Air-Dry Feed Containing 90 Percent Dry Matter)

| | | | | | Daily Nutrients per Animal | | | | | | | |
| | | | | | Digestible Energy | | | | | | | |
Body Weight (lb)	Average Daily Gain[b] (lb)	Daily Feed per Animal (lb)	Total Protein (lb)	Digestible Protein (lb)	For Maintenance (kcal)	Per Pound of Gain (kcal)	Total (kcal)	TDN[c] (lb)	Ca (gm)	P (gm)	Carotene[d] (mg)	Vitamin A (IU)
					Calves Finished as Short Yearlings							
400	2.3	11.8	1.3	1.0	6600	3900	15600	7.8	20	15	22	8850
600	2.4	16.4	1.8	1.3	9000	5300	21700	10.9	20	17	31	12300
800	2.2	19.4	1.9	1.5	11200	6600	25700	12.9	20	18	37	14600
1000	2.2	23.0	2.3	1.7	13300	7800	30500	15.3	21	21	44	17300
					Finishing Yearling Cattle							
600	2.6	17.5	1.8	1.3	9000	5300	22800	11.4	20	17	33	13100
800	2.7	22.3	2.2	1.7	11200	6600	29000	14.5	20	20	42	16700
1000	2.6	25.8	2.6	1.9	13300	7800	33600	16.8	23	23	49	19400
1100	2.3	25.8	2.6	1.9	14200	8400	33500	16.8	23	23	49	19400
					Finishing Two-Year-Old Cattle							
800	2.8	23.3	2.3	1.7	11200	6600	29700	14.9	22	22	44	17500
1000	2.9	28.2	2.8	2.1	13300	7800	35900	18.0	26	26	54	21200
1200	2.7	31.0	3.1	2.3	15200	9000	39500	19.8	28	28	59	23300

[a] *Nutrient Requirements of Beef Cattle*, Sub-Committee on Beef Cattle Nutrition, National Research Council, revised 1963.

[b] Average daily gain for finishing cattle is based on cattle receiving stilbestrol. Finishing cattle not receiving stilbestrol gain 10 to 20 percent slower than the indicated values.

[c] The TDN was calculated from digestible energy by assuming 2000 kcal of DE per pound of TDN.

[d] The carotene requirement was calculated from the vitamin A requirement assuming 400 IU of vitamin A per milligram of carotene.

Table 112
Nutrient Requirements of Feeder Cattle Expressed as Percentage Composition of Air-Dry Rations[a]

Body Weight (lb)	Average Daily Gain[b] (lb)	Daily Feed per Animal (lb)	Total Protein (%)	Digestible Protein (%)	Digestible Energy (kcal/lb)	TDN[c] (%)	Ca (%)	P (%)	Carotene[d] (mg/lb)	Vitamin A (IU/lb)
				Calves Finished as Short Yearlings						
400	2.3	11.8	11.0	8.2	1325	66	0.37	0.28	1.9	750
600	2.4	16.4	11.0	8.2	1325	66	0.27	0.23	1.9	750
800	2.2	19.4	10.0	7.5	1325	66	0.23	0.21	1.9	750
1000	2.2	23.0	10.0	7.5	1325	66	0.20	0.20	1.9	750
				Finishing Yearling Cattle						
600	2.6	17.5	10.0	7.5	1300	65	0.25	0.21	1.9	750
800	2.7	22.3	10.0	7.5	1300	65	0.20	0.20	1.9	750
1000	2.6	25.8	10.0	7.5	1300	65	0.20	0.20	1.9	750
1100	2.3	25.8	10.0	7.5	1300	65	0.20	0.20	1.9	750
				Finishing Two-Year-Old Cattle						
800	2.8	23.3	10.0	7.5	1275	64	0.21	0.21	1.9	750
1000	2.9	28.2	10.0	7.5	1275	64	0.20	0.20	1.9	750
1200	2.7	31.0	10.0	7.5	1275	64	0.20	0.20	1.9	750

Percentage of Ration or Amount per Pound of Feed

[a] *Nutrient Requirements of Beef Cattle*, Sub-Committee on Beef Cattle Nutrition, National Research Council, revised 1963.
[b] Average daily gain for finishing cattle is based on cattle receiving stilbestrol. Finishing cattle not receiving stilbestrol gain 10 to 20 percent slower than the indicated values.
[c] The TDN was calculated from digestible energy by assuming 2000 kcal of DE per pound of TDN.
[d] The carotene requirement was calculated from the vitamin A requirement assuming 400 IU of vitamin A per milligram of carotene.

cattle. As in the previous NRC tables, ration requirements are shown both on a daily requirement basis and a percentage composition basis. Morrison's standards are shown in Table 113.

STARTING CATTLE ON A FEED OF GRAIN

Feeder cattle coming directly off grass or a stocker ration consisting solely of roughages undergo a rather serious physiological shock if they

Table 113
Morrison's Feeding Standards for Finishing Cattle[a]

Class of Cattle and Weight (lb)	Requirements per Head Daily							
	Dry Matter (lb)	Digestible Protein (lb)	Total Digestible Nutrients (lb)	Calcium (gm)	Calcium (lb)	Phosphorus (gm)	Phosphorus (lb)	Carotene (mg)
Calves finished for baby beef								
400	9.6–12.1	1.05–1.15	7.4– 8.6	20	0.044	15	0.033	25
500	11.3–13.8	1.14–1.26	8.8–10.2	20	0.044	16	0.035	30
600	13.2–15.8	1.26–1.37	10.2–11.8	20	0.044	17	0.037	35
700	14.8–17.5	1.39–1.52	11.6–13.2	20	0.044	18	0.040	40
800	16.7–19.3	1.52–1.68	12.6–14.4	20	0.044	18	0.040	45
900	17.7–20.3	1.64–1.82	13.5–15.5	20	0.044	18	0.040	50
Finishing yearling cattle								
600	15.0–17.6	1.18–1.32	10.7–12.3	20	0.044	17	0.037	35
700	16.5–19.1	1.36–1.52	12.7–14.3	20	0.044	18	0.040	40
800	17.8–20.4	1.52–1.68	14.1–15.9	20	0.044	19	0.042	45
900	18.9–21.7	1.64–1.82	15.4–17.2	20	0.044	20	0.044	50
1,000	20.0–23.0	1.71–1.91	16.0–18.0	20	0.044	20	0.044	55
1,100	21.0–24.0	1.76–1.96	16.5–18.5	20	0.044	20	0.044	60
Finishing 2-year-old cattle								
800	19.6–22.2	1.46–1.62	14.1–15.9	20	0.044	20	0.044	45
900	20.7–23.5	1.53–1.78	14.6–17.4	20	0.044	20	0.044	50
1,000	22.0–25.0	1.65–1.85	16.5–18.5	20	0.044	20	0.044	55
1,100	24.0–27.0	1.70–1.90	17.0–19.0	20	0.044	20	0.044	60
1,200	24.0–27.0	1.70–1.90	17.0–19.0	20	0.044	20	0.044	65

[a] Taken by permission of The Morrison Publishing Company, Ithaca, N.Y., from *Feeds and Feeding*, 22nd edition, by Frank B. Morrison.

are abruptly started on a high-energy ration. This is mainly because the rumen microfloral population consists of specific organisms that are adapted to the use of feeds high in crude fiber. Rumen pH drops drastically as a result of the high level of acetic acid produced from fermentation of soluble sugars and starches present in the high-concentrate ration, retarding the growth of certain microflora and thus causing a serious imbalance in the microfloral population. Upwards of 3 weeks are required to reestablish a normal status in the rumen, and feed intake and weight gains are seriously reduced. If the shock just referred to is imposed upon the shock of weaning and exposure to the variety of viruses found in saleyards and trucks or trains, death losses in starting calves on feed can run as high as 5 to 10 percent.

To further complicate matters, comparatively few of the feeder cattle shipped into the feedlots have been fed any grain. Consequently, they must be taught to eat it. In the farm feedlots this can be done easily by putting about a pound of grain per head in the feed trough and feeding 4 to 6 pounds of hay or silage on top of it. In eating the leaves and chaff from the bottom of the trough the cattle will also pick up the grain and will quickly learn to eat it. If it is desired to get the cattle on a full feed of grain rather promptly, the grain should be increased by about 1 pound per head daily until they are eating 1 pound of grain per 100 pounds live weight, after which time increases of 1 pound a head every third day for 2-year-old steers, 0.5 pound every third day for yearlings, and 0.25 pound every third day for calves are recommended. All this time, of course, the cattle should have as much roughage in the form of hay or silage as they will eat. If all the cattle do not appear hungry when they are fed, the hay or silage should be sufficiently reduced to ensure that all animals come to the feed trough promptly and begin eating when the grain is fed.

If the grain ration is increased gradually as suggested above, 2-year-old cattle will attain a minimum full feed of grain (1.8 percent of their live weight) about 30 days after being started on feed, yearlings in about 40 days, and calves in about 50 days. Further increases, of course, should be made as the cattle increase in weight and become better adjusted to a heavy feed of grain.

Starting cattle on feed in commercial feedlots that have their own feed processing plants is another matter. In this instance the roughage portion of the ration—usually chopped hay, cottonseed hulls, or some other bulky by-product feed—is mixed with coarsely prepared grain, molasses, and a fortified and medicated protein supplement and offered free-choice several times daily. The first feed offered may contain 50 to 60 percent roughage, with reductions being made at weekly intervals or so until the roughage content is down to 5 or 10 percent. The feedbunks should never be completely empty when this method is used.

For cattle that are to be fed on self-feeders, complete mixed starter rations such as those fed in commercial feedlots will serve well, if the feed is not too bulky to feed down in the feeders. Some farmer-feeders use a bulky grain such as oats or barley or ground ear corn as their starter concentrate and additional ground cobs are sometimes used. If cattle are being fed on pasture, they voluntarily consume an adequate amount of roughage to prevent digestive difficulty.

Obviously the weight of the cattle will increase more rapidly than their capacity for feed. Consequently, it is usually necessary to reduce the roughage from time to time in order to maintain the level of grain consumption at at least 1.8 percent of the live weight. (Calves, because of their rapid growth, may increase in size sufficiently to permit appropriate increases of grain without reducing the roughage below the amount being eaten when they attained a minimum full feed.) However, reducing the roughage to a weight below 0.5 percent of the live weight is seldom justified, since further decreases in the roughage level fed will not result in increases

FIG. 70. A heavy feed of corn silage with supplements and limited grain during the preceding winter cheapened the overall cost of gains on these steers being full-fed in drylot during the summer for the fall market. (Corn Belt Farm Dailies.)

in concentrate intake. In calculating roughage consumption the cobs (20 percent) in corn-and-cob meal and the estimated dry weight of the cobs and stover in corn silage should, of course, be considered as air-dry roughage. On a dry basis, corn silage consists of about 50 percent grain and 50 percent cob and stalk.

HIGH-ENERGY VERSUS MEDIUM-ENERGY FINISHING RATIONS

Researchers have found that increasing the energy content of finishing rations by removing or reducing the roughage portion of the ration improves feed efficiency, at least with respect to pounds of feed required per unit gain. This improved efficiency does not hold when the calculations are made on the basis of TDN or calories. When roughage calories are more economical than concentrate calories—and they often are in the corn and sorghum growing areas if the entire plant is harvested as silage—it may not always be wisest to feed the ration that requires the least pounds of feed to produce a pound of liveweight gain. Farmer-feeders, and even commercial feeders, who grow or buy silage are interested in pounds of gain produced per acre of corn or sorghum grown or purchased. Thus it should be profitable to review several experiments studying the effect of level of silage and grain in the rations fed finishing cattle.

An Indiana experiment summarized in Table 114, as well as another Indiana trial summarized in Table 85, Chapter 12, show that corn silage can be used to reduce the cost of steer gains without seriously affecting rate of gain and carcass grade. Remembering that about 15 percent of the wet weight of corn silage is composed of corn grain itself, it can be seen that grain intake, between lots, does not differ greatly. Note that in both trials, as corn grain level in the ration increased, pounds of TDN required per unit gain also increased. Thus at the levels of silage fed, the roughage portion did not decrease efficiency. In these two tests silage, and thus energy level, was held constant throughout the trail. Unfortunately, all calves in Table 114 were fed the same length of time. The gains of the calves on high silage would surely have been reduced had they been fed until they attained the same slaughter weight as those in the other three lots, but the carcass grades would also have been improved.

Feeders often mistakenly believe that only older, more mature cattle can make good use of medium-energy rations such as those high in corn silage. The comparisons between calves and yearlings described in Table 85 illustrate that this is not necessarily true. Note the reduction in pounds of feed required per pound gain in both age groups as energy level in

Table 114

Effect of Levels of Corn and Corn Silage in Steer Calf Finishing Rations (480-Pound Steer Calves Fed 205 Days)[a]

Level of Corn Grain Fed	1.5 Pounds	One-Third Full Feed	Two-Thirds Full Feed	Full Feed
Number of steers fed	20	20	20	20
Average daily gain (205 days) (lb)	2.01	2.28	2.31	2.31
Daily feed (lb)				
Corn grain	1.5	4.4	8.7	13.1
Supplement A[b]	2.0	2.0	2.0	2.0
Corn silage	40.0	35.0	29.0	14.0
Feed per pound gain (87% dry-matter) (lb)	8.6	8.2	9.0	8.6
TDN per pound gain (lb)	5.5	5.4	6.3	6.4
Carcass grades	8 C, 12 G	13 C, 7 G	17 C, 3 G	17 C, 3 G
Feed cost per pound gain (cents)[c]	13.4	13.5	16.0	17.2

[a] Indiana Cattle Feeders' Day Report, 1966.

[b] A fortified protein supplement formula described in Chapter 17.

[c] Feed prices used, per ton: corn, $40; Supplement A, $80; corn silage, $8.

the ration increased, but also, when feed required is expressed as TDN, both age groups required more TDN on the high-energy rations. More interestingly, the increase in TDN requirement was even smaller for the calves than for the yearlings (55 versus 69 pounds TDN), indicating that calves can make excellent use of corn silage rations. Farmer-feeders who feed these types of rations would incur a nonfeed cost of 4 to 5 cents per pound gain, which must be added to the very economical feed costs shown to arrive at the total cost of gain.

Another test that shows that limited or medium-energy levels may be more profitable than higher-energy rations is the Iowa test summarized in Table 115. As in the Indiana test, all cattle were fed on a time-constant basis—that is, all lots were sold at once. Had the heavy-silage lot been fed another 30 to 40 days to equalize final weights, both gain and feed efficiency would have been reduced slightly, but live grade would have improved, resulting in higher selling prices. The profit picture might not have improved, especially if the cattle were sold on a retail-cut basis, as the extra finish would only have reduced cutability.

From an industry standpoint, the silage programs had the following advantages:

1. More live weight and carcass beef were produced per acre of corn fed.

Table 115
Comparison of Performance of Steers on Various Corn Plant Combinations (265-Day Feeding Period for All Lots)[a]

Type of Ration	Ear Corn	Half Ear Corn, Half Silage	Corn Silage
Number of steers	24	24	24
Weight data (lb)			
Average initial weight	573	572	575
Average final weight	1,235	1,267	1,170
Average daily gain	2.51	2.63	2.25
Market to market average daily gain, 301 days (lb)[b]	2.25	2.35	2.02
Daily feed consumption (lb)			
Ground corn[c]	20.2	12.1	3.7
Corn silage[c]	—	7.1 (24.3)	11.0 (37.7)
Iowa 80 premix[d]	0.5	0.5	0.5
Total	20.7	19.7	15.2
Estimated corn grain	16.2	15.7	9.3
Feed required (lb) to produce:			
One pound of live weight	8.2	7.5	6.8
One pound of retail beef	23.0	20.7	17.9
Beef production per acre of 100-bu corn			
Liveweight basis (lb)	768	890	1,307
Retail-cut basis (lb)	275	321	496
Feed cost per pound liveweight gain (cents)	15.0	14.1	11.3
Feed + 10 cents per day per steer (cents) to produce:[e]			
One pound liveweight gain	19.0	17.9	15.8
One pound of retail beef	52.8	49.5	41.6
Profit per steer if sold on:[e]			
Live basis @ $23, $22.75, $22.50 per cwt	$ 9.90	$15.21	$19.05
Retail-cut basis @ $66.85 per cwt	$ 9.90	$19.45	$33.07
Cattle profit per acre of 100-bu corn if sold on:[e]			
Live basis at above prices	$11.31	$19.30	$41.83
Retail-cut basis at above prices	$11.31	$24.68	$72.62

[a] Adapted from Iowa Agricultural Experiment Station, AS Leaflet R69.

[b] Includes 36 days of feeding in a pretest conditioning period.

[c] On 89-percent dry-matter basis, with wet weight of silage in parentheses.

[d] High-urea fortified supplement described in Chapter 17.

[e] Cattle cost $25 per cwt delivered. Feed costs used, per ton: corn, $40; silage, $8; supplement, $120.

2. More profit was made per steer and per acre of corn fed.

3. Retail cutout was improved because of less waste fat trim.

4. A greater number of steers were fed per acre of corn.

A question that many cattle feeders ask is whether a combined stocker-feeder program in which only corn silage and supplement are fed during the stocker program, followed by a shorter finishing period on a higher-energy ration, would not combine the economy of silage feeding with superior carcass merit. The Illinois station conducted such a study, and the results are shown in Table 116. Heavy steer calves were fed on silage rations for either 112, 168, or 224 days and finished on a full feed of cracked shelled corn, soybean meal, and limited corn silage, to about 1,050 pounds and estimated choice live grade. Another lot was fed a full feed of corn from the outset as a control. As in the Iowa and Indiana studies, the higher-energy rations produced the most rapid but costliest gains. The medium-energy or high-silage cattle graded higher in the Illinois test, in contrast to other tests mentioned. This may be explained by the fact that all cattle were fed to approximately equal weights and, as the lot fed only silage for the longest period gained somewhat more slowly, they were older when slaughtered, especially in comparison with the lots on full feed from the start and those fed heavy silage for only 112 days. Marbling score, the chief determinant of carcass grade, is highly related to age, and even though the slower gaining steers had a quarter-inch less back fat, their marbling score was significantly higher.

As in the Iowa test, profit per steer was higher, but profit per acre of corn fed was still more favorable for the lots on a heavy feed of silage. These findings perhaps are most important for farmer-feeders, who feed cattle in the first place mainly to provide a better market for their crops and who feed only one drove of cattle per year. Nonfeed costs on these farms are not increased appreciably by extending the feeding period from 7 months to 10. In a commerical lot this additional 90 days could add from $5 to $10 in nonfeed costs per head to the overall cost of finishing a steer and, in the large-volume operation, this is especially important.

A point not to be overlooked in favor of the medium levels of energy in the finishing ration is cutability of carcass. The British breeds, especially, have for 50 years or more been selected for early maturity and, if fed the high-energy rations from an early age onward, they often reach choice grade at 850 to 950 pounds. Thus they become inefficient and produce wasty carcasses at weights that the industry cannot afford to live with. Lower grades of cattle such as "Okies" or Holsteins can make use of the higher-energy rations more advantageously, as discussed in Chapter 13.

In summary, it appears that corn silage can satisfactorily make up

Table 116

Effect of Length of Heavy Silage Feeding Period on Performance and Carcass Merit of Feeder Steers (10 Steers per Lot)[a]

Heavy Silage Feeding Phase				
Days on heavy silage	0	112	168	224
Total gain in period (lb)	—	210	272	372
Average daily gain (lb)	—	1.86	1.68	1.66
Average daily and total ration in period (lb)				
Corn silage	—	34.1–3,819	35.8–6,0143	4.1–7,638
Soybean meal	—	1.5–168	1.5–252	1.5–336
Feed cost per cwt gain in period[b]	—	$12.29	$14.22	$13.87
Gain per ton of silage (lb)	—	110	91	95
Gain per acre of corn fed (lb)[c]	—	1,980	1,638	1,710

Grain Feeding Phase				
Days on full feed of grain	196	112	105	63
Average daily gain in period (lb)	2.48	2.71	2.05	1.54
Average daily and total ration in period (lb)				
Corn, cracked shelled	12.7–2,489	14.6–1,635	13.9–1,459	12.8–806
Soybean meal	1.5–294	1.5–168	1.5–157	1.5–94
Corn silage	15.9–3,116	18.7–2,094	18.8–1,774	14.0–882
Feed cost per cwt gain in period	$15.84	$16.44	$21.07	$25.06

Combined Silage-Grain Feeding Phases				
Total days to finished weight	196	224	273	287
Average final weight (lb)	1,063	1,095	1,066	1,047
Average daily gain (lb)	2.48	2.29	1.79	1.63
Total feed consumed per steer				
Corn, cracked shelled (bu)	44.5	29.4	26.0	14.4
Total corn (bu)[d]	50.7	41.2	41.6	31.4
Corn silage (tons)	1.56	2.96	3.90	4.26
Soybean meal (lb)	294	336	409	430
Corn acreage utilized per steer[c]	0.64	0.53	0.54	0.42
Steers fed per acre of corn	1.56	1.89	1.83	2.38
Steer gain per acre of corn fed (lb)	810	971	896	1,116
Feed cost per cwt gain[b]	$15.84	$14.52	$17.25	$16.21
Dressing percent	62.7	61.0	61.8	60.2
Carcass grade	G+	G+	C−	C
Fat cover over 12th rib (in.)	0.63	0.59	0.63	0.39
Return over steer and feed cost[e]	$11.53	$10.77	$19.41	$8.32
Return over steer and feed costs per acre of corn fed	$17.99	$20.36	$35.52	$31.61

[a] Adapted from Illinois Agricultural Experiment Station, *Illinois Research*, Winter 1963.
[b] Feed costs used, per ton: corn silage, $10; soybean meal, $80; shelled corn, $40.
[c] Corn silage yield estimated at 18 tons or 80 bushels per acre.
[d] Includes corn in silage, assumed to equal 15 percent of the weight of the silage.
[e] Based on $26 stocker steer cost and carcass prices ranging from $34.60 to $37.00 per hundredweight.

one-third to one-half of a complete, mixed ration fed to finishing cattle, especially the lower grades, or it can be used as the principal ingredient in the ration for 150 to 200 days before finishing on a high-energy ration.

In the event the spread in fed cattle prices again becomes large, the somewhat reduced finish on the cattle fed large amounts of corn silage may penalize users of this excellent feeding program. The likelihood of a very wide spread in prices between high good and choice cattle occurring with consistency is so remote that this factor should not deter any feeder who is debating the question of whether corn silage will have a place in cattle finishing in the future.

COMMERCIAL FEEDLOT HIGH-ENERGY RATIONS

The principles demonstrated in the discussion of energy levels in the rations fed by farmer-feeders in general also hold when it comes to choosing the preferred formula for commercial feedlot rations. Several differences, however, should be noted. The commercial feeder prefers to feed—and in fact almost must feed—a complete mixed ration. His automatic feed-processing and mixing plant and his unloading trucks or wagons are all geared to handling and delivering to the bunk, rations in completely mixed and proportioned form. As already mentioned, this type of feedlot generally feeds a lower-grading kind of steer that responds more satisfactorily to high-energy rations, especially in terms of type and amount of finish or cutability produced. Another difference is that the commercial feeder feeds for a shorter period than his counterpart in the farm situation.

The Arizona station, noted for its nutrition research in commercial feedlot rations, has compared one lot of cattle that were fed so-called 1954 rations with other lots fed modern or 1965 rations that contained 80 or 90 percent concentrates. The rations and performance data are shown in Table 117. The rations for all lots contained 40 percent roughage for the first 56 days.

The obvious differences between the modern and the 1954 rations are in method of grain processing, additions of tallow, vitamin A, and stilbestrol, and reduction in roughage content, at least with respect to the 90 percent concentrate ration. Protein, calcium, phosphorus, and fiber content were similar, yet the modern ration with 80 percent concentrate produced 32 percent more gain on 17 percent less feed than did the 1954 ration. The feed cost savings amounted to $2.10 per hundredweight gain or $8.40 in producing 400 pounds of gain.

This experiment, like those previously reviewed from the Corn Belt stations, shows that about 80 percent concentrates or a concentrate to rough-

Table 117
Comparison Between 1954 Ration and 1965 or Modern 80 and 90 Percent Concentrate Rations[a]

Type of Ration	80 Percent Concentrate		90 Percent Concentrate			1954 Ration	
Number of days fed	56	84	56	42	42	56	113
Ration ingredients (lb)							
Ground alfalfa hay	20.0	5.0	20.0	10.0	2.0	25.0	20.0
Cottonseed hulls	20.0	15.0	20.0	15.0	8.0	15.0	—
Steam-processed milo	44.7	63.3	44.6	58.9	73.6	—	—
Dry-rolled milo	—	—	—	—	—	49.3	73.8
Cottonseed pellets	4.5	5.5	4.5	5.0	5.0	4.5	—
Molasses	5.0	5.0	5.0	5.0	5.0	5.0	5.0
Tallow	4.0	4.0	4.0	4.0	4.0	—	—
Dicalcium phosphate	0.5	0.5	0.5	0.5	0.5	0.3	0.3
Urea	0.8	0.6	0.8	0.5	0.6	0.4	0.4
Salt	0.5	0.5	0.5	0.5	0.5	0.5	0.5
Ground limestone	—	0.6	—	0.6	0.8	—	—
	100.0	100.0	100.0	100.0	100.0	100.0	100.0
Vitamin A premix (gm)	10	10	10	10	10	—	—
Stilbestrol	+	+	+	+	+	−	−
Performance data							
Number of steers	16		16			16	
Average initial weight (lb)	626		633			628	
Average daily gain (shrunk 5%) (lb)	3.0		2.9			2.2	
Average daily feed (lb)	23.7		23.3			20.5	
Feed per cwt gain (lb)	786		813			924	
Feed cost per ton	$53.68		$56.25			$50.50	
Feed cost per cwt gain	$20.48		$21.38			$22.58	

[a] Adapted from Arizona Cattle Feeders' Day Report, 1967.

age ratio of 4:1 is apparently ideal with respect to rate of gain, feed efficiency, physiological well-being, and carcass merit. Under unusual economic circumstances, even greater concentrate levels may be indicated, but this situation is the exception rather than the rule.

Before closing the subject of level of concentrate or energy in the ration, it should be mentioned that there is a very real problem with respect to liver condemnations at slaughter time, often amounting to more than 50 percent, when all-concentrate or 90 percent concentrate rations are fed. The economic loss is substantial. A highly inflamed rumen wall condition, known as rumen parakeratosis, also occurs and surely interferes with absorption of nutrients from the rumen. The physiology of these two conditions is not well understood, but they are believed to be associated with the lowered pH in the rumen. Including 75 milligrams of terramycin or aureomycin per day in the rations of cattle fed all-concentrate rations reduces the incidence of abscessed liver. Founder and chronic bloat also occur in 10 to 15 percent of cattle fed such rations, depending on the degree of care with which the cattle are started on feed.

Chapter 15

CARBONACEOUS CONCENTRATES and THEIR USE in the FINISHING RATION

With respect to source, concentrates are of two kinds: (1) the grains of farm-grown crops and (2) commercially manufactured feedstuffs. With respect to composition, concentrates may be divided into two groups: (1) carbonaceous feeds, or those containing a relatively high percentage of energy and a low percentage of protein, and (2) nitrogenous feeds, or those especially rich in protein. In general the farm-grown grains belong in the carbonaceous class, whereas a majority of the commercial feedstuffs are nitrogenous.

Carbonaceous feeds, also referred to as carbohydrate feeds, as their name implies, are rich in chemical compounds that contain a relatively large amount of carbon. The more important of these compounds are starches, sugars, and fats. All these materials, when consumed by animals, are utilized chiefly for maintenance, muscular energy, and conversion to body fat, which is stored in various parts of the body but especially in the superficial layers of muscle found just beneath the skin. Carbonaceous concentrates are thus of prime importance in finishing cattle for the market and generally make up the major portion of finishing rations.

COSTS OF FARM GRAINS

Prices paid for feed grains by commercial cattle feeders, or prices used in calculating costs by farmer-feeders who may produce their own grain supply, vary considerably from area to area. As with most commodities, the prevailing price for a feed grain is a matter of supply and demand and transportation or freight costs. Government price support programs also are a factor, especially in years of average or smaller than average acreages and yields. Figure 71 shows that feed grain prices in the western Corn Belt and in the Sorghum Belt are well below the national average. This circumstance undoubtedly has contributed greatly to the phenomenal increase in cattle feeding in Nebraska, western Iowa, Colorado, and the Panhandle area. One might project that the western states will reduce their feeding operations and that the southern states will not assume an important role as a cattle-finishing area because of the noncompetitive feed prices in these areas. It is more economical to move surplus cattle to surplus feed areas, but partially offsetting this apparent saving is the cost of moving the fed cattle or carcass beef to consumer centers at either end of the country. Drastic shifts in freight rates can quickly alter the relative competitive position of certain areas, favorably or unfavorably—for instance, reductions in rates on freight shipments to the South in 1967 substantially reduced feed grain prices in that region, relatively speaking.

Other factors such as climate, labor costs, feeder cattle supply, and related factors, of course, all enter into this question but, because feed costs amount to about 85 percent of all costs in the cattle-finishing program, relative feed prices must be given a high priority when assessing regional suitability for cattle feeding.

IMPORTANT ENERGY SOURCES

The important carbonaceous concentrates used in cattle feeding are corn, grain sorghum, oats, barley, wheat, beet pulp, and molasses. Although the use of some of these feeds is limited to certain sections, a brief discussion of all of them will be attempted, along with a discussion of other feeds that are less commonly used.

CORN

In corn (*Zea mays*) is found the explanation of the remarkable development attained by the beef cattle industry of the United States, especially in earlier years and of course specifically in the Corn Belt. Other countries have cattle of as good or better breeding; other regions have as good or more favorable climate; other nations have more and better grasslands; but no other country has such an abundant supply of corn available for cattle feeding. As mentioned in Chapter 2, United States corn production

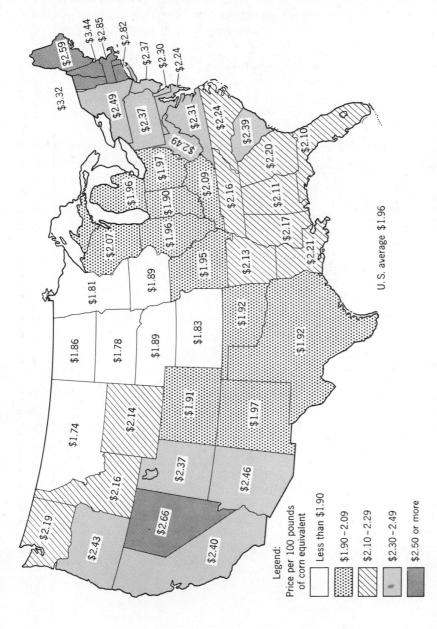

FIG. 71. Feed grain prices received by growers, per 100 pounds of corn equivalent, for the crop years 1957–1961. (W. W. Pawson, Cattle Feeding Research Workshop, Denver, Colorado, 1964.)

Legend:
Price per 100 pounds
of corn equivalent

Less than $1.90
$1.90 – 2.09
$2.10 – 2.29
$2.30 – 2.49
$2.50 or more

U. S. average $1.96

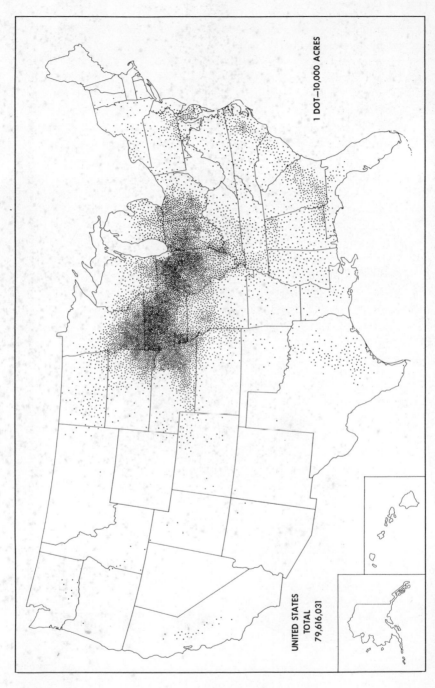

UNITED STATES
TOTAL
79,616,031

1 DOT—10,000 ACRES

FIG. 72. Corn acreage harvested for all purposes in 1959. Corn is easily the most important cattle-finishing feed in the United States, especially in the eastern half of the country. (U.S. Department of Commerce.)

consistently amounts to some 4 billion bushels annually, of which 15 percent is consumed by beef cattle.

Corn is unsurpassed as a feed for finishing cattle. It is highly palatable, being eaten readily by cattle of all ages. It meets the requirements for a high-energy feed, being high in digestible carbohydrates and fat—the two nutrients that normally furnish most of the energy in finishing rations. Corn is easily stored and easily prepared for consumption. It is produced in great abundance and, consequently, is readily obtainable. In fact, it may be said that corn combines all the essentials of a valuable cattle feed, save one. It is noticeably deficient in protein. For this reason corn should not be fed alone, but should be supplemented with feeds that are high in protein, such as the oilseed meals and alfalfa hay.

In all sections where corn is extensively grown (see Fig. 72) it should, and normally does, form the major portion of the concentrate of the finishing ration. It is seldom possible for Corn Belt feeders to use other carbonaceous concentrates to any great extent without materially increasing the cost of the ration or lowering its efficiency, or possibly both. Grain sorghums approach corn in this respect in the southern and western fringes of the Corn Belt and in the Sorghum Belt itself—that is, the Panhandle of Oklahoma and Texas, and adjoining parts of Kansas, Colorado, and New Mexico.

NEW VERSUS OLD CORN

In view of the large amount of corn placed in storage under the government loan and purchase agreement, some of which is carried over for 2 or more years, the question arises concerning the feeding value of such corn in comparison with corn that has been recently harvested. Old corn contains less moisture than new corn and consequently is a more concentrated feed. Assays of corn of different ages show a loss of carotene with increase in storage time. This is probably of little significance if plenty of high-quality legume hay is fed with the corn, but it might be of great importance if the corn is relied on as the principal source of carotene. Feeding tests such as those summarized in Table 118, comparing corn stored for one or more years with recently harvested corn, show no significant difference between old and new corn for finishing cattle.

Another test, comparing still older corn with new corn, was conducted at the Ohio station. They fed steers for 84 days on either the old or new corn, both hammermill-ground, and quality mixed hay, soybean meal, and minerals. The steers in one lot fed each type of corn received additionally 20,000 IU of vitamin A. As may be seen in Table 119, the old corn pro-

Table 118

Relative Value of Old and New Corn for Finishing Cattle (210-Day Feeding Period)[a]

	4-Year-Old Corn	3-Year-Old Corn	2-Year-Old Corn	1-Year-Old Corn	New Corn
Corn data					
Relative hardness of kernels (%)	164	160	174	150	100
Moisture (%)	10.4	10.9	12.4	12.9	16.7
Vitamin A (IU per pound)	821	809	869	1,330	1,720
Crude protein (14% moisture basis) (%)	9.8	8.5	8.6	8.3	8.0
Cattle data					
Average initial weight (lb)	735	736	732	732	738
Average daily gain (lb)	2.20	2.15	2.10	1.97	2.15
Average daily ration (lb)					
Shelled corn (14% moisture)	16.3	15.0	16.1	15.7	14.9
Protein supplement	1.0	1.0	1.0	1.0	1.0
Alfalfa hay	4.0	4.0	4.0	4.0	4.0
Feed per cwt gain (lb)					
Shelled corn (14% moisture)	741	696	766	799	693
Protein supplement	45	47	48	51	47
Alfalfa hay	181	187	190	203	186
Shrinkage (%)	2.0	1.4	3.2	2.3	1.9
Dressing percent	63.7	63.8	63.9	63.5	62.8

[a] Iowa AH Leaflet 160.

Table 119

Comparison of Old and New Corn, with Supplemental Vitamin A, for Finishing Steers[a]

	1960 Corn		1954 Corn	
	Vitamin A	Control	Vitamin A	Control
Average daily gain (lb)	2.11	1.94	2.08	1.98
Average daily ration (lb)	18.8	18.4	19.1	18.5
Feed per cwt gain (lb)	893	952	915	938
Feed cost per cwt gain	$17.61	$18.66	$18.02	$18.40
Carcass grade	G+	G+	G+	G+
Dressing percent	61.0	61.8	61.2	61.8
Blood vitamin A (mcg/100 ml)	39.4	26.1	41.7	26.2

[a] Ohio Farm and Home Research, 1962.

duced results comparable to those of new corn. What appeared to be a gain response to supplemental vitamin A in both treatments was apparently not real as the higher dressing percentages of the unsupplemented steers offset the differences in liveweight gains. The old corn contained only 0.08 milligrams of carotene per pound compared with 0.49 for the new corn. Apparently the hay provided all the vitamin A activity required by these cattle.

The field sheller or corn combine is undoubtedly responsible for a great increase in the feeding of high-moisture new corn, especially if storage facilities such as the airtight Harvestore are available. Such high-moisture grain undergoes fermentation if stored in large quantities and has all the characteristics of silage. For this reason it is discussed more thoroughly under the topic of ear-corn silage.

VALUE OF SOFT CORN

Often a field of late-planted corn fails to mature before the first killing frost, and about once in 8 or 10 years an unusually early frost damages a considerable percentage of the crop over a large portion of the Corn Belt. Frost of course kills the green, growing stalk and halts the translocation of the food nutrients from stalk to ear, and the transfer of water from ear to stalk where it normally evaporates through the leaves. As a consequence the ears fail to dry out, and they contain an abnormally high percentage of moisture, as much as 40 percent not being uncommon. Such corn is commonly termed "soft." In practice, soft corn is corn that contains too high a percentage of moisture to be stored in the ear without danger of heating. Heating is likely to occur when the amount of moisture in the kernel in newly harvested ear corn exceeds 25 percent.

Soft corn is also immature and is not to be confused with mature corn that has a high moisture content. Because soft corn is unmarketable, it is retained on the farm and fed. The feeding value of such corn has long been a matter of controversy, and no one will deny that it is inferior to sound corn. However, results obtained at experiment stations where the two feeds have been compared indicate that soft corn is a better feed for finishing cattle than is commonly supposed. On the basis of dry-matter content, soft corn is as efficient as mature corn in producing gains. However, the gains from soft corn are somewhat less rapid, because its greater bulk results in a smaller consumption of dry matter in comparison with mature corn.

Corn that is abnormally high in moisture will freeze hard during severe weather so that cattle can scarcely eat it. For this reason soft corn should, if possible, be fed during the fall and early winter. To obtain as great a consumption of corn as possible, the roughage allowance should be limited to 4 or 5 pounds daily per 1,000 pounds live weight. Silage is less satisfactory than hay as roughage for cattle fed on soft corn.

SOFT EAR-CORN SILAGE

One problem connected with the utilization of soft corn is the matter of storage. This difficulty can be avoided by picking the ears soon after the first killing frost and making them into ear-corn silage, which is superior to the same corn fed as ear corn, as shown in Table 120. Making silage of the entire stalk is not always successful as the frosted leaves often will have been blown off.

HIGH-MOISTURE EAR-CORN SILAGE

The results of an early, pioneering experiment with the feeding of soft corn, harvested as ear-corn silage, are reported in Table 121. The

Table 120
Value of Soft Corn for Finishing Cattle[a]

Two-Year-Old Steers, Fed 80 Days	Soft Corn		Mature Corn from Previous Year
Form of Corn Fed	Ear Corn	Ear-Corn Silage	Ear Corn
Average percent moisture	35.82	59.65	13.61
Average daily gain (lb)	3.20	3.35	3.52
Average daily ration (lb)			
Ear corn	31.48	—	25.59
Linseed meal	1.89	1.89	1.89
Ear-corn silage	—	45.43	—
Alfalfa hay	2.70	2.35	2.70
Feed per pound gain (lb)			
Ear corn	9.84	—	7.27
Linseed meal	0.59	0.56	0.54
Ear-corn silage	—	13.56	—
Alfalfa hay	0.84	0.70	0.77
Corn (dry matter) per pound gain (lb)	6.42	5.34	6.28
Dressing percent	60.0	60.7	60.6
Cattle and hog gains per acre (lb)	247.0	292.0	—

[a] Illinois Bulletin 313.

Table 121
Ear-Corn Silage for Finishing Calves[a]

Equal Areas of Corn Harvested and Fed	Ear-Corn Silage	Ear-Corn Silage	Ground Ear Corn
Date harvested	Sept. 4–9	Sept. 24–25	Oct. 25–30
Moisture in grain when harvested (%)	53.4	37.5	Under 25
Average daily gain (lb)	2.23	2.12	2.10
Average daily ration (lb)			
Corn component	29.4	27.3	13.3
Cottonseed meal	1.6	1.6	1.6
Alfalfa hay	2.0	2.0	2.0
Oat straw	—	—	2.9
Total days feed from area	3,018	3,153	3,329
Calculated total gain to be credited area (lb)	6,721	6,686	6,987
Selling price per cwt	$15.70	$15.65	$15.35
Dressing percent	61.1	61.4	60.4
Return per bushel of corn obtained from the area allowed to mature	$1.19	$1.11	$1.07

[a] Illinois Experiment Station, Mimeographed Report, 1928–29, Calf Feeding Experiments.

favorable response to the feeding of corn harvested in this form leads one to ask why this method of harvesting and feeding corn need be limited to those occasional years when corn fails to mature.

If high-moisture ear-corn silage will produce the same amount of gain in cattle as will fully matured ear or shelled corn from an equal area of land, there would seem to be several advantages to be gained by harvesting the corn as ear-corn silage. First, it permits the harvesting of a considerable portion of the corn crop during favorable autumn weather. Second, it stores the corn in a suitable form for feeding to cattle of all ages without further preparation, eliminating the expense and extra handling of feed involved in grinding or shelling. Third, it is possible to pasture stalk fields while they are still green and palatable, or the green stover may be made into green stover silage. Or, if the green stover is not needed, the stalk may be plowed under in time for the field to be sown to wheat.

The method of making ear-corn silage is relatively simple. The ears, with or without the attached husks, are gathered with a mechanical picker without husking rolls and are hauled to the silo to be ground or chopped and blown into the silo. It is unnecessary to tramp the silage in the silo, because the weight of the corn and the absence of much bulky material will cause the silage to settle almost as rapidly as it is made. However, it is necessary to have someone in the silo to see that the silage is evenly distributed, unless a mechanical spreader is used. Otherwise the cut corn tends to pile up in the center while the husks and cobs accumulate along the walls.

If the corn is overly ripe and contains insufficient moisture to ensure thorough packing and fermentation, a small amount of water should be added through a pipe or hose inserted into the blower of the cutter. The filled silo should be sealed either by running sufficient green stover or other forage through the cutter to make a layer approximately 3 feet deep on top of the ear corn, or by covering it with a plastic silo cap. When ear-corn silage is sealed in this manner there is little spoilage. The airtight or oxygen-free silo is of course ideal for storing such silage.

Because ear-corn silage is heavy in proportion to its volume, a relatively small silo will hold sufficient silage to feed a carload of cattle, even though they receive little other feed. If the silo is of the conventional type and also quite large in diameter, a small drove of cattle will require insufficient silage daily to prevent spoilage in warm weather. A silo 12 by 40 feet holds approximately 120 tons of ear-corn silage, or enough to finish 40 to 50 head of cattle during an average-length feeding period.

Research work by the Iowa and Indiana stations is reported in Table 122. These studies deal with corn that was mature to the extent that it contained only 32 percent moisture or less. These studies indicate, as did the Illinois studies with immature ear-corn silage, that the dry matter of high-moisture ear corn is superior to that of mature dry corn when stored as ear-corn silage, as faster gains are produced on less dry matter. A probable explanation is that the cob portion of the ear-corn silage is more efficiently utilized by the rumen microflora as a result of the fermentation it undergoes in the silo. On a dry basis, ear-corn silage consists of approximately 20 percent cob. It should be noted in Table 122 that dry-matter intake was not increased in the high-moisture corn lots. Consequently, it cannot be said that the corn is more palatable, although it would seem so upon observing the apparent relish with which cattle eat it.

Experimental data on the feeding value of ensiled high-moisture shelled corn are discussed in Chapter 24, which deals with preparation of feeds. The subject of ensiled ear corn is introduced here only in connection with the subject of ensiled soft ear corn, which appropriately belongs here.

ARTIFICIALLY DRIED HIGH-MOISTURE CORN

Extensive data are unavailable concerning the effect of artificial drying on the feeding qualities of corn. For farmers who have no silo for storing high-moisture corn but who wish to practice field shelling, artificial drying

Table 122
Value of Mature High-Moisture Ear-Corn Silage Compared with Dry Ground Ear Corn for Finishing Cattle[a]

Experiment Station	Indiana				Iowa		
Year	1956		1957		1957		
Days on Test	117		126		119		
	High-moisture	Dry ground	High-moisture	Dry ground	High-moisture	High-moisture + stilbestrol	Dry ground
Moisture and additives (%)	32.2	17.7	32.5	15.5	31	31 + stilbestrol	14.5
Number of animals	10	16	36	35	36	18	36
Sex	Steers	Steers	Heifers	Heifers	Steers	Steers	Steers
Initial weight (lb)	958	960	467	466	804	805	802
Pounds gain	299	272	279	275	354	327	364
Average daily gain (lb)	2.56	2.33	2.21	2.18	2.98	3.17	3.05
Ground ear corn per day (lb)[b]	20.6	22.1	11.88	13.45	20.0	20.4	22.9
Total feed per day (lb)[b]	23.3	25.8	18.43	20.22	24.3	24.7	27.2
Ground ear corn per cwt gain (lb)[b]	807	951	555	617	675	643	750
Total feed per cwt gain (lb)[b]	953	1,111	860	927	819	778	889
Cost per pound gain (cents)	22.0	25.0	17.4	18.4	14.8	14.1	16.1
Advantage of high-moisture corn (cents)	3.0		1.0		1.3	2.0	

[a] Indiana Feeders' Day Reports, 1956 and 1957; Iowa Feeders' Day Report, 1957.
[b] All high-moisture corn values converted to equivalent moisture for comparison.

387

Table 123
Results of High-Temperature Drying of High-Moisture Field-Shelled Corn When Fed to Beef Steers[a]

Type of Corn Fed	Field-Dried 16%-Moisture Corn	High-Temperature Dried[b] High-Moisture Corn
Number of calves	9	10
Average initial weight (lb)	556	554
Average final weight (lb)	858	859
Average daily gain (lb)	2.39	2.42
Average daily ration (lb)		
Corn silage	16.8	17.3
Shelled corn	9.1	9.1
Alfalfa hay	2.1	2.1
Soybean meal	1.3	1.3
Feed cost per cwt gain ($)[c]	16.88	16.81

[a] Illinois Cattle Feeders' Day Report, 1955.

[b] Corn contained 30% moisture and was kiln-dried at 180°F until it contained 16% moisture.

[c] Feed prices used were: corn silage, $12 per ton; shelled corn, $1.40 per bushel; soybean meal, $80 per ton; alfalfa hay, $20 per ton.

seems a practical alternative. A problem presents itself in the heavy corn-growing areas because the capacity of the flow-type dryers commonly operated by the elevators is insufficient to keep up with the large amount of corn that suddenly needs drying when field shelling commences in earnest. Either farm-size or larger portable "batch" dryers ordinarily relying on forced air, heated to temperatures ranging from 180°F upward, are being used to take care of this emergency. The data shown in Table 123 indicate that drying 30 percent moisture content mature corn at 180°F did not reduce its feeding value for cattle, but such high temperatures slightly reduce the feeding value of such corn for swine and poultry. Few cattle feeders feed their corn exclusively to cattle. Therefore the drying method should be carefully considered if there is a shift to more and more field shelling of corn with high moisture content.

EFFECT OF VARIETY OF CORN ON FEEDING VALUE

During 1930–1940, when hybrid corn was rapidly displacing the open-pollinated varieties, the feeding value of the different hybrids in relation to open-pollinated corn was widely discussed. Some hybrids were claimed to be so hard that they were eaten with difficulty and consequently the kernels were swallowed with only a little chewing. Other varieties were believed not only to be softer but also to possess an aroma and flavor that made them more palatable.

A discussion of this topic is unimportant today, because open-pollinated corn has been replaced by hybrid corn for all practical purposes. Plant breeders have continued to do research in breeding "tailor-made" hybrids for specific purposes. For example, special hybrids have been developed that produce more forage tonnage and therefore are ideally suited for silage, although such "silage hybrids" fail to produce more beef gains per acre, primarily because of the lower yield of grain. Other hybrids have been produced that are 50 percent higher in protein content than ordinary corn. Feeding trials such as the one reported in Table 124 indicate that protein supplements may be eliminated from finishing rations in the future if plant breeders improve the yield of the high-protein corn hybrids. Hybrids have

Table 124

Value of High-Protein Corn for Finishing Cattle (175 Days)[a]

Ration	Lot 7 Regular Corn	Lot 8 Regular Corn Plus Soybean Meal	Lot 9 High-Protein Corn
Protein content of corn (%)	8.8	8.8	13.2
Initial weight (lb)	928.0	924.5	926.5
Final weight (lb)	1,244.5	1,310.0	1,319.0
Total gain (lb)	316.5	385.5	392.5
Average daily gain (lb)	1.81	2.20	2.24
Average daily ration (lb)			
Shelled corn	13.1	13.4	14.5
Soybean meal	0.3[b]	1.5	0.3[b]
Corn silage	24.9	25.1	25.0
Feed eaten per cwt gain (lb)			
Shelled corn	726.0	607.0	644.0
Soybean meal	18.0	70.0	15.0
Corn silage	1,377.0	1,138.0	1,114.0
Feed cost per cwt gain ($)	30.85	28.19	26.68
Selling price per cwt ($)	33.50	34.00	34.25
Return over cost of cattle and feed ($)	−16.16	3.58	12.44
Total corn consumed per steer, including corn in silage (bu)	51.8	52.7	56.0
Return realized per bushel of corn fed ($)	1.37	1.75	1.90

[a] Illinois Cattle Feeders' Day Report.

[b] Fed to equalize the protein fed in Lot 8.

also been developed that contain twice the usual amount of oil, and there are other hybrids that contain high levels of sucrose or sugar in the stalks.

FEEDING CORN TO BREEDING ANIMALS

Although corn ranks high as a feed for finishing slaughter cattle, its use as a feed for breeding animals is often severely criticized. It is stated that, owing to its high starch and fat content, corn is likely to produce such an accumulation of fat around the reproductive organs that their normal function is impaired. Also it is thought that the great strain put on the system by the digestion and assimilation of such a highly concentrated feed, and the elimination of the oxidized products, may seriously and perhaps permanently impair the function of some of the organs. This condition, commonly referred to as "burnt up by too much corn," is regarded as particularly likely to occur in young bulls and heifers.

No one denies that breeding animals should not be kept in unduly high condition, but it has not been established that it is necessary to withhold corn from breeding animals. Rather, the general use of corn in moderate quantities for cattle of all ages and descriptions by many successful breeders would indicate that corn, when properly supplemented and wisely fed, has an important place in the ration of the breeding herd. However, because corn is low in protein and mineral content, there is greater need for proper supplementary feeds when corn is fed to breeding cattle than when fed to finishing steers.

GRAIN SORGHUMS

There has been a great impetus in the growing of the grain-type sorghums with the development of sorghum hybridization in the 1950s. The increase in sorghum acreage can be explained by three reasons: (1) corn, wheat, and cotton acreage allotments stimulated interest in grain sorghum as a replacement crop; (2) great progress in developing combine-type hybrid sorghums with yields of 150 to 200 percent more grain than nonhybrid varieties; and (3) the increase in acreage brought under irrigation in the regions particularly adapted to sorghum growing and now known as the Sorghum Belt.

Sorghums of all kinds are still being grown to some extent throughtout the southern half of the country, as seen in Figure 73, but the major portion of the acreage grown exclusively for grain, often called milo by cattle feeders, is located in the Panhandle area of Texas and Oklahoma

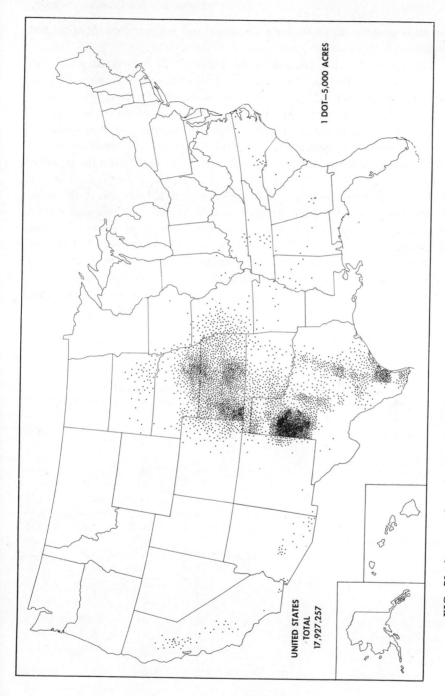

1 DOT—5,000 ACRES

UNITED STATES
TOTAL
17,927,257

FIG. 73. Acreage of sorghums grown for all purposes in 1959. (U.S. Department of Commerce.)

with some overlap in southwestern Kansas and eastern New Mexico and Colorado.

The newer hybrids not only yield higher with respect to grain, but they are also adapted to harvesting with the combine because of their shorter stalks. A problem that continues to occupy the attention of plant breeders is that of late maturity. Each year thousands of acres are damaged by frost before maturity is reached. The development of open-panicled or "sprangly-headed" hybrids has helped to reduce the moisture in the grain so as to improve the storage qualities, but high moisture is still a major problem.

A discussion of recommended varieties and hybrids has little point here because so much variation exists among sections of the country with respect to rainfall or irrigation water supplies, length of growing season, soil fertility, and drying conditions. Local or state crop specialists should be consulted for recommendations as to suitable sorghum varieties and hybrids and latest improvements, which are being made each year. Yields of 150 bushels and more per acre are not uncommon in modern grain sorghums, and yields consistently compare favorably with corn.

Table 125
Feedlot Results for Finishing Heifers with Sorghum Grain, Sorghum Grain and Corn, or Sorghum Grain, Corn, and Wheat (112 Days)[a]

Number of heifers per lot	10	10	10	10	10	10
Grain (%)						
Sorghum	100	75	50	25	—	33
Corn	—	25	50	75	100	33
Wheat	—	—	—	—	—	33
Average initial weight (lb)	607.5	608.5	607.0	608.5	607.0	610.0
Average final weight (lb)	897.0	936.5	945.0	946.5	935.0	936.0
Average daily gain (lb)	2.58	2.93	3.10	3.02	2.93	2.91
Average daily ration (lb)						
Grain	17.0	17.6	17.5	16.2	16.2	16.6
Supplement	1.0	1.0	1.0	1.0	1.0	1.0
Alfalfa hay	2.0	2.0	2.0	2.0	2.0	2.0
Prairie hay	1.7	1.7	1.7	1.7	1.7	1.7
Feed per cwt gain (lb)						
Grain	658	600	565	538	553	570
Supplement	39	34	32	33	34	34
Alfalfa hay	77	68	65	66	68	69
Prairie hay	67	59	56	57	59	59
Feed cost per cwt gain	$16.55	$15.42	$14.96	$14.86	$15.69	$15.77

[a] Kansas Livestock Feeders' Day Report, 1967.

In chemical composition, grain sorghums in general are quite similar to shelled corn. An exception is protein content, which is higher in grain sorghums—roughly 11 percent compared with 9 percent in corn. The two feeds are approximately equal with respect to total digestible nutrients or energy content, and both are low in calcium and phosphorus. The carotene content of grain sorghums is very low. Fortification of sorghum rations with carotene or vitamin A concentrate is essential in finishing rations that do not also contain at least 4 to 6 pounds of excellent-quality legume hay or an appreciable amount of good silage. High-carotene hybrids appear promising in the developmental stages and may eventually offset the lack of this nutrient in grain sorghums.

Sorghums should always be processed for feeding to cattle because they are hard-seeded and much of the sorghum grain otherwise passes through the digestive tract in an incomplete stage of digestion. Swine do not do a good job of recovering sorghums from voided catttle feces. The effect of preparation on the feeding value of sorghums is discussed in Chapter 24.

An experiment at the Kansas station, reported in Table 125, made a comparison of grain sources in a finishing ration for heifers, using straight sorghum, a combination of sorghum and corn, straight corn, and a combination of sorghum, corn and wheat. All grains were ground without heat treatment and were full-fed with 2 pounds of alfalfa hay and 1 pound of protein supplement daily. Some prairie hay was fed for the first 56 days. The general conclusions were:

1. The straight ground sorghum grain ration produced significantly slower gains.

2. Additions of corn to the ration improved rate of gain and feed efficiency.

3. Addition of wheat did not affect rate of gain but did improve feed efficiency.

4. Mixtures of grain seemed to be more acceptable to the animals over a longer time.

5. Carcass grade and carcass characteristics were not affected by type of grain in the ration.

The steam-processing and rolling of milo has improved its feeding value so that it compares quite favorably with corn and even surpasses barley, at least with respect to rate and cost of gain. The Arizona test summarized in Table 126 shows that such processing enhances the nutritional properties of milo. Data in Chapter 24 show that improved digestibility is the major reason for this effect. The same treatment also improves

Table 126

Comparison of Milo and Barley in Dry-Rolled and Steam-Processed Form When Fed to Finishing Steers (140 Days)[a]

Type of Grain	Milo		Barley	
Method of Processing	Dry-Rolled	Steam-Processed	Dry-Rolled	Steam-Processed
Number of steers	10	10	10	10
Average initial weight (lb)	559	551	567	567
Average daily gain (lb)	2.90	3.20	2.90	3.11
Average daily feed (lb)	23.4	24.7	20.6	22.0
Feed per cwt gain (lb)	848	812	749	743
Feed cost per ton	$49.62	$49.62	$57.29	$57.29
Feed cost per cwt gain	$21.04	$20.15	$21.46	$21.28
Major ration components (%)				
Ground alfalfa	5.0		5.0	
Cottonseed hulls	15.0		10.0	
Milo or barley	68.4		74.9	
Cottonseed pellets	4.5		3.0	
Molasses	5.0		5.0	
Urea	0.6		0.5	
Minerals and salt[b]	1.5		1.6	
Total	100.0		100.0	
Crude protein (%)	11.50		11.40	
Calcium (%)	0.48		0.35	
Phosphorus (%)	0.33		0.35	
Crude fiber (%)	10.40		10.20	

[a] Adapted from Arizona Cattle Feeders' Day Report, 1965.

[b] Vitamin A was added to both rations.

barley but not nearly to the same extent. The inference is that with the modern processing methods used especially by the commercial feedlots in the Southwest, milo is an excellent cattle-finishing feed that is highly competitive with corn and even excels the other grains commonly used today.

Those feeders on a smaller scale who cannot justify the investment in steam-processing equipment can still obtain satisfactory results from finely ground milo, especially if fed with small amounts of any type of silage, cottonseed hulls, or ground hay.

The problem of high moisture content in sorghum grain can be solved, as it is with corn, by ensiling it at about 30 percent moisture, either as whole grain or as head chops. This subject is discussed further in Chapter 24.

OATS

Formerly oats were not important in beef cattle feeding, but after the loss of the market for oats for feeding horses they were used rather extensively for cattle and still are in some parts of the country. For finishing purposes oats are too high in crude fiber and too low in digestible nutrients to be fed alone, but when mixed with other grains such as corn or sorghum, which are higher in energy and less bulky, they give very satisfactory results. The tough, fibrous hull and small kernel make it advisable to grind or roll oats for older cattle, as they swallow a considerable percentage of the kernels with insufficient chewing if the oats are fed whole. Grinding oats is usually inadvisable for calves, unless the grain with which the oats are fed is also ground; in this case grinding insures a better mixture of the two feeds.

Both whole and ground oats are more palatable than corn for calves, as indicated by the relative amounts eaten in the Kansas experiment reported in Table 127, where the two grains were fed free-choice in separate self-feeders. The slight difference in palatability was probably because the oat grains were bulkier.

FIG. 74. Hybrid grain sorghums are increasing in importance as a source of high-energy concentrate for finishing rations.

Feeding experiments indicate that oats more nearly approach the value of shelled corn for feeding heavier cattle than for feeding calves. In tests conducted at the Indiana station with 2-year-old steers the substitution of ground oats for one-third of the shelled corn ration slightly decreased the feed required per 100 pounds of gain. However, in none of the seven experiments with calves shown in Table 128 did oats have as high feeding value as shelled corn. On the average, 118 pounds of oats replaced 100 pounds of shelled corn, or 2 bushels of oats replaced approximately 1 bushel of shelled corn. One objection to feeding oats to young feeder cattle is that oats tend to make the cattle grow rather than fatten, actually a reflection of their lower energy value. Oats are approximately equal to ground ear corn when fed at a level not exceeding half of the concentrate portion of the ration.

Oats may constitute as much as one-half or two-thirds of the grain ration while the cattle are becoming accustomed to a full feed. Because oats contain less energy and are bulkier than corn, their use at this time lessens the danger that some of the steers will overeat and become foundered. However, the oats should gradually be reduced as the feeding progresses, until they form only 20 to 30 percent of the grain ration, or they may be eliminated altogether, during the last third of the finishing

Table 127

Relative Palatability of Shelled Corn and Oats When Fed Free-Choice to Beef Calves[a]

	Lot 1		Lot 2	
	Shelled Corn	Whole Oats	Ground Shelled Corn	Ground Oats
Average daily consumption (lb)				
First 28 days	1.96	3.18	1.83	3.45
Second 28 days	1.96	5.85	2.50	5.02
Third 28 days	4.56	4.57	4.96	4.35
Fourth 28 days	4.70	6.32	6.48	5.46
Last 26 days	2.44	8.53	4.80	7.33
Average 138 days	3.14	5.65	4.10	5.09
Average initial weight (lb)	412		409	
Average final weight (lb)	738		729	
Average daily gain (lb)	2.36		2.32	

[a] Kansas Cattle Circular 38B.

Table 128
Comparison of Corn with Corn and Oats for Finishing Cattle

	Shelled Corn			Shelled Corn and Oats		
	Daily Gain (lb)	Grain per cwt Gain (lb)	Selling Price per cwt	Ratio of Corn to Oats	Daily Gain (lb)	Grain per cwt Gain (lb)
Two-year-old steers						
Indiana	2.50	593	$12.90	2:1[a]	2.60	564
Indiana	1.88	584	11.50	2:1[a]	2.16	508
Indiana	2.15	592	7.45	2:1[a]	2.18	586
Calves						
Illinois	2.00	594	8.50	4:1	1.96	603
Illinois	2.06	447	8.65	2.5:1	2.02	472
Nebraska	2.47	453	9.50	2:1	2.40	494
Minnesota	2.32	580	9.60	4:1	2.19	602
Oklahoma	1.95	469	8.75	1[a]:1[a]	1.95	495
Kansas	2.11	444	6.75	0:1[b]	2.02	442
Kansas	2.14	476	7.00	1.3:1	2.13	516
Average of 7 lots of calves	2.15	495	$8.40	—	2.09	518

[a] This grain fed ground.
[b] Whole oats alone fed first 100 days and shelled corn alone last 100 days.

period. Obviously the relative prices of corn and oats should be considered carefully in determining the extent to which oats should be used in the finishing of cattle for market. As already mentioned, feeding large amounts is unprofitable unless the price of oats per bushel is less than half that of corn.

Oats are somewhat higher in protein and mineral content than corn, and for this reason they are especially valuable for breeding stock. When used in mixtures with other grains, they are highly regarded by herdsmen who fit cattle for shows and sales. For such animals the oat hulls are in no way objectionable, since they give bulk to the grain ration.

BARLEY

Large amounts of barley are used in cattle feeding in the northern and northwestern states and in Canada. At these latitudes corn often fails to mature, and barley, with its shorter growing season, is raised in its place. Barley is also grown to a limited extent along the western edge of the Corn Belt, where a grain crop is desired that will mature before the arrival of hot dry weather.

Barley, like oats, has a kernel surrounded by a tough, heavy hull that materially lessens its digestibility and makes it somewhat unpalatable unless it is processed. The kernel itself is rather hard and flinty in texture and does not "chew up" as easily as corn, making the grinding, crushing, or rolling of barley all the more necessary if satisfactory results are to be obtained.

In chemical composition barley falls midway between oats and corn. Because of its higher protein and mineral content it is a somewhat better-balanced feed than corn, but its lower percentage of fat and greater amount of fiber lessens its energy value. Many comparisons have been made between shelled or ground corn and ground barley as feeds for finishing cattle. Some of the results of these studies are shown in Table 129.

Barley may vary considerably in quality, which may be highly important as shown by the daily gains of three lots of steers fed shelled corn, native barley, and northern-grown barley by the Illinois station. The gains were 3.08, 2.81, and 3.35 pounds per head, respectively. Because weather conditions for the production of high-grade barley are almost opposite those required for a good grade of corn, it is not surprising that results vary widely in studies of the relative value of these two grains.

Table 129
Barley versus Corn for Finishing Cattle

	Steer Calves[a]		Yearling Steers[b] (4 Years)		Steer Calves[c]	
	Ground Shelled Corn	Ground Barley	Ground Shelled Corn	Ground Barley	Shelled Corn	Ground Barley
Daily gain (lb)	2.11	1.96	2.02	2.00	2.34	2.21
Feed per cwt gain (lb)						
Grain	570	613	379	384	476	476
Protein concentrate	47	51	—	—	64	68
Corn silage	—	—	—	—	414	431
Hay	176	199	841	824	119	114

[a] Kansas Cattle Circular 36A.
[b] Oregon Bulletin 528.
[c] Minnesota Bulletin 300.

Although feeding tests show considerable variation in the relative value of corn and barley for finishing cattle, most of them indicate that ground barley is less palatable than shelled corn, that it has a tendency to cause bloat, and that barley-fed cattle usually sell for less than corn-fed cattle when marketed. In view of these facts, the feeding of barley by itself is confined largely to those sections where barley is extensively grown and consequently is relatively cheap.

In the Corn Belt proper, barley should always be fed mixed with corn or corn and oats to lessen the frequency of bloat. In such mixtures barley has a feeding value fully equal to that of corn, but when fed alone it is only about 90 percent as valuable as corn on a pound basis. Or it may be said that 5 bushels of barley have about the same feeding value for finishing cattle as 4 bushels of shelled corn.

ROLLED BARLEY WITHOUT HAY

Barley contains about 15 percent hulls and thus is said to contain its own built-in roughage supply. It is not surprising, in view of increasing costs of processing roughages, that commercial feeders especially have become interested in using rolled barley and fortified supplement as a complete feed for finishing cattle. The South Dakota station compared a mixture of 80 percent rolled corn and 20 percent ground alfalfa hay with dry-rolled barley in finishing yearling cattle. A supplement was designed that equalized the fiber, protein, calcium, and phosphorus fed to all cattle. Results are summarized in Table 130. In brief, the corn-alfalfa ration resulted in slightly more rapid gains but with about equal efficiency. As a result of this and other tests by this station, they recommend that barley be supplemented with about 10 percent hay for most satisfactory performance. Their recommendations are intended largely for farmer-feeders, and the all-barley ration should not be ruled out for the commercial feedlots in or near the barley-growing areas.

WHEAT

When prices of feed grains work upward, wheat prices often appear competitive with the more commonly used feeds. This situation may occur as often as once in 4 or 5 years. Many tests have been conducted at various experiment stations, and, not unexpectedly, such major wheat-producing states as Oklahoma, Kansas, and Nebraska have been most active in the field. These researchers found that wheat was a more satisfactory feed when used as a partial substitute for other grains. Table 131 was prepared

Table 130
Rolled Barley Compared with Corn and Alfalfa Hay Rations[a]

	Trial 1 (154 Days)		Trial 2 (204 Days)	
Basal ration (%)				
Rolled shelled corn	80	—	80	—
Ground alfalfa hay	20	—	20	—
Dry-rolled barley	—	100	—	100
Number of steers	31	31	40	40
Average initial weight (lb)	798	798	700	695
Average daily gain (lb)	2.22	1.97	2.47	2.37
Average daily ration (lb)				
Basal ration	19.8	15.5	23.1	20.1
Supplement[b]	1.0	2.0	1.0	2.0
Hay[c]	—	0.2	0.3	0.3
Total	20.8	17.7	24.4	22.4
Feed per cwt gain (lb)	938	897	961	950
Carcass grade	C−	G+	C	C
Condemned livers	7	8	3	14

[a] Prepared from data in South Dakota Agricultural Experiment Station Bulletin 539.
[b] Supplements contained stilbestrol, vitamin A, and minerals plus enough protein to equalize the rations when fed at the rates shown.
[c] Hay used in getting cattle safely on full feed.

Table 131
Comparative Value of Wheat, Corn, and Sorghum When Wheat Is Fed at 50 Percent or Less of the Ration[a]

Wheat Compared with Corn				Wheat Compared with Sorghum		
If Corn Price Is:		Feeder Can Pay for Wheat:		If Sorghum Price Is:	Feeder Can Pay for Wheat:	
(bu)[b]	(cwt)	(bu)[b]	(cwt)	(cwt)	(bu)[b]	(cwt)
$1.20	$2.14	$1.54	$2.57	$1.80	$1.40	$2.33
1.25	2.23	1.61	2.68	1.90	1.48	2.47
1.30	2.32	1.67	2.78	2.00	1.56	2.60
1.35	2.41	1.73	2.88	2.10	1.64	2.73
$1.40	$2.50	$1.80	$3.00	$2.20	$1.72	$2.87

[a] Adapted from "Is This the Year for You to Feed Wheat?" by Ovid Bay, *Farm Journal*, Western Edition, June 1965.
[b] Weights used per bushel: corn, 56 pounds; wheat, 60 pounds.

after reviewing the recent tests with wheat, especially those conduted at the Fort Hays, Kansas, Agricultural Experiment Station.

Results secured at the Kentucky and Kansas stations indicate that ground wheat may successfully be fed alone if it is spread evenly over a rather liberal feed of corn silage. Evidently, the silage reduces the sticky nature of the ground wheat by supplying more bulk and by compelling the cattle to eat the wheat more slowly.

RYE

Rye is usually regarded as a feed for swine and is ordinarily fed to beef cattle only when its price is considerably below that of corn. Rye grains are small and hard and, of course, should be ground. Like wheat, rye gives much better results when fed with other grains, in which case ground rye may be considered equal to ground wheat.

GRAIN MIXTURES

Experienced feeders of purebred cattle greatly prefer a mixture of several grains to corn alone for fitting cattle for the show and sale ring. On the other hand, the men who feed steers for the market usually feed corn or sorghum alone unless these grains are scarce and high in price compared with other grains. Feeding experiments indicate that grain mixtures usually are sufficiently better than single grains such as corn or barley to justify their use, even though additional labor is required in their preparation. An exception appears to be a mixture of corn and oats for calves, which, as shown in Table 128, usually produces smaller gains and less finish than corn alone. Another exception is a mixture of ground oats and barley, which probably contains too much fiber and is too bulky for young animals.

Table 132 gives comparative values of oats, barley, and sorghum grain in terms of various values placed on corn. Adjustments for differences in feeding value and for different weights per bushel were taken into account.

BEET PULP

In the manufacture of sugar from beets, great quantities of wet beet pulp result as a residue of the sugar-extraction process. Thousands of cattle

Table 132

Approximate Value of Oats, Barley, and Sorghum Grain in Terms of Price Levels for Corn When Fed to Beef Cattle[a]

	Value of Other Grains When Fed		
If Corn Price per Bushel Is:	Oats per Bushel	Barley per Bushel	Sorghum Grain per 100 Pounds
$1.00	$0.48	$0.76	$1.65
1.10	0.53	0.83	1.81
1.20	0.58	0.91	1.98
1.30	0.63	0.98	2.14
1.40	0.68	1.06	2.31
1.50	0.73	1.14	2.47
1.60	0.78	1.21	2.64
1.70	0.83	1.28	2.79

[a] Courtesy Malcolm Clough and Ralph Jennings, U.S.D.A.

and tens of thousands of sheep and lambs are finished annually on this material in the beet-growing sections of the West. Because of the large amount of moisture that it contains—approximately 90 percent—wet beet pulp should be regarded as a diluted carbonaceous roughage similar to corn silage and is more fully discussed in connection with that subject in Chapter 19.

A considerable quantity of the beet pulp produced at the sugar refineries is dried with waste steam to produce dried beet pulp, which may be bagged and sold as plain beet pulp or mixed with molasses to produce dried molasses beet pulp. Approximately 100 pounds of the plain dry pulp are obtained from each ton of sugarbeets processed. Four hundred pounds of molasses are mixed and dried with about 1,600 pounds of dried pulp to make a ton of dried molasses beet pulp.

In chemical composition dried beet pulp is a carbonaceous concentrate. Because of its bulk it produces slightly faster gains when mixed with corn or ground barley than when fed alone. Mixtures of one part dried beet pulp and two parts corn, or equal parts of dried pulp and shelled corn, were equal in all respects to a full feed of shelled corn in three tests conducted at the Nebraska station. Plain dried pulp gave as good results as dried molasses pulp when fed with ground barley in two tests made at the Colorado station. (See Table 133.)

Table 133
Value of Dried Beet Pulp for Finishing Cattle

	Colorado[a]				Nebraska[b]		
	Ground Barley	Dried Molasses Beet Pulp	Ground Barley and Plain Beet Pulp	Ground Barley and Dried Molasses Beet Pulp	Ground Shelled Corn	Ground Shelled Corn, 2 Parts; Dried Beet Pulp, 1 Part	Ground Shelled Corn, 1 Part; Dried Beet Pulp, 1 Part
Average initial weight, (lb)	713	723	348	349	549	550	546
Average final weight (lb)	1,120	1,120	746	746	1,064	1,059	1,060
Average daily gain (lb)	2.26	2.20	2.04	2.03	2.19	2.17	2.19
Average daily ration (lb)							
Grain	10.6	—	3.6	3.6	10.9	7.2	5.4
Dried beet pulp	—	10.7	3.4	3.4	—	3.6	5.4
Protein concentrate	2.1	2.1	1.0	1.0	1.0	1.0	1.0
Silage	15.7	15.6	9.2	9.2	21.2	17.3	16.6
Legume hay	7.3	7.7	4.2	4.2	1.9	1.9	1.9
Dressing percent	63.0	63.8	62.5	62.5	59.9	60.5	60.2

[a] Colorado Bulletin 422.
[b] Nebraska Bulletin 359.

Usually the cost of beet pulp is considerably higher than the price of corn, milo, or barley; consequently, it is not fed to beef cattle to any extent except in the vicinity of sugar refineries where it is purchased direct from the mills. Frequently small amounts are used in fitting cattle for show. One quart of dried beet pulp moistened with 1 quart of diluted feeding molasses or plain water and allowed to swell overnight will yield about 2 quarts of soft, succulent feed, which may be fed alone or mixed with other concentrates.

MOLASSES

A large quantity of low-grade molasses is produced each year as a by-product of the sugar-refining industry. Frequently it is shipped in tank cars or river barges to the locality where it is to be fed or mixed with other ingredients in a complete feed.

There are four kinds of feeding molasses—cane, beet, corn, and wood. Most of the cane molasses is made in the South where it is commonly called "blackstrap." Beet molasses is produced principally in the sugarbeet areas of the West. Corn molasses is available in relatively small amounts for feeding purposes. It is a by-product of the corn milling industry in the manufacture of corn sugar. Wood molasses, also called holocellulose molasses, is a by-product of the pressed board or hardboard industry.

Formerly cattle feeders showed a strong preference for cane molasses, claiming that the beet variety was less palatable and much more laxative. More extensive use of beet molasses, however, has convinced feeders that there is little difference between the two kinds. The use of corn molasses in beef cattle feeding has been too limited to permit a definite statement about its feeding value. However, in three tests where it was used at the Illinois and Nebraska stations, no superiority over blackstrap was disclosed. Because of its high viscosity, corn molasses is difficult to handle during cold weather. Indiana studies have recently shown that wood molasses is equal to cane molasses in every respect studied except free-choice palatability. When included in mixed rations, even this problem disappeared.

Feeding experiments indicate that 1 or 2 pounds of molasses usually can be supplied to full-fed cattle without appreciably affecting the consumption of the other feeds. In theory this intake of additional nutrients should result in increased rate of gain, but in only a few of the experiments where small amounts of molasses have been fed has the increase been significant. In fact, in 10 out of 25 experiments where not more than 5 pounds of molasses were fed daily, the average gains were no larger than those of the check lots. As molasses increased the cost of the ration and on the average lessened instead of increased the selling price of the cattle, the use of small quantities for the purpose of accelerating gains and producing a quick finish appears unjustified. (See Table 134.)

Molasses has a definitely useful place in cattle feeding when grain is scarce. By feeding molasses in amounts large enough to replace one-third to two-thirds of the usual grain ration, a sufficient number of cattle may be finished to utilize the farm roughage supply to best advantage. Although cattle fed only grain during such a year probably will return a higher profit per head than those fed a limited amount of grain and considerable molasses, the larger number it is possible to feed by using molasses will probably return larger profits to the individual farmer and to the industry as a whole.

Feeders who make large profits on molasses-fed cattle during a year of feed scarcity should not be misled as to its feeding value in terms of corn or sorghum. When fed to the extent of one-third or more of the grain ration, molasses is approximately 80 percent as efficient as shelled corn or sorghum in producing a given amount of gain. Another way of expressing the value of molasses is to say that 75 pounds of molasses are equal to 1 bushel or 56 pounds of corn or about 60 pounds of sorghum. Therefore, unless 75 pounds of molasses (or 6.5 gallons, as 1 gallon of molasses weighs 11.7 pounds) can be bought for less than a bushel of corn or 60 pounds of sorghum, it is not an economical replacement for

Table 134
Value of Small Amounts of Molasses for Finishing Cattle

Age of Cattle	Calves		Yearlings		2-Year-Olds		All Tests	
Number of Tests	12		6		7		25	
	Check	Mo-lasses	Check	Mo-lasses	Check	Mo-lasses	Check	Mo-lasses
Average molasses fed (lb)	—	1.4	—	2.6	—	3.0	—	2.2
Average grain eaten (all lots) (lb)	9.9	10.1	13.8	12.1	15.9	13.5	12.5	11.6
Average grain eaten (unrestricted lots) (lb)	9.9	10.1	15.1	14.8	17.6	15.2	12.6	12.1
Average daily gain (lb)	2.02	2.15	2.53	2.42	2.79	2.72	2.36	2.38
Water drunk daily (lb)	54.0	56.3	—	—	48.8	47.2	52.3[a]	53.6[a]
Feed consumed per cwt gain (lb)								
Grain	491	471	547	496	567	496	509	484
Protein Concentrate	98	92	77	80	108	110	96	95
Corn silage	333	311	201	256	840	852	553	549
Hay	90	86	215	215	118	125	129	137
Molasses	—	55	—	124	—	115	—	88

[a] Average of 17 comparisons made of water consumption.

Table 135

Value of Molasses as a Substitute for Much or All of the Grain Ration[a]
Yearling Steers—150 Days

Method of Feeding Molasses	Shelled Corn[b]	Cane Molasses — Poured on Silage	Ground Corn 25% Ground Oats 25% Molasses 50% — Poured on Grain and Silage	Ground Corn 25% Ground Oats 25% Molasses 50% — All Feeds Except Silage Machine Mixed	Ground Corn 50% Ground Oats 50% Molasses — Self-Fed
Average initial weight (lb)	753	754	753	750	752
Average daily gain (lb)	2.51	2.08	2.57	2.51	2.24
Average daily ration (lb)					
Corn or corn and oats	13.9	—	6.8	5.9	6.8
Molasses	—	14.3	7.2	6.1	5.6
Soybean meal	1.9	2.9	2.6	2.2	2.6
Corn silage	20.0	22.7	20.0	20.0	20.0
Alfalfa hay	2.0	2.0	2.0	2.9[c]	2.0
Feed per cwt gain (lb)					
Corn or corn and oats	516	—	265	234	278
Molasses	—	642	282	243	251
Soybean meal	74	129	100	88	104
Corn silage	796	1,021	778	796	893
Alfalfa hay	80	96	78	117	89
Cost of gain per cwt	$15.09	$14.85	$13.64	$13.25	$14.76
Shrinkage in shipment (lb)	41	66	54	72	59
Return above cost of cattle and feed	$69.35	$41.09	$61.88	$54.50	$47.73
Price per ton at which molasses would have been as profitable as corn at $1.12 and oats at 40 cents a bushel	—	−$3.25	$11.20	−$7.42	−$26.20
Water drunk daily (lb)					
Aug. 5–19	92	108	—	—	—
Aug. 22–Sept. 6	86	—	—	119	—

[a] Illinois Agricultural Experiment Station Mimeographed Report.
[b] Grain fed to molasses lots was ground to absorb the molasses.
[c] Hay was ground and mixed with grain to help absorb the molasses.

large amounts of these grains. Furthermore, when the lower selling price of the molasses-fed cattle is considered, the actual economic advantage of the molasses may disappear altogether. (See Table 135.)

A substitution of 3 to 5 percent of molasses for grains is commonly made in complete mixed cattle rations. The chief purpose of such molasses additions is to reduce the dust problem associated with finely pulverized feeds. If the molasses costs no more than grain per hundredweight, it is a good addition. Even higher levels up to 10 percent may be profitably used in preconditioning or starter rations. Such rations are highly palatable to calves that are unaccustomed to eating anything other than forages. Molasses serves as a good binder if used at the rate of 5 to 10 percent in the making of certain range supplements in the form of pellets or cubes.

SUGARBEETS

Sugarbeet pulp has already been discussed and the use of beet tops is discussed in Chapter 8. In recent years, as new areas of irrigated land have been brought into production and the acreage of sugarbeets has increased, questions are being raised concerning the feeding value of the fresh beets themselves. The Arizona station conducted trials at their Yuma branch for 2 years to evaluate fresh beets when fed as an addition to a good 85 percent concentrate finishing ration. Table 136 contains the pertinent data. The beets were topped and then dug daily and chopped through a conventional field chopper just prior to being fed. The Arizona workers concluded that cattle receiving the beet-supplemented rations performed comparably with those on the control rations, with no adverse effect on carcass quality. The beets had a feed replacement value of $10 to $14 per ton. Obviously if large quantities are to be fed, specialized harvesting and processing equipment must be developed.

MISCELLANEOUS CARBONACEOUS CONCENTRATES

From time to time, especially when the price of grain is high, cattle feeders become interested in possible substitutes that are ordinarily used only in the locality where they are produced. Among such feeds are dried citrus pulp and dehydrated potatoes.

Dried Citrus Pulp. The rapid increase in production of canned and frozen citrus fruit juices since the early 1940s has made available large quantities of orange and grapefruit pulp for livestock feeding. Much of this pulp is fed in fresh form in the vicinity of canning establishments,

Table 136
Effect of Feeding Chopped Sugarbeets in Addition to an 85 Percent Concentrate Ration[a]

	Without Sugarbeets	With Sugarbeets
Number of steers	32	32
Average initial weight (lb)	704	692
Average final weight (lb)	1,009	1,013
Average daily gain (lb)	2.51	2.66
Average daily mixed ration (lb)	19.4	15.5
Average fresh beets per day (lb)	—	21.2
Average beet dry matter per day (lb)	—	4.2
Total dry matter per day (lb)	19.4	19.8
Mixed ration per cwt gain (lb)	812	622
Wet beets per cwt gain (lb)	—	839
Dry beets per cwt gain (lb)	—	168
Total dry ration per cwt gain (lb)	812	790

[a] Adapted from data in Arizona Cattle Feeders' Report, 1963. (Two trials at the Yuma station.)

but an appreciable amount is dried and bagged for use elsewhere. Since the canning and freezing of citrus juice is an expanding industry, the present production of dried pulp may be but a fraction of the future supply.

Dried citrus pulp compares favorably with ground ear corn in total digestible nutrients but is much lower in digestible protein. Although it has a slightly bitter taste it is eaten readily by cattle and has proved to be a good concentrate for finishing cattle when properly supplemented with a protein concentrate. In two Florida trials, steers full-fed dried pulp without any corn consumed about 1 pound of pulp per 100 pounds live weight but gained more slowly than steers fed snapped corn or a mixture of snapped corn and dried citrus pulp. (See Table 137.)

Dehydrated Potatoes. A considerable percentage of both Irish and sweet potatoes grown commercially during an average year are cull tubers that are too small or misshapen to be sold for human consumption. Many of them are fed raw to cattle and hogs in the vicinity where they are grown, but some are dried and sold as dehydrated potatoes, which are a much more valuable feed than the fresh form. Dehydrated sweet potatoes are regarded more highly than dried Irish potatoes, because they are more palatable. Neither type is of much economic importance as a cattle feed except in potato-growing areas, but they may be purchased by cattle feeders elsewhere when grain is scarce and abnormally high in price.

Dehydrated potatoes, like dried citrus pulp, compare favorably with corn in total digestible nutrients, but are noticeably lower in protein. They are less palatable than corn and when fed as the only concentrate are not eaten in sufficient amounts to produce a satisfactory rate of gain. In two tests made at the Oklahoma station, steer calves full-fed dried sweet potatoes ate considerably less feed and gained more slowly than the check lot fed ground shelled corn. However, a third lot fed equal parts of ground corn and dried sweet potatoes consumed more concentrates and made faster gains than the check lot. Even in this third lot the substitution of dried

Table 137
Value of Dried Citrus Pulp and Dehydrated Sweet Potatoes for Finishing Cattle

	Florida (Average 2 Trials)			Oklahoma (Average 2 Trials)		
	Ground Snapped Corn	Dried Citrus Pulp	Ground Snapped Corn, 2 lb Dried Citrus Pulp	Ground Shelled Corn	Dried Sweet Potatoes	Ground Corn, Dried Sweet Potatoes
Initial weight (lb)	608	590	607	507	506	508
Final weight (lb)	892	850	883	862	825	877
Total gain (lb)	284	260	276	355	319	369
Average daily gain (lb)	2.37	2.17	2.30	2.14	1.92	2.22
Average daily ration (lb)						
Ground snapped or shelled corn	10.6	—	2.0	11.3	—	6.2
Corn substitute	—	7.9	9.0	—	9.0	6.2
Protein concentrate	2.9	2.9	2.9	1.5	1.9	1.7
Sorgo silage	—	—	—	9.5	8.8	9.5
Hay	5.8[a]	5.7[a]	5.8[a]	1.0[b]	1.0[b]	1.0[b]
Feed per cwt gain (lb)						
Ground snapped or shelled corn	455	—	87	528	—	279
Corn substitute	—	365	322	—	478	279
Protein concentrate	124	135	127	71	100	79
Silage	—	—	—	445	478	430
Hay	245[a]	264[a]	250[a]	59[b]	53[b]	45[b]

[a] Carpet and Bermuda grass hay.
 Alfalfa hay.

potatoes for half of the corn ration reduced the net profit the first year because the dried potatoes cost $19 a ton more than corn, and the second year because the potato-fed cattle sold for 50 cents a hundred less than those fed ground shelled corn. (See Table 137.)

FATS IN FINISHING RATIONS ᐧ

Fats and oils have a gross energy value approximately 2.25 times that of carbohydrates. In recent years fats have accumulated in surplus quantities because of closer trimming of pork and beef carcasses in the packing plants, and the advent of synthetic detergents with a resultant decrease in demand for fats by the soap industry. Research with poultry has amply demonstrated that if fats can be bought at prices not much higher than those paid for carbohydrates, more economical broiler rations can be formulated with the use of such fats.

Aside from the energy value of fats, the settling of dust arising from finely ground components of the ration such as ground alfalfa hay, dehydrated alfalfa, or finely ground grain is an important factor. The lubrication provided for feed grinders, mixers, augers, and pelleting dies, where used, lengthens the life of expensive feed-processing equipment.

Arizona workers recommend 4 percent fat in finishing rations fed in summertime particularly. When this level of fat is used, energy level in the ration can be increased and fiber level can be reduced. Arizona workers believe this largely accounts for the fact that cattle in their summertime tests maintain the same conversion rates as their winter-fed cattle, contrary

Table 138
Effect of 4 Percent Fat Addition to Barley or Milo
Rations Containing 11 Percent Crude Protein
(112 Days)[a]

Type of Grain	Barley		Milo	
	Average Daily Gain (lb)	Feed/Cwt Gain (lb)	Average Daily Gain (lb)	Feed/Cwt Gain (lb)
No fat	3.05	742	2.61	812
4 percent fat	3.32	690	2.97	710
Improvement (%)	8.8	7.0	13.8	12.6

[a] Arizona Cattle Feeders' Day Report, 1963.

Table 139
Effect of Level of Animal Fat in Finishing Steer Rations[a]

Fat Levels (%)	0	5	10
Number of animals	16	16	16
Average initial weight (lb)	774	770	776
Average daily gain (lb)	2.01	2.57	2.27
Feed per pound gain (lb)	11.8	9.0	9.9
Carcass grade	G+	C−	C−
Dressing percent	61.4	61.4	60.5
Separable fat, 9th, 10th, 11th rib (%)	27.8	32.1	30.2

[a] V. R. Bowman, et al., *Journal of Animal Science*, 16:833.

to the usual 10 to 15 percent reduction in feed efficiency in summer-fed rations not containing fat. One of their many experiments with 4 percent fat added to an 80 percent milo or barley ration fed in summertime is summarized in Table 138.

Several stations have compared animal fats with vegetable fats and have found no consistent differences for or against any type of fat so long as it was protected from rancidity by use of an antioxidant additive.

The level of fat to include in finishing rations has also been studied extensively. Nevada workers compared 0, 5, and 10 percent levels of fat in a medium-energy ration and concluded that a 10 percent level was probably somewhat high but that 5 percent improved performance significantly, as may be seen in Table 139. It is an accepted practice for commercial feedlots to include 2 to 5 percent fat in all their rations, depending on costs of the fat.

Chapter 16

PROTEIN REQUIREMENTS of FEEDER CATTLE and HOW to SUPPLY THEM

One of the first discoveries made by early investigators of animal feeding was the need of the animal body for the complex organic compounds called proteins. The distinctive characteristic of proteins is that they contain nitrogen, an element indispensable to all animal life. Nitrogen is necessary for the building and repair of nearly all the tissues that make up the animal body. Muscle, the skin and its modifications, and the connective tissues consist almost entirely of protein materials that have been elaborated and built up from the nitrogenous compounds in the feeds consumed by the animal. A second significant feature of protein feeds is that they are rich in phosphorus, an element that is highly necessary for the development of bone.

Apparently, it is not possible for ration protein in excess of immediate rumen bacterial and body requirements to be stored as protein to satisfy future needs. Instead, after the immediate needs for protein are satisfied, the nitrogen from the excess is excreted in the urine and lost, except for whatever fertilizer value it may have. The non-nitrogenous fraction of the excess ration protein can be utilized as energy in meeting maintenance requirements, or it can be converted to fat and stored, as happens when carbohydrate or other feed components are consumed in excess of maintenance requirements. Since protein feedstuffs are relatively expensive, it is evident that their use for the production of fat and energy is not economical. It is therefore important for the cattle feeder to know the protein requirements of feeder cattle of different ages and under different conditions, in order to supply the optimum amount of this nutrient without feeding it in excess of the need.

PROTEIN REQUIREMENTS

In Chapter 6 it is shown that protein requirements for beef cattle are expressed on a quantitative rather than qualitative basis. That is, protein is protein to a ruminant, provided that the sources are of equal digestibility, and it matters little if the protein is deficient in certain amino acids. It is further shown that nonprotein nitrogen sources, such as urea, can be used to satisfy at least part of the protein needs of cattle.

In the nutrient requirement tables in Chapters 8, 10, and 14, protein requirements are expressed in four ways: (1) percentage of total protein in the ration, (2) percentage of digestible protein in the ration, (3) daily requirement of total protein, and (4) daily requirement of digestible protein per head. There are still other ways of expressing protein requirements for feeder cattle, some of which are actually more convenient to use in practice than are the tables referred to above, although they may be somewhat less exact.

METHODS OF EXPRESSING PROTEIN REQUIREMENT IN FINISHING RATIONS

Total Protein Requirement. Finishing rations should contain between 10 and 11 percent total protein, or they should furnish between 1.3 and 2.9 pounds of total protein daily, depending on the weight of the animal. When computing daily total protein requirements for a drove of feeder cattle it is often more convenient, although less accurate, to express them on the basis of amount required daily per 1,000 pounds live weight of animal. In this situation 2.5 to 3 pounds may be used as the requirement. Actually calves require somewhat more protein per unit of weight than do older cattle, but because they usually consume more feed per unit of weight, the percentage protein in the ration does not need to be increased as much for calves as one might think.

Digestible Protein Requirement. The digestible protein in a ration is, of course, that portion of the total protein that is actually digested and made available to the bacterial flora in the paunch or to the animal itself. Naturally it is less than the total protein figure, and the extent to which it is less depends on the quality of the ration. As shown earlier, digestible protein for cattle represents approximately 60 percent of the

FIG. 75. Legume-grass mixtures such as this one, consisting largely of red clover and timothy, produce hay that can supply appreciable amounts of protein in the rations of beef cattle. (University of Illinois.)

total protein in high-roughage rations and 75 percent of the total protein in more concentrated rations such as those fed to finishing cattle.

Fiber content of a protein concentrate or of a ration is a fairly good guide for evaluating the digestibility of the protein in the ration, but this rule is not infallible. Digestible protein requirements, when expressed as percent digestible protein in the ration, range from 7.5 to 8.2 percent, with the younger cattle again having the higher requirement. When requirements are expressed as daily requirements of digestible protein, 1 to 2.2 pounds are required daily per head, or approximately 2 to 2.5 pounds per 1,000 pounds of live weight.

Daily Protein Concentrate or Legume Roughage Equivalent. This expression refers to the supplemental protein needed to balance the grain or carbonaceous concentrate portion of the finishing ration. Corn, for example, when full-fed contributes only about half of the protein required by cattle being fed finishing rations. If the roughage portion of the ration is nonleguminous, the protein content of the ration will not be improved by the roughage. Consequently all finishing rations consisting of carbonaceous concentrates and nonlegume roughage require supplemental sources of protein. Generally speaking, 2 pounds of a conventional high-protein concentrate (35 to 50 percent total protein) are required daily to balance finishing rations if no leguminous roughage is included. Each pound of air-dry legume roughage such as alfalfa or clover hay reduces the protein concentrate requirement by about 0.25 pound. If legume silage is fed, it should be reduced to an air-dry basis by dividing the amount consumed by 3 in determining its contribution of supplemental protein. Legume green chop can be reduced to an air-dry basis by dividing the amount fed by 8. Roughages consisting of mixtures of grasses and legumes generally supply only half as much supplemental protein as straight-legume roughages. Consequently each pound of such air-dry roughage contributes the equivalent of 0.125 pound of protein concentrate.

Ratio of Protein Concentrate to Grain. Protein concentrates are often mixed with the grain portion of the ration prior to feeding; for example, when self-feeding or when feeding with automatically metered mixing equipment. In these circumstances it is desirable to know what proportions of protein concentrate and grain to mix in order to insure correct daily consumption of protein concentrate. In other words, a given ratio is maintained between the weight of corn and the weight of the protein feed used. The advantage of such a method is its simplicity. The feeder is much more likely to know how much corn he is feeding than he is to know how much the steers really weigh. Its chief disadvantage is that the same ratio is likely to be maintained throughout the feeding period. The ration is therefore apt either to be low in protein at the beginning or higher than need be at the end. This problem, however, can be easily overcome by gradually widening the ratio as the feeding period progresses. By the time about two-thirds of the feeding period is complete, hay or pasture intake will have been reduced to such an extent that more supplemental protein will be needed, and thus the ratio will narrow again. Based on such a plan, the ratios in Table 140 will be found fairly satisfactory.

Table 140
Ratio Between Nitrogenous Concentrate and Corn for Full-Fed Steers

Ration	Period	Two-Year-Olds	Yearlings	Calves
Corn–nonlegume roughage	1st third	1:6	1:5	1:4
	2d third	1:8	1:7	1:6
	Last third	1:7	1:6	1:5
Corn–legume hay	1st third	None	None	1:10
	2d third	None	1:10	1:8
	Last third	1:10	1:8	1:7

If ground ear corn is fed instead of shelled corn, the ratio of protein concentrate to grain should be narrowed somewhat, but if grains that are higher in protein content, such as sorghum or barley, are used, then the ratio should be widened. When full-feeding grain to cattle on legume pasture, the ratio may be widened still further, even to 1:15 during the early pasture season, narrowing to 1:10 or 1:12 as the pasture matures.

SOURCES OF PROTEIN

While all ordinary feedstuffs contain some protein, the amount furnished by the cereal grains, usually the principal component of finishing rations, is so small that other feeds containing a relatively high percentage of protein must be added if satisfactory results are to be obtained. Because of their nitrogen content, protein feedstuffs are spoken of as "nitrogenous feeds," and are divided according to their nature into nitrogenous concentrates and nitrogenous roughages. In the main, the concentrates consist principally of the by-products that result from the milling of cereal grains and from the extraction of oil from seeds that have a high percentage of fat. The more commonly known feeds of this class are linseed (flax) meal, cottonseed meal, soybean meal, gluten (corn) meal, and wheat bran.

The nitrogenous roughages are represented by the different legume hays and silage and are, of course, entirely farmgrown except for dehydrated legumes. Clover, alfalfa, soybean, cowpea, and lespedeza are the principal legume hays used in cattle feeding. Dehydrated alfalfa, usually fed to cattle in pellet form, remains a roughage though ground and pelleted. In connection with nitrogenous roughages, green legume forage such as alfalfa, red, alsike, and sweet clover pastures should be mentioned. Steers with access to such grazing obtain a large percentage of the protein needed from these pasture crops.

Ideally, the major portion of the protein needed by beef cattle should be furnished in the form of legume roughage grown on the farm where the cattle are fed, because of the price relationships shown in Table 141. This also makes it possible to realize one of the purposes for which cattle are kept, namely to furnish a means of marketing legume crops without losing much of the nitrogen that the crops secure from the air. This especially applies to the farmer-feeders in the Midwest, but of course is not applicable to commercial feedlots, which must buy all or most of their feeds.

Despite the large amounts of high-protein feeds produced in this country, they usually fall far short of being enough to balance the enormous tonnage of grains, straw, stubble, and low-protein mill feeds that are used

Table 141
Relative Cost of Digestible Protein in Farmgrown Legume Hay and in Protein Concentrates[a]

	Digestible Nutrients per Ton		Approximate Farm Value per Ton	Approximate Value of Carbohydrates and Fat[b]	Net Cost of Digestible Protein	Net Cost of Digestible Protein per Pound
	Protein (lb)	Carbohydrates and Fats (lb)				
Alfalfa hay	218	796	$ 20.00	$10.87[c]	$ 9.13	$0.042
Clover hay	144	892	20.00	11.95[c]	8.05	0.056
Soybean meal (50%)	928	660	100.00	13.20	86.80	0.094
Soybean meal (43%)	738	834	80.00	16.68	63.32	0.086
Cottonseed meal (41%)	666	768	90.00	15.36	74.64	0.112
Linseed meal (37%)	652	908	80.00	18.16	61.84	0.095
Wheat bran	266	1,072	50.00	21.44	28.56	0.056

[a] Calculated from Tables of Digestible Nutrients in Morrison's *Feeds and Feeding*, 22nd edition.

[b] In determining the money value of the digestible carbohydrates, an arbitrary value of $0.02 per pound was assumed. This is the approximate cost of the digestible carbohydrates in corn at $1.12 per bushel:

1 bushel (56 lb) corn contains:

3.7 lb digestible protein @ 8 cents	=	$0.296
41.1 lb digestible carbohydrates @ 2 cents	=	0.822
Total value of nutrients in 1 bushel of corn	=	$1.118

[c] Arbitrarily decreased one-third because of the higher percentage of fiber in roughages.

417

annually in meat, egg, and milk production. Swine and poultry are unable to use legume roughages to any extent; high-protein feeds must therefore be included in their rations regardless of price, if a satisfactory level of production is to be maintained.

PROOF OF THE NEED FOR A PROTEIN CONCENTRATE

In discussing the matter of establishing the need for a protein concentrate under practical feeding situations, the type of roughage fed is of primary importance. For this reason research data are presented that take the different types of roughage into account.

When the Roughage Is Leguminous. Many feeders consider that cattle receiving a ration of grain and legume hay or silage have little or no need for a protein concentrate. However, most experimental feeding trials show that the addition of a small amount of an oil meal to such a ration usually results in a noticeable increase in the average daily gains. Whether the use of such material proves to be financially profitable depends on the relative costs of the protein concentrate and the feeds that it displaces or saves, as well as on the amount of premium that highly finished cattle command on the market. Under normal conditions the use of a protein supplement is not justified during the first half of the feeding period while the cattle are consuming large amounts of legume roughage. During the last half, however, a small amount of protein concentrate usually is advisable, since the amount of hay or silage eaten at this time is seldom enough to furnish the amount of protein required to maintain the proper ratio between protein and carbohydrates for the most effective action of the rumen bacteria.

Apparently one of the benefits derived from adding a protein concentrate to a ration of grain and legume roughage during the last half of the feeding period is its stimulating effect upon appetite, which tends to become sluggish as cattle approach market finish. Most of the protein concentrates are highly palatable. Consequently, not only is the protein concentrate itself consumed, but the cattle often eat more grain and hay than they would eat if the protein feed were omitted. Faster and more efficient daily gains are the result.

The results usually obtained from adding a protein concentrate to a corn-legume roughage ration are well illustrated by two of the three experiments reported in Table 142. In these trials the use of the protein feeds brought about a slight increase in the consumption of corn and produced slightly larger daily gains. In all three trials, however, the feed costs were increased to the extent that the use of the supplement would have

Table 142
Effect of Adding a Protein Concentrate to a Ration of Corn and Legume Hay for Two-Year-Old Steers

	Iowa		Nebraska		Nebraska (Average 2 Trials)	
Legume Hay Fed	Clover Hay		Alfalfa Hay		Alfalfa Hay	
Supplement Fed	None	Linseed Meal	None	Linseed Meal	None	Cottonseed Meal
Average daily gain (lb)	2.33	2.56	2.31	2.26	2.32	2.57
Average daily ration (lb)						
Shelled corn	19.9	22.0	17.9	17.6	17.2	17.7
Protein concentrate	—	1.5	—	1.7	—	1.7
Legume hay	8.9	7.1	10.5	8.7	11.0	10.4
Feed per cwt gain (lb)						
Concentrates	857	919	775	856	735	732
Legume hay	381	278	455	388	496	422
Net return per head (including swine gains)	$3.48	$5.48	$13.71	$7.44	$23.02	$26.63

been unprofitable had not the cattle receiving it sold for approximately 25 cents a hundred more than those that were fed only corn and legume hay.

Somewhat better results may be expected in calves and yearlings, since their need for protein is greater than that of older cattle. (See Table 143.) Also younger cattle have less capacity and therefore do not ordinarily eat enough legume roughage to secure as much protein from that source as they need. Note that in both experiments, when a protein concentrate was added, total concentrate consumption increased.

It should not be inferred, however, that the profit realized from feeding calves is always increased by the use of purchased concentrates. The gains made are larger, but probably more costly. Under normal market conditions, however, the superior condition and grade produced by the protein concentrate usually result in an advance in selling price sufficient to cover the increased cost of gains with something left over to add to the profit.

When Only Part of the Roughage Is Leguminous. Unless nearly all of the roughage portion of the ration consists of a good grade of legume roughage, the feeding of a nitrogenous concentrate is usually advisable. Sometimes an unusual demand for the common protein feedstuffs forces

Table 143
Need of Calves for a Protein Concentrate with Corn and Alfalfa

Supplement Fed	Illinois None	Illinois Cottonseed Meal	Kansas None	Kansas Linseed Meal	Nebraska None	Nebraska Cottonseed Meal
Average daily gain (lb)	1.97	2.17	2.29	2.37	2.41	2.54
Average daily ration (lb)						
Shelled corn	12.4	11.0	10.5	9.9	11.1	10.6
Protein concentrate	—	2.2	—	1.7	—	1.8
Hay	6.2	6.1	10.3	9.8	4.4	4.6
Feed per cwt gain (lb)						
Concentrates	621	605	457	488	460	486
Hay	319	283	450	413	182	180
Net return per head	7.41	8.86	28.89	27.13	12.07	13.33

Table 144
Need for a Protein Concentrate When Only Part of the Roughage Is Legume Hay

Roughages Fed	Indiana (Calves) (Average 3 years) Corn Silage and Clover Hay		Minnesota (Calves) Corn Silage and Alfalfa Hay		Indiana (2-Year-Olds) (Average 3 years) Corn Silage and Soybean Hay	
Supplement Fed	None	Cottonseed Meal	None	Linseed Meal	None	Cottonseed Meal
Average daily gain (lb)	1.78	2.17	2.02	2.32	2.35	2.46
Average daily ration (lb)						
Shelled corn	10.8	9.7	13.3	13.5	14.3	12.7
Protein concentrate	—	1.5	—	1.9	—	2.3
Corn silage	9.0	10.2	4.8	4.3	24.1	23.6
Legume hay	2.8	2.9	1.8	1.3	4.4	4.5
Feed per cwt gain (lb)						
Shelled corn	626	467	660	580	611	518
Protein concentrate	—	72	—	80	—	95
Corn silage	510	484	215	187	1,030	961
Legume hay	163	138	91	57	198	183
Net return per steer	$7.93	$14.49	$10.72	$13.36	$8.68	$6.66

their price so high that their use materially increases the cost of gains, but the increase in selling price that results from the better condition and finish of the cattle is usually sufficient to increase the net profit.

Obviously, less oil meal is needed where clover or alfalfa makes up a considerable part of the roughage portion of the ration than where carbonaceous roughages, such as corn, sorghum silage, or prairie hay predominate. Table 144 illustrates this type of feeding situation. In all of the experiments reported, the amounts of clover and alfalfa eaten were small. This situation usually exists when corn or sorghum silage is fed, but often it does not exist when legume hay and corn or sorghum stover or oat straw are the roughages used. Should the consumption of legume hay equal 6 or 8 pounds per day, the amount of protein concentrate needed would be appreciably less than the quantities indicated in the table.

When Grain Is Full-Fed on Legume Pasture. In the Corn Belt many feeder cattle are full-fed ground ear corn, usually in self-feeders, on legume or legume-grass rotation pastures during the peak grazing season. Table 145 shows the effect on rate and cost of gain, and daily consumption of feed, of adding a protein concentrate to a full feed of ground ear corn. Whenever protein concentrate was included, more corn and less pasture was consumed. The feeding of protein concentrate throughout the summer, rather than during the late summer only, proved profitable because of the higher rate of gain and a slightly lower cost per hundredweight of gain.

DIFFERENT AMOUNTS OF PROTEIN CONCENTRATE COMPARED

In the early years of experimental beef cattle feeding, the practice was to feed 2 to 3 pounds of protein concentrate per head daily to 2-year-old steers after they were on full feed, even though they were fed considerable legume hay. For example, the average daily ration of a drove of 2-year-old steers fed at the Indiana station during the winter of 1910–1911 was approximately 23 pounds of shelled corn, 3.33 pounds of cottonseed meal, and 10 pounds of clover hay after the first 60 days. Such amounts of protein concentrates were soon found to be too large for the most economical gains, and they were gradually reduced. In fact, the results of many feeding experiments later indicated the need for less and less protein concentrate in the ration of finishing cattle. Today less than half the amount fed in the Indiana experiment mentioned previously is recommended for mature steers fed a liberal amount of legume hay.

Table 145

Performance of Yearling Steers Self-Fed Ground Ear Corn on Legume-Grass Pasture with or without Supplemental Protein[a]

	Protein Supplement[b]	No Protein
Number of steers	15	15
May 7–July 30, 1957, 84 days		
Average initial weight (lb)	691.3	696.7
Average daily gain (lb)	2.83	2.17
Average daily feed consumption (lb)	17.11	13.65
Cost per cwt gain ($)	16.85	14.67
July 31–August 27, 28 days		Protein Added[b]
Average initial weight (lb)	928.7	879.3
Average daily gain (lb)	2.33	2.95
Average daily feed consumption (lb)	13.45	23.55
Cost per cwt gain ($)	14.40	18.25
Summary—entire 112-day period (May 7–August 27, 1957)		
Average final weight (lb)	994.0	962.0
Average daily gain (lb)	2.70	2.31
Average daily feed consumption (lb)	16.20	16.13
Cost per cwt gain ($)	15.12	15.87

[a] Illinois Cattle Feeders' Day Report, 1957.

[b] Soybean meal added in the ratio of 1:12.5.

Results of other experiments with younger cattle are shown in Table 146. It has, of course, long been known that young cattle require more protein concentrate in proportion to their weight than do older cattle. However, feeding experiments show that there is little difference in the requirements per head among cattle of different ages when all are fed appropriate amounts of roughage of the same type, that is, legume or nonlegume. Knowing this has greatly simplified the feeding of cattle because the same thumb rules for supplying protein may be applied to cattle of all ages.

The preceding discussion and the information presented in the accompanying tables indicate that a rule for feeding protein concentrates to full-fed cattle need take into account only one factor—the amount of legume roughage consumed daily. Since 4 pounds of legume hay contain approximately the same amount of digestible protein as 1 pound of high-quality protein concentrate, the following simple rule should be sufficiently accurate for practical feeding operations:

Table 146
Value of Different Amounts of Protein Concentrates for Feeder Calves (Fed in Drylot)

Critical Feeds	Oklahoma[a] Average of 4 years (Calves) Cottonseed Cake Alfalfa Hay			Ohio[b] Average of 2 years (Calves) Mixed Supplement Mixed Hay, Corn Silage			Kansas[c] Cottonseed Cake Alfalfa Hay			
Average Protein Concentrates per Day, lb	0.5	1.0	1.5	0.8	1.6	2.4	0.5	1.0	1.5	2.0
Average daily gain (lb)	1.98	2.08	2.17	1.99	2.20	2.21	1.98	2.06	2.07	2.12
Feed per cwt gain (lb)										
Concentrates	588	593	579	609	593	599	515	519	539	550
Dry roughage	51	48	46	85	77	77	101	97	97	94
Silage	346	331	316	351	319	317	462	442	438	432
Profit per head ($)	22.64	26.45	31.74	17.55	25.25	24.81	7.79	9.17	6.62	6.28

[a] Oklahoma Bulletin B-428.
[b] Ohio Bimonthly Bulletins 179 and 186.
[c] Kansas Circular 105.

> To cattle fed no legume hay
> Feed 2 pounds of protein meal per day,
> But for each pound of hay you feed
> One-fourth pound less of meal they'll need.

The application of this simple rule to full-fed cattle nearly always results in their getting enough digestible protein to meet the requirements given in the requirement tables presented earlier.

Today's progressive cattle feeders are ensuring that their finishing rations contain adequate vitamin A, calcium, phosphorus, and trace minerals. In addition, stilbestrol is generally used and, as indicated elsewhere, the use of this hormone-like compound promotes growth and therefore may increase the need for supplemental protein in the ration. Table 147 presents a summary of data collected by the Nebraska station in a study in which four levels of protein were fed to yearling steers consuming ground ear corn, alfalfa hay, and an amount of soybean meal required to supply the

Table 147
Performance of Steers Fed a Ground Ear-Corn Ration with Varying Levels of Protein[a]

	Protein Level			
	9%	10%	11%	12%
Number of steers[b]	20	20	20	20
Average daily ration (lb)				
Ground ear corn	18.6	18.4	17.6	18.4
Alfalfa	2.0	2.0	2.0	2.0
Supplement	0.6	1.0	1.5	1.9
Average daily gain (lb)	2.28	2.24	2.30	2.47
Feed per cwt gain (lb)				
Ground ear corn	816	820	774	749
Supplement	26	46	64	78
Hay	87	89	86	80
Total	930	956	924	908
Carcass grade[c]	20.2	20.4	20.5	20.7
Dressing percent[d]	59.5	59.7	59.4	59.7

[a] Nebraska Feeders' Day Report, 1965.
[b] Initial weight of steers was 730 lb. Two lots of 10 head in each treatment. Trial conducted May 7 to September 19, 1963, for total of 135 days.
[c] Carcass grade: high choice = 21, average choice = 20.
[d] Dressing percent = (hot carcass weight less $2\frac{1}{2}\%$) ÷ final experimental weight.

level of protein being studied. In addition the ration was completely fortified with vitamin A, minerals, and stilbestrol.

As shown in Table 147, daily gain improved slightly but, because total intake of ration increased slightly, the added gain was undoubtedly due to the added energy intake. Feed efficiency improved somewhat with increasing protein level, but not enough to pay the cost of the protein supplement. Carcass grade and dressing percent were not affected. The Nebraska workers concluded that although 9 percent total protein in the ration of yearling cattle was generally adequate, increasing the supplement enough to raise the protein level to 10.5 percent would be good insurance because the protein content of grains and roughages varies considerably. This would seem to substantiate the National Research Council protein requirement of 11 percent for calves and 10 percent for yearlings.

PROTEIN CONCENTRATES AS SUBSTITUTES FOR GRAIN

Sometimes cottonseed or soybean meal is cheaper than, or at least as cheap as, grain in certain unusual circumstances. At such time cattlemen are likely to feed large amounts of these concentrates in an attempt to cheapen the ration by replacing part of the grain. A study of Table 148

Table 148

Value of Feeding an Excess Amount of Protein Concentrate to Replace Part of the Corn Ration of Finishing Cattle

	Illinois (2-Year-Old Steers)			Illinois (Calves)		Oklahoma[a] (Calves)	
	Soybean Meal			Cottonseed Meal		Cottonseed Cake	
Average Protein Concentrate per Day	2.3	3.9	6.4	1.6	4.2	2.0	7.0
Average daily gain (lb)	2.75	2.93	2.89	2.44	2.57	2.23	2.24
Feed per cwt. gain (lb)							
Shelled corn	595	534	442	461	324	451	225
Protein concentrate	85	113	220	67	162	88	312
Total concentrate	680	647	662	528	486	539	537
Dry roughage	150	141	143	82	317	515	514
Silage	—	—	—	333	78	—	—
Net return per head ($)	14.08	18.72	19.42	29.52	28.15	20.67	13.68

[a] Average of three tests.

will disclose that protein concentrates fed in excess of the amount needed for their protein content will as a rule replace their weight of grain in producing a pound of gain. The energy of the protein concentrate is used for the same purposes as any other source of energy as already mentioned.

Occasionally one still hears about "protein poisoning." The data in Table 148 surely demonstrate that such a possibility is remote, at least when the protein concentrate being fed is entirely of plant origin. Under certain conditions there is danger of toxicity from excess consumption of protein concentrates containing high levels of urea, a nonorganic form of nitrogen. This is another matter and is discussed in Chapter 17.

Chapter 17

The PRINCIPAL PROTEIN CONCENTRATES USED in CATTLE FEEDING

Almost all of the common protein concentrates that are used in cattle feeding are by-products of the cereal and vegetable-oil milling industries. Bran and gluten meal are obtained from the cereal mills, whereas cottonseed meal, linseed meal, and soybean meal are by-products of the oilseed-processing industry. Table 149 gives a breakdown of protein concentrate supplies for all livestock feeding. The percentage of any one protein concentrate source that is fed straight, as contrasted to being fed as an ingredient in a mixed supplement, is difficult to determine. Estimates vary, but probably in the neighborhood of 60 percent of all protein concentrates fed to cattle are fed as a part of a commercially prepared supplement. This percentage is increasing because the protein concentrate is being used as a convenient carrier for numerous desirable additives such as vitamins, antibiotics, and hormones.

Nonprotein nitrogenous materials such as urea are serving as extenders of the protein concentrate supply for ruminants by their use as partial substitutes for the oil meals. A review of the characteristics, processing methods, and comparative value of the principal protein concentrates used in cattle rations should assist feeders in making the proper choice from among the various concentrates available. However, changing price conditions (due largely to variations in supply), quality of the remainder of the ration being fed, and differences in processing methods used, all tend to make it unwise to set forth hard and fast rules as to the relative value of protein concentrates.

Table 149

Supply of Major Protein Feedstuffs for All Classes of Livestock[a]

Protein Feeds	Amount Fed in United States (in Thousands of Tons)			Wholesale Price per Ton at Leading Markets ($) 1959–1963 (5-Year Average)	
	1944	1954	1964		
High-protein feeds					
Soybean meal	3,627	5,428	9,236		68.80
Cottonseed meal	1,982	2,405	2,680		63.00
Linseed meal	459	488	306		62.60
Copra meal	42	182	101		68.90
Peanut meal	102	18	99		60.90
Grain protein feeds					
Gluten feed and meal	918	1,034	1,406	Feed	39.50
				Meal	67.00
Distillers' dried grains	634	251	409		59.60
Brewers' dried grains	217	238	295		44.80
Other by-product feeds					
Wheat millfeeds	5,488	4,567	4,766	Bran	37.40
				Midlings	38.10
				Shorts	39.00
Rice millfeeds	150	303	395		34.00
Alfalfa meal	922	1,324	1,536		47.90

[a] U.S.D.A. Economic Research Service Statistical Bulletin 410, 1967.

SOYBEANS AND SOYBEAN MEAL

Although soybeans were almost unknown in many parts of the United States before 1920, they now constitute one of the major crops, especially in the Corn Belt and the Delta section of the old Cotton Belt, as shown in Figure 76. Approximately 750 million bushels of soybeans are harvested annually in the United States, 90 percent of which are sold to milling companies that extract the valuable oil and sell the residue to commercial feed companies and farmers as soybean meal. The remainder are processed for industrial and human food purposes. Soybeans and soybean products are also an important export commodity.

Soybeans not only contain a high percentage of protein, but also a high percentage of oil. The fat content varies from 15 to 20 percent, depending somewhat on the variety of beans.

Soybean oil is a valuable commercial product, being used in the manu-

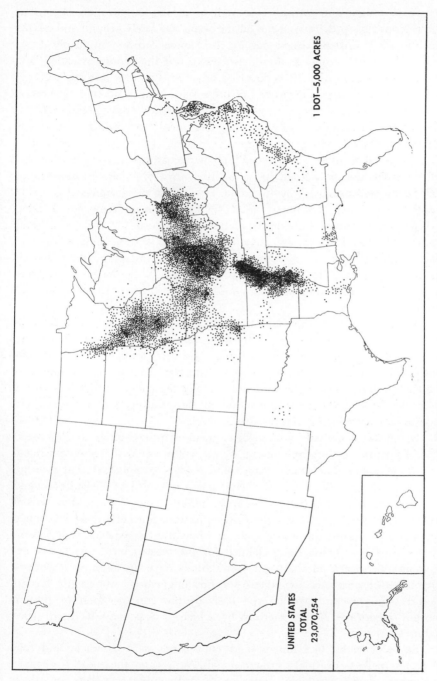

1 DOT—5,000 ACRES

UNITED STATES
TOTAL
23,070,254

FIG. 76. Acreage of soybeans grown in the United States for all purposes in 1959. (U.S. Department of Commerce.)

429

facture of paints and varnishes as well as in the preparation of various edible products. In extracting the oil the beans are finely ground and heated and the oil is either pressed out by mechanical presses or dissolved out with a chemical solvent. If the oil is pressed out the residue is called "old process" soybean meal; if it is dissolved out, the residue is termed "solvent" or "new process" meal. Two types of presses are used in making old process meal—hydraulic and expeller presses. Consequently, old process meal is often called "hydraulic meal" or "expeller meal," according to the type of press used in extracting the oil.

The meal is subjected to very high temperatures in the screwlike expeller presses and as a result has a slightly burnt or "toasted" appearance and flavor not possessed by hydraulic or solvent meal unless it is given a special "toasting" treatment after the oil has been removed. Toasted soybean meal is more valuable than untoasted meal for both swine and poultry but has little if any advantage for beef cattle. In fact, there is some evidence that the high temperatures used during toasting depress the digestibility of the protein slightly, but this is no longer a problem of much practical importance, since most soybeans are now solvent-processed.

Soybean meal has the highest protein content of any feed that is available in quantity for beef cattle feeding, namely, from 41 to 50 percent. Because of its ready availability and usually comparatively low price per unit of protein, it is the most common "straight" protein concentrate purchased by midwestern cattle feeders to supply the protein needed by feeder cattle. Soybean meal is less important in the Southwest and West where cottonseed meal is highly competitive in price.

In chemical composition, soybean meal is quite similar to cottonseed meal. Experiments indicate that it is practically equal to cottonseed meal in feeding value. For many years some feeders complained that soybean meal was too laxative for cattle being fed a full feed of shelled corn and legume hay, but with the adoption by milling companies of the solvent process such complaints have almost disappeared. Formerly there was much discussion regarding the relative merits of hydraulic, expeller, and solvent meals and the superior value claimed for the toasted meals. Extensive experiments at several of the Corn Belt stations have disclosed no important advantage of one over the others for finishing cattle. Apparently feeders should buy soybean meal on the basis of the cost per unit of protein content regardless of the method by which it was manufactured. (See Table 150.)

Soybean meal, like linseed and cottonseed meal, is made and sold as cubes, pellets, or finely ground meal. As a rule, the meal is cheaper and more readily available. However, pea-size cubes or pellets are much

Table 150
Effect of Method of Processing on Value of Soybean Meal for Finishing Cattle

	Illinois				Iowa		Illinois	
	Calves				Yearling Steers (Average 2 trials)		Steer Calves (Average 3 trials)	
	Hydraulic		Expeller					
	Maximum Temperature, F		Maximum Temperature, F				Old Process[a]	New Process
	180°	220°	200°	300°	Expeller	Solvent		
Average daily gain (lb)	1.99	1.89	1.86	1.90	1.78	1.87	2.06	2.00
Average daily ration (lb)								
Shelled corn	9.2	8.8	8.4	8.6	10.5	10.6	10.2	10.1
Protein concentrate	1.4	1.3	1.2	1.3	1.3	1.3	1.5	1.4
Corn silage	8.1	8.1	8.1	8.1	12.0	12.0	4.9	4.9
Legume hay	2.0	2.0	2.0	2.0	1.6	1.6	2.7	2.7
Feed per cwt gain (lb)								
Shelled corn	461	468	451	454	594	564	493	504
Protein concentrate	68	69	67	67	76	68	72	73
Corn silage	406	427	435	425	677	642	226[b]	231[b]
Legume hay	100	106	108	105	91	86	134[b]	137[b]

[a] The old process meal was made by the hydraulic method in two trials and by the expeller method in the third trial.

[b] Corn silage fed in two of the three trials. In the third trial alfalfa was the only roughage fed.

better for feeding with shelled corn, and pellets or cubes 1 to 2 inches in size are recommended for use in the range area where the feed is frequently scattered on the ground.

High-protein soybean meal (50 percent) has assumed considerable importance and is likely to increase in this respect. In the usual solvent-processing methods soybean hulls are virtually all removed, leaving a meal that contains slightly more than 50 percent protein. Usually some of the hulls are added back to the meal so that the protein content is reduced close to whatever the guarantee calls for, usually about 44 percent. The swine and, especially, the poultry feed industries are willing to pay a premium for the lower fiber-content 50 percent protein meals because of their higher energy content. Because of this demand from the mixed-feed industry for the 50 percent meal, and also because soybean processors have developed a fairly good market for their hulls for industrial uses, some processors have discontinued production of the once customary 42 to 44 percent concentrate. The relative feeding value of the higher protein concentrates has not been thoroughly tested with cattle but the results of one Illinois study

FIG. 77. The cotton fields of the South and West constitute an important source of protein concentrate for cattle in the form of cottonseed meal. (International Harvester Company, Chicago, Illinois.)

would seem to make it a safe assumption that these concentrates should be bought on a cost per unit of protein basis, as is the case with other protein concentrates. The lower fiber content does not assume quite the same importance in cattle feeding as it does in feeding poultry and swine because cattle are normally fed higher fiber-content rations and the slightly reduced fiber content of the total ration resulting from the use of a 50 percent protein concentrate is of little importance.

COTTONSEED AND COTTONSEED MEAL

Despite the fact that the acreage of cotton being grown in the United States is only about half of what it was before the initiation of crop control acreage allotments in the 1930s, cottonseed products remain extremely important to both the finishing and the cow-calf programs. Figure 78 shows that cotton production is now concentrated in the Mississippi Delta region, Texas, and in the irrigated areas of Arizona and California. Beef cattle consume about 40 percent of the 2.5 million ton annual production of cottonseed meal and pellets, and nearly all of the 1.3 million tons of cotton-

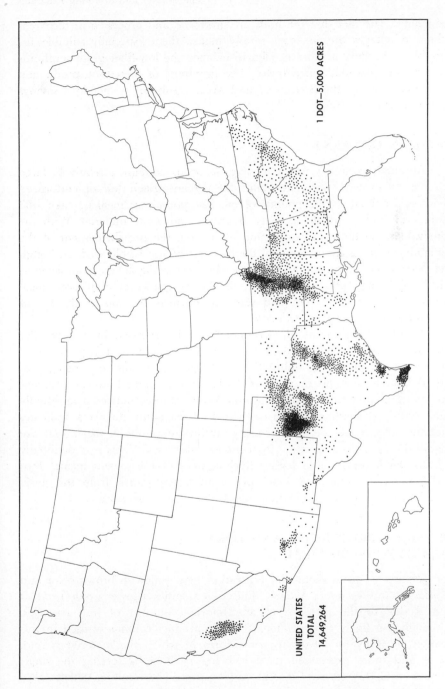

UNITED STATES
TOTAL
14,649,264

1 DOT—5,000 ACRES

FIG. 78. Acreage of cotton harvested in the United States in 1959. (U.S. Department of Commerce.)

seed hulls. The fact that the hulls are produced and processed in the areas where roughages are in short supply makes them especially suitable for providing the bulk or fiber needed to balance the low-fiber grain sorghums used as carbonaceous concentrates. The proximity of the cotton production to the feedlots in the Southwest and West results in tremendous savings in transportation costs.

PROTEIN CONTENT

Although cotton is not a legume, its seeds contain a relatively large amount of protein, normally about 20 percent. Each ton of cottonseed yields approximately 322 pounds of oil, 931 pounds of meal or cake, 480 pounds of hulls, and 267 pounds of linters and waste material. With the removal of the hulls and extraction of the oil, the protein content of the remaining meal is approximately double that of the whole seed. Although the percentage of protein varies somewhat according to the completeness with which the hulls are removed, the better grades of cottonseed meal contain 41 percent protein and stand at or near the top of the list of high-protein feedstuffs fed to cattle.

About 51 percent of all cottonseed is still processed by the expeller process, although a changeover to solvent extraction is taking place in the cottonseed processing industry, with 47 percent of the total being processed by that method in the late 1960s. A few mills, however, still use the hydraulic process. The protein and oil content of cottonseed meal usually vary slightly depending on the method used. Experimental work indicates that the solvent method of processing cottonseed results in meals that are slightly lower in feeding value than the expeller meals. This is undoubtedly due to the lower fat and higher fiber content of the solvent meals. Two Oklahoma tests showed a $7.62 per head larger return from the steers fed hydraulic-processed meal than from those fed solvent meals.

PHYSICAL PROPERTIES AND FORM
OF COTTONSEED MEAL

Cottonseed meal should have a rather light yellowish brown color and a pleasant nutlike odor. A dark, dull color signifies a lower grade product and is due to the presence of an abnormal number of hull particles. Although the finely ground "meal" is the product commonly used in this country, it is by no means the only form in which the material is sold. More and more cottonseed meal is being pelleted by forcing the finely ground meal through steel dies of varying sizes. The resulting pellets range from 0.125 to 0.75 inch in diameter and from 0.25 to 1.5 inches in length.

The small pellets are satisfactory for feeding with shelled or coarsely ground corn. The larger pellets, commonly referred to as range cubes, are popular in the range area where cake or cubes are frequently fed to cattle during the winter by scattering them on the dry ground. Often the cottonseed meal is mixed with 10 to 20 percent of its weight of ground alfalfa to obtain a cake that is a valuable source of carotene. Molasses is often added to the mixture in amounts sufficient to produce a pellet that does not crumble during shipment and handling.

COTTONSEED MEAL POISONING

In the past, cattle that were fed cottonseed meal over an extended period sometimes became unthrifty in appearance, with harsh coats and inflamed eyes. Some animals even, in time, became blind. Such trouble was rather common in the South where large quantities of cottonseed meal were often fed along with cottonseed hulls or other low-grade roughages for rather long periods, and some feeders came to believe that cottonseed meal was not a safe feed for cattle that were to be fed for several months. It has now been established that a number of cases of so-called "cottonseed meal poisoning" were due to a deficiency of carotene or vitamin A. The characteristic symptoms of the disease have been produced by feeding rations low in this vitamin, and they have disappeared entirely when an adequate amount of vitamin A was provided.

If legume hay of good quality, or silage, is included in the ration containing cottonseed meal as a protein supplement, it is doubtful if carotene or vitamin A supplementation is essential. This is especially true if the feeder cattle are not from an extremely droughty area. Table 151 shows the results of a 3-year study dealing with this subject.

The so-called cottonseed meal poisoning referred to above is not to be confused with the gossypol poisoning produced by feeding large amounts of cottonseed meal to swine. Evidently this chemical, which is present in cottonseed meal processed according to standard procedures, does not affect cattle.

LINSEED MEAL

Linseed meal, linseed oil meal, or simply oil meal, as it was once called, is the product that results from the extraction of oil from flaxseed. Flax, like cotton, is not a legume but produces a seed containing a high percentage of protein and oil. When the oil is extracted, the percentage of protein in the residue is further increased. The average protein content of flaxseed is approximately 17 percent, whereas that of the meal is slightly more than twice this amount.

Flax is grown on a much smaller scale than cotton or soybeans in the United States. Altogether only about 300,000 tons of flaxseed are produced annually, over half of which are grown in Minnesota and North Dakota. This amount is insufficient to meet the demand for linseed oil and an additional 800,000 tons of flaxseed are imported, principally from Argentina. Before World War II nearly half of the linseed cake produced in the United States was exported to Europe, but since the war the amount exported has been negligible.

All flaxseed except that needed for planting is sold to processing plants

Table 151

Effect of Adding a Crude Carotene Concentrate to the Cottonseed Meal Fed to Feeder Steer Calves (Three Trials, 174 Days)[a]

	Cottonseed Cake	Cottonseed Meal Pellets + Carotene[b]
Total number of steers	29[c]	30
Average weights (lb)		
Initial	496	495
Final	870	868
Average daily gain	2.15	2.14
Average daily ration (lb)		
Ground shelled corn	11.12	11.43
Protein supplement	1.50	1.50
Alfalfa hay	1.00	1.00
Sorghum silage	6.59	6.58
Salt	0.04	0.04
1–1–1 mineral mixture	0.04	0.04
Feed cost per cwt gain ($)	19.22	20.63
Financial results ($)		
Selling price per cwt	30.08	29.92
Total value per steer (3% shrink)	253.88	251.93
Initial cost per steer	145.38	145.08
Feed cost per steer	71.89	74.77
Total cost per steer	217.27	219.85
Return per steer	36.61	32.08
Average carcass grade[d]	C	C

[a] Oklahoma Bulletin B-428.
[b] Crude carotene concentrate was added to the cottonseed meal to supply 13.7 mg of carotene per pound of supplement fed.
[c] One steer was removed in the second trial and not included in these data.
[d] Carcass grades obtained only in the third trial.

where the oil is extracted by either the expeller or the solvent method. Almost none of the whole seed is used for feeding purposes.

Many cattle feeders, especially those feeding for the higher market grades, feed some linseed meal in the belief that sleeker haircoats with resultant higher selling prices are obtained. The alleged effective agent, mucin, is a gelatinous material, covering the outer hull of the flaxseed, which is removed from the hull in extraction. While it is true that linseed meal contains mucin, experimental work done at the Iowa station did not satisfactorily prove that either haircoat or carcass quality is affected by the use of mucin.

PHYSICAL PROPERTIES OF LINSEED MEAL

In appearance linseed meal is grayish brown in color and somewhat coarser in texture than cottonseed meal. Like cottonseed and soybean products, it is made in various degrees of fineness, ranging from finely ground meal to pellets. The pea-size pellet has been increasingly popular and has largely displaced the finely ground product in many sections of the country.

Linseed meal, because of its marked adhesive qualities under pressure, is especially easy to pellet. Consequently, more of it is processed and sold in this form than is the case with cottonseed meal. Practically all of the linseed meal exported is in the form of cake.

LINSEED MEAL AND COTTONSEED MEAL COMPARED

Theoretically, cottonseed meal is somewhat more valuable than linseed meal since it is from 5 to 10 percent higher in protein. Most commercial feeders, however, prefer linseed meal if it costs no more per ton because of its supposed beneficial effect on the general health of the cattle. This is probably brought about by the tendency of the linseed meal to produce a slightly laxative effect, whereas cottonseed meal tends to make cattle somewhat constipated. Cottonseed meal is often considered superior to linseed meal when used in connection with laxative feeds such as silage, alfalfa hay, or grass.

GLUTEN FEED

Gluten feed is a by-product of corn and is produced in considerable quantities by corn wet millers. It consists of the outer layers of the corn kernel, which are separated from the starch particles in the wet milling processing of corn. Sometimes the outer hull is separated from the underlying gluten layer, which is then sold as gluten meal.

Gluten feed contains about 25 percent protein whereas gluten meal has approximately 40 percent. Both these feeds are used more extensively for dairy cattle than for beef cattle.

Gluten feed is decidedly inferior to the oil meals in finishing rations when used as the sole protein supplement, but it can be successfully substituted for up to one-half of the oil meals. Gluten meal, on the other hand, is almost equal to the oil meals but, as is true of gluten feed, the best use of gluten meal is as a partial substitute for the more commonly used oil meals. Equal parts of gluten meal and linseed meal gave results approximately equal to linseed meal alone in finishing calves at the Kansas station. (See Table 158.) Since gluten feed contains less protein than cottonseed or linseed meal, it should be purchased at a correspondingly lower price per ton.

WHEAT BRAN

Large quantities of wheat bran are produced annually by the flour mills of North America. Although comparatively little of this material is used in the finishing of beef cattle, feed prices occasionally are such as to make it the cheapest source of protein available. Because of its bulk and high percentage of fiber, bran is not an especially good feed for cattle being finished for market. Its rather pronounced laxative effect when fed in large quantities is also unfavorable to its extensive use for cattle finishing. On the other hand, these very qualities commend it as a feed for breeding animals and for young cattle intended for the breeding herd. When mixed with the common farm-grown grains, wheat bran adds bulk and lightness to the ration as well as generous quantities of phosphorus, an element greatly needed by pregnant cows and young, growing bulls and heifers.

Bran differs from most of the nitrogenous concentrates already discussed in having considerably less protein. For this reason it must be fed much more liberally than the oil meals to add the same amount of protein to the ration, but the entire protein requirement of the ration should ordinarily not be supplied with bran alone because this amount of bran would make the ration unduly bulky and laxative. Results of feeding trials in which bran was used are shown in Table 152. Bran contains somewhat more carbohydrate than the oil meals, a fact that should be considered in determining the relative economy of these feeds. For practical purposes, 2 pounds of bran may be said to have the same feeding value as 1 pound of any of the oil meals.

Table 152

Comparative Value of Wheat Bran and Cottonseed Products for Steers[a]

Supplement Fed	Corn Full-Fed 140 Days		Corn Two-Thirds Full-Fed 112 Days	
	Wheat Bran	Cottonseed Meal	Wheat Bran	Cold Pressed Cottonseed Cake
Approximate ratio, corn to supplement	78:22	90:10	58:42	82:18
Initial weight (lb)	973	988	778	743
Average daily gain (lb)	1.76	2.11	1.55	1.59
Feed per pound gain (lb)				
Concentrates	14.19	10.83	7.19	6.93
Prairie hay	—	—	5.14	4.94
Corn stover	5.06	4.21	—	—

[a] Nebraska Bulletins 116 and 132.

Bran is most extensively used for finishing cattle in the highly specialized cattle-feeding sections of the eastern states where both protein concentrates and carbonaceous feeds must be purchased in rather large quantities. Under such conditions a feed such as bran, which contains both protein and carbohydrates, finds considerably more favor than it does in the Corn or Sorghum Belts where an adequate supply of carbohydrates is available in farm-grown grain.

BREWERS' AND DISTILLERS' GRAINS

In the pre-Prohibition era the expended grains of the liquor industry were commonly fed in the form of wet mashes and "slops" to cattle located near the distilleries. Now, however, they are usually dried, bagged, and sold as brewers' and distillers' dried grains.

Brewers' grains, made principally from barley, contain about twice the protein and fiber but only 90 percent of the total digestible nutrients of the original grain. Because of their bulky nature they are seldom fed to beef cattle but are used principally in the manufacture of mixed feeds for dairy cows. In a test at the Illinois station, brewers' dried grains proved to be a much less valuable source of protein for beef calves than soybean meal. (See Table 153.)

Table 153
Value of Brewers' and Distillers' Dried Grains for Feeder Cattle

	Brewers' Grains Calves, Illinois		Distillers' Dried Grains and Solubles 2-Year-Old Steers, Iowa		Heavy Calves, Nebraska		
	Soybean Meal	Brewers' Grains (Barley)	Linseed Meal	Distillers' Grains (Corn)	Linseed Pellets	Distillers' Grains with Solubles	No Protein Con- centrate
Protein content of supple- ment (%)	44.7	31.5	—	—	34.8	25.9	—
Weight of supplement per bushel (lb)	31	18	—	—	—	—	—
Average daily gain (lb)	2.06	1.88	2.36	2.21	1.88	1.68	1.35
Average daily ration (lb)							
Shelled corn	9.2	7.2	10.5	9.2	12.7	10.4	11.3
Protein concentrate	1.3	3.0	1.5	2.0	1.7	2.6	—
Corn silage	7.4	7.4	35.5	32.1	—	—	—
Legume hay	2.0	2.0	1.2	1.4	5.7[a]	5.1[a]	6.0[a]

[a] Prairie hay fed in Nebraska experiment.

Distillers' dried grains are obtained from corn, rye, wheat, or grain sorghum used in the manufacture of beverage and industrial alcohol. Those from corn and wheat are much higher in protein and total digestible nutrients than those from rye or grain sorghum and consequently are much more valuable per ton as livestock feed.

In the disposal of the distillery slops after distilling off the alcohol, the solid particles of grain are screened out and dried to make distillers' dried grains. The liquid portion, containing the water-soluble nutrients and very fine particles of grain, is condensed by evaporation and dried to form distillers' dried solubles. This product has been widely publicized as an excellent source of the B vitamins, which are so important in the feeding of poultry and swine. As cattle have little need for these vitamins, distillers' dried solubles are seldom fed to beef cattle except when mixed with distillers' dried grains. This mixture is called distillers' dark grains or distillers' dried grains with solubles.

Distillers' dried grains may be regarded as a satisfactory substitute for the oil meals in the rations of finishing cattle if sufficiently large amounts are fed to furnish the proper amount of protein. Feeding tests indicate that 1 ton of distillers' dried grains will replace about 1,500 pounds of soybean or cottonseed meal and 10 bushels of shelled corn if fed at the rate of 2 to 3 pounds per head daily. Feeding larger amounts usually is uneconomical unless the dried grains are no higher in price per pound than shelled corn.

UREA AND OTHER NONPROTEIN
NITROGENOUS MATERIALS

As discussed in Chapter 6, rumen bacteria can utilize nitrogen from nonfeed sources such as urea and ammonia in the synthesis of bacterial protein which, in turn, can be digested and used by the ruminant in meeting its own protein requirement. The nonprotein nitrogenous substances mentioned are produced synthetically in larger quantities each year and are now in a very favorable competitive position, as to price, with the protein concentrates.

Because of its hydroscopic nature, urea is prepared for use as feed by the addition of substances that prevent caking in storage. Urea contains 46.7 percent nitrogen but dilution with these substances reduces the nitrogen in feed urea to approximately 42 to 45 percent. Since the crude protein content of a feed is determined by multiplying the percentage nitrogen content by 6.25, the crude protein content of feed-grade urea is 262 to 281 percent. Cottonseed meal is only 43 percent protein; thus approximately 6.1 pounds of 43 percent cottonseed meal are required to furnish as much nitrogen as is present in 1 pound of urea feed. Consequently, if the nitrogen of urea feed were utilized by cattle to the same extent as the nitrogen in cottonseed meal, 1 pound of urea would replace 6.1 to 6.5 pounds of cottonseed meal as far as the protein of the ration is concerned.

When administered to cattle in solution in the form of a drench, urea is highly toxic; as little as 0.25 pound of urea administered to an adult cow directly into the paunch causes death in 40 to 90 minutes. The toxicity is believed to be the result of rapid conversion of urea nitrogen to ammonium carbamate which is absorbed directly into the bloodstream from the paunch. However, if urea is fed to cattle by thoroughly mixing it with the grain ration or with silage, no ill effects are noted. Coating the urea particles so as to make them more slowly available also provides some protection against toxicity.

Early studies of urea as a possible protein substitute indicated that it would be utilized to a higher degree if it were fed with a readily available carbohydrate, such as molasses, so that both nitrogen and a source of energy would be simultaneously available to the rumen bacteria. Feeding trials at the Oklahoma and Iowa stations indicate that the addition of molasses may not be needed by cattle that are fed large quantities of carbohydrates in the form of farm grains. (See Table 154.) It is possible that molasses favors the utilization of urea by stocker cattle fed mainly low-grade roughages such as ground corn stover and corn cobs, although here the principal value of molasses may be as an appetizer and a source of minerals.

Table 154
Value of Urea as a Protein Substitute for Feeder Cattle

	Oklahoma[a] Calves, 167 Days Average 2 Trials			Iowa[b] Yearlings, 175 Days Average 2 Trials			
	Cottonseed Meal	1/4 Urea N 3/4 Cottonseed Meal N	1/2 Urea N 1/2 Cottonseed Meal N	Protein Concentrate	1/2 Urea N 1/2 Protein N	Urea	Urea plus 1 lb Molasses
Average daily gain (lb)	2.14	2.11	2.17	2.26	2.32	2.24	2.22
Protein supplement (lb)	1.5	1.5	1.5	1.25	0.75	0.20	0.20
Urea intake daily[c]	—	(0.06)	(0.12)	—	(0.10)	(0.20)	(0.20)
Protein content of supplement (%)	41.8	46.7	47.3	—	—	—	—
Feed per cwt gain (lb)							
Shelled corn	528	511	518	662	681	701	686
Protein supplement[c]	71	71	69	61	40	9	9
Molasses	—	—	—	—	—	—	45
Legume hay	47	49	46	214	213	226	217
Silage	445	457	436	—	—	—	45
Feed cost per cwt gain	$25.92	$26.30	$25.53	$21.92	$21.06	$20.13	$21.39
Cost of protein supplement per ton	92.50	98.05	98.05	—	—	—	—

[a] Oklahoma Miscellaneous Publications 11 and 13.
[b] Iowa Mimeo. AH Leaflets 176 and 179.
[c] Urea included in weight of protein supplement.

Urea is finding its greatest acceptance in protein supplements fed with high-concentrate complete rations for finishing cattle. An experiment conducted by Oklahoma workers at Panhandle Agricultural and Mechanical College, Goodwell, illustrates that urea makes a satisfactory substitute for the entire amount of cottonseed meal required to bring a high-milo ration up to 12 percent crude protein. The rations were designed to be equal in nitrogen and mineral content. Results after 143 days on feed are shown in Table 155. Note that dehydrated alfalfa additions did not improve this milo ration. Gain and feed efficiency were comparable in all lots and the resulting greater return over feed costs is due to the lower cost of the urea-containing rations. Urea supplied 27.6, 23.5, and 13.1 percent of the nitrogen in the three urea-supplemented lots.

Apparently the concern with toxicity and poor utilization of urea nitrogen in earlier experiments was not well founded. Undoubtedly the various components in the complex urea-containing supplements used in the Indiana trials summarized in Table 156 eliminated the palatability and toxicity problems and enhanced the utilization of the very high levels of

Table 155

Performance of Beef Steers Fed Urea-Containing 12 Percent Protein Rations (143 Days)[a]

Supplement Fed	Cotton-seed Meal	Cotton-seed Meal + Dehy-drated Alfalfa	Urea	Urea + Dehy-drated Alfalfa	Cotton-seed Meal + Urea + Dehy-drated Alfalfa
Number of animals	23	21	23	22	18
Ration ingredients (%)					
Ground milo	87.50	84.75	96.25	91.50	87.75
Dehydrated alfalfa	—	5.00	—	5.00	5.00
Cottonseed meal	8.10	7.00	—	—	3.20
Urea	—	—	0.98	0.84	0.46
Mineral premix	4.40	3.25	2.77	2.66	3.59
Total	100.0	100.0	100.0	100.0	100.0
Average initial weight (lb)	716	714	714	722	721
Average daily gain (lb)	2.48	2.35	2.38	2.40	2.34
Average daily feed consumed (lb)	20.0	19.7	19.8	19.8	19.5
Feed per pound gain (lb)	8.06	8.39	8.31	8.23	8.35
Return per steer over feed costs	$6.08	$8.44	$13.51	$14.98	$11.60

[a] Adapted from data in Oklahoma Cattle Feeders' Day Report, 1967.

Table 156

Comparison of Purdue Supplement A and High-Urea Supplements
(Yearling Steers, Initial Weight, 640 Pounds; 184 Days)[a]

Supplement Designation	Supplement A	Purdue 64	Purdue 80	Purdue 96
Protein equivalent (%)	32	64	80	96
Ingredient formula				
Macronutrients per 1,000 lb				
Soybean meal (lb)	650	—	—	—
Cane molasses (lb)	140	140	140	140
Dehydrated alfalfa meal (lb)	140	510	407	306
Urea (42% nitrogen) (lb)	—	210	279	347
Bonemeal (lb)	52	104	130	155
Salt (lb)	18	36	44	52
Total	1,000	1,000	1,000	1,000
Micronutrients per 1,000 lb				
Cobalt carbonate (gm)	2	4	5	6
Zinc oxide (gm)	625	1,250	1,563	1,865
Vitamin A (IU, millions)	10	20	25	30
Vitamin D (IU, millions)	1.5	3	4	5
Ration fed and performance				
Supplement per head per day (lb)	2.0	1.0	0.8	0.67
Corn silage (lb)	16.9	16.9	17.0	17.0
High-moisture ground ear corn (lb)	12.5	14.5	13.9	14.8
Average daily gain (lb)	2.39	2.32	2.29	2.32
TDN per cwt gain (lb)	524	548	542	553
Cost per pound gain (cents)	14.6	14.3	14.0	14.2

[a] Adapted from Indiana Cattle Feeders' Day Report, 1967.

urea used. The Iowa station has also formulated an 80 percent protein-equivalent supplement that is being used successfully. These high-urea supplements are usually being fed with rations that contain silage or ensiled high-moisture grain.

There is substantial evidence that the addition of sulfur, in the form of flowers of sulfur or Glauber's salt, to high-urea supplements improves cattle performance. A nitrogen to sulfur ratio of 15:1 is suggested. The sulfur apparently enhances urea conversion to amino acids by the rumen microflora.

Although much research is still being done on the subject of urea, the following paraphrase of statements by Indiana researchers[1] summarizes

[1] W. M. Beeson, and T. W. Perry, "Formulating and Feeding High Urea Supplements," Purdue University Research Progress Report 249, 1966.

the current recommendations concerning the essentials for optimum utilization of high-urea supplements:

1. There should be a readily available source of energy such as molasses or grain.

2. Adequate levels of calcium and phosphorus must be supplied. High-urea rations are usually deficient in both calcium and phosphorus.

3. Special attention should be given to supplying the proper level of trace minerals, especially cobalt and zinc.

4. Sulfur may become a limiting factor for the synthesis of the amino acids methionine and cystine by rumen microorganisms. Experimental evidence indicates that nitrogen-sulfur ratio should not be wider than 15:1.

5. Dehydrated alfalfa meal should be used as a source of unidentified factors for microsynthesis of protein. High-urea supplements formulated to supply 90 percent or more of the nitrogen from urea should contain 36 percent or more of dehydrated alfalfa meal for maximum performance on a wide variety of rations.

6. To improve palatability, 3.5 percent salt should be added to high-urea supplements.

7. Fortification with the proper levels of vitamin A and feed additives, such as stilbestrol and antibiotics, is essential to meet recommended daily allowances and to balance the ration completely.

8. Urea should be free-flowing and mixed homogeneously throughout the formula.

9. The maximum intake and desired level of the supplement to be fed should be made clear to the person who will be using the supplement. Mixing the supplement with grain, silage, or total ration is recommended so that each animal obtains the correct amount of supplement.

10. The highest-quality ingredients should be used in a high-urea supplement, avoiding filler feeds such as ground corn cobs, oat hulls, or screenings.

There is no conclusive evidence that supplements that contain urea or similar nitrogenous materials are superior to those that do not. Therefore premium prices should not be paid for such supplements. In fact, because these nitrogen sources are usually cheaper, the feeder should expect to buy them for less than would be paid for supplements containing oil meals alone as protein sources.

Biuret, a nitrogenous material produced by heating urea, is less toxic than urea and thus shows promise as a supplement. Ammonia, when mixed with molasses, apparently is satisfactory as a nitrogen source but cases of supersensitivity in steers have been reported by Kansas and Oklahoma workers when such supplements were used. Morea, a commercial product containing mainly urea, molasses, phosphoric acid, and ethanol or ethyl alcohol as a ready source of available energy, is being self-fed by some feeders with reported success but experimental data are inconclusive on this point. Diammonium phosphate shows promise as a combined nonprotein nitrogen and phosphorus source.

ALFALFA MEAL

Dehydrated alfalfa meal made from leafy alfalfa cut in the prebloom stage of maturity contains 18 to 22 percent protein and may therefore be regarded as a protein concentrate. Because the meal is usually ground extremely fine, it cannot easily be mixed with coarse feeds and is much too fine and dusty to be fed alone. However, when formed into pellets about the size of corn kernels it is an excellent protein supplement to feed with shelled or ground corn. As its protein content is only about half that of cottonseed or soybean meal, approximately twice as much must be fed to supply a given amount of protein.

The demand for dehydrated alfalfa meal for poultry and swine feeding is so great that it is usually too high-priced to make it as economical a source of protein as linseed, cottonseed, or soybean meal for beef cattle, at least on the basis of protein content alone. Feeding tests conducted at the Nebraska station suggest that dehydrated alfalfa may have a much higher feeding value than its protein content would indicate. (See Table 157.) Alfalfa meal made from green leafy alfalfa that was dehydrated immediately after cutting contains abundant carotene. Alfalfa meal makes a real contribution as a source of highly available phosphorus. It should be noted that dehydrated alfalfa pellets had a much lower feed replacement value in the second experiment with calves than in the first experiment with yearlings.

MIXED SUPPLEMENTS

A question in the minds of most smaller-scale cattle feeders concerns the economy of commercial mixed supplements. Undoubtedly the fact that such supplements are a convenient way to supply all needed additives to rations of grain and roughage has much to do with their increasing use. As a rule they are pelleted and bagged in convenient 50-pound bags, or delivered in bulk with special discounts. Use of the protein supplement as the carrier for stilbestrol, antibiotics, vitamin A, or trace minerals, all

Table 157
Value of Dehydrated Alfalfa Pellets as a Protein Concentrate for Beef Cattle[a]

	Finishing Yearling Steers					Finishing Steer Calves		
	1.5 lb Soybean Meal	1 lb Soybean Meal, 1 lb Dehydrated Alfalfa Meal	0.5 lb Soybean Meal, 2 lb Dehydrated Alfalfa Meal	3 lb Dehydrated Alfalfa Meal	1.5 lb Dehydrated Alfalfa Meal	Linseed Pellets	Dehydrated Alfalfa Pellets	No Protein Concentrate
Average daily gain (lb)	2.32	2.52	2.62	2.71	2.47	1.88	1.98	1.35
Average daily ration (lb)								
Ground ear corn	17.6	18.2	17.8	18.4	18.7	—	—	—
Shelled corn	—	—	—	—	—	12.7	11.8	11.3
Protein concentrate	1.5	2.0	2.5	3.0	1.5	1.7	3.3	—
Corn silage	11.3	11.8	11.6	11.9	11.5	—	—	—
Prairie hay	—	—	—	—	—	5.7	5.0	6.0
Feed per cwt gain (lb)								
Corn	757	720	678	677	756	673	596	838
Protein concentrate	64	79	95	110	61	91	166	—
Corn silage	487	466	442	438	465	—	—	—
Prairie hay	—	—	—	—	—	306	254	441
Dressing percent	61.6	63.7	60.8	62.5	59.8	59.6	59.4	[b]

[a] Nebraska Cattle Progress Reports 190 and 194.
[b] This lot was not finished sufficiently for slaughter.

447

Table 158
Protein Mixtures versus Single Protein Feeds for Feeder Cattle

| | Wisconsin[a] (Average of Three Trials) | | Kansas[a] (Average of Three Trials) | | | | | | |
	Linseed Meal	Linseed Meal, ½; Cottonseed Meal, ½	Cottonseed Meal	Linseed Meal	Corn Gluten Meal	Cottonseed Meal, ½; Linseed Meal, ½	Cottonseed Meal, ½; Corn Gluten Meal, ½	Linseed Meal, ½; Corn Gluten Meal, ½	⅓ Each, Cottonseed, Linseed, Corn Gluten Meal
Average daily gain (lb)	2.41	2.41	2.18	2.29	2.20	2.33	2.23	2.36	2.35
Feed per cwt gain (lb)									
Shelled corn	343	354	425	418	407	416	420	402	404
Protein concentrate	83	73	45	43	45	42	44	42	42
Corn silage	504	505	368	350	344	338	352	354	356
Legume hay	114	115	91	87	91	85	90	85	85

[a] Mimeographed reports of cattle feeding experiments.

of which may be desirable ration ingredients under certain conditions, makes this a popular and usually economical method of feeding the supplements.

In addition to these built-in conveniences in commercial mixed supplements, it is sometimes claimed that the mixture or variety of protein sources improves the quality of the supplement. It is sometimes further claimed that the 12 to 18 percent of protein present in a mixed feed is in reality more efficient than the much larger amount of protein in the oil meals fed alone, owing to the lack of certain essential amino acids in the ration when a single protein supplement is fed. This argument fails to recognize that the protein compounds eaten by cattle are broken down and resynthesized by the rumen microflora before digestion and assimilation. Consequently the "quality" of protein fed appears to be unimportant in beef cattle rations. This view is supported by results of experiments conducted at the Wisconsin and Kansas stations, in which none of the protein mixtures used proved significantly superior to linseed meal. (See Table 158.)

DRY ROUGHAGES and
THEIR USE in
FINISHING RATIONS

Roughages differ from concentrates principally in their fiber content. Most concentrates are very low in fiber, with few of the common ones having more than 10 percent. Roughages, on the other hand, have a large amount of fiber, particularly when in the nonvegetative or mature stage. Hay averages about 28 percent and straw approximately 38 percent of fiber when cut and harvested at the appropriate stage and time.

Fiber consists largely of cellulose, hemicellulose, and lignin, all complex, relatively insoluble compounds that form the walls of plant cells. In young, immature plants the cell walls are comparatively thin and the percentage of fiber, and especially lignin, is relatively small. With the approach of maturity, however, the cell walls thicken, and there is a great increase in the fiber content. More important, with approaching maturity the nondigestible lignin portion increases fastest, lowering the available nutrient content still further.

In chemical composition, fiber is a carbohydrate—that is, it is essentially like starches and sugars. However, because of its relative insolubility it is only partly utilized as a food nutrient by domestic animals. Thus the first important step in the digestion of crude fiber, at least for the nonlignin portion, is the softening of the fibrous tissue that occurs when it absorbs large quantities of water in the rumen. The fiber is then partly converted to less complex, nutritionally available compounds—principally volatile fatty acids—through fermentation brought about by the microorganisms in the rumen. Because of the capacity and structure of the digestive organs and their symbiotic relationship with rumen microflora, cattle are more efficient utilizers of roughage than nonruminant farm animals. The methods by which roughages are broken down in the rumen are discussed in greater detail in Chapter 6.

CLASSIFICATION OF ROUGHAGES

Roughages, like concentrates, may be classed as carbonaceous or nitrogenous, depending on the percentage of protein that they contain. Carbonaceous roughages include hay and pasture from the grasses, straw from the cereal grains, and stalks and leaves of corn and the sorghums. Nitrogenous roughages include the forage from legume crops.

Roughages may also be divided into dry roughages and green or succulent materials. For dry roughages the plants are cut when almost mature and are allowed to cure—that is, to lose moisture through evaporation until the roughage contains only 15 percent or so of moisture. For green roughages, immature green crops are pastured by the animals, or the freshly cut green material is fed to cattle before it withers. In silage the freshness or succulence of green roughages is preserved by storage immediately after cutting in specially designed, airtight silos.

FUNCTION OF ROUGHAGE IN THE FINISHING RATION

Roughage plays three main roles in the cattle finishing ration: (1) it contributes to the total nutritive value of the ration, (2) supplies bulk, and (3) serves as a source of minerals and vitamins.

To Provide a Portion of the Required Nutrients. Although most of the gain made by cattle in the feedlot is credited to the concentrate portion of the ration, the part played by the roughage component should not be overlooked. On farms where cattle are fed mainly for the purpose of utilizing unmarketable roughages, the efficiency with which the roughage is used in the production of gains is the factor that often determines the success of the feeding venture.

In general, roughage is most important during the early part of the feeding period. It is then that the cattle have the greatest appetite for such material, and large quantities can be fed with little likelihood of causing the cattle to overeat or go off feed. Except for short-fed cattle, which, of course, should be got on a full feed of grain in the shortest time possible, roughage may well compose at least half of the ration for the first 4 to 6 weeks. Starting with a ration composed almost entirely of roughage, concentrates should be added gradually until the cattle are on a full feed of grain by the end of approximately 6 weeks, depending

on the age and condition of the cattle and relative cost of roughage nutrients. At this time the concentrate content of the ration should range from
80 to 100 percent for steers that are to be fed until they reach choice
grade.

With a longer feeding period roughage is often used alone for the
first few weeks, especially with cattle that are thin and empty when they
arrive at the feedlot. With feedstuffs at ordinary prices, such a plan
tends to reduce the total cost of feed without materially affecting the total
gains made.

Although the ratio maintained between grain and roughage at different
stages of the finishing process varies widely in practical feeding operations,
the ratios given in Table 159 are fairly representative of feeding programs
in the Corn Belt area in this respect. Possibly more hay was fed than
would be used if hay were scarce and had to be purchased, as it must
be by most commerical feedlots, and this of course explains to a large
extent why operators of custom lots design rations that are considerably
lower in roughage.

To Furnish Bulk to the Ration. Owing to the great capacity and
peculiar structure of their digestive systems, cattle are capable of consuming
and utilizing a considerable amount of roughage. It is being demonstrated
daily that cattle can exist on an exclusive concentrate diet, but feeding
experiments as well as practical experience demonstrate that only the experienced feeder who, for economic reasons, is almost forced to reduce the
roughage content of finishing rations, should attempt the feeding of all-
concentrate rations.

Table 159

**Approximate Ratio of Grain to Roughage at Different Stages
of the Feeding Period**

	Ratio of Grain to Roughage (Air-Dry Basis)			
Division of Feeding Period	Large Amount of Roughage Available	Amount of Roughage Limited	Long Feeding Period (Over 200 days)	Short Feeding Period (Approximately 90 days)
1st third	1:4	1:1	2:3	2:1
2nd third	2:3	3:1	3:2	3:1
Last third	3:2	4:1	3:1	5:1
Average for entire period	2:3	5:2	3:2	3:1

Roughage has formed the principal, if not the only, feed of cattle under natural conditions for countless generations. Through processes of natural selection cattle have developed a digestive system, with its population of microorganisms, that functions best when the organs are moderately distended by bulky materials. Cattle naturally only partially masticate their food while eating. When swallowed the food goes into the paunch where it absorbs large quantities of water and begins the process of fermentation through the action of the rumen microflora. It is then regurgitated into the mouth in balls or boluses weighing about 0.25 pound and thoroughly chewed by the resting animal before it is again swallowed for further digestion. Roughage is essential in this process of rumination. Cattle fed on an exclusive concentrate diet spend comparatively little time chewing their cuds. Hence the grain that they eat, if unprocessed on the one hand or ground too finely on the other, is imperfectly masticated and therefore not so thoroughly digested as it would be if part of the ration consisted of roughage.

In furnishing bulk to the ration, roughage tends to lighten or dilute the contents of the alimentary tract so that the particles of concentrates are completely exposed to the rumen microflora and their enzymatic action. Although it is not yet established scientifically, it appears that a certain minimal amount of fiber, or at least bulk, is required to prevent the irritation of the rumen walls. When rations containing less than 10 percent of roughage are fed, rumen parakeratosis occurs in a majority of the animals. Abscessed livers, generally condemned in packing plants by federal inspectors, also are common. Fortunately, although again scientific explanation is lacking, including at least 75 mg of the tetracycline antibiotics in the daily ration prevents the liver malady.

To Furnish Minerals and Vitamins. A much larger concentration of minerals and vitamins occurs in the leaves and stems of plants than in the seeds. Consequently roughages are a better source of calcium, trace minerals, and vitamins A and D than are the farm grains. Cattle fed an abundance of high-quality roughage seldom show symptoms of mineral or vitamin deficiency, whereas such symptoms are occasionally seen in cattle that are fed heavy grain rations with limited amounts of low-grade roughage or no roughage at all.

THE AMOUNT OF ROUGHAGE TO FEED

The role of roughage in the finishing ration has received much attention in recent years, mainly because it is the main deterrent to the design and operation of economical automatic feeding systems. The automation of the processing and feeding of concentrates is relatively simple and economical. Thus if roughages could be entirely eliminated from the finishing ration, greater economies, especially in the category of nonfeed costs, could be effected.

One can find published results of successful feeding of 100 percent concentrate rations, but just as often one can find reports of complete failure. Apparently variations in level of management and in amount and kind of fortification in the supplement fed with the high-grain rations account for the wide difference in degree of success.

Barley and ear corn, two highly important finishing feeds, apparently have sufficient built-in roughage content in the form of hulls and cob so that it is comparatively easy to design acceptable rations based on either of these concentrates. Test results from the North Dakota and South Dakota stations show that 1 to 4 pounds of hay added to barley rations increased rate of gain about 0.20 pound daily but did not reduce costs of gain.

Shelled corn and milo are another matter, as they are both quite low in fiber. Observations in the commercial feedlots of the Southwest and West suggest that the lowest practical level of roughage in milo rations is in the neighborhood of 10 percent. When tallow is included in the high-concentrate ration at the level of 3 to 5 percent, slightly higher levels of roughage—up to 20 percent—can be expected to produce the same results as lower roughage content rations, with the bonus of fewer physiological disturbances.

When less than 10 percent roughage is used in finishing rations, it apparently matters little what the roughage is, as the overall fiber and energy content is not greatly changed regardless of which roughage is fed. It is desirable to leave this low level of roughage in its bulkiest form consistent with its ability to fit into an automated system. For example, shredding of hay is preferred to extremely fine grinding in this situation, provided the feed mixer and unloading wagon can properly handle the shredded hay.

Another point with respect to all- or high-concentrate rations is the comparative dressing percentage of the finished cattle. In trials comparing levels of roughage, when gains and feed efficiency are calculated on the basis of carcass gain rather than liveweight gain, the higher-concentrate rations produce more favorable results. Cattle on such rations also shrink less in shipment and thus should probably be sold on a grade and yield basis if at all possible.

The few, exceptional trials where all-concentrate rations have been successfully fed should encourage researchers to continue in their efforts to eliminate roughage completely and without risk from cattle finishing rations.

ROUGHAGES COMMONLY USED IN FINISHING RATIONS

Roughage, especially in the form of hay, is grown and harvested throughout the country. The roughage crops vary in species, quality, and yield, depending on rainfall, soil fertility, and choice of the farmer in some instances (see Fig. 79). The great variety of roughages used in finishing rations may be divided into two broad classifications—leguminous and non-leguminous. The leguminous roughages are usually considered to be a good source of protein, minerals, and vitamins but only average to low in energy or total digestible nutrients. On the other hand, the nonleguminous rough-ages, such as corn or sorghum silages, used in finishing rations are generally good to excellent sources of energy, especially from a cost standpoint. But the nonleguminous roughages are poor sources of proteins, minerals, and vitamins, and rations using this group of roughages exclusively must be supplemented with rather completely fortified supplements.

LEGUME ROUGHAGES

According to Morrison,[1] legume roughages have the following advantages over other roughages as a livestock feed:

1. They lead in yield of palatable hay per acre.
2. They are the richest of all common forages in protein content.
3. Their protein helps correct the deficiencies in the proteins of cereal crops.
4. Legume forages are the highest in calcium content among all farm-grown feeds.
5. Legume forage excels in vitamin A (carotene) value.
6. Field cured legume hay is rich in vitamin D.
7. Legume forage is rich in other vitamins besides carotene and vitamin D.
8. Legumes increase the yield and protein content of grasses growing in a legume-grass mixture.
9. Legumes are highly important in maintaining soil fertility.

Cutting at the proper stage and curing under good conditions are essential for the above qualities to be present in hay or silage made from leguminous forages, as discussed later in this chapter.

[1] F. B. Morrison, *Feeds and Feeding,* 22nd edition.

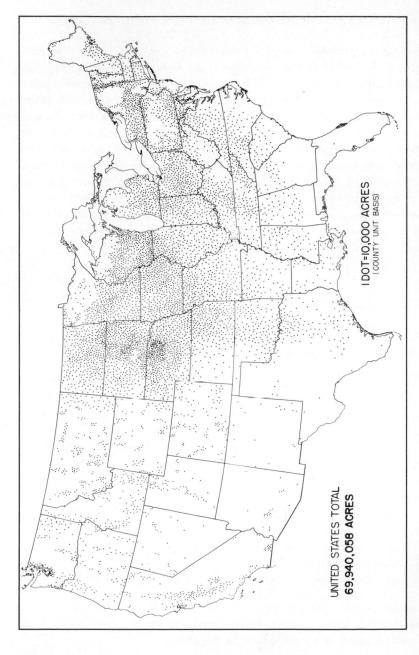

UNITED STATES TOTAL
69,940,058 ACRES

1 DOT=10,000 ACRES
(COUNTY UNIT BASIS)

FIG. 79. Hay is cut throughout the United States, with the possible exception of Florida. The hay acreage in the Corn Belt has declined to some extent during the last decade. (U.S. Department of Commerce.)

ALFALFA HAY

Alfalfa is the most important tame forage crop grown and is the standard by which all others are judged. The development of wilt-resistant and winter-hardy varieties has caused almost universal adoption of this crop as the preferred forage, with its closest competitor among the legume forages being red clover.

Often the market price of alfalfa hay is 20 to 30 percent above that of clover, indicating its superiority in the minds of most feeders. This superiority is due to its higher protein content and freedom from dust and mold. Western-grown alfalfa is seldom damaged by rain and usually retains its bright green color and does not spoil in the stack or bale. Hay made farther east, on the other hand, is often subjected to rain, which leaves it discolored, dusty, and sometimes moldy. Obviously such hay is less palatable and less nutritious than hay put up under ideal weather conditions.

Occasionally alfalfa hay is criticized because of its somewhat laxative nature when fed in large quantities and because it sometimes causes acute bloating. Both these effects are more pronounced in calves than in older cattle. The problem is lessened when alfalfa hay is fed with ground ear corn instead of shelled corn.

Table 160

Comparison of Clover and Alfalfa Hay When a Surplus Amount of Protein Is Supplied (2-Year-Old Steers)[a]

Average of 4 Trials	Corn Cottonseed Meal Clover	Corn Cottonseed Meal Alfalfa	Corn Cottonseed Meal Clover Silage	Corn Cottonseed Meal Alfalfa Silage
Average daily gain (lb)	2.38	2.27	2.32	2.35
Average daily feed (lb)				
Shelled corn	16.2	15.8	12.6	12.5
Cottonseed meal	2.9	2.9	2.9	2.9
Legume hay	12.1	12.7	2.9	2.6
Corn silage	—	—	27.6	28.9
Average feed per cwt gain (lb)				
Corn	679	700	542	530
Cottonseed meal	122	128	126	125
Hay	509	558	126	111
Silage	—	—	11.90	12.28

[a] Indiana Bulletin 245.

There has been much discussion of the relative value of clover and alfalfa hays as roughage for finishing cattle. Unfortunately most of the comparisons have involved the feeding of sufficient nitrogenous supplements to make balanced rations without the additional protein furnished by the legume hays. Obviously under such conditions it would hardly be expected that the difference in protein content of the hays would be apparent.

This is the situation in the Indiana experiments summarized in Table 160. In all lots involved in these comparisons 2.5 pounds of cottonseed meal were fed per thousand pounds live weight, an amount that fully met the protein requirements of the cattle even though no legume hay was included in the rations. It is only reasonable to believe that alfalfa would make a more favorable showing if no protein concentrate were fed, or if it were so limited as to make the protein contained in the legume roughages of real need to the animals.

This method of feeding was duplicated by the Illinois, South Dakota, and Wisconsin stations in the feeding trials reported in Tables 161 and 162. In all these comparisons of red clover and alfalfa the superiority of alfalfa is well established.

CLOVER HAY

Ordinarily, the term "clover hay" is used to refer to red clover or, more likely, a mixed hay in which red clover, a legume, is mixed with timothy or some other grass. Red clover and timothy comprise the backbone of most rotation mixtures in cropping systems used in the Corn Belt that call for leaving the legume-grass seeding down for only one or two years. When longer rotations are used, alfalfa and brome are apt to be substituted. Clover hay or mixed clover-grass hay is preferred by some feeders to alfalfa hay because it is less laxative. Since a protein concentrate is often fed, even when a legume roughage is used, because of its palatability factor, the lower protein content of mixed hay is not serious. Clover or mixed hay is more palatable than alfalfa hay. When used along with silage it tends to cause a more normal state in the digestive tract. Obviously the relative proportion of clover and grasses affects the value of this type of hay with respect to protein, mineral, and vitamin content.

MISCELLANEOUS LEGUME HAYS

Red clover is only one of several different kinds of clover, all of which are used for hay to some extent. Alsike clover is usually considered more

Table 161
Comparison of Clover and Alfalfa Hay for Feeder Cattle

	Illinois Experiment Station (2-Year-Old Steers—126 Days)		South Dakota Experiment Station (Yearling Steers—91 Days)	
	Ear Corn Corn Silage Alfalfa Hay	Ear Corn Corn Silage Clover Hay	Corn Silage Alfalfa Hay	Corn Silage Clover Hay
Average daily gain (lb)	2.38	2.05	2.49	2.29
Average daily ration (lb)				
Ear corn	16.1	16.0	—	—
Corn silage	25.3	26.7	58.3	58.1
Legume hay	4.3	2.0	4.0	3.5
Feed per cwt or lb gain (lb)				
Ear corn	675	783	—	—
Corn silage	1,065	1,302	23	25
Legume hay	181	100	1.6	1.5

Table 162
Comparison of Clover and Alfalfa Hay When a Limited Amount of Protein Supplement Is Fed (Yearling and Two-Year-Old Steers—Average of 4 Trials)[a]

Average Time Fed, 161 Days	Lot 1 Clover Hay	Lot 2 Alfalfa Hay
Average daily gain (lb)	2.19	2.22
Average daily ration (lb)		
Corn	8.5	8.9
Cottonseed meal	1.5	.9
Legume hay	5.3	5.3
Corn silage	28.0	28.0
Feed per cwt gain (lb)		
Corn	394	407
Cottonseed meal	70	42
Legume hay	250	242
Corn silage	1,292	1,264

[a] Wisconsin Agricultural Experiment Station, mimeographed reports of calf feeding trials.

a pasture than a hay crop because of its short growth and consequently low yield. However, when cut and harvested it produces an exceedingly fine-textured hay that is greatly relished by cattle. In chemical composition it has slightly more protein than red clover but is somewhat lower in energy.

Mammoth clover resembles red clover in general appearance but is taller, coarser, and somewhat later-maturing. Although it yields heavily, the hay is often rather coarse and unpalatable. It contains the lowest percentage of protein found in any of the more common legume crops used for hay.

Lespedeza was grown extensively as a hay and pasture crop in the southern and southeastern states in the 1940s and 1950s, but less is being grown each year because of low yields. The improved annual varieties, Kobe and Korean, grow to a foot or more in height and yield about 2 tons of hay per acre. Lespedeza hay cut at the right stage of maturity and properly cured is an excellent roughage and compares favorably in feeding value with alfalfa, as shown in Table 163. However, if it is allowed to become overly ripe, the stems become extremely tough and wiry and are neither palatable nor easily digested. Such overripe hay is sometimes deceptive as to its quality because it often has a good green color and contains an abundance of leaves. *Serecia lespedeza,* a perennial variety, is much lower in value than the annual varieties and can hardly be recommended for use in finishing rations.

Sweet clover is not a satisfactory hay crop, because it is generally too coarse and woody to make a high grade, palatable hay. Its coarse,

Table 163

Value of Lespedeza Hay for Feeder Cattle

	Illinois		Illinois		Missouri	
	2-Year-Old Steers 84 Days		Yearling Steers 53 Days		Calves 112 Days	
Hays Compared	Alfalfa	Lespedeza	Soybean	Lespedeza	Alfalfa	Lespedeza
Average daily gain (lb)	2.99	3.02	2.52	2.54	1.92	1.69
Feed per cwt gain (lb)						
Shelled corn	609	603	578	505	279	319
Protein concentrate	39	39	40	38	23	27
Legume hay	244	241	208	203	494	443
Corn silage	—	—	—	—	113	129
Dressing percent	60.2	61.5	57.6	56.3	—	—

rank growth makes it exceedingly difficult to handle and adds to the labor of baling it or putting it into the stack or mow. With favorable soil and weather conditions a crop of fair quality hay may be secured in the fall from sweet clover sown in oats or wheat the previous spring. Such hay is quite similar to alfalfa in color and texture and cattle eat it readily. Close clipping of the first-year growth in late September is unlikely to cause serious damage to the stand.

NONLEGUME HAYS

Hays made from grasses may differ to some extent in palatability and feeding value in comparison with legume hays, but, generally speaking, tame grass hays are similar to each other in chemical composition regardless of species. Factors such as stage of cutting, curing methods, and yield affect the value of grass hays more than does the species. Thus it is advisable to grow the adapted and recommended grasses for a given locality, cutting them for hay when they are in their most nutritious state.

It has been well demonstrated that mixtures of legumes and grasses produce a greater tonnage of higher quality forage than do grasses seeded alone when grown in areas where such mixed seedings are adapted. It seldom pays to grow straight-grass meadows except, of course, in areas such as those where likelihood of erosion precludes growing anything but perennial grass meadows, or where rainfall is apt to limit yields of tame meadows to an unprofitable level.

Grass hays, if cut at any but the very early stage, resemble cereal straw in composition, being low in protein, phosphorus, and carotene. Early-cut grass hay from well fertilized meadows, on the other hand, may equal or approach the composition of legume or mixed hay.

TIMOTHY AND ORCHARD GRASS

Many years ago timothy was the standard roughage for finishing cattle, but today most up-to-date feeders apologize for its use. From the standpoint of production, timothy has several advantages. It thrives reasonably well on a wide variety of soils; it is a perennial and thus does not require frequent reseeding; the hay is very easily cured and is usually quite free of dust and mold.

As a finishing feed for cattle, however, timothy has few advantages as may be seen in Table 164. Practically its only positive feature apart from its brightness and quality is its nonlaxative properties. Steers or calves

Table 164

A Comparison of Red Clover, Timothy, and Mixed Hay
for Feeder Cattle[a]

2-Year-Old Steers	Red Clover Hay	Mixed Hay	Timothy Hay and Oat Straw
Protein content of hay (%)	10.3	7.7	5.0 (timothy)
			3.5 (straw)
Average daily gain (lb)	2.47	2.28	1.99
Average daily feed (lb)			
Shelled corn	20.3	21.0	20.0
Cottonseed meal	2.0	2.5	3.0
Hay	8.7	8.5	6.0
Oat straw	—	—	0.3

[a] Iowa Bulletin 253.

receiving this roughage are seldom troubled with scours. In total digestible nutrients, timothy is somewhat lower than the common legume hays, and its protein content is less than one-third and one-half that of alfalfa and clover, respectively.

Orchard grass is a recommended grass for pasture and hay in many parts of the Corn Belt and the cooler sections of the southeastern states. It yields well and, when used as a pasture crop, provides both early and late grazing, thereby extending the grazing season. Hay is often made from surplus spring growth of this excellent grass. The Tennessee station conducted a trial in which they compared orchard grass hay containing 10.1 percent crude protein with alfalfa hay containing 14.3 percent. Supplemental protein was reduced in the alfalfa lot to equalize crude protein intake. Performance was good on orchard grass, but it did not measure up to alfalfa hay in either rate or cost of gain (see Table 165).

PRAIRIE HAY

Prairie hay made from native grass meadows in the Plains states resembles timothy in chemical composition and general value as a cattle feed. It is usually much cheaper than timothy, however, especially on the farms and ranches where it is produced. In the western states great quantities of prairie hay are made, and in such areas prairie hay may well be combined with alfalfa, especially in feeding yearling steers. It should be remembered that prairie hay varies greatly in the species of grasses and other plants that it contains. Occasionally it has a large percentage of needle grass or other material that irritates the mucous lining of the cattle's mouths. Such hay should of course be avoided.

Prairie hay is nonlaxative and does not tend to produce bloat. This fact together with its freedom from mold makes it a popular roughage to feed cattle at stockyards and on the show circuit. Table 166 shows that prairie hay, when properly supplemented, has a place in the finishing ration. In the future, prairie hay will be used even more for cows and stockers than presently, as less and less roughage is included in finishing rations.

Hays made from the cereal crops and from Sudan or Johnson grass have a feeding value for finishing cattle comparable to timothy or prairie hay and usually are also better utilized by breeding cattle or stockers.

Table 165

Orchard Grass Hay versus Alfalfa Hay in Finishing Rations for Yearling Steers (86 Days) [a]

	Corn, Cottonseed Meal, and Orchard Grass Hay	Corn, Cottonseed Meal, and Alfalfa Hay
Number of steers	14	14
Average initial weight (lb)	787	800
Average final weight (lb)	1,004	1,051
Total gain (lb)	217	251
Average daily gain (lb)	2.52	2.92
Average daily ration (lb)		
Ground ear corn	18.7	20.3
Cottonseed meal	2.0	1.5
Hay	4.7	4.7
Feed per cwt gain (lb)		
Ground ear corn	742	695
Cottonseed meal	79	51
Hay	186	161
Total	1,007	907
Feed costs per head daily (cents) [b]	51.6	55.4
Feed cost per pound gain (cents)	20.5	19.0
Slaughter grade	G+	G+

[a] Tennessee Agricultural Experiment Station H-72-7-3, 1961.

[b] Feed costs per ton: ear corn, $40; cottonseed meal, $65; orchard grass hay, $32; alfalfa hay, $40.

Table 166

Value of Prairie Hay for Feeder Cattle

	Nebraska Bulletin 93 84 Days			Nebraska Bulletin 100 Average of 3 Experiments	
	Ear Corn, Prairie Hay	Ear Corn, Prairie Hay 50%, Alfalfa 50%	Ear Corn, Alfalfa Hay	Shelled Corn, Prairie Hay	Shelled Corn 90%, Linseed Meal 10%, Prairie Hay
Daily gain (lb)	1.20	2.01	2.06	1.51	2.18
Feed per cwt gain (lb)					
Corn	787	470	460	1,171	808
Protein concentrate	—	—	—	—	90
Hay	1,516	1,047	1,075	521	387

FACTORS AFFECTING HAY QUALITY

Even though hay usually makes up considerably less than one-fourth of the total feeds fed to finishing cattle, the quality of hay, within a kind of hay, may materially affect performance. Data such as those shown in Table 167 effectively demonstrate this point.

There is considerable disagreement as to when hay should be cut to obtain the greatest feeding value. It is generally agreed that early cutting favors higher protein content, finer texture, less fiber, and higher digestibility. Many believe, on the other hand, that late cutting results in greater tonnage, more total digestible nutrients, and usually more favorable weather for field curing of the first cutting. The last point is undoubtedly true, but data such as those shown in Table 168 disprove the others.

Method of curing and storage also may materially affect hay quality. Data on this subject are plentiful and those in Table 169 are typical. Naturally the amount of rainfall and cloudy weather affect curing time; therefore this factor is a greater consideration in some areas than others.

Table 170 shows the effect of stage of maturity upon digestible protein content, and Table 171 shows the effect of method of harvesting and storing upon carotene content.

The method of cutting hay is of concern today, as several kinds of machines are available for cutting and field-conditioning hay. The Kansas station conducted a trial with weaner heifer calves fed limited rolled sorghum grain in addition to alfalfa hay, processed by various methods, according to appetite. The various treatments applied to the hay were:

    1. Control—raked and baled.

    2. Crushed—crushed with one smooth steel roll and one spiral-grooved rubber roll, raked, and baled.

    3. Rotary cut—cut, lacerated, and windrowed in one operation by means of a 12-foot trail-behind twin-rotor rotary mower, and baled.

    4. Swathed-crimped—swathed and crimped by means of a 12-foot self-propelled windrower with crusher-crimper attachment, and baled.

Table 167

Value of Quality Hay When Fed with a Full Feed of Ground Ear Corn to Feeder Cattle[a]

	Trial 1 (259 Days)		Trial 2 (??4 Days)	
	Poor Hay	Good Hay	Poor Hay	Good Hay
Number of steers in lot at start	12	10	14	14
Number of steers in lot at close	12	10	14	14
Average weight at start of test (lb)	636	630	472	475
Average weight at close of test (lb)	1,049	1,132	889	965
Average daily gain (lb)	1.60	1.94	1.86	2.19
Average daily ration				
Ground ear corn (lb)	13.6	15.0	11.1	12.4
Supplement (lb)	1.5	1.5	1.5	1.5
Hay (lb)	1.9	3.3	2.1	2.8
Minerals (oz)	2.0	1.9	1.3	1.3
Salt (oz)	0.5	0.6	0.6	0.7
Feed per cwt of gain (lb)				
Ground ear corn	852.0	774.0	598.0	569.0
Supplement	94.0	77.0	80.0	68.0
Hay	117.0	169.0	112.0	127.0
Minerals	8.0	6.0	4.0	4.0
Salt	2.0	2.0	2.0	2.0
Ground ear corn plus supplement				
(lb)	946.0	851.0	678.0	637.0
Cost per cwt of gain ($)	23.14	21.99	20.50	19.69
Dressing percent	59.97	63.04		

[a]Ohio Research Bulletin 732.

Table 168

Effect of Maturity at Cutting on Composition and Yield of Alfalfa (3 Years)[a]

Stage of Maturity	Bud	$\frac{1}{10}$ Bloom	Full Bloom and $\frac{1}{10}$ Bloom[b]	Full Bloom
Composition data				
Percent protein	21.3	20.5	19.6	18.4
Digestible dry matter (%)[c]	63.2	61.1	59.5	57.8
Percent leaves	54.9	52.9	52.8	51.1
Yield data				
Average number of cuttings per year	5.0	4.3	4.0	3.7
Dry matter yield, tons per acre	4.18	4.39	4.32	4.35
Protcin yield, tons per acre	0.89	0.90	0.85	0.80
Digestible dry matter, tons per acre	2.64	2.68	2.57	2.51

[a] Nebraska Beef Cattle Progress Report, 1968.
[b] First cutting at full bloom, remainder at $\frac{1}{10}$ bloom.
[c] Determined by *in vitro* technique.

5. Wafered—cut with a flail-type cutter, field-dried to about 15 percent moisture in windrows, and wafered with a commerical wafering machine.

Results of the Kansas experiment are shown in Table 172. Rotary-cut hay proved to be least desirable, undoubtedly due to leaf and particle shedding at baling time. Simultaneous swathing and crimping of the hay

Table 169

Effect of Method of Curing and Storage on Quality Factors in Hay[a]

Method of Curing and Storage	Hours in Swath	Leaf Loss (%)	Dry Matter Loss (%)	Storage Dry Matter Loss (%)	Dry Matter Left to Feed as Standing Crop (%)
Field cure—rain damage	108	60	40	5	60
Field cure—no rain	54	35	20	5	79
Barn dried—no heat	29	25	20	10	81
Barn dried—heat added	29	25	15	3	85
Wilted silage	8	15	20	18	83

[a] U.S.D.A. Bureau Animal Industry-Inf. 142.

Table 170

Digestible Protein Content of Alfalfa and Grasses in Relation to Maturity[a]

	Stage of Growth	Digestible Protein[b] (%)
Alfalfa	Immature	17.0
	After bloom	5.4
Kentucky bluegrass	Before heading	15.0
	After bloom	2.7
Orchard grass	Before heading	13.0
	After heading	4.9
Timothy	Pasture stage	13.9
	In seed	2.2
Mixed grasses	Immature	10.3
	At haying stage	4.7

[a] California Extension Service Circular 125.
[b] On 15 percent moisture basis.

produced a maximum response, equal to the more expensive wafering process. The excellent performance by the heifers on the swathed-crimped hay, aside from the reduced curing time and labor saved, undoubtedly accounts for the rapid rise in popularity of the self-propelled swather-crimper in the commercial hay-growing areas where hay is often sold on an analysis basis.

Table 171

Carotene Content of Roughages When Cut, Stored, and Fed[a]

Roughage	Carotene (micrograms per gram dry matter)		
	Cut	Stored	Fed
Early silage	354	341	140
Barn-dried hay	361	190	20
Field-cured hay	211	66	9
Late silage	216	238	84

[a] Cornell Feed Service No. 35.

Table 172

Performance of Weaned Heifer Calves Fed Various Field-Conditioned
Alfalfa Hays (93 Days)[a]

	Control	Crushed	Rotary Cut	Swathed-Crimped	Wafered
Number of heifers per lot	10	10	10	10	10
Initial weight (lb)	438	441	442	443	442
Average daily gain (lb)	1.10	1.18	1.05	1.30	1.28
Average daily ration (lb)					
Alfalfa hay	11.8	13.1	11.3	11.9	13.0
Rolled sorghum grain	3.5	3.5	3.5	3.5	3.5
Feed per cwt gain (lb)					
Alfalfa hay	1,072.7	1,110.2	1,076.2	915.4	1,015.6
Rolled sorghum grain	318.2	296.6	333.3	269.2	273.4
Total feed required per cwt gain (lb)	1,390.9	1,406.8	1,409.5	1,184.6	1,289.0
Feed cost per cwt gain[b]	$16.46	$16.44	$16.76	$14.00	$15.08

[a] Kansas Agricultural Experiment Station Bulletin 460, 1963.
[b] Feed costs: alfalfa, $20 per ton; rolled sorghum grain, $1.80 per cwt.

OTHER NONLEGUMINOUS DRY ROUGHAGES

Corn stover, once a fairly important roughage used in finishing the 2- and 3-year-old steers commonly fed in the past, is no longer used except as bedding, or possibly as silage for dry cows. Other stovers are used to a small extent in areas where grown.

SORGHUM STOVER

The stover of the grain sorghums is essentially like corn stover in chemical composition and feeding value. However, because the stalks are somewhat finer-textured, a greater percentage of the stalk is ordinarily eaten. When fed alone or with other dry carbonaceous roughage, sorghum stover is not very satisfactory for finishing cattle. When combined with legume hay or silage, much better results are secured. Even when legume hay constitutes half or more of the total roughage ration, some nitrogenous concentrate should usually be fed as all stovers are low in protein. Sorghum stover, like corn stover and the grass hays, is more valuable in rations for cattle other than those on finishing rations.

FIG. 80. A two-row self-propelled chopper can harvest up to 500 tons per day of a forage sorghum crop. Grain content of the silage made from some varieties of forage sorghum may be as high as 40 percent, on a dry-matter basis. (John White, New Mexico State University.)

CEREAL STRAWS

Of the straws of the common cereal grains, oat straw is the most valuable as a cattle feed. When cut with a binder at the optimum time and stored before it is damaged by weather, oat straw has a feeding value close to that of timothy hay. But when it is raked up after a combine, it has little chaff or leaves and often has been damaged by sun and rain. In this condition it is not highly palatable, and comparatively small quantities will be eaten. Oat straw is, of course, a carbonaceous roughage and should be fed to finishing cattle only when other constituents of the ration furnish the necessary protein. Feeding a small quantity of oat straw tends to reduce the bloat and scouring occasionally noticed in cattle receiving a heavy feed of alfalfa hay or barley. Also, added to a heavy silage ration it satisfies, at a very low cost, the desire of the cattle for a dry roughage material.

Barley straw is somewhat inferior to oat straw for feeding purposes, principally because most varieties of barley are bearded, and the beards sometimes cause the cattle's mouths to become sore.

Wheat straw has the lowest feeding value of any of the common cereal straws and should be regarded as an emergency feed for finishing cattle, to be fed only when ordinary roughages are scarce and high-priced. Satisfactory gains were obtained in an experiment at the Nebraska station using yearling steers that were fed a full feed of shelled corn, cottonseed cake, and equal parts of wheat straw and alfalfa hay.

When straw is fed to finishing cattle, it is usually fed according to appetite. In the days when cereal grains were cut with a binder and threshed, cattle frequently were permitted access to the strawstack, which afforded shelter as well as feed. Now that the grain is combined the straw usually is fed in the form of baled straw in mangers and bunks. Since it is a low-grade, inexpensive feed, it is usually fed in much larger amounts than the cattle will eat, and the refused portion is removed from the bunks every few days for bedding.

CORN COBS

Corn cobs have long been regarded as an important roughage for grain-fed cattle when fed in the form of ground ear corn. Their value was believed to lie chiefly in the volume and bulk which they gave the ration rather than in the digestible nutrients supplied. Cobs are hard and woody and will not be eaten alone even by starving cattle—a fact that has caused cattlemen to consider them almost worthless as a source of feed nutrients. Nevertheless, digestion trials have shown that cobs contain 45 pounds of digestible nutrients per hundredweight, or about the same as oat straw. Cobs have usually saved both grain and roughage in producing 100 pounds of gain in experiments comparing ground ear corn with shelled corn. This shows either that cobs have a definite feeding value or that their presence in the ration favors the utilization of the other components.

Experiments at the Ohio, Iowa, and Nebraska stations, in which ground cobs were fed in addition to those present in gound ear corn, show clearly that corn cobs are a valuable roughage material for finishing cattle. (See Table 173.) In the Ohio and Iowa tests, where both corn and hay were fed according to appetite, the added cobs replaced grain rather than roughage in both the daily ration and in feed consumed per hundred pounds of gain. But in the Nebraska experiments, where the roughage fed was sorghum silage, the addition of cobs had little effect on the consumption of corn on a shelled basis but greatly reduced the consumption of silage. These results indicate that corn cobs are a better companion roughage for low-grade hay, which is eaten in small amounts, than for highly palatable hay and silage. It is possible that the additional bulk imparted to

Table 173

Value of Corn Cobs as a Feed for Finishing Cattle

	Ohio Mimeo. Series 52 Calves and Yearlings (Average 3 Trials)			Iowa Mimeo. AH Leaflet 165 Yearlings (Average 2 Lots Each)			Nebraska Bulletin 396 Yearlings (Average 3 Trials)		
	Shelled Corn	Ground Ear Corn	Ground Ear Corn + Cobs	Shelled Corn	Ground Ear Corn	Ground Ear Corn + Cobs	Shelled Corn	Ground Ear Corn	Ground Ear Corn + Cobs
Average daily gain (lb)	1.92	1.92	1.85	2.38	2.43	2.23	2.42	2.35	2.13
Average daily ration (lb)									
Shelled corn	12.4	11.1	9.4	15.2	14.3	13.0	13.5	13.5	13.1
Corn cobs	–	2.5	4.2	–	2.7	5.4	–	3.1	6.2
Protein concentrate	2.0	2.0	2.0	1.5	1.5	1.5	1.5	1.5	1.5
Hay	4.3	4.3	4.2	5.1	3.6	3.6	–	–	–
Sorgo silage	–	–	–	–	–	–	27.3	16.7	6.1
Total daily consumption of air-dry roughage (including cobs) (lb)	4.3	5.8	8.4	5.1	6.3	9.0	8.2[a]	8.1[a]	8.0[a]
Ratios of corn to roughage (including cobs)	2.9:1	1.6:1	1.1:1	3.4:1	2.6:1	1.4:1	1.6:1	1.7:1	1.6:1
Feed eaten per cwt gain (lb)									
Shelled corn	631	552	478	657	651	576	559	574	613
Corn cobs	–	129	220	–	123	240	–	131	291
Protein concentrate	103	103	108	64	66	67	62	63	70
Hay	221	223	226	219	163	158	–	–	–
Sorgo silage	–	–	–	–	–	–	1,126	712	286

[a] Sorgo silage reduced to 15 percent moisture basis.

the Ohio and Iowa rations by the ground cobs resulted in more complete digestion of the corn and protein concentrate.

COTTONSEED HULLS

In the cotton-growing areas, huge quantities of cottonseed hulls are produced in the processing of cottonseed at the oil mills. The 500 pounds of hulls resulting from processing 1 ton of cottonseed add up to 1.3 million tons of hulls produced annually. As they are low in protein and high in fiber, their feeding value would appear to be low compared with most other roughage materials. They are bulky and obviously cannot be economically transported far from their source. Thus they are usually fed in Texas and the western states where cotton production is most highly concentrated.

The Arizona station has obtained excellent results from the substitution of cottonseed hulls for chopped alfalfa hay in their 80 percent concentrate ration, especially when they included 4 percent tallow in the ration. Table 174 gives the results of one of their studies, and it will be noted that the steers on regular hulls consumed the most feed and gained fastest, indicating possibly that the hulls were responsible for a more desirable physiological status in the rumen. The fact that the hull rations were not so efficiently converted to gain indicates a lower available energy value, and thus the price paid for hulls probably must be at least one-third lower than that for good alfalfa hay. Feeders who feed complete mixed rations

Table 174
Comparison of Alfalfa and Two Types of Cottonseed Hulls in Finishing Rations (140 Days)[a]

Roughage Source	Alfalfa	Regular Cottonseed Hulls	Delinted Cottonseed Hulls
Alfalfa in ration (%)	20	5	5
Hulls in ration (%)	—	15	15
Number of steers	16	16	16
Average initial weight (lb)	591	595	589
Average daily gain (lb)	2.70	3.02	2.88
Average daily feed (lb)	20.1	23.7	23.8
Feed per cwt gain (lb)	746	786	832
Feed cost per ton	$54.58	$53.68	$53.68
Feed cost per cwt gain	$19.87	$20.48	$21.66

[a] Adapted from Arizona Cattle Feeders' Day Report, 1967.

Table 175
Low-Quality Roughages in Rations for Growing Calves (125 Days)[a]

	Cottonseed Hulls	Oat Straw	Cotton Burrs	Corn Cobs
Number of calves	22	22	22	22
Average initial weight (lb)	506	514	499	505
Average daily gain (lb)	1.78	1.92	1.58	1.85
Feed intake per head per day (lb)	16.4	16.5	16.0	17.3
Feed per cwt gain (lb)	922	862	1,013	937

[a] Adapted from North Carolina State College A.I. Report 73, 1961.

like to include some hulls, because hulls require no processing and make the ration bulkier, and cattle seem to stay on feed better.

A comparative trial conducted by North Carolina investigators, reported in Table 175, shows the relative values of several lower-grade roughages when included in rations containing considerably less energy than the ration used by the Arizona workers. The rations used by the North Carolina group contained 61.5 percent of the roughages being compared, so that differences in roughage value are more likely to be expressed. Oat straw appeared to be slightly superior to hulls or cobs. Cotton burrs were definitely inferior. If these roughages were fed at lower levels, such as 10 or 15 percent, it is doubtful if it would matter which one were fed. Therefore, when low-grade roughages are used with high-concentrate rations, the one available at lowest cost should be used, provided that any required processing is comparable in cost. Hulls mix readily with the remainder of the ration, whereas the others, especially cobs and straw, require grinding or chopping, a rather expensive operation.

SOYBEAN HULLS AND FLAKES

As swine and poultry nutritionists and feed manufacturers have formulated rations that are higher and higher in energy content, and thus lower in fiber, they have demanded a soybean meal with little hull content. Consequently a potentially important roughage source has become available for beef cattle feeding in the form of soybean hulls. Soybean hulls contain about 10 percent crude protein, 40 percent nitrogen-free extract, and 35 percent crude fiber, a composition not greatly different from that of the better grass hays. Studies have shown that flaking the soybean hulls increases

Table 176

Feeding Value of Soybran Flakes Fed Beef Heifers (137 Days)[a]

Type of Substitution	Roughage Replacement		Ear-Corn Replacement	
Feeds Fed	Hay, Silage, Corn	Hay, Silage, Corn, Soybran Flakes	Hay, Silage, Corn	Hay, Silage, Soybran Flakes
Number of heifers	16	16	24	24
Average initial weight (lb)	458	468	455	458
Average daily gain (lb)	1.51	1.78	1.38	1.52
Average daily ration (lb)				
Ground ear corn	4.0	4.0	4.0	—
Soybran flakes	—	5.0	—	4.0
Hay crop silage	15.0	7.5	15.0	15.0
Mixed hay	5.0	2.5	5.0	5.0
Salt (oz)	0.5	0.5	0.5	0.5
Minerals (oz)	0.4	0.4	0.4	0.4
Feed per cwt gain (lb)				
Ground ear corn	258	223	287	—
Soybran flakes	—	279	—	261
Hay crop silage	988	419	1,078	977
Mixed hay	329	140	359	326
Salt	2	2	2	2
Minerals	2	2	2	2

[a] Adapted from Ohio Agricultural Experiment Station Series No. 122, 1961.

the digestibility of the dry matter of the hulls from about 60 percent for ground raw hulls to 70 percent for flaked hulls. Undoubtedly the heat treatment involved in the flaking process is responsible for the improved digestibility. Soybean hulls are an example of a roughage with high fiber content that has a rather high digestibility, the reason being that the major portion of the fiber is cellulose, which is digestible by beef cattle if it is not highly bound by indigestible lignin.

The Ohio station conducted two trials with soybean flakes, also called soybran flakes. The results are shown in Table 176. In one trial they substituted 5 pounds of flakes for 7.5 pounds hay-crop silage and 2.5 pounds hay, or, in other words, they substituted the flakes for a like amount of air-dry roughage. In the other trial they substituted 4 pounds of flakes for 4 pounds of ground ear corn. The Ohio workers concluded that the flakes more than replaced the feeding value of the hay and even proved to be more valuable than ear corn. Thus toasted soybean or soybran flakes have a value comparable to ear corn, oats, or barley, and yet they have the bulky characteristic of a roughage.

RICE HULLS

In some of the southern states, notably Arkansas, Louisiana, and Texas, large quantities of rice hulls result from the rice milling industry. The composition of rice hulls is comparable to that of low-quality hay, and they thus would appear to offer possibilities as an economical source of roughage, since there is little other commercial outlet for rice hulls. The Arkansas station conducted a trial with yearling steers fed a grain mix consisting of 80, 10, and 10 percent of ground shelled corn, crimped oats, and cottonseed meal, respectively, plus vitamin A, minerals, and salt, plus either prairie hay or rice hulls according to appetite. The data are summarized in Table 177. Performance was comparable in all respects and, with the lower price prevailing, the rice hull steers produced gains at $2.55 lower cost per hundredweight gain. Neither bloat, scours, nor other ill effects resulted from feeding rice hulls as the only roughage source.

Table 177

Performance of Steers Fed Rice Hulls or Prairie Hay as Roughage Sources (84 Days)[a]

Roughage Fed	Prairie Hay	Rice Hulls
Number of steers	8	10
Average initial weight (lb)	711	692
Average daily gain (lb)	2.93	2.86
Average daily ration (lb)		
Grain mixture	20.6	17.7
Prairie hay	3.4	—
Rice hulls	—	4.3
Total	24.0	22.0
Feed per pound gain (lb)	8.5	7.9
Feed cost per cwt gain[b]	$19.79	$17.24
Carcass grades	1 C, 7 G	1 C, 9 G

[a] Adapted from *Arkansas Farm Research*, Vol. 12, No. 4, 1963.
[b] Feed costs per ton: prairie hay, $26; rice hulls, $8.

MISCELLANEOUS ROUGHAGE SUBSTITUTES

In many cattle-feeding areas, roughage nutrients are costlier than concentrate nutrients. Add to the higher cost of roughage nutrients the processing and storage costs, and the problem of feeding by automated methods, and one can see why feeders and researchers alike are casting about for every conceivable substitute for roughages in finishing rations.

Oyster Shells. Several stations, notably Nebraska, have studied the use of about 2.5 percent hen-size oyster shells as a substitute for the entire roughage in finishing rations. They have achieved reasonable success, and their results have been confirmed by some, though not all, stations. The high calcium level fed when oyster shells are used apparently must be compensated for by also increasing the phosphorus level in order to keep the calcium-phosphorus ratio in balance. The scabrous character of the shells serves the same physiological function as the "roughness" quality of hays.

Sand. The addition of 2 percent sand to a completely mixed high-concentrate ration improved performance of steers at the Iowa station. In six experiments the improvement was small but consistent, with a 5 percent improvement in both rate of gain and feed efficiency. Sand did not improve conventional rations containing some hay; thus sand apparently serves the same function as the oyster shells mentioned above.

Poultry Litter. Broiler-house litter has been successfully used as both a roughage and a nitrogen source in feeding trials at the Arkansas and Virginia stations. Such litter contains upward of 30 percent protein equivalent and, depending on type of bedding used, may contain 15 to 30 percent crude fiber. Wood shavings apparently result in a less desirable litter, for cattle feeding at least, than cottonseed hulls or peanut hulls which are also sometimes used as litter.

Poultry litter can satisfactorily make up at least 25 percent of the total ration. As the nitrogen is in the form of nonprotein nitrogen, it cannot be expected to be utilized quite as well as natural protein nitrogen. The litter should be ground in a hammermill before mixing. Commercial cattle feeders in the broiler-producing areas of the Southeast are making quite extensive use of litter as an ingredient in their complete feeds.

Cattle Manure. Cattle manure itself is being tested, and it appears that if the manure from cattle on grower rations is dried and mixed in a complete high-concentrate ration at the level of 5 to 15 percent, it serves as a good substitute for roughage.

Chapter 19

SILAGE as a FEED
for BEEF CATTLE

The first silage experiment reported in a United States Government publication (1875) was that of Professor Manly Miles of the University of Illinois, who made studies of cornstalk and broomcorn silages stored in pit silos. However, extensive use of silage in beef cattle feeding did not begin until about 1910. Before that time it was generally believed that silage was a feed mainly for dairy cows, and it was thought that its succulent nature would produce a marked diarrhea, an undesirable condition for steers on full feed. But the results obtained by using it at agricultural experiment stations and in feedlots of progressive cattlemen were so favorable that the feeding of silage to feeder cattle soon became a common practice throughout the Corn Belt.

By no means has the use of silage been confined to the Corn Belt. Although it is true that corn was the first and is still the principal crop used in silage making, many other crops are ensiled in those parts of the country where corn cannot be grown successfully. In the arid West and Southwest the sorghums as well as corn furnish an enormous tonnage of silage, especially where grown by benefit of irrigation. Large amounts of silage are also now made from alfalfa, oats, field peas, clover, cowpeas, soybeans, rye, Sudan grass, and other farm crops in various sections of the country. Indeed, it appears likely—if, in fact, it is not already true—that silage will be the principal roughage used in beef production both in the Corn Belt and throughout the United States.

PRINCIPLES OF MAKING SILAGE

The making of silage involves choosing the harvesting time that results in the greatest yield of digestible nutrients, and then processing and storing the crop in such a way that losses are lowest. It is well to remember that a silage cannot be better than the crop that was harvested to make it. In fact, it is unfortunately true that much silage is considerably less valuable than the original crop from which it was made. This is due, first of all, to excessive losses during fermentation and from top spoilage and, second, to unfavorable aroma and conditions that reduce consumption to low levels. An understanding of the chemical changes that forage undergoes during ensiling should serve to emphasize the important steps necessary in making good silage. These chemical changes occur in the following order, with much overlapping:

1. Respiration of the plant cells of the chopped forage particles continues for a few hours, consuming much of the trapped oxygen and producing carbon dioxide and heat as end products. Plant enzymes contained in the forage are also active during this phase.

2. Lactic and acetic acids are produced by anaerobic bacteria, provided that sufficient soluble carbohydrate material is available for the fermentation to proceed. This process continues for several days to several weeks, depending on the rate of acid production. When a pH of about 4 is reached, fermentation practically stops, and the silage undergoes little further change. The length of time required to reach the desired pH determines, to a large degree, the amount of energy lost because of fermentation.

3. If the proper pH is not reached because of insufficient soluble carbohydrate in the forage, butyric acid-producing bacteria become active, further reducing the energy content of the silage. A foul-smelling, slimy, unpalatable silage results from this type of fermentation. Furthermore, the butyric acid-producing organisms may attack the proteins, converting them to volatile fatty acids and ammonia.

In summary, there are two important essentials in making high-quality, palatable silage with minimum nutritive losses from excessive fermentation: (1) pack the silage well so as to exclude oxygen, thereby reducing cell respiration and enzyme activity; (2) if grain is not a part of the silage plant, add some type of soluble carbohydrate material to promote acetic and lactic acid production and to inhibit the activity of butyric acid-producing bacteria.

EXTENT OF LOSSES IN MAKING SILAGE

For convenience of discussion, silage losses may be divided into (1) seepage losses, (2) fermentation losses, and (3) top spoilage. Extent of seepage loss depends on the amount of moisture present (moisture within the plant itself and added moisture in the form of rain or snow), on pressures exerted, and on whether the excess moisture, containing soluble materials, may drain or seep from the silo. Seepage losses can be eliminated or reduced by (*a*) wilting the freshly cut forage to approximately 30 percent dry matter while it lies in the freshly cut swath or in a loose windrow, or (*b*) by cutting forage only after it has matured sufficiently to contain above 28 to 30 percent dry matter, or (*c*) by adding an absorbent material such as ground ear corn or ground corn cobs. Seepage losses of nutrients are not as high, on a percentage basis, as the apparent loss in weight, as most of the liquid seeping from a silo is water. Nevertheless, seepage losses of actual nutrients may run as high as 10 percent of the dry matter, mostly in the form of soluble carbohydrate material and valuable silage acids, in large tower silos where much pressure is exerted. In trench, bunker, or stack silos, seepage losses are usually lower, especially if rainwater is diverted. Under ideal conditions no seepage need occur. The possibility of seepage should be allowed for by locating the silo so as to provide drainage for the liquid that sometimes results.

The extent of fermentation losses, usually the greatest source of loss in other kinds of silage besides corn and sorghum, may go as high as 25 percent of the total dry matter, but ranges from 5 to 10 percent in most silages. The extent depends on availability of a soluble carbohydrate for bacterial fermentation or conversion to the volatile fatty acids, lactic and acetic, or the addition of mineral acids themselves, and the effectiveness with which air is excluded from the silage.

Top spoilage losses vary greatly and are the most apparent of all losses because they are easily observed. The extent of these losses, which may range from practically zero to as high as 20 to 25 percent, depends on how well the fresh material is packed, how fine the forage is chopped, whether rain or snow falls on the surface of the silage, how much is fed off the exposed surface each day, and whether a covering material of some sort is used.

Field losses, due to loss of dry leaves as a result of overmaturity or excessive wilting, may be an important loss in legume silages in some situations.

Total losses from all sources may be expected to run from an average of 10 to 15 percent for corn or sorghum silages stored in good tower silos to an average 15 to 25 percent for straight legume or legume-grass silages. Use of oxygen-free storage or airtight silos will reduce losses by at least one-half, while total losses may go as high as 35 percent in poorly located and constructed surface silos. Figure 81 shows the effect of moisture content and type of silo on the magnitude of dry matter losses in legume-grass silage.

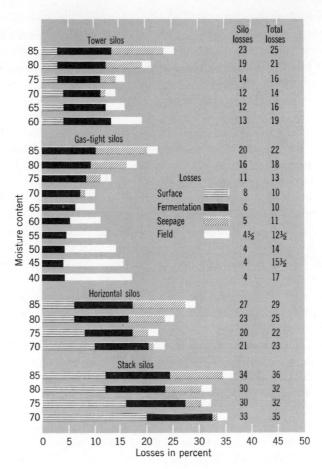

FIG. 81. Effect of moisture content on dry matter losses in legume-grass silages.
(A. T. Hendrix, Agricultural Research Service, U.S.D.A.)

PRESERVATIVES

Preservatives, additives, or conditioners are added to legume and grass silages for one or all of four reasons: (1) to hasten the natural production of lactic and acetic acids so as to lower the pH and prevent undue fermentation losses, (2) to accelerate the lowering of the pH, thus preventing the more adverse forms of fermentation, (3) to correct certain nutritional deficiencies that may be inherent in the silage, and (4) to improve the odor and palatability of the silage.

Preservatives may be divided into two broad categories: first, the carbonaceous or carbohydrate materials such as molasses and corn and, second, the mineral acids, chief of which is sodium metabisulfite. Good silage can be made, and much is made, without preservatives, but those who are inexperienced in making silage, especially grass and legume silages, will do well to investigate their use.

The economy of preservatives is difficult to determine for each and every farm. The carbonaceous preservatives can be justified without question because they are recovered in the silage as a carbonaceous concentrate with approximately 85 percent of their energy or feeding value being retained. On the other hand, such preservatives as sodium metabisulfite, while doing an excellent job of reducing losses in the silo and of improving palatability in some cases although not all, make no further contribution to the ration. Application of sodium metabisulfite and similar material presents a mechanical problem because the amount added is extremely small.

Table 178 gives recommended levels of preservatives when the material is cut with a direct-cut field chopper. If wilting is practiced, the use of preservatives is questionable. Additional amounts of cob, up to 200 pounds, are often used to reduce seepage of high-moisture silages. Corn or other high-energy feeds may be added up to 10 percent of the weight of the silage in order to balance a finishing ration with respect to energy content.

Numerous experiments have recently been conducted in which 0.5 percent of either limestone or urea or both have been added to fresh-cut whole-plant corn silage as the silo was filled. These 10-pound-per-ton additions raise the pH from about 3.5 to about 4.25 and increase the protein

Table 178
Rates of Preservative Application with Different Forages

	Suggested Pounds per Ton of Crop		
Kind of Crop	Molasses[a]	Ground Ear Corn[b]	Sodium Metabisulfite
Grasses	40–60	100–150	5
Cereals	25–50	75–100	5
Legume-grass mixtures	50–100	125–150	7.5
Legumes	75–100	150–200	10
Corn and sorghum	None	None	None

[a] If dried molasses preservatives are used, reduce amounts by 50 percent.
[b] If ground shelled corn is used, reduce amounts about 10 percent.

content on a dry basis from about 7.5 to about 9 percent. Silage intake has usually increased along with a 5 to 8 percent increase in rate of gain, and with improved feed efficiency of about the same magnitude. Some further bonuses are a reduction in acid erosion of concrete silo walls, a higher carotene content, and an economical source of supplemental protein. Responses have been greatest when grower or medium-energy finishing rations have been fed.

SEALING THE SILO

Many farmers make no attempt to seal their silos but allow a natural seal of rotted and moldy silage to form at the top, through which little air gains access to the silage below. This is a wasteful practice because the amount of spoiled silage, which must be discarded when the silo is opened, may easily represent 10 to 15 percent of the total weight of green forage stored in tower silos, and considerably more in bunker or trench silos. This loss of feed may be greatly reduced by leveling the surface of the silage, packing the silage thoroughly to exclude the air from the top layer, and covering it with a sheet of plastic weighted down with 4 or 5 inches of earth or limestone, or even old tires. Ordinary agricultural limestone makes an especially good seal on uncovered silos because of the tight, impervious crust that it forms after being wet by a heavy rain. Well-packed silage covered with a plastic seal should not shrink in weight more than 3 to 5 percent because of top spoilage.

KINDS OF SILOS

The most common kind of silo found in the Corn Belt is the upright or tower silo, 16 to 20 feet in diameter and 40 to 60 feet high. Concrete staves are the most popular building material, although many monolithic concrete silos and a few glazed tile silos erected many years ago are still serviceable and are being used.

The metal silo is increasing in number, especially the glass-coated steel silo called the Harvestore, which has the feature of being airtight. Thus the ensiled forage loses little carotene or other nutrients as a result of the controlled fermentation. The chopped forage is blown into the silo through an opening in the roof and is removed at the bottom by a specially designed bottom unloader. A valve in the top attached to an inflatable "breather bag" under the roof serves to maintain the same pressure within the silo as exists on the outside.

Other manufacturers now market metal silos of various designs, but in general the airtight feature is found in all of them. The method of unloading is the major difference. A big advantage of the bottom-unloading silos is that they may be refilled from the top while feeding from the bottom, allowing for season-long harvesting.

Upright silos made of metal, concrete, or similar material are high-priced, and there have been various attempts to provide less expensive structures for the storage of silage. Perhaps the most common of the less costly forms is the trench silo, which consists of a long straight-sided ditch, 10 to 30 feet wide, constructed on a 5- to 10-degree slope in order that surface water seeping into the trench may drain out at the lower end. An important advantage of the trench silo, besides its inexpensive construction, is the low cost of filling it. No special equipment is required at the silo, since the trucks or unloading wagons that are used to haul the crop from the field harvester are driven into the silo, where they dump and spread the chopped forage with little or no hand labor.

Driving the truck or tractor and wagon over the silage during filling assists in packing the material, but further packing with a tractor alone is recommended. The principal disadvantages of the trench silo are that the trench tends to widen each year through the crumbling of the walls and that it is sometimes difficult to maintain roads to the silo over which the silage may be hauled to the cattle during bad weather.

FIG. 82. Large-volume trench, bunker, or surface silos and automatic loading equipment, coupled with unloading-mixer wagons and fenceline bunks, reduce nonfeed costs for larger feedlots. (Western Livestock Journal.)

FIG. 83. Filling a bunker silo, using a self-unloading truck. Proper leveling and packing reduce spoilage losses. (John Deere, Moline, Illinois.)

Except in regions of light rainfall, the walls and floors of trench silos should be made of concrete to prevent serious erosion damage by heavy rains during the spring and summer when the silos are empty. Bunker silos, consisting of either concrete or wooden walls and a concrete floor at ground level, are equally satisfactory, and mud usually is not so much a problem.

Many farmer-feeders and especially larger feedlots are having fairly good results from storing silage in large, carefully made stacks in the open, with or without retaining walls, without protection to either the tops or sides of the piles. The losses encountered from spoilage in these bunker or surface silos are claimed to be less than the annual interest and deprecia-tion charges on a conventional silo. This method of storage is commonly used for preserving pea vines, cannery refuse, beet tops, and other silage materials of low feeding value, but corn and sorghum silages are also being so stored by very large commercial feedlots. The surface silo may well prove to be the best method of preserving grasses and legumes because, as some feeders claim, spoilage costs are lower than silo costs. Labor costs of feeding from surface silos are lower because tractor-mounted scoops are used to fill automatic unloading trucks or wagons.

Large pits, resembling gravel or sand pits with no concrete used in their construction, may be found in the Southwest and West where rainfall is low. These make very economical and satisfactory silos.

COSTS OF MAKING SILAGE

The total costs of making and feeding silage may be grouped under the following items: (1) depreciation of silo, filling, and feeding equipment; (2) interest on investment in silo and equipment; (3) maintenance and repair of silo and equipment; (4) taxes; (5) silage losses, including field losses; (6) net cost of preservative; (7) labor cost of filling silo and feeding silage.

Such costs naturally vary considerably, depending principally on the type of silo and size of operation. The larger capacity silos have a real advantage in that fixed costs, such as those for filling and feeding machinery, can be spread over more tons of silage or larger numbers of cattle fed. Table 179 shows how size and type of silo affect estimated silo costs and cost of silage losses per ton.

The estimated total cost per net ton fed—that is, the estimated total cost from standing crop to feedbunk per ton actually fed—is shown in Table 180. Some operators do this job more cheaply, but some also have higher costs, mainly because losses are greater and inefficiencies in operating expensive silo-filling equipment are more frequent. Silo-filling equipment is owned cooperatively in many areas, and neighbors often trade work at silo-filling time. The difficulty with this arrangement is that the desirable time for harvesting a crop may arrive simultaneously on all farms in the neighborhood. The custom-harvesting of silage crops is on the increase.

SILAGE IN THE FINISHING RATION

In all cases where cattle are given a liberal amount of grain, any silage fed should be considered a part of the roughage ration. Its use should therefore be in accordance with the well-recognized rules for the feeding of roughage discussed in Chapter 18.

Because silage often contains some corn or other concentrate, men inexperienced in its use often make the mistake of treating it as a concentrate. Beginning with a small amount of silage, they gradually increase the allowance per day throughout the feeding period when, as a matter of fact, the reverse of this procedure usually should be followed. In view of the fact that silage contains little corn or other concentrate in proportion to its weight and bulk, it is manifestly impossible for steers to consume any great amount of grain and silage at the same time. The logical procedure is to feed the maximum amount of silage during the first part

Table 179
Estimated Cost per Ton Capacity of Silos[a]

Type of Silo	Average Initial Cost ($) per Ton Capacity		Estimated Life (yr)	Annual Cost[b] per Ton (% of initial cost)	Annual Cost ($) per Ton Capacity —Silo Only		Estimated Silage Losses (%)	Annual Silo Costs[c] per Gross Ton ($)	
	100-Ton	200-Ton			100-Ton	200-Ton		100-Ton	200-Ton
Tower									
Wood stave	12.00	10.00	25	8.7	1.04	0.87	14	1.74	1.57
Monolithic concrete	12.00	9.00	35	7.4	0.89	0.67	14	1.59	1.37
Concrete stave	13.00	10.00	30	7.8	1.02	0.78	14	1.72	1.48
Clay tile stave	22.50	17.50	45	6.5	1.46	1.04	12	2.06	1.64
Galvanized steel	17.50	13.50	40	6.7	1.17	0.90	12	1.77	1.50
Gastight steel	32.00	25.10	40	7.0	2.24	1.76	5	2.49	2.01
Horizontal									
Unlined trench	1.00	0.80	5	26.0	0.26	0.21	23	1.41	1.36
Wood wall trench	6.00	4.50	15	11.5	0.69	0.52	20	1.69	1.52
Concrete trench	7.60	7.00	25	8.5	0.65	0.60	20	1.65	1.60
Tilt-up horizontal	6.00	5.20	25	8.5	0.51	0.44	20	1.51	1.44
Wood wall bunker	7.50	5.00	20	9.7	0.72	0.38	20	1.72	1.38
Miscellaneous									
Welded wire, paper	1.00	0.80	3	36.8	0.47	0.38	20	1.47	1.38
Plastic film envelope	1.50	1.40	1	103.5	1.56	1.45	5	1.81	1.70
Stack	0.05	0.04		103.5	0.05	0.04	33	1.70	1.69

[a] A. T. Hendrix, Agricultural Research Service, U.S.D.A.
[b] Includes depreciation, maintenance, interest on investment, and taxes.
[c] Includes spoilage loss with green forage crop valued at $5 per ton.

Table 180

Summary of Estimated Costs of Harvesting, Storage, Storage Losses, and Feeding of Silage by Different Methods per Net Ton[a] (Silo Capacity, 200 Tons)[b]

Type of Silo	Silo Cost per Year, Net Ton ($)	Storage Losses per Net Ton ($)	Filling Costs per Net Ton ($)	Silo, Storage Losses, and Filling Costs per Net Ton ($)	Estimated Feeding Costs per Net Ton ($)	Total Costs, Standing Crop to Feedbunk ($)
Tower						
Wood stave	1.01	0.81	2.65	4.47	1.05	5.52
Monolithic concrete	0.78	0.81	2.65	4.25	1.05	5.30
Concrete stave	0.91	0.81	2.65	4.38	1.05	5.43
Clay-tile stave	1.18	0.68	2.69	4.45	1.05	5.50
Galvanized steel	1.02	0.68	2.69	4.30	1.05	5.35
Gastight steel	1.85	0.26	2.69	4.41	1.09	5.50
Horizontal						
Unlined trench	0.27	1.49	2.69	4.46	0.48	4.94
Wood wall trench	0.65	1.25	2.59	4.48	0.08	4.56
Concrete trench	0.75	1.25	2.59	4.58	0.08	4.66
Tilt-up horizontal	0.55	1.25	2.59	4.39	0.08	4.47
Wood wall bunker	0.48	1.25	2.59	4.32	0.08	4.40
Miscellaneous						
Welded wire, paper	0.48	1.25	3.00	4.72	0.48	5.20
Plastic film envelope	1.52	0.26	2.53	4.32	0.48	4.80
Stack	0.06	2.46	3.58	6.10	0.48	6.58

[a] A. T. Hendrix, Agricultural Research Service, U.S.D.A.
[b] Forage crop estimated value $5 per ton in field.

487

of the feeding period, decreasing the amount gradually, so as to permit a larger consumption of concentrates.

EFFECT OF SILAGE ON GRAIN CONSUMPTION

Except when fed in very small quantities, the use of silage lessens slightly the consumption of grain. This is probably because well preserved silage is more palatable than most dry roughages and thus competes more actively for the appetite of the steer, and because the concentrate present in the silage naturally tends to replace part of that in the grain ration. The use of corn silage in moderate amounts decreases the consumption of the corn fed as concentrate by approximately 10 percent of the weight of the silage fed, whereas the shelled corn content of the silage usually is 15 to 17 percent of the weight of silage eaten. Consequently the feeding of silage in moderate amounts usually results in a somewhat larger total consumption of corn than is realized from a nonsilage ration. This is illustrated in Table 181. Naturally the feeding of silages other than corn or sorghum will not increase grain intake but may in fact reduce it slightly.

AMOUNT OF SILAGE TO FEED

Silage varies so much in moisture content that it is impossible to lay down definite rules as to the amount that should be fed. Some men put

Table 181
Effect of Silage on the Consumption of Grain (Two-Year-Old Steers)

Indiana Experiment Station		Number of Trials Averaged	Average Daily Consumption		Corn Present in Silage (lb)[a]	Total Corn Consumed (lb)
			Shelled Corn (lb)	Corn Silage (lb)		
I.	Shelled corn, cottonseed meal Clover hay, corn silage	10	14.2	25.1	3.8	18.0
	Shelled corn, cottonseed meal Clover hay	10	17.1	—	—	17.1
II.	Shelled corn, cottonseed meal Corn silage	5	14.5	29.9	4.5	19.0
	Shelled corn, cottonseed meal Clover hay	5	17.5	—	—	17.5

[a] Assuming that 15 percent of weight of silage is corn.

the crop into the silo when it is comparatively green, whereas others wait until it is fairly mature. Naturally the greener silage contains more water. Also, it frequently happens that silage in a given silo varies greatly in moisture content from top to bottom. This is particularly likely in a large silo that requires several days to be filled or in a silo containing silage made from rather dry forage to which considerable water was added. Unless the flow of water is carefully regulated, some of it filters down through the silage, leaving the material at the top only moderately moist while that near the bottom is saturated. Consequently, any recommendation as to the amount of silage to feed should be regarded only as an estimate.

The best index to follow in determining how much silage to feed is the appetite of the cattle. Knowing the approximate amount of grain that should be consumed at a given stage of the feeding period, one may feed as much silage as will not decrease appreciably the consumption of grain. Less than this amount may, of course, be fed but the use of greater amounts is almost certain to result in smaller daily gains and lower finish.

Although from the standpoint of the cattle there is no minimum amount of silage to feed, there are certain practical objections to using this material in small quantities. The principal items in the cost of silage, as mentioned earlier, are the labor, machinery, and storage charges. These items decrease rapidly per ton as the amount of silage made increases. Consequently silage is a much cheaper feed when put up in large quantities. Also, considerable labor is involved in the feeding of silage compared with most other roughages, but this labor decreases per ton with the amount of silage fed. Thus if silage is to prove an economical feed, it must be fed in relatively large quantities. Except in special situations, if silage is used it should furnish 75 to 100 percent of the roughage on an air-dry basis.

Owing to the high water content of silage, inexperienced feeders find it somewhat difficult to determine its dry roughage equivalent. Because of the great variation in moisture this can only be approximated, but for practical feeding purposes it may be roughly estimated by dividing the weight of fresh silage by 3. This means that cattle fed 30 pounds of silage are consuming approximately the same amount of roughage as cattle eating 10 pounds of hay.

SILAGE VERSUS DRY ROUGHAGE

When corn or sorghum silage is supplemented with the proper amount of nitrogenous concentrate to balance the ration, it compares favorably as a roughage with clover or alfalfa hay, as shown in Table 182.

Its principal advantages over these dry roughages are its certainty

Table 182

Comparison of Corn Silage and Legume Hay as Roughages for Steers on Full Feed

	Number of Trials Averaged	Daily Gain (lb)	Feed per Cwt Gain			
			Corn (lb)	Cotton-seed Meal (lb)	Silage (lb)	Hay (lb)
I. Indiana: 2-year-olds						
Corn, cottonseed meal, corn silage	5	2.42	600	115	1,240	—
Corn, cottonseed meal, clover hay	5	2.35	742	118	—	456
II. Illinois: calves						
Corn, cottonseed meal, corn silage	5	2.08	476	70	377	96
Corn, alfalfa hay	5	1.97	525	73	—	254

of supply—the total failure of corn or sorghum to make a silage crop is extremely rare—and the opportunity to utilize a large amount of roughage that has no market value. The chief disadvantages of silage compared with legume hay are the comparatively large amount of labor and equipment required for its harvest and storage, and the necessary cash outlay for commercial feedstuffs to furnish the protein needed to balance the ration. Comparatively little additional protein is needed with clover hay as the roughage, and still less with alfalfa. Corn and sorghum silage, however, require a daily allowance of 1 to 2 pounds of protein concentrate per steer if satisfactory results are to be obtained from their use. Addition of urea, as a nitrogen source, to corn silage at silo-filling time is an alternative already mentioned.

Whether corn or sorghum silage or legume hay will prove most satisfactory for a particular feeder depends both on his method of feeding and handling his cattle and on the amount of legume hay and other dry roughages available. As said elsewhere, corn or sorghum silage is unsurpassed as a roughage for lower-grade, heavy steers that are to be fed only a short time before being sent to market. It is also well suited for cattle of any grade that are to be finished in a leisurely manner over a long period.

Legume hay, on the other hand, is preferred by many feeders for putting a choice finish on higher-grade steers within 120 to 140 days and for finishing calves within a period of 6 to 8 months. Legume hay is much more satisfactory than silage when the cattle are fed their grain in self-feeders. By using large racks or mangers, hay can be self-fed by filling the mangers as needed. Silage, on the other hand, must be fed daily.

Self-feeding surface silos are fairly satisfactory for feeding stockers but generally prove unsatisfactory for finishing cattle. The cost of constructing a silo and the outlay for harvesting and feeding machinery amount to an expense that is justified only when a reasonably large tonnage of silage is made and used. Otherwise the machinery and storage charge per ton of silage is so high that silage ceases to be an economical feed. Obviously silage is a more satisfactory feed for the man who feeds 100 or more cattle per year than for the farmer who feeds only 15 to 25 head.

FEEDING SILAGE TO CALVES

Some feeders who have found corn or sorghum silage a satisfactory feed for yearling steers consider it too bulky for finishing calves. However, in some respects it is more useful for feeding animals of this age than for older cattle because calves are fed for a much longer time and thus have greater need for variety and succulence in their rations. By starting calves on nearly a full feed of silage, 12 to 15 pounds a head, and adding grain and protein concentrate as the calves grow and attain greater capacity, the calves are gradually got up to a full feed with almost no risk of their going off feed or becoming foundered.

On most farms the silo is empty by April or May, and the change to dry roughage results in an increased consumption of grain at the time of year when the appetite of full-fed cattle is usually slowed by the onset of hot weather. Should the supply of silage last until calves are finished, the amount should be reduced to 4 or 5 pounds a day with the arrival of summer or about June 1.

The proper amount of silage for calves is a somewhat disputed question. The amount to feed depends greatly on size and age of the calves and length of the feeding period. The Illinois station had excellent success feeding 8 pounds of silage and 2 pounds of legume hay per head daily for the first 6 or 7 months and 6 pounds of silage and 2 pounds of hay thereafter. However, in experiments in which different amounts of silage were compared, equally good gains were secured by feeding 16 pounds of silage and 3 pounds of hay during the first 140 days and approximately half these amounts thereafter. The calves fed in this manner maintained their feed consumption and rate of gain much better during the summer than the other lots. Consequently this plan of feeding appears to be satisfactory for calves that are to be carried into late summer and fall.

VALUE OF A DRY ROUGHAGE WITH SILAGE

Although satisfactory results may be expected from a ration consisting of shelled corn, a protein concentrate, and corn or sorghum silage, feeding experiments indicate that the ration is improved by adding a small amount of dry roughage such as hay or straw. The amount of such roughage eaten is usually only 2 or 3 pounds a day, but it appears to have a noticeable effect on the rate of gain, especially if the dry roughage fed is clover or alfalfa hay. These legumes are rich in minerals and vitamins, both of which are likely to be deficient in a ration in which corn or sorghum silage is the only roughage. Inasmuch as calves have greater need for minerals and vitamins than do older animals, their response to the addition

Table 183

Value of a Dry Roughage with a Heavy Silage Ration

Number of Trials Averaged	Rations Compared	Feed per Pound Gain					
		Daily Gain (lb)	Shelled Corn (lb)	Cotton-seed Meal (lb)	Legume Hay (lb)	Oat Straw (lb)	Corn Silage (lb)
4 Indiana (two-year-olds)	Shelled corn, cottonseed meal, corn silage	2.37	5.97	1.14	—	—	12.66
	and Shelled corn, cottonseed meal, clover hay, corn silage	2.45	5.74	1.12	1.41	—	11.16
3 Indiana (two-year-olds)	Shelled corn, cottonseed meal, clover hay, corn silage	2.41	5.67	1.16	1.09	—	10.37
	and Shelled corn, cottonseed meal, oat straw, corn silage	2.45	5.67	1.15	—	.56	10.59
2 Illinois (calves)	Shelled corn, cottonseed meal, corn silage	2.03	5.09	1.02	—	—	7.49
	and Shelled corn, cottonseed meal, alfalfa hay, corn silage	2.20	4.79	.96	1.53	—	4.64

of a legume hay is usually greater. If no legume hay is available, calcium should be supplied in the form of ground limestone.

If the dry roughage fed is straw, hulls, or average to low grade hay, it may be kept before the cattle with little danger of their eating enough to interfere with their consumption of grain or silage. However, if it is good-quality legume hay, it should be fed at the rate of 2 or 3 pounds a head, preferably about noon after the morning's feed of grain and silage has been eaten. Table 183 summarizes the results of several tests which demonstrate the value of small amounts of dry roughage.

FEEDING MOLDY AND FROZEN SILAGE

There is seldom any trouble from feeding small quantities of moldy silage to beef cattle. Moldy grain is another matter and should not be fed.

During extremely cold weather a considerable amount of frozen silage is encountered in regions where temperatures drop to 0°F or below, especially in silos where little feed is removed each day. Silages above 70 percent in moisture content freeze more solidly than drier silages. This frozen material should be removed from the walls of the silo as soon as it is possible to knock or pry it loose. If the pieces are small and not numerous, they may be piled in the center of the silo after the morning's feed, where they will often thaw before night. With a large quantity of frozen silage, however, or with the temperature much lower than freezing, this method is impractical. Instead, the frozen silage should be piled just outside the silo where it can be watched and fed as soon as it is reasonably well thawed. If left longer, it becomes moldy and unfit for use.

The presence of small quantities of frozen silage in the ration is unlikely to cause trouble, but the feeding of large amounts is highly inadvisable. Not only is such material unpalatable and difficult for the cattle to eat, but it is likely to cause serious digestive disorders. Excessive scouring is one of the common aftereffects of feeding frozen silage.

The question is often asked as to how long one must wait after filling a silo before it is safe to begin feeding from it. Many successful feeders begin feeding immediately, thus eliminating the top spoilage that occurs with waiting. However, the silage in the upper layer is usually quite warm and lacks the typical mildly acid taste of properly fermented silage; consequently it is not readily consumed by the cattle. Waiting a week or so insures that most of the fermentation has occurred, the silage will have cooled off, and cattle will start off on the new and strange—to them, at least—feed with far greater relish.

EFFECT OF SILAGE ON SHIPPING AND SALE

Some feeders hold the opinion that silage-fed cattle undergo a much greater shrink between feedlot and market than cattle fed dry roughage. For this reason they sometimes remove the silage from the ration a few days before shipping and supply a dry roughage in its place. However, records kept of the actual shrinkage undergone by silage-fed and nonsilage-fed cattle, respectively, show that this practice is unwise because the silage-fed cattle in reality shrank less than hay-fed cattle. It was true that some silage-fed cattle lost more weight in transit, but they were more inclined to take on a greater fill after they were unloaded. Silage cattle have a reputation at the market for being unusually good "drinkers." Whereas this kind of fame probably does not work to their advantage when it comes to selling price, it does indicate that their owners get good weights. When sold and weighed at the feedlot, silage cattle may have a slight advantage because of greater fill. Selling on grade and yield eliminates the question entirely.

In former years when dry roughage was the orthodox feed for finishing cattle, the few silage-fed steers received at the market were regarded with considerable suspicion. It was feared they would resemble grass-fattened cattle in having a low dressing percentage and dark-colored flesh of rather poor quality. Comparisons of the carcasses of such cattle with those of animals finished on dry roughage disclosed no important differences in amount or quality of finish traceable to silage. As the number of silage-finished cattle increased, packers and shippers bought them as a matter of course and frequently found them superior to any dry-roughage-fed cattle included in their purchases. Naturally any discrimination against such cattle soon disappeared, and they were bought on the basis of their actual worth.

Today a large percentage of grain-fed steers received at large mid-western livestock markets have been fed silage, and a buyer seldom stops to inquire whether a given drove has had silage. If the cattle show unusual fill, the buyer may suspect that they have been fed silage, but whether this is the cause of their condition makes little difference to him. He estimates their dressing percentage, judges their worth on the rail, and tries to buy them on that basis.

EFFECT OF SILAGE ON PROFITS

No one ration or plan of cattle feeding is more profitable than another under all conditions. However, if an attempt were made to classify rations on the basis of the net profit realized from their methodical use over a period of years, those containing moderate and fairly liberal amounts of silage probably would be at the top. This is partly because silage-fed cattle have a smaller feed bill than those fed only dry roughage. Then, as a rule, silage-fed cattle develop more uniformly, and a smaller percentage are affected by founder and bloat. Because of their attractive appearance when marketed they frequently sell higher than cattle of similar quality that were fed for the same length of time, but on rations not containing silage.

It is not without significance that nearly all Corn Belt experiment stations have for many years used a ration containing corn silage as a check or standard with which new and little-tried feed combinations are compared. This practice is the result of obtaining better returns from feeding corn silage than from feeding dry roughages in literally scores of experiments made over a period of more than 60 years.

If cattle feeders in the areas where high yields of corn or grain sorghums can be obtained with relative economy fail to use silage made from these feeds rather heavily in their finishing programs, they are ignoring probably their greatest competitive advantage over feeders in other parts of the country who must rely almost exclusively on high-concentrate rations. In fact, taking advantage of this economical roughage source may be the one factor that will keep Corn Belt feeders in a sufficiently competitive position to remain an important segment of the cattle feeding industry.

POISONOUS GAS IN NEWLY FILLED SILOS

It has long been known that gases collect above the silage in tower silos during or immediately after filling. These gases, mostly carbon dioxide, have not been found particularly dangerous. Within the last few years, however, a more ominous situation has been recognized, and numerous fatalities from inhaling poisonous gas have been reported among men working in silos, especially in the main corn-growing states. The toxic gas has been identified as nitrogen dioxide. It seems to develop most readily in corn silage grown in years when drought damages corn that has had heavy applications of nitrogen fertilizer. Under these conditions the plants evidently absorb more nitrates from the soil than they are able to use. The excess nitrate is released during the ensiling process and, through a series of chemical reactions, is changed into nitrogen dioxide or nitrous oxide gas. Symptoms of "silo fillers' disease" are severe coughing and burning or choking pains in the throat and chest. The condition usually arises as a person enters a silo during filling or within 48 hours after filling. A physician should be consulted if such symptoms occur.

Scientists from the U.S. Department of Agriculture recommend these safety precautions in filling all tower silos:

1. Run the blower for 10 minutes before going into a partly filled silo during the silo-filling process. Always keep the blower running while you are inside.

2. Be alert for irritating odors. Nitrogen dioxide is heavier than air and collects near the surface of the silage. The gas tends to settle in the silo chute and around the base of the silo.

3. Watch for yellowish brown fumes—they are a sign of nitrogen dioxide gas. If the silo is dark, use a flashlight so that you can see.

4. Keep children and animals out of the silo and away from it during filling.

5. After the silo is filled, wait at least a week before going inside if the blower cannot be operated. Do not let children stay near the silo. If necessary, use a temporary fence to help keep them safe.

COMPARISON OF VARIOUS SILAGES FOR BEEF CATTLE

To the man who lives in the oldest cattle-feeding area, the Corn Belt, the term "silage" usually means ensiled corn. However, outside the region where corn can be successfully grown, various other crops are used for silage purposes. Since the tonnage of corn silage made and fed greatly exceeds that made from other silage materials, it is convenient to compare these other kinds of silage with corn silage as a standard.

CORN SILAGE AS THE PRINCIPAL FEED FOR FINISHING CATTLE

A full feed or varying high levels of corn silage, supplemented with an appropriate amount of protein concentrate and grain, results in economical gains at low cost, as discussed under the subject of limited grain rations in Chapter 14, and particularly as shown in Tables 114, 115, and 116. If the silage is made from high-yielding corn, thin cattle make a noticeable improvement in condition during the first half of the feeding period on silage alone, with no additional corn. Usually it is important to feed as little as 2 or 3 pounds of hay per day, or no hay at all, to obtain the consumption of as much silage as possible. This, of course, means that

1 to 2 pounds of protein concentrate must be fed daily, as well as a mineral supplement.

Cattle with sufficient quality to sell near the top of the market, if they have the necessary finish, usually return more profit if they are fed a high-concentrate ration during at least the last half of the feeding period. Heavy or long yearling steers are more satisfactory for such a plan of feeding than younger cattle, as they attain a satisfactory finish for their grade in a shorter time. Gains made by cattle on limited grain rations are relatively low after 120 to 130 days.

AT WHAT STAGE SHOULD CORN BE ENSILED?

It has generally been thought that the best-quality silage is produced by cutting the corn when the kernels have hardened to the point where the interior has a stiff doughlike consistency, but a large portion of the stalk and most of the leaves are still green. Silage made from immature corn is likely to be sour and "sloppy." Moreover, the cutting of very green corn entails a waste, since the total amount of food nutrients in the corn plant continues to increase until the ears are fully mature. On the other hand, it is difficult to make silage from corn that is almost ripe. Unless sufficient water is added to soften the dry stalks and leaves and permit packing them so firmly that all air is excluded, silage made from overripe corn is very likely to mold.

A measured area of corn ensiled in the undented or "roasting-ear" stage from August 12 to 14 at the Ohio station proved to be much lower in feeding value than an equal acreage ensiled September 5 to 8, when the corn was well dented. Samples taken of the corn at the time of harvest showed that 17 percent less dry matter per acre was obtained from the early-harvested corn. Moreover, considerable juice leached out of the silo after this corn was stored. As a result only 755 pounds of gain per acre were obtained from the immature silage compared with 940 pounds per acre from the well-dented corn. The cattle fed the immature silage were not so well finished and were valued 25 cents per hundred lower than the steers fed the riper silage.

In a recent study the Ohio station—along with numerous others, it might be added—investigated the possibility of allowing corn to mature beyond the dent stage (the usual time for cutting corn for silage) to permit the crop to store the absolute maximum of nutrients before harvesting. The corn plant normally continues to transfer nutrients to the kernels until the grain moisture content has been reduced to about 30 percent. However, when the plant has reached this stage of maturity only the upper leaves are green, and the moisture content of the entire plant, including ears, has been reduced from the normal silage-cutting moisture level of 70 percent to less than 50 percent. Obviously this drier silage is higher in energy content on an as-fed basis. Thus nutritionists wanted to know if cattle might consume more nutrients per day when such drier silage was fed. If so, rate of gain should increase, and this, combined with a known slightly higher nutrient yield per acre, should result in more beef gains per acre of corn harvested as silage.

The results of the Ohio test are shown in Table 184. The table also supplies data that can be used for another comparison between all-silage, silage-and-corn, and all-corn rations. To balance and equalize all of the rations with respect to protein and minerals, all lots were fed equal amounts

Table 184

A Comparison of Corn Silages Harvested at Two Stages of Maturity (98 Days)[a]

Ration Plan Stage of Corn Silage	Silage[b]		Silage + Ear Corn		Ear Corn	
	Normal	Mature	Normal	Mature		
Number of heifers	14	14	14	14	14	13
Average initial weight (lb)	480	482	482	491	489	453
Average daily gain (lb)	1.54	1.66	1.95	1.97	2.12	2.33
Average daily ration (lb)						
Corn silage	26.3	16.6	10.0	7.7	—	—
Silage dry matter	7.5	9.2	2.9	4.3	—	—
Alfalfa hay	3.0	3.0	3.0	3.0	3.0	3.0
Ground ear corn	—	—	8.0	8.0	12.9	14.3
Soybean meal	1.0	1.0	1.0	1.0	1.0	1.0
Salt	0.04	0.05	0.04	0.04	0.05	0.05
Minerals	0.04	0.04	0.04	0.04	0.04	0.04
Feed per cwt gain (lb)						
Corn silage	1,703	1,005	512	389	—	—
Silage dry matter	489	559	147	216	—	—
Alfalfa hay	194	181	153	152	141	129
Ground ear corn	—	—	409	404	610	612
Soybean meal	65	60	51	50	47	43
Salt	2	3	2	2	2	2
Minerals	3	2	2	2	2	2
Total air-dry feed per cwt gain (lb)	802	910	779	848	802	788

[a] Adapted from Ohio Agricultural Experiment Station Series No. 129, 1963.
[b] Dry matter content of silages: normal, 28.7 percent; mature, 55.6 percent.

of hay, soybean meal, and minerals, in addition to the silage and/or corn. The heifers eating the more mature silage consumed more silage dry matter and gained faster in the comparison between lots on high silage rations, but required considerably more feed because of poorer utilization of the more fibrous stalk portion of the ration. Thus it is doubtful whether the slight increase in nutrients stored in the more mature plant is enough to offset the depressed utilization that results from late cutting. The chances of spoilage are increased, of course, by the drier silage. Thus there appears to be no real reason for cutting corn at later than dent stage or drier than 65 to 70 percent moisture.

Table 185 shows data from an Illinois study in which the effect of stage of maturity and dry matter content are related to ear and leaf content of a ton of corn forage.

Table 185

Ear-Corn Content and Leaf-Stalk Hay-Equivalent in One Ton of Corn Forage at Various Stages of Development[a]

Dry-Matter Content of Forage[b] (%)	Ears[c] (bu)	Leaves and Stalks[d] (lb)	Dry-Matter Content of Forage[b] (%)	Ears[c] (bu)	Leaves and Stalks[d] (lb)
15	0.2	308	24	3.1	312
16	0.5	309	25	3.4	312
17	0.8	309	26	3.8	313
18	1.1	309	27	4.1	313
19	1.5	310	28	4.4	314
20	1.8	310	29	4.8	314
21	2.1	311	30	5.1	315
22	2.5	311	31	5.4	315
23	2.8	312	32	5.7	315

[a] Illinois Bulletin 576.

[b] In the absence of dry-matter determinations, dry-matter content of forage may be estimated on the basis of stage of development, as follows:

Ears beginning to form	15%	Early dent	25%
Kernels forming	17	Well dented	28
Early milk	20	Kernels hardening, most leaves green	30
Late milk	23	Kernels hardening, fewer leaves green	32

[c] On the basis of 15 percent moisture, 70 pounds of ears per bushel.

[d] Hay-equivalent value, on the basis of 15 percent moisture.

The Oklahoma Extension Service makes the following recommendations as to the most desirable time to cut forage crops for silage:

Corn: when grain is in the dent stage.
Sorghums: when grain is in the dough stage.
Alfalfa: one-fourth to one-half bloom stage.
Sweet clover: early bloom stage.
Other clovers: one-fourth to one-half bloom stage.
Cereals: soft dough to hard dough stage.
Vetch: full bloom stage.
Sudan: early heading stage.
Grasses: bloom to early heading stage.

Weather conditions, of course, often determine the time of making
silage. During periods of severe drought it may happen that the corn must
go into the silo with the base of the stalk and the lower leaves brown
and dry and the ear and upper portion of the stalk still green. Whereas
first-class silage cannot be made under such conditions, to delay longer
would permit the drying up of the entire plant.

VARIATION IN FINENESS OF CHOP

Although the subject of degree or fineness of chop and type of chopper
for corn silage might more appropriately be discussed in Chapter 24, which
deals with feed preparation, the subject is introduced here because the
trials to be described also include data on effect of stage of maturity on
feeding value of corn silage. Fineness of chop is obviously related to degree
or firmness of packing in the silo, hence to the rate of fermentation and
possibly even the end products of fermentation. There is also the possibility
of an interrelationship between degree of maturity and fineness of chop
of corn silage with respect to quality or feeding value of the final product.

The Michigan station conducted a comprehensive experiment in which
corn from the same field was chopped on three dates—in mid-September,
mid-October, and mid-November—and stored in two separate silos on each
date. The corn in one silo was chopped fine ($\frac{3}{8}$ inch) and the other
was chopped medium fine ($\frac{1}{2}$ to $\frac{3}{4}$ inch). The six silages were fed to
heavyweight calves for 180 days. In addition to the respective silages, the
calves received about half a feed of ground shelled corn and supplement
mixture (1 percent of body weight daily). In addition to supplemental
protein the supplement contained appropriate levels of calcium, phosphorus,
antibiotic, vitamins A and D, salt, and stilbestrol. The data of greatest
interest are contained in Table 186.

Table 186

Comparison of Fine- and Medium-Chopped Corn Silage Harvested in September, October, and November (180 Days)[a]

	Fine	Medium	Fine	Medium	Fine	Medium
Agronomic data						
Harvest date	Sept. 13		Oct. 17		Nov. 14	
Average dry matter (%)	28.2		48.2		59.6	
Adjusted yield of silage, tons[b]	17.3		15.2		13.5	
Bushels corn per ton of silage[b]	4.3		4.9		5.5	
Fineness of chop	Fine	Medium	Fine	Medium	Fine	Medium
Pounds dry matter per cubic foot of silage	13.4	11.1	12.0	10.0	11.9	11.0
Change from fine to medium chop (%)	−16.9		−16.9		−8.0	
Cattle performance						
Number of steers	18	18	18	18	18	18
Average initial weight (lb)	539	538	539	538	538	539
Average daily gain (lb)	2.89	2.85	2.78	2.63	2.78	2.69
Average daily ration (lb)						
Corn silage	33.0	32.7	19.2	19.3	17.1	18.2
Corn + supplement	8.33	8.17	8.10	8.02	8.04	7.81
Total[c]	19.5	18.5	19.0	18.5	19.5	19.1
Feed per cwt gain (lb)[c]	678	650	689	703	702	710
Feed cost per cwt gain	$11.42	$11.17	$11.59	$12.02	$11.74	$11.84

[a] Adapted from data in Michigan Beef Cattle Day Report, 1967.
[b] 30 percent dry matter.
[c] 85 percent dry matter.

There was no frost prior to the September harvest date, but both of the other cuttings were repeatedly frosted. The November cutting was also subjected to a 9-inch snowfall. Water was added to the November cutting but not to the others.

The total tons of dry matter produced per acre were lower with each successive cutting, no doubt because of leaf loss after the frosts and progressive leaching of soluble stalk components. The increase in bushels of corn grain per ton of silage with increasing maturity is a reflection of the loss of stalk weight already mentioned and a slight increase in corn grain weight itself. The fine-cut silage packed better, as reflected by the 8 to 17 percent greater density in the silo. This means that about 10 percent more fine-cut silage can be stored in the same silo space as medium-cut silage.

The pooled data for the harvest date comparisons show that the cattle fed September-cut silage gained 0.17 and 0.13 pound per day faster than those fed October- and November-cut silage, respectively. The feed required per hundredweight gain was 664, 691, 706 pounds for the three lots, in order of harvest date. Carcass grades and selling price also favored earlier dates. With respect to harvest date, the September-cut silage was superior in every respect, with each month's wait being increasingly less desirable.

The pooled data for fine versus medium cutting show that differences in cattle performance were small but slightly in favor of fine cutting. The later the harvest date, the greater the advantage from fine cutting. When the possibly greater power requirements for fine cutting are considered, chopping corn silage finer than one-half inch at normal harvesting time is unwarranted.

EFFECT OF CORN YIELDS ON VALUE OF SILAGE

It would seem that silage made from hybrid corn yielding 80 to 100 bushels per acre would have a much higher feeding value per ton than the silage made from open-pollinated corn yielding considerably less. Also it would seem that silage made during a dry year would have a low feeding value because of the low yield of grain. However, corn yields have a much greater effect on the yield per acre of silage than on the feeding value per ton. (See Table 187.) Although the yields of both dry corn and silage at the Illinois station have varied widely from year to year, the corn content per ton of silage has remained rather constant, especially for the hybrid corn varieties.

Silage made and fed at the Nebraska station during the extremely dry years 1934 and 1936 produced gains on stocker calves and yearlings as large as those secured in previous tests with normal corn silage. The corn from which the silage was made grew stalks 5 to 7 feet high. Only a few ears filled, and those were immature and watery. Part of the corn was relatively immature when cut and part of it required water as it went into the silo. The yield of silage per acre was very low, but apparently many of the feed nutrients that would have gone into the ears were still present in the stalks and leaves and thus were available to the cattle.

Numerous studies have shown that fertilizers, which increase tonnage of silage, also increase bushel yields to about the same degree. In general, then, silage made from highly fertilized fields has a feeding value, per ton, equal to but no higher than ordinary silage. Planting rates above 30,000 seeds per acre may reduce the grain content of silage somewhat, but tonnage yields will be high and total yield of nutrients per acre can

Table 187
Effect of Yield of Corn on Corn Content of Silage[a]

	Yield of Corn per Acre (bushels)[b]	Yield of Silage Corn per Acre (tons)	Corn per Ton of Silage (bushels)
1. Open-pollinated corn			
1925	72.9	8.84	8.24
1926	44.7	8.71	5.13
1927	68.5	14.73	4.65
1928	78.9	12.62	6.25
1929	48.3	9.70	4.98
Average	62.7	10.92	5.85
2. Hybrid corn			
1946	92.2	13.29	6.64
1947	66.9	10.51	6.37
1948	106.5	16.68	6.39
1949	91.0	12.63	7.20
1950	52.0	9.35	5.38
Average	81.7	12.49	6.40
Percentage increase	30.3	14.4	9.4
3. Dry years			
1934 (dry year)	35.0	6.38	5.49
1935 (normal rainfall)	57.1	10.25	5.57
1936 (dry year)	33.9	7.51	4.50

[a] Illinois Agricultural Experiment Station, unpublished data.
[b] 14 percent moisture.

be increased by this method, if the necessary fertility is applied and if moisture during the growing season is adequate.

VARIETIES OF CORN FOR SILAGE

Although certain corn hybrids yield an enormous tonnage of silage owing to their rank growing tendencies, such silage is not especially valuable for finishing cattle because of the low percentage of grain present. Silage made from such "silage hybrids" may be suited to dairy cattle that need a highly succulent ration, but the ordinary corn varieties that produce a high yield of sound grain make a much more satisfactory silage for finishing cattle. Not only does it require smaller amounts of additional concentrates when fed to animals that are being finished, but it also produces more growth and development when fed to heifers and stockers that are being carried through the winter.

The Minnesota station conducted a trial with stocker calves in which they compared regular corn silage with silage made from a "high-sugar" hybrid corn and with silage made from one of the popular new sorgo-Sudan hybrids. All lots were fed a full feed of the particular silage, limited ground ear corn, supplement, and about 2 pounds of hay. Separate metabolism trials showed that the two corn silages had similar digestibility and TDN values, but that the sorgo-Sudan silage was 20 percent lower in this respect. The feeding trial data reported in Table 188 indicate, however, that regular corn silage was significantly superior to high-sugar silage, and even more

Table 188

Performance Data on Calves Fed Regular Corn Silage, High-Sugar Corn Silage, or Sorgo-Sudan Silage (133 Days)[a]

Silage Fed	Regular	High-Sugar	Sorgo-Sudan
Agronomic data			
Wet yield per acre, tons	7.8	10.6	9.7
Dry matter (%)	36.2	26.0	28.7
Dry matter yield per acre (lb)	5,646	5,490	5,555
Price per ton, as fed[b]	$9.63	$7.50	$7.94
Value per acre, as fed	$75.28	$73.20	$74.06
Feedlot data			
Number of calves	18	17	18
Average initial weight (lb)	492	500	495
Average daily gain (lb)	1.86	1.68	1.37
Average daily feed (lb)[c]			
Silage	9.0	8.7	7.9
Supplement	1.2	1.2	1.2
Ground ear corn	3.8	3.8	3.8
Brome-alfalfa hay	2.0	2.0	2.0
Total	16.0	15.7	14.9
Feed cost per cwt gain	$13.24	$14.26	$17.00
Silage dry matter per cwt gain (lb)	424	458	516
Gain per acre of silage fed (lb)	1,332	1,199	1,077

[a] Adapted from Minnesota Beef Cattle Feeders' Day Report, 1966.

[b] Assumed a price of $8 per ton of 30 percent dry matter silage, or 1.33 cents per pound dry matter. Prices for silages fed were adjusted accordingly, assuming the same value for each pound of dry matter.

[c] 88 percent dry matter basis.

so to the sorgo-Sudan silage. As yields were comparable, and all low for that matter, regular corn again appears to be the best type of corn to grow for silage.

Corn breeders are developing hybrids with grain that is higher in both protein and oil content. As yet, yields are below those of regular corn varieties, but in feeding trials with both stockers and feeders at the Illinois station, protein supplements were successfully omitted from the ration. Thus if yields can be made comparable, these hybrids appear promising indeed. It should be mentioned that the high-lysine corn varieties recently developed, which, being higher in this essential amino acid, are especially valuable in formulating swine and poultry rations, cannot be expected to have higher value for cattle because individual amino acids are not required as such.

VALUE OF FROSTED CORN FOR SILAGE

The silo offers the best way to utilize corn that has been frosted before the ears are mature. If put into the silo immediately after the frost, the silage has practically the same feeding value as it would had the corn been ensiled just before the frost. If, however, the corn is allowed to stand until the leaves are withered and dry, many of the leaves are blown away and the quality of the silage is greatly reduced. A heavy rain on frosted corn also causes much damage by leaching out some of the soluble feed nutrients. It is usually necessary to add some water to frosted corn to make it pack well in the silo.

SORGHUM SILAGE

The various forage sorghums, such as Tracy, Sart, Atlas sorgo, and many new hybrids, are quite satisfactory silage crops and are grown extensively for this purpose in the semiarid and irrigated parts of the Southwest. The hybrid forage sorghums now in use far surpass those used prior to the 1960s. Yields of silage per acre are usually larger for forage sorghums than for corn, especially in years of low rainfall during the growing season or when both are grown on irrigated acreage. Should early frosts occur, less damage is suffered by the sorghums than by corn because the leaves are not so easily lost.

Silage made from the forage sorghums is somewhat lower than corn silage in both protein and fat and is higher in crude fiber. With few exceptions, feeding experiments comparing corn and sorghum silages point to

Table 189

Comparisons Between Corn, Sorghum, and Sorghum-Sudan Silages
Fed to Finishing Cattle[a]

Type of Silage	156-Day Trial		184-Day Trial		
	Corn	Sorghum-Sudan	Corn	Oat	Sorghum
Average initial weight (lb)	730	738	451	450	450
Average daily gain (lb)	2.76	2.19	2.12	1.70	1.66
Average daily ration (lb)					
Silage	48.4	43.0	25.7	22.4	24.5
Ground ear corn	—	—	5.9	5.7	5.7
Cracked shelled corn	12.7	12.7	—	—	—
Supplement	2.5	2.5	1.9	1.9	1.9
Feed per cwt gain (lb)					
Silage	1,755	1,962	1,213	1,315	1,481
Ground ear corn	—	—	279	333	342
Cracked shelled corn	212	268	—	—	—
Supplement	90	113	90	112	115
Feed cost per cwt gain[b]	$15.33	$17.63	$13.92	$14.97	$15.78

[a] Adapted from data in South Dakota Beef Cattle Field Day, 1967.

[b] Costs of silage, per ton: corn, $8; sorghum-Sudan, $7.93; oat, $6; sorghum, $6.

the superiority of corn silage. Although the difference is not great, it is
sufficient to indicate that corn is the more valuable silage crop in areas
where corn can be grown competitively, except during unusually dry
weather. Under irrigation conditions, the added yield of sorghums as silage
makes an acre of forage sorghum equal to an acre of corn.

It is rather difficult to make direct comparisons between corn silage
and silage made from the newer hybrid forage sorghums and sorghum-
Sudan hybrids because each station doing the testing usually uses a different
hybrid or cross, generally bred or selected for the particular climatic conditions prevailing in the state where recommended. Table 189 shows the
results of tests conducted in South Dakota. The results are similar to,
but not necessarily the same as, those obtained in other sorghum growing
states. It is apparent that, as yields of both corn and sorghum varieties
or hybrid number have improved through breeding and better cultural
practices, the grain content of corn silage has improved faster than the
grain content of sorghum silage, in relation to yield of silage expressed
as tons per acre. Therefore it appears that more grain must be added
to forage-sorghum silages, and especially the sorghum-Sudan hybrid silages,

than formerly if they are to be competitive with corn silage. This means that the spread in price or value between corn and sorghum silages is widening with improved yields.

HAY-CROP SILAGE

This term refers to silage made from alfalfa, clover, bromegrass, timothy, and other crops that are ordinarily grown, either in pure stands or in mixtures, for hay. Other designations for this kind of silage are legume-grass silage, grass silage, or merely alfalfa, red clover, or bromegrass silage, if one of these species comprises most or all of the freshly cut forage. Whether it is better to harvest these crops as silage rather than cut them for hay depends on a number of factors. In general, the advantages of ensiling over harvesting for hay are as follows:

1. The crop is less seriously damaged by rains during the harvest period.
2. A much smaller percentage of the leaves is lost; leaves are the most valuable portion of the forage.
3. Considerably more protein is obtained per acre, mainly because fewer leaves are lost, but partly because some crops, principally the grasses,

FIG. 84. Direct-cutting an alfalfa-brome mixture for silage. (John Deere, Moline, Illinois.)

are cut earlier for silage than for hay and thus the protein content is higher.

4. The carotene present at cutting time is preserved much better in silage than in hay.

5. Hay-crop silage is somewhat more palatable and digestible than hay; consequently it may compose a larger percentage of the ration of finishing cattle without seriously reducing the rate of gain.

6. Silage is more suited to mechanized feeding operations.

The principal disadvantage of using legumes and grasses for silage rather than for hay is the higher cost of harvesting. This added cost is represented chiefly by the expensive field harvesters, the silo-filling equipment, and the silos themselves. The use of surface or horizontal silos reduces the cost considerably. If, by chance, silage-making equipment is already available for making corn silage it can, of course, be used for making hay silage with little additional expense.

MAKING HAY-CROP SILAGE

Much more care and judgment are needed in making high-quality hay-crop silage than in making corn silage. Cutting the forage when it is either too ripe or too green, allowing it to lie in the swath or windrow too long before being ensiled, or failing to add the proper amount of preservative, if required, to the forage, may result in unpalatable silage of low feeding value.

Two types of field cutters are available. One type mows and chops the forage in a single operation, usually called direct-cutting. The other type picks up the forage from the windrow after it has been cut with a mower and windrowed with a side-delivery rake, or has been cut and swathed with a swather-crimper. If the first type of harvester is used, the forage should be cut when somewhat riper than is advisable for the second type, as little moisture is lost by the forage during handling. Cutting the forage when too green with this harvester results in a wet, "sloppy" silage that is unpalatable and of low feeding value because of its high moisture content. If the moisture content is unduly high (75 to 82 percent), a considerable percentage of the feed value is lost in the juices that seep through the doors and walls of the upright silo or from the ends of trenches and bunkers. Free water in a silo, whether or not seepage occurs, is to be avoided because it is highly destructive to the silo, first, through the great hydrostatic pressure exerted on the walls and, second, through the constant corrosive action of the silage acids on the walls of concrete and metal silos.

FIG. 85. One method of adding ground shelled corn as a preservative to direct-cut alfalfa. (University of Illinois.)

Most authorities agree that the best quality of conventional hay-crop silage is made from forage that contains 65 to 70 percent moisture or 30 to 35 percent dry matter at the time it is put into the silo. If it is desired to fill the silo when the moisture content is higher, the forage should be allowed to lie in the swath or windrow to permit evaporation of the excess water. When the forage is mowed, raked, and chopped with a field cutter from the windrow without waiting for it to dry, on a warm sunny day the moisture content drops 3 or 4 percent between the time the forage is cut and the time it is stored in the silo. Mowed forage, lying in the swath, loses about 5 percent moisture per hour on a sunny day.

HAYLAGE

The term "haylage" is in common usage among researchers, farmers, and cattle feeders alike. The product actually is little different from legume or grass-legume silage except in moisture content. Most haylage is harvested with a swather-crimper and left in the swath until it has dried down to 40 to 50 percent dry matter. Haylage has been called "wet hay" or "dry silage." Apparently because of its higher dry-matter content it is generally voluntarily consumed at a higher rate than silage made from the same crop, when intake is expressed on a dry-matter basis. Haylage also is not so acid as silage and has a less pungent, less pronounced butyric acid odor. Because of its higher intake, it apparently can be used as a partial or complete substitute for protein supplement, as seen in Table 190. However, it can also be seen from the "level of haylage" study in the table that this product cannot be substituted for the corn in the ration.

The storage of haylage presents some special problems, as it is fluffy and difficult to pack. It is best stored in an airtight metal silo, but can be successfully stored in conventional concrete, open-top silos if the walls are in perfect condition and if the top is sealed with plastic and green chop, limestone, or soil for ballast. Note that the studies covered in Table 191 show no real differences between haylages stored in the two types

Table 190
Haylage as a Protein and Roughage Source in Cattle-Finishing Rations[a]

Type of Study	Level of Supplement		Level of Corn		Level of Haylage			
Length of Feeding Trial	182 Days		168 Days		112 Days			
Initial weight of all cattle (lb)	557		760		525			
Number of cattle per lot	40	40	40	40	10	10	10	10
Average daily gain (lb)	2.32	2.29	1.96	2.62	1.68	2.06	2.20	2.30
Average daily feed (lb)								
Haylage[b]	9.2	9.2	16.1	9.7	14.7	11.6	8.1	6.1
Corn[c]	11.6	11.2	7.6	14.7	3.1	6.2	9.3	10.3
Supplement A	—	1.0	1.5	1.5	—	—	—	—
Feed per pound gain (lb)	9.0	9.5	12.4	9.9	10.6	8.7	7.9	7.1
Feed cost per cwt gain[d]	$13.42	$15.90	$20.67	$17.97	$14.19	$14.57	$12.88	$12.14

[a] Adapted from data in Indiana Cattle Feeders' Day Report, 1966.
[b] 87 percent dry-matter basis.
[c] 15.5 percent moisture.
[d] Feed prices used: haylage, $24 per ton of 85 percent dry-matter material; corn, $40 per ton; supplement A, $80 per ton.

Table 191
Effect of Type of Storage on Making Haylage[a]

Days Fed and Initial Weight (lb)	196 Days—525 Pounds				168 Days—760 Pounds	
Type of Roughage	Chopped Hay	Hay-lage	Hay-lage	Silage	Hay-lage	Hay-lage
Type of Storage	Mow	Gas-tight Silo	Open Silo	Open Silo	Gas-tight Silo	Open Silo
Number of cattle per lot	20	20	20	20	40	40
Average daily gain (lb)	2.12	2.08	1.92	1.88	2.26	2.32
Average daily feed (lb)						
Roughage[b]	5.8	5.8	5.6	5.7	12.5	13.3
Corn[c]	11.4	11.7	10.9	11.2	11.2	11.1
Supplement A	1.0	1.0	1.0	1.0	1.0	1.0
Feed per pound gain (lb)	8.5	8.9	9.1	9.8	11.3	11.4
Feed cost per pound gain[d]	$15.91	$16.52	$16.92	$17.66	$18.32	$18.14

[a] Adapted from Indiana Cattle Feeders' Day Report, 1966.
[b] 87 percent dry-matter basis.
[c] 15.5 percent moisture.
[d] Feed prices used, per ton: roughage, $24; corn, $40; supplement A, $80.

of upright silos. Also note that haylage is comparable to hay but somewhat superior to conventional legume-grass silage made from the same crop.

RECONSTITUTED HAY

Some cattle feeders who have an airtight silo built into an automated feeding system and who thus are not equipped to feed hay will, after the silo has been emptied of conventional haylage, sometimes refill it during the winter or early spring with chopped hay to which water is added. By doing this they need not change rations during the feeding period for a particular drove of cattle. Such chopped hay apparently undergoes fermentation similar to haylage and is slightly superior to the hay from which it was made.

FEEDING VALUE OF HAY-CROP SILAGE

Several experiments have compared the feeding value of silage made from grasses and legumes with that of hay made from the same field and also with corn silage grown and harvested the same year. Extensive tests

Table 192

Comparison of Corn Silage, Legume Silage, and Legume Hay for Finishing Cattle

	Pennsylvania (Average 3 Trials)			Illinois (1 Trial)			Michigan (Average 4 Trials)		
	Corn Silage	Al- falfa Silage	Al- falfa Hay	Corn Silage	Al- falfa Silage	Al- falfa Hay	Corn Silage	Al- falfa[a] Silage	Al- falfa[a] Hay
Average initial weight (lb)	627	627	630	852	854	852	491	493	488
Average daily gain (lb)	2.18	2.15	2.06	2.87	2.55	2.45	1.81	1.93	1.81
Average daily ration (lb)									
Corn	12.6[b]	12.6[b]	12.6[b]	16.3	15.5	16.5	2.0[c]	9.2	9.2
Protein concentrate	1.5	—	—	2.3	1.0	1.0	1.6	1.0[d]	1.0[d]
Silage	15.1	20.1	—	22.0	21.7	—	28.0	25.0	—
Legume hay	—	—	8.3	2.0	2.0	6.8	2.0	—	9.5
Feed per cwt gain (lb)									
Shelled corn	581[b]	586[b]	609[b]	568	610	671	111	473	506
Protein concentrate	68	—	—	80	39	41	87	9	10
Silage	688	938	—	766	853	—	1,532	1,293	—
Alfalfa hay	—	—	399	70	79	279	136	—	504
Dressing percent	60.3	60.0	59.8	59.3	59.5	59.4	—	—	—

[a] A mixture of alfalfa and red clover fed in two trials.
[b] Ground ear corn fed in Pennsylvania trials.
[c] Average for entire period but fed during only last half.
[d] Average for last 30 to 60 days, during which time it was fed.

were made at the Michigan and Pennsylvania stations. At each station three lots of cattle were fed to compare corn silage, alfalfa silage, and alfalfa hay, respectively. At the Pennsylvania station all lots of cattle were fed the same amounts of ground ear corn, while the silages and hay were fed according to appetite. In the Michigan tests no grain was fed to the corn silage lots during the first half of the test, and about one-third of a full feed during the last half. Grain was fed to the legume hay and silage lots in amounts to make these cattle gain as much as those fed corn silage. Results of these two series of experiments are given in Table 192.

It will be noted that rations based on silage made from alfalfa or other legume crop required the addition of as much as 1 pound of grain per day per hundredweight in order to produce gain and finish comparable to that produced by a full feed of good quality corn silage and hay plus protein supplementation.

MISCELLANEOUS SILAGE MATERIALS

Almost any green material can be ensiled successfully if it is tightly packed in a well-built silo so that no air pockets remain and if sufficient carbohydrate material is present to promote fermentation. Naturally some crops are better suited for silage purposes than others. In addition to having a sufficiently high carbohyrate content to produce enough organic acids to arrest the action of putrefying bacteria before they cause decomposition, a good silage material must have a physical texture that produces an abundance of fine particles to ensure close packing. All plants that have an abundance of leaves in proportion to the amount of coarse, woody stems are good silage materials in this respect. A silage crop should also produce a heavy yield per acre. Otherwise the money or time spent in cutting, hauling, and land rental will be so great that the cost of making the silage will be uneconomical.

Few of the silages discussed below have been used extensively in either practical or experimental beef cattle feeding. However, their yield and other qualities make them suitable silage crops for those regions where they can be successfully grown.

Sudan Grass. Sudan grass belongs to the same group of plants as the grain sorghums. It differs from the sorghums principally in being finer-textured and in having a comparatively light yield of seed. When seeded in rows or broadcast in fertile soil it grows 6 to 8 feet tall and yields an enormous tonnage of green forage per acre. Sudan grass intended for silage should be planted in rows to encourage better development of heads and leaves and greater seed production. It should be cut when the seeds are in the dough stage, although very good silage can be made at almost any stage of maturity. In the southern states where the growing season is long, both the first and second growths of Sudan grass are available for silage. The newer sorghum-Sudan hybrids are intermediate between Sudan and forage sorghum in feeding value, but at least 50 percent higher-yielding.

Table 193
Oat and Pea Silage for Finishing Cattle[a]

Full Feed of Grain Plus:	Prairie Hay	Oat and Pea Silage Prairie Hay	Sunflower Silage Prairie Hay
Average daily gain (lb)	1.79	2.48	2.06
Feed per pound gain (lb)			
Grain mixture	5.87	4.06	5.49
Hay	9.24	0.96	1.15
Silage	—	13.24	15.00
Linseed meal	0.55	0.40	0.48

[a] University of Alberta, Canada, Mimeographed Report.

Oat and Pea Silage. A mixture of oats and field peas is often grown for forage purposes in the northern United States and in Canada. If the supply of such forage is greater than needed for pasture, its successful harvest for other purposes is often a perplexing problem because oats and peas do not usually mature at exactly the same time. The silo offers a satisfactory way of harvesting such a crop. The carbohydrates of the oats ensure successful preservation of the pea vines which, if ensiled alone, might spoil because of their high protein content. Silage made from oats and peas is both a palatable and nutritious feed for beef cattle, as shown by Table 193.

Oat Silage. Oats are grown throughout the country, but they seldom return a satisfactory profit to the farmer, because of their low yield and low market price. Their yield is often greatly reduced by the sudden advent of hot dry weather a few days before they are ripe enough to cut. This hazard can be avoided, in part at least, by cutting the oats while they are still green and making them into silage. Oats that are to be ensiled should be cut when the kernels are in the dough stage.

In the northern and northwestern states, where oats are grown much more successfully than corn, oat silage may well take the place of corn

FIG. 86. The income from an oat crop, when harvested as silage and fed to stockers or cows, may easily be doubled as compared with the usual harvesting methods. (University of Illinois.)

Table 194
Oat Silage for Finishing Cattle

	Two-Year-Old Steers				Yearlings	
	University of Alberta, Canada (Average of 3 Trials)		Illinois Experiment Station		Illinois Experiment Station	
	Oat Silage (lb)	Sunflower Silage (lb)	Oat Silage (lb)	Corn Silage (lb)	Oat Silage (lb)	Hay-Crop Silage (lb)
Average daily gain (lb)	2.48	2.28	2.88	3.08	2.09	2.29
Feed per cwt gain (lb)						
Grain	408	525	568	507	535	504
Protein concentrate	37	41	81	72	54	4
Hay	129	161	70	65	—	—
Silage	1,300	1,562	6,08	622	1,126	1158

silage in beef cattle rations. Extensive tests at the Illinois station and others show that oat and other cereal silages used in finishing rations have a feeding value comparable to that of hay-crop silage but somewhat below that of corn silage (see Table 194). Perhaps even more important than the increased income from the oat crop as a result of harvesting as silage is the improvement in stand obtained from the legume-grass seeding that usually accompanies the oat crop. A good clipping of hay from the new legume-grass seeding during the first fall is not uncommon when the oat nurse crop has been removed as silage instead of by the usual combining procedure.

Millet Silage. Formerly several varieties of millet were commonly grown by Corn Belt farmers as emergency hay crops. Millet is still occasionally sown in cornfields where the stand of corn has been ruined by floods or insects too late in the season to warrant replanting. Millet is a rank-growing crop and produces a high yield of hay per acre. The quality of hay is poor, however, because of its coarse texture and the numerous hard seeds found in the heavy heads.

Experiments show that ensiling millet greatly improves its value as a feed. In the silo the seeds absorb sufficient water to make them soft and more easily eaten by cattle. Because little labor is involved in growing and harvesting millet, and because of the high yields and the fact that it matures in time to allow it to be seeded after a crop of wheat or early oats is removed, millet silage can be produced at very low cost per ton. In feeding value, however, it is distinctly inferior to corn silage, at least when fed as the principal ingredient in the ration.

Canning Refuse. The pea vines, cobs and husks from sweet corn, and beet tops that accumulate around large canning factories are quite generally used for finishing beef cattle. Frequently these materials are allowed to accumulate in huge stacks, which in time become "natural silos" through the exclusion of air from the interior by the decomposition of the material at the surface. Not only the top but also the sides of the pile may be spoiled for a depth of 2 to 3 feet, depending on the nature of the material and the care with which it was stacked.

Some farmers living in the vicinity of canneries realize the feeding value of this refuse and are willing to buy it and store it in modern silos. Often it is returned to the owners of the land on which the crops were grown in partial payment for use of the land by the canneries, or it is returned in partial payment for the canning crop purchased.

Cattle feeding is quite general in the vicinity of canneries, both because of the large amount of cheap feed available and because the supply of manure helps keep the soil at maximum productivity. Owing to the wide variation in silages resulting from the different crops, stages of cutting, and methods of storing, no definite rules for feeding can be given. In general, better returns are secured by feeding them in combination with grain and hay than by feeding them alone.

Wet Beet Pulp. In the sugarbeet-growing areas of Colorado and other beet-producing states, large quantities of wet beet pulp are used in finishing cattle for market. Whether it is fed directly from the mill or from a silo, its feeding value compares favorably with corn silage. In feeding trials at the Colorado station 1 ton of corn silage replaced 3,980 pounds of wet beet pulp, but required 43 pounds more beet molasses, 33 pounds more cottonseed cake, and 74 pounds more alfalfa hay.

Potato Silage. Silage made from surplus or cull potatoes compared favorably with corn silage as a feed for beef calves in tests of these feeds at the Colorado station. Each ton of potato silage replaced 2,642 pounds of corn silage, 42 pounds of barley, and 7 pounds of linseed cake, but required 476 pounds more alfalfa hay.

Chapter 20

FINISHING CATTLE
on PASTURE

In the farmer-feeder situation, the contribution of pasture, as such, to the finishing ration may vary all the way from being the only feed used to none. The first circumstance occurs when older feeder cattle, such as 2-year-old or older steers or slaughter cows, are sold for slaughter directly from grass. The degree of condition attained naturally is seldom sufficient from such slaughter cattle to grade above standard or low good. At the other extreme are cattle that are fed finishing rations in drylot where the sources of roughage consist of harvested hays or silage, if any roughage is fed at all. As pastures are seldom used by commercial feedlots for finishing cattle, the material in this chapter is applicable only to farmer-feeders, some of whom are found in all parts of the United States.

Pasture makes its greatest contribution in the finishing programs that are conducted during the summer. In this circumstance various combinations of pasture and concentrates are used, depending on the relative amounts of pasture to be harvested and sold through feeder cattle and the amounts of grain or concentrate available, the time at which cattle are to be sold, and the availability and skill of the labor being used.

With the steady rise in tillable land values in recent decades, less and less pasture is planted if the soil can grow cash crops successfully. Thus many farmer-feeders are forced into winter or year-round drylot feeding, particularly in the prairie sections where all the land is tillable. For the grain farmer who must keep most of his land in cultivated crops, winter feeding has certain advantages, as will presently be discussed. But for that man who, because of the large size of his farm or the presence of rolling or timbered land, has a considerable area of pasture, summer feeding will always have an appeal.

ADVANTAGES OF SUMMER FEEDING ON PASTURE

The main reasons in favor of summer feeding on pasture may be summarized as follows:

1. Summer gains on pasture are usually cheaper than winter gains in drylot because (*a*) less grain is eaten per pound of gain; (*b*) pasture is a cheaper form of roughage than harvested hay; (*c*) less labor is required in feeding and caring for cattle, as the labor involved in feeding roughage is eliminated, and cattle fed grain on pasture are usually fed by means of self-feeders, whereas hand feeding is often used in drylot.

2. Ordinarily larger daily gains are secured in the summer than in the winter over feeding periods of equal length. Cattle are likely to be more comfortable in summer than in winter because weather conditions are more uniform and feedlots are dry. In addition, summer rations are on the whole superior to winter rations, as explained in the next paragraph.

3. Cattle fed on pasture during the summer usually receive a better-balanced ration than cattle fed during the winter in drylot. Fresh-pasture forage is an excellent source of protein, minerals, and vitamins, and cattle fed on pasture are not often deficient in these important nutrients.

4. No investment in buildings to afford shelter is required.

5. Swine following cattle on pasture make larger gains and show a lower death loss than those following cattle in drylot.

6. The manure produced is spread on the fields by the cattle themselves, avoiding loss of fertility through leaching and heating, and saving much labor.

7. Summer-fed cattle are commonly marketed during the late summer and fall when well-finished cattle are usually higher priced than at any other season of the year. Winter-fed cattle, on the other hand, are marketed in the late winter and spring when prices paid for fed cattle are relatively lower.

DISADVANTAGES OF SUMMER FEEDING ON PASTURE

Summer feeding of cattle on pasture has certain disadvantages, which are more likely to apply to the small farmer with a quarter-section of land or less than to the man who owns several hundred acres.

1. The land used for pasture may return a larger gross cash income if planted in crops.

2. An adequate supply of feed in the form of grass is uncertain, owing

to the possibility of unfavorable weather. Winter killing or a late freeze after germination may result in complete failure of rotation pasture. Drought and insect or hail damage are always a threat.

3. Grains, especially corn and milo, are relatively high-priced during summer and early fall.

4. The farmer has less time to devote to cattle in summer than in winter.

5. Flies and extremely hot weather may cause cattle much discomfort.

6. If permanent pastures are used, the manure is dropped on the same fields year after year.

7. Shade and water are hard to provide in temporary and rotation pastures.

8. Feeder cattle are scarce, high in price, and the grade and weight desired are hard to obtain, if spring purchases are made.

9. Summer pasture feeding programs are not so well adapted to use of laborsaving equipment such as may be used in drylot.

10. Weather damage to grain in feeders or bunks is possible because of rain and windstorms, which are more prevalent in summertime.

11. Swine are less apt to utilize completely the grain voided in manure dropped on pasture.

DRYLOT VERSUS PASTURE FOR SUMMER FEEDING

A large number of cattle are fed in drylot for the late summer and fall markets. However, a majority of these cattle are purchased in the fall and have been fed considerable grain by the arrival of spring. Consequently they usually carry too much flesh to be fed on pasture during the summer and therefore are kept in the drylot until they are ready for market. Only steers of strictly choice grade justify such a long feeding period, but if they have sufficient conformation and finish to sell near the top of the market, they usually are more profitable than when marketed earlier, because higher prices are paid for choice and prime steers during late summer and fall.

Although it is generally agreed that cattle in good, thrifty feeder condition in the spring will make faster and more economical gains when fed on pasture than in the drylot, it is often claimed that these advantages are more than offset by the lower prices received for pasture-fed cattle when marketed. This opinion is well supported by numerous feeding experiments in which prices paid for the lots fed on pasture have usually been 25 cents to a dollar less per hundredweight than for those fed in drylot.

Table 195
Comparison of Drylot and Pasture for Summer-Fed Steers

	Full-Fed in Drylot				Full-Fed on Pasture			
			Feed per Cwt Gain (lb)				Feed per Cwt Gain (lb)	
	Daily Gain (lb)	Shelled Corn per Day (lb)	Concen-trate[a]	Hay	Daily Gain (lb)	Shelled Corn per Day (lb)	Concen-trate[a]	Hay
2-year olds								
Illinois	2.12	19.7	927	411	2.00	19.9	992	—
Kentucky (3-year-average)[b]	1.81	10.1	608	908	2.07	10.1	531	—
St. Joseph stockyards	3.08	18.2	671	288	2.76	18.6	763	112
Average 5 trials	2.13	13.6	684	685	2.19	13.8	670	20
Yearlings								
Kansas	1.76	11.5	706	242[c]	2.10	13.0	667	55[c]
Ohio	1.88	12.8	787	401[c]	2.36	12.7	625	—
Ohio	1.92	15.0	840	331[c]	2.13	15.0	758	—
Missouri	1.85	13.4	804	121	2.18	15.5	780	—
Missouri	2.32	12.9	611	229	2.20	12.3	613	—
Missouri	2.46	15.4	680	135	2.13	12.7	653	—
Nebraska (average of 3 trials)	2.19	16.1	743	187	2.31	15.7	686	112
Illinois	2.45	14.1	618	204	2.34	12.3	483	58
Average of 10 trials	2.12	14.3	728	222	2.24	14.1	665	45

[a] Grain plus protein concentrate.
[b] Mixed pasture used in this experiment.
[c] Includes some silage reduced to dry roughage equivalent.

Buyers defend these prices by saying that cattle fed on pasture yield less beef in proportion to their live weight, that their carcasses have a higher shrinkage, and that the beef is poor in color, with the lean being too dark and the fat being yellow, instead of white as found in cattle fed in drylot. Although none of these claims except that relating to the fat color has been supported by slaughter and carcass studies made by impartial investigators, the fact remains that market buyers look with disfavor upon pasture-fed cattle, even though they show good quality and finish.

Grass-fed cattle can be distinguished while alive from drylot cattle by their rough, dry, sunburned hair and the greenish color of their feces. Removing the cattle from pasture about 2 weeks before marketing will correct the color of the feces, but even a month of drylot feeding usually does not improve their hair sufficiently to escape some price discrimination. There is no improvement in appearance of the hair unless the drylot adjoins a barn or shed as protection from the sun. Selling on grade and yield will eliminate the possibility of price discrimination for pasture-fed cattle.

Although the financial statements on comparable droves of cattle frequently show a larger net return over feed costs for drylot than for pasture feeding, it should not be assumed that this system of feeding is necessarily better under all conditions. Such financial statements seldom take into account the relative farm costs of harvested roughages and pasture, the relative amounts of labor used in feeding, or the comparative quantities of manure recovered and returned to the land. All these items are very much in favor of the pasture-fed cattle.

All things considered, the net difference between pasture and drylot feeding is not great. The summarized data in Table 195 give pertinent information on this subject. It should be noted that some of the earlier tests reported dealt with nonlegume pastures but, if anything, the use of legume or legume-grass pastures favors pasture feeding.

In the last analysis, the amount of available pasture is the factor that usually determines whether cattle purchased in the spring or carried through the winter in stocker condition are fed on pasture or in the drylot. If the topography of the farm or the crop rotation results in a large acreage of pasture the cattle, in all likelihood, will have been purchased mainly to utilize the otherwise surplus pasture. Thus, in the Corn Belt at least, feeding on pasture is usually the result of a system of farming rather than a separate project undertaken because of advantages peculiar to the enterprise itself.

FIG. 87. Full-fed steers self-fed on rotation pasture. Whether to self-feed on pasture in the summer, rather than in drylot, is a matter for each farmer to decide after weighing the advantages against the disadvantages. (Corn Belt Farm Dailies.)

SELECTING CATTLE FOR SUMMER FEEDING

Cattle to be finished during the summer may be purchased any time between September or October of the preceding year and the date when it is desired to turn them on pasture. Purchases made in the fall consist largely of calves and light yearlings, which will make considerable growth during the winter before the period of heavy feeding begins. Spring purchases, on the other hand, are often mature steers or yearlings with sufficient flesh to insure their being in choice slaughter condition after a feeding period of 4 or 5 months.

Regardless of their age and time of purchase, cattle to be fed during the summer should be selected with considerable care. Cattle from droughty areas that are light but healthy and of good to choice grade usually work best in this program, especially if summer feeding is preceded by a stocker program in the winter.

In the Winter Wheat Belt, literally millions of calves are bought in the fall for grazing the wheat fields from fall to spring. Unfortunately, moisture conditions necessary for adequate growth on the wheat to graze all of these calves happen only occasionally. Thus many of the calves are prematurely sold in the Wheat Belt from January to March, and even in good years most of them are sold by May 1. These calves all make ideal cattle for feeding in summertime. Steers are usually more suitable than heifers for summer feeding, mainly because so many of the heifers will be in calf.

Steers bought in the spring for summer feeding should have a fair amount of flesh; otherwise it is difficult to get them into choice grade by the end of the grazing season. In selecting steers with plenty of flesh one should avoid cattle that have been "warmed up" on corn or, still worse, poor-doing steers cut from droves that were fed grain all winter. Most experienced feeders prefer cattle that have never had any grain. Two-year-old western hay-fed steers or yearlings wintered in the Corn Belt principally on corn silage make ideal cattle for summer feeding. While each spring thousands of half-finished, corn-fed steers are sent back to the country for further feeding, the gains made by such animals are usually more expensive than those made by cattle unaccustomed to a heavy grain diet.

VARIETIES OF PASTURE

Before the mid-1930s bluegrass was the pasture most commonly used in the summer feeding of steers. Even though legume-grass rotation pastures have taken over as the most important pasture for summer feeding, blue-grass is still considered by many to be an especially valuable forage for grain-fed cattle because it does not tend to cause scouring or bloat. It is ready to graze much earlier in the spring than are most of the legume forages, and its firm sod withstands trampling much better than most other pasture crops.

During the spring and early summer, bluegrass is palatable and nutritious, but after ripening its seed in midsummer it becomes more or less dormant, especially during a dry season. At this stage it is not so palatable, and cattle getting a full feed of grain eat comparatively little of the grass.

It is this tendency to go dormant that reduces the carrying capacity of bluegrass and lessens its value in a summer feeding program. True, with the coming of fall rains and cooler weather, it starts growing again and often furnishes considerable grazing during September and October. The common practice is to remove the cattle to the drylot before this growth begins, however, lest the new, green grass interfere with the grain consumption. The fall growth of bluegrass pastures is utilized to greater advantage by newly purchased stocker calves or yearlings.

Along the western and northern borders of the Corn Belt, bluestem, bromegrass, and orchard grass pastures are used extensively for summer-fed steers. All these pasture forages are similar to bluegrass in composition, and similar gains may be expected from them. Recently bromegrass has been grown extensively in the Corn Belt where it has given much better results than bluegrass, particularly when mixed with alfalfa.

Many longtime cattle feeders look with disfavor upon straight legume pastures for grain-fed cattle, believing that they produce both scouring and bloat. However, comparisons of these forages with bluegrass have shown that legume pastures are capable of producing so much more gain per head and per acre that they must be rated as valuable pasture crops for cattle despite these objections. Both red clover and alfalfa, either alone or in mixtures, have given much larger and more economical gains than bluegrass in experiments summarized in Table 196. Fairly good results have been obtained with sweet clover, but the short grazing season of the second year's growth makes it necessary to transfer the cattle to other pasture or to the drylot about the middle of August.

SMALL GRAIN PASTURES

Opinions differ as to the best way to use winter small-grain pastures. In the Winter Wheat Belt of Kansas, Oklahoma, and Texas, calves are most often grazed without supplemental feeding except during inclement

Table 196
Value of Legume Forages for Steers Full-Fed on Pasture

Yearling Steers Initial Weight 550 @ 650 lb	Nebraska May 5–Nov. 3 182 Days			Illinois May 4–Oct. 23 172 Days			Illinois May 6–Nov. 15 193 Days		
	Drylot	Alfalfa Pasture	Native Pasture	Blue-grass Pasture	Red Clover Pasture	Sweet Clover Pasture	Brome-grass Pasture	Alfalfa Pasture	Mixed Clover Pasture
Date turned on pasture	—	May 5	May 5	May 5	May 5	May 5[b]	May 6	May 6	May 13
Date turned off pasture	—	Sept. 22	July 28	Oct. 23	Oct. 23	Oct. 23	Sept. 23	Sept. 23	Sept. 23
Days on pasture	—	140	84	172	172	172	140	140	133
Steers per acre	—	3	2	2	2	2[b]	2.4	2.4	2.4
Days in drylot	182	42	98	None	None	None	53	53	53
Average daily gain on pasture (lb)	2.67[a]	2.70	2.49	2.16	2.47	2.27	2.25	2.44	2.28
Average daily gain in drylot	2.19[a]	2.60	2.51	—	—	—	2.57	2.36	2.27
Average daily gain total period (lb)	2.41	2.68	2.50	—	—	—	2.34	2.42	2.28
Feed per cwt gain (lb)									
Shelled corn	659	605	661	658	663	648	574[d]	553[d]	609[d]
Protein concentrate	—	65	141	45	44	44	25	24	27
Alfalfa hay	289	29	18	—	—	—	58	57	62
Hog gains per steer (lb)	—	—	—	67	67	58	—	—	—

[a] Average daily gain in drylot before and after July 28.
[b] Spring seeding of sweet clover used after August 7, one steer per acre.
[c] Clover mixture: approximately 80% sweet clover and 20% red clover.
[d] Weight of ground ear corn.

524

weather. Because of the low dry-matter content of such pastures, some operators add a source of dry roughage or even a high-energy feed to increase the total intake of nutrients—and, it is hoped, to improve performance.

The Oklahoma station conducted a test in which half of the calves grazed wheat pasture alone and the other half had free access to a high-concentrate mixture consisting of 77 percent ground milo, 8 percent molasses, and 15 percent chopped alfalfa hay. The results are summarized in Table 197. The calves that were offered supplemental feed consumed 10 pounds per head daily but gained only 0.61 pound more per head daily than calves consuming no supplemental feed. This added gain would not offset the cost of the supplemental feed.

In the Southeast, winter small grains are often grown specifically for winter pasture. The question arises as to how best to utilize these pastures and at the same time produce a steer ready for slaughter. The Georgia station studied this problem at its Coastal Plain station at Tifton and the results of the test are shown in Table 198. They used five management systems as follows:

1. Drylot—on ground snapped corn, cottonseed meal, and Coastal Bermuda hay.

2. Started in drylot on above ration, then transferred to grazing plus limited ground snapped corn.

3. Fall maintenance feeding followed by winter grazing and limited feeding on oat pasture.

Table 197
Response of Heifer Calves to High-Energy Feed on Wheat Pasture (114 Days)[a]

	Wheat Pasture Only	Wheat Pasture + High-Energy Feed
Number of heifers	20	20
Average initial weight, 11/21/67 (lb)	481	482
Average final weight, 3/14/68 (lb)	637	708
Total gain (lb)	156	226
Average daily gain (lb)	1.37	1.98
Feed consumed per head (lb)		
Hay (during snow cover)	135	90
Concentrate feed	—	1,216

[a] Oklahoma State University, Animal Science Research Progress Report, Miscellaneous Publication 80, 1968.

Table 198

Systems of Utilizing Winter Grazing in Finishing Yearling Steers[a]

	Drylot Feeding	Drylot + Pasture	Oats Pasture	Rye Pasture	Limited Oats Pasture
Number of steers	10	10	10	10	10
Total days fed	147	147	223	223	223
Days maintenance feeding	—	—	76	76	76
Days in drylot	147	76	35	49	35
Days on pasture	—	71	112	98	112
Average initial weight (lb)	714	716	710	716	730
Average final weight (lb)	1,024	1,061	1,094	1 084	1,104
Average daily gain, finishing period (lb)	2.10	2.35	2.42	2.39	2.44
Average daily ration (lb)[b]	25.36	14.45	6.69	10.00	14.74
Ground snapped corn	17.73	10.08	3.75	6.07	12.66
Cottonseed meal	2.50	1.29	0.60	0.90	0.60
Coastal Bermuda hay	5.13	3.08	2.34	3.03	1.48
Acres grazing per lot	—	7.00	10.00	10.00	5.00
Feed cost per cwt gain	$24.04	$20.28	$18.65	$22.06	$20.36
Average dressing percent	60.2	59.2	58.3	58.8	59.2
Average carcass grade	G	G−	G	G−	G

[a] Adapted from data in Georgia Agricultural Experiment Station Circular N.S. 31, 1962.
[b] Based on 147-day finishing period.

4. Same plan as 3, except rye pasture was used.

5. Same plan as 3, except grazing was for only 5 hours daily and ground snapped corn was full-fed.

The pastures were seeded in October after 100 pounds of actual nitrogen was applied to the soil per acre. A charge of $40 per acre was made for the pasture. Grazing began in mid-December and continued for 112 days. A short, 35-day, drylot feeding period followed for steers on treatments 3, 4, and 5 to bring them to comparable slaughter weights. All of the pasture treatments produced more rapid and more economical gains, with the most profitable being the lot on oat pasture continuously plus limited grain and hay. Carcass grades were comparable for all treatments.

It appears that systems making full use of the small-grain pasture with only enough supplemental feed to provide intakes at the level of about 1 percent of body weight offer the most profitable way to utilize such pastures in finishing programs. Short periods of drylot feeding at the end of the grazing season in the spring are recommended.

TURNING STEERS ON PASTURE

Better results are secured if grass or legume-grass pastures are allowed to make a fairly good start before the cattle are turned onto them. The first growth that appears is high in moisture and very low in energy content. By April 20 to May 1 bluegrass is usually mature enough to be used in the central part of the Corn Belt. In the famous bluegrass sections of Kentucky and Iowa the grazing season opens about 10 days earlier and later, respectively.

An exception to these dates are steers that have received a heavy feed of grain during the winter. Better results are usually obtained if such cattle are turned onto the pasture in late March or early April as soon as the grass begins to grow. In this way they become accustomed to the grass so gradually that they are not apt to lose their appetite for corn. When full-fed cattle are turned onto a heavy growth of grass their grain consumption temporarily falls off, sometimes to only half of what it was in the drylot. Although the cattle usually regain their appetite for grain by the end of the third or fourth week, gains made during such a transition period are far from satisfactory.

Steers should never be turned onto straight-legume or legume-grass rotation pastures before the pastures are 6 to 8 inches high. Grazing too early results in reduced forage production through the remainder of the grazing season. Such pastures usually reach the desired height by April 15 to May 10 in the midwestern cattle-finishing areas, with earlier dates in the South and the West.

BLOAT

Use of legume and legume-grass rotation pastures has resulted in an increase in the incidence of pasture bloat. This type of bloat is often fatal and is not to be confused with the troublesome, but less dangerous, common feedlot bloat. Great economic losses are incurred during some years because of deaths and poor performance of bloated animals. Accurate data are difficult to obtain, but the figure often used is $100 million lost due to bloat per year in the United States alone. The annual loss from bloat is estimated at one death per 200 head of cattle. Of even greater importance, however, is the loss resulting from failure to use these nutritionally more valuable pastures instead of grass pastures simply because of the farmer's fear of bloat.

The causes of legume bloat are still under investigation, but the picture is becoming clearer. In almost all cases the paunch becomes markedly distended on the left side, due to the formation of a stable frothy mass in the left dorsal portion of the rumen. Trapped within this mass is a large quantity of rumen gas, mostly carbon dioxide, which cannot easily be eliminated by the usual belching process. Gas continues to be produced faster than it can be expelled, and pressure builds up until the animal dies of apparent suffocation. The actual cause of death is not fully established, but suffocation currently seems the most plausible explanation.

The important discovery that appears to be the best explanation for the real problem, the stable froth, was made by a group of Wisconsin investigators. They believe that certain plants, notably alfalfa, are bloat-producing because they contain critical amounts of a plant enzyme called pectin methyl esterase or PME. The enzyme, when present in sufficient quantity, reacts with pectin, a carbohydrate always present in forages, and changes it to pectic acid and alcohol. Pectic acid reacts with calcium, a valuable mineral in legumes, in the rumen to produce the sticky, foamy mass that traps the rumen gas, preventing its escape and finally resulting in the fatal buildup of pressure.

A number of antifoaming agents are available, and many of them are effective if administered at a sufficient and a constant rate. These agents, called surfactants, generally act by reducing the surface tension of the otherwise stable froth, permitting the gas to escape at a rate fast enough to prevent serious consequences. Among the products being used for this purpose are animal fats, lecithin, vegetable oils, mineral oils, detergents, turpentine, silicones, glycerol, and—one that looks especially promising—poloxalene, a synthetic polymer. The effectiveness of this product was first announced by a Kansas group and it has now been approved by the Food and Drug Administration for use in livestock feeds as a bloat preventive.

A test conducted at Kansas compared two methods of administering poloxalene with a third lot receiving no treatment, using steers that were grazing immature alfalfa pasture. Results are summarized in Table 199. One group received poloxalene from a molasses-salt block containing 30 grams per pound of block. No other salt was provided. Another group received poloxalene in a self-fed ground milo preparation to which "Bloat Guard," a commercial premix containing 53 percent of the active agent, was added at the rate of 10 grams per pound of milo. This group and the control group also received molasses-salt blocks to which no poloxalene was added. The steers were rotated between treatments every 14 days to eliminate the possibility of anatomical differences in the steers.

The poloxalene in the molasses-salt block most successfully prevented bloat. No steers died, but many severe cases of bloat were produced, so

Table 199
Effectiveness of Poloxalene for Controlling Bloat in Steers Grazing Immature Alfalfa (42 Days)[a]

Treatment[b]	Poloxalene in Feed	Poloxalene Block	Control
Average number of animals	4	4	4
Number of bloat ratings[c]	336	336	336
Cases of bloat	48	20	98
Highest bloat score	3	2	3+
Average bloat score[d]	0.69	0.54	1.38

[a] Adapted from Kansas Livestock Feeders' Day Report, 1967.
[b] All 12 animals received each treatment, being rotated at 14-day intervals.
[c] All 4 animals on each treatment were rated twice daily for 42 days.
[d] A scale ranging from 0 for no bloat to 4 for severe bloat was used.

that there was no question that bloat-producing forage was available.

Bloat is also common in cattle fed in drylot on almost any combination of alfalfa hay and barley. Several stations have tested poloxalene in the prevention of bloat under these circumstances and here, too, it apparently is effective so long as at least 10 grams of the compound are consumed daily.

Some old and some new recommendations for preventing and controlling bloat through management other than medication are as follows:

1. Permit cattle to fill on grass or hay before first turning out on pasture.

2. Allow cattle free access to dry forage or a grass pasture adjoining the legume pasture.

3. Leave cattle on the legume pasture 24 hours a day rather than penning them at night.

4. Use pasture mixtures that provide no more than 50 percent of legume forage.

5. Mix additional cob or oats with ground ear corn or other grains fed on legume pasture.

TREATMENT OF BLOAT

There are almost as many methods for the treatment of bloat as there are recommendations for preventing it. The presence of bloat is easily recognized by a pronounced swelling of the left flank. So long as the distention causes the animal no great discomfort, there is no need for alarm, as recovery usually occurs with treatment. However, the bloated animal should be kept under observation, because the condition occasionally worsens very quickly.

If bloating persists or proceeds to a more advanced stage, treatment by medication or otherwise may be necessary. Treatment consists in administering compounds that will break down the stable froth that is preventing the gas from escaping. As already mentioned, numerous products are available from commercial sources. Also, either 1 pint of mineral oil, or 2 ounces of aromatic spirits of ammonia, or 2 ounces of turpentine, diluted with 1 pint of cold water often brings relief.

A simple way of relieving acute bloat in animals that are easily caught and not critically advanced in bloat is to force the escape of the gas through a 6-foot length of smooth ½- or ⅝-inch rubber hose, one end of which is inserted far back into the animal's mouth and carefully pushed down the esophagus and into the paunch. With a little practice there is no difficulty in getting the end of the hose over the trachea and into the esophagus.

If the formation of gas proceeds despite the remedies administered

(a)   (b)

FIG. 88. (a) A badly bloated steer. (b) Trocar and cannula for tapping critically bloated animals. (O. M. Franklin Serum Company, Denver, Colorado.)

or after use of the stomach tube, it is necesssary to tap the paunch to permit the gass to escape. This is done by means of a trocar, a sharp-pointed instrument encased in a cannula or sheath. The point of the trocar is placed at a spot on the animal's left side equidistant from the last rib, the hipbone, and the transverse processes of the lumbar vertebrae. The handle is then struck sharply with the palm of the hand in the direction of the animal's right knee. Then, while the cannula is kept in place, the trocar is withdrawn, permitting the gas to escape slowly. This procedure is a last resort and should only be used in extreme cases.

DAY VERSUS NIGHT GRAZING

Many farmers who full-feed their cattle during the summer confine them in the drylot overnight and turn them onto the pasture during the day. The object of this procedure is to induce the cattle to eat as much grain as possible by holding them in the feedyard.

As cattle that are continually on pasture spend more time grazing at night than during the day, especially in hot weather, it would appear that confining them in the lot during the day, where they have access to water and shade, and turning them onto the pasture at night might be a better procedure.

These two methods of using pastures were compared at the Illinois station. A third and fourth lot, fed in drylot and on pasture, respectively, were also included in the test. Grazing at night proved to be much the best plan of utilizing the pasture. Observations of the cattle showed that those confined in the drylot during the day and turned onto the pasture at night usually grazed steadily for about 3 or 4 hours after being turned onto pasture, whereas the cattle pastured during the day usually sought the protection of an artificial shade soon after being put into the field. The principal difference observed between the cattle pastured at night and those left on pasture continually was that the night-pastured cattle spent much time during the day lying in their well-bedded shed, while the cattle continually in the pasture stood most of the time under their sunshade. (See Table 200.)

GRAIN RATIONS FOR PASTURE-FED STEERS

Either corn or milo may be fed to cattle on pasture. Their concentrated form makes them combine well with a bulky roughage such as pasture. Ground ear corn is too bulky and unpalatable for cattle fed on nonlegume

Table 200

A Comparison of Night and Day Grazing for Full-Fed Cattle[a]

	Confined in Drylot	On Pasture at Night	On Pasture During Day	On Pasture Continually
Average daily gain (lb)	1.74	2.20	1.99	2.00
Average daily ration (lb)				
Shelled corn	12.3	13.6	12.4	13.7
Soybean meal	1.0	1.0	1.0	1.0
Clover hay	4.7	1.9	1.8	0.2[b]
Pasture, acres per steer	—	0.25	0.25	0.50
Feed per cwt gain (lb)				
Shelled corn	706	617	625	686
Soybean meal	55	45	49	49
Clover hay	272	86	93	11
Pasture days	—	11	12	24
Return per head above feed costs ($)	14.00	25.72	20.58	16.89

[a] Illinois Mimeographed Report.

[b] Hay fed in drylot 6 days before marketing.

pastures but is better than shelled corn for cattle on sweet clover or alfalfa because the cob particles tend to reduce the prevalence of scours and bloat (see Table 201).

Owing to the relatively high percentage of protein in young grasses or legumes, the need for a nitrogenous concentrate is not so urgent with pasture-fed cattle as it is with those in drylot. Except for slightly higher daily gains resulting from a larger consumption of the more palatable ration, there is no important advantage in feeding protein supplements to full-fed steers that have access to abundant high-quality green pasture. However, bluegrass, bromegrass, and all legume forages contain a much lower percentage of protein in midsummer than they do in the spring. Moreover, they are less palatable and consequently are consumed in smaller amounts. For these reasons a protein concentrate should always be fed with the grain on grass pasture beginning about July 1, and on legume pastures about August 1, unless unusually favorable weather conditions delay the maturing of the forage beyond the usual time.

The need for a protein concentrate during the summer and fall by cattle fed on bluegrass pasture is well shown by Table 202. Many feeders feed protein supplement from the start of the grazing period to make certain that they do not wait until the pasture is too dry before starting it. In

Table 201

Shelled versus Ground Ear Corn for Cattle Full-Fed on Pasture

| | Mixed Pasture[a] | | | Bluegrass[b] | | Alfalfa[b] | |
	Shelled Corn	Ground Shelled Corn	Ground Ear Corn	Shelled Corn	Ground Ear Corn	Shelled Corn	Ground Ear Corn
Average weight (lb)	660	667	667	703	709	705	708
Days fed	190	190	190	133	133	133	133
Average daily gain (lb)	2.47	2.40	2.19	2.38	2.08	2.31	2.32
Average daily ration (lb)							
Shelled corn	16.6	15.9	15.3[c]	15.0	13.2[c]	12.8	11.4[c]
Protein concentrate	—	—	—	1.0	1.0	1.0	1.0
Alfalfa hay	2.2	2.4	2.3	—	—	—	—
Feed per cwt gain (lb)							
Shelled corn	673	663	654[c]	599	635[c]	554	491[c]
Protein concentrate	—	—	—	42	50	43	43
Alfalfa hay	90	98	105	—	—	—	—
Gain of hogs per steer (lb)	—	—	—	63	23	66	38

[a] Nebraska Mimeographed Circular 140.
[b] Illinois Mimeographed Report.
[c] Shelled basis.

Table 202

Value of a Protein Supplement at Different Stages of the Grazing Period for Steers Fed on Bluegrass Pasture

| | First Period May 5 to June 30 | | Second Period July 1 to October 20 | | Total Period May 5 to October 20 | |
	Average Daily Gain	Feed Per Cwt Gain	Average Daily Gain	Feed Per Cwt Gain	Average Daily Gain	Feed Per Cwt Gain
Missouri[a] (average 4 trials)						
Corn alone	2.26	613	1.76	1,581	1.97	940
Corn and cottonseed meal	2.27	612	2.05	1,332	2.12	833
Missouri (average 3 trials)						
Corn only first 56 days / Corn 8:cottonseed meal 1, last 112 days	2.46	330	2.33	771	2.37	637
Corn 8:cottonseed meal 1, 168 days	2.44	354	2.30	806	2.36	671
Corn only first 56 days / Corn 12:cottonseed meal 1, last 112 days	2.47	330	2.23	826	2.31	667
Corn 12:cottonseed meal 1, 168 days	2.64	324	2.38	776	2.47	600

[a] First period, May 1 to July 31; second period, August 1 to December 1.

such cases, and especially if the pasture is legume-grass or straight legume, feeders may widen the usual 10–12:1 ratio of ground ear corn to protein concentrate to 12–15:1 during at least the first 2 months of the grazing season.

The success achieved by including several additives with the supplements fed to finishing cattle being fed in drylot has tempted many feeders who feed grain on grass to use such fortified supplements. The rate of gain is almost always increased to some extent, but when the cost of the fortified supplement is added to the cost of grain also being fed, the cost per pound of gain invariably comes out higher, rather than lower as one might expect from an increase of gain. Stilbestrol and phosphorus are apparently the only two additives that can be relied on to reduce cost of gain in steers being fed on pasture, assuming that protein is adequate, of course.

TURNING HALF-FINISHED CATTLE ONTO PASTURE

Cattle that have received a fairly liberal ration during the winter, so that they are half-finished or better at the opening of the grazing season, should be finished in a drylot rather than on pasture. To turn such animals onto pasture will probably result in a marked decrease in grain consumption for the first 3 or 4 weeks while they are becoming accustomed to pasture. This decreased grain consumption, together with the "washy" character of the spring pasture, results in very moderate gains for the first 4 to

Table 203
Drylot versus Pasture for Finishing Half-Finished Cattle[a]

Place Finished	First Trial		Second Trial		Third Trial		Average 3 Trials	
	Drylot	Pasture	Drylot	Pasture	Drylot	Pasture	Drylot	Pasture
Weight in spring (lb)	880	880	870	875	827	809	859	855
Average daily gain (lb)								
First month	2.06	0.66	2.16	1.61	2.37	1.34	2.20	1.20
Second month	1.90	2.05	1.58	1.41	1.35	1.66	1.61	1.71
Third month	1.01	1.00	1.38	1.56	1.35	1.38	1.25	1.31
Total—90 days	1.65	1.24	1.71	1.53	1.69	1.46	1.68	1.41
Grain eaten per day (lb)								
First 10 days	14.90	8.22	14.55	10.27	16.00	14.50	15.15	11.00
Total—90 days	17.04	11.84	16.17	14.46	16.74	15.92	16.65	14.07
Grain per pound gain (lb)								
First month	7.60	13.94	7.06	6.62	6.96	11.02	7.21	10.53
Total—90 days	10.28	9.55	9.46	9.45	9.89	10.88	9.88	9.96

[a] Indiana Bulletin 142.

6 weeks. In fact, it is not unusual for cattle under such conditions to show an actual weight loss when weighed 2 or 3 weeks after leaving the drylot.

If it is necessary to put half-finished cattle on pasture because of a scarcity of dry roughage, they should be turned onto pasture early in the spring when the grass first starts to grow, and their dry roughage should be continued until the pasture is fairly mature. In this way the cattle become accustomed to the change in their ration very gradually and are unlikely to go off feed. (See Table 203.)

GREEN CHOP IN DRYLOT FINISHING RATIONS

Modern methods of harvesting forages for silages are being adapted to daily cutting of forage for feeding in drylot in the fresh state as green chop. This system of utilizing legume-grass or legume forage is especially suited to operators of feedyards and to large farm feedlots for the following reasons:

1. Carrying capacity of pasture is doubled because of more complete utilization of the crop.
2. The quality of the forage is higher by at least one government grade, over the same forage harvested and fed as hay, if harvested at the proper stage.
3. The forage is more uniform in quality.
4. Ration changes and adjustments can be made more accurately and conveniently.
5. Less fence, shade, and watering equipment are required.
6. The cattle expend less energy because walking about the pasture is eliminated.
7. Fewer cases of bloat occur, although green chop is not a sure cure for bloat.
8. Flies are easier to control in drylot than on pasture.
9. The process of checking cattle for disease, bloat, and other problems is simplified.
10. Weeds and insect damage to pastures are easier to control.
11. Puddling and packing of pasture soil is lessened.
12. Haircoats are in better condition in cattle fed in drylot.

Problems arise in this method of feeding, some of which are serious enough to cause some feeders who have tried green chop to abandon the plan. The disadvantages of green chop are:

1. Prolonged wet weather makes harvesting difficult or impossible.
2. Labor requirements are increased, especially in smaller operations.

One study found that 0.4 man-hour of labor was required per ton of fresh forage harvested.

3. Equipment costs are high; machinery costs amounted to $1.75 per ton in one study. If silage-making equipment is already on hand, however, this extra item of expense is small.

4. Large numbers must be on feed—at least 100 head or more—to keep per-head costs of labor and machinery low.

5. Scouring is prevalent with this type of feed.

Table 204

Comparison of Pasture and Green Chop When Steers Were Grazed or Fed Green Chop Without Grain from May to September (120 Days) and Full-Fed in Drylot for 75 Days (Average of 2 Years)[a]

	Grazing		Green Chop	
	No Supplement	Supplement	No Supplement	Supplement
Number of steers	16	15	16	15
Average initial weight (lb)	778	782	807	801
Average final weight (lb)	1,183	1,182	1,186	1,176
Average daily gain (lb)	1.86	1.84	1.95	1.97
Average feed consumed per steer daily				
Pasture, brome-alfalfa (acres)	0.85	0.85	—	—
Clippings, brome-alfalfa (lb)[b]	—	—	58.2	55.0
Ground ear corn (lb)	8.4	7.8	7.7	7.1
Hay, brome-alfalfa (lb)[c]	0.4	0.4	0.9	0.9
Supplement (lb)	0.6	1.8	0.4	1.8
Feed cost per cwt gain ($)	20.00	22.10	19.70	21.40
Selling price per cwt ($)	23.16	23.45	23.36	23.10
Margin per steer over feed costs ($)	22.13	16.79	25.25	14.83
Estimated beef per acre (lb)	192	161	283	256
			1.5 tons hay	1.5 tons hay
Returns per acre of pasture ($)	41.00		46.42	

[a] Iowa Miscellaneous Publication AH 693.

[b] Daily consumption during pasture season was 79 pounds per steer in the first year and 90 pounds in the second.

[c] Daily consumption during finishing period was 3 pounds per steer in both years.

6. Forage quality of roughage consumed may become poor if the crop becomes fibrous because of advanced maturity. Steers on pastures selectively graze only the new growth, thereby consuming a more nutritious roughage.

7. A system of feeding green chop is confining, in that fresh forage must be chopped daily.

Results of the Iowa experiments reported in Tables 204 and 205 show what may be expected from steers being managed under two different programs for utilizing pasture forage. In general, the contribution to the ration by pasture forage was about twice as great when it was harvested as green chop instead of pasture, in a steer program in which no grain was fed for 120 days followed by a 75-day full feed of grain in drylot for both lots. (See Table 204.) Profits per steer were no greater, but more steers could be fed per acre of pasture.

In a comparison between steers full-fed on pastures and steers fed limited grain plus pasture forage harvested as green chop, the pasture made a substantial contribution toward total feed needs when harvested and fed as green chop. Profits per steer fed were noticeably higher than when the pasture was grazed, especially if supplement was not fed. (See Table 205.)

FINISHING CATTLE ON GRASS ALONE

Comparatively few cattle are grazed without grain in the Corn Belt with the expectation of selling them for slaughter in the fall. There is grassland in many parts of the country, notably the Flint Hills section of Kansas and Oklahoma, where older steers will reach choice grade on grass. Other cattle programs are proving more profitable, so that this system of finishing cattle is passing out of the picture in the United States, but it is quite common in Australia, New Zealand, and Argentina. The most common practice where no grain is fed on pasture is to buy heavier cattle in the spring or at the start of the grass season and carry them to fall, or the end of the best grazing, then sell them as long 2- or even 3-year-old steers ready for slaughter. A few long-aged Mexican and Canadian steers are exported to the United States just for this purpose each year.

Choice-quality 2-year-old steers often attain enough finish on excellent pasture alone to grade low to average choice as slaughter cattle. However, most feeders who graze steers of this age and weight prefer to feed for 30 to 50 days in drylot to improve the condition sufficiently for the steers to grade still higher. Ordinarily the 10 to 15 bushels of corn or sorghum required are more than paid for by the higher selling price received. Choice

Table 205

Comparison of Pasture and Green Chop When Steers Are Full-Fed on Pasture or Limited-Fed Grain and Green Chop in Drylot for 135 Days (Average of 2 Years)[a]

| | Grazing plus Full-Feed of Corn | | Green Chop plus Limited Corn[b] | |
	No Supplement	Supplement	No Supplement	Supplement
Number of steers (2 yr)	16	16	16	16
Average initial weight (lb)	814	820	820	816
Average final weight (lb)	1,123	1,162	1,125	1,130
Average daily gain (lb)	2.32	2.56	2.33	2.39
Average daily feed consumed				
Pasture, brome-alfalfa (acres)	0.25	0.25	—	—
Clippings, brome-alfalfa (lb)	—	—	51.6	49.2
Ground ear corn (lb)	20.8	20.1	13.6	13.4
Hay, brome-alfalfa (lb)	—	—	0.1	0.1
Supplement (lb)	—	2.0	—	2.0
Feed cost per cwt gain ($)	23.40	23.60	20.00	22.70
Selling price per cwt ($)	24.18	24.88	24.58	24.18
Margin per steer over feed costs ($)	21.29	28.41	35.30	25.52
Estimated beef per acre (lb)	14	6	398	310

[a] Iowa Miscellaneous Publication AH 693.

[b] Ground ear corn—5 pounds per steer daily during first month, 10 pounds second month, 15 pounds third month, and a full feed of ground ear corn thereafter.

steers such as the fleshy 2-year-olds in Figure 89 are ideal for this short feeding program.

LIMITED GRAIN RATIONS FOR PASTURE-FED CATTLE

Feeding a limited grain ration to cattle on pasture is unusual. Unless they are moved from the pasture to the drylot each day for feeding, some of the animals may not be at the bunks when the grain is fed and therefore will not get any. The more common practice is to feed no grain at all during the first 2 or 3 months while the pasture is palatable and nutritious, and to supply a full feed during the late summer and fall when grazing conditions are less favorable. This plan has the advantage of utilizing the pasture when it is at its best and of supplying abundant digestible nutrients

FIG. 89. Choice stocker steers on grass without grain, in Iowa. When grazed in areas other than the range areas, such steers as these are placed in the feedlot for 30 to 50 days before going to market. (American Hereford Association.)

during the period when cattle on even a limited feed of grain will eat relatively little of the dry, unpalatable pasture. Another advantage of deferring the feeding of grain to the last half of the summer is that much labor is saved.

It has already been said that the feeding of some form of dry feed on legume pastures reduces the tendency to bloat. Ground ear corn or oats, fed in limited amounts, is quite effective for this purpose and at the same time serves as an energy source and as a moisture-absorbing agent in the digestive tract. This absorbent effect slows the passage of the feeds consumed, which conceivably should improve digestion and absorption of nutrients. Thus if the pasture consists largely of succulent legumes, the limited feeding of ground ear corn or oats, which contain considerable fiber, seems justified, although admittedly this recommendation is not based on extensive research.

AREA OF PASTURE PER STEER

Pastures vary so much in productivity that no definite statement can be made as to their carrying capacities. When cattle receive a full feed of grain, only one-third to one-half as much pasture is required as when the cattle are finished on pasture alone. Some feeders use a minimum of pasture, feeding as many as 80 yearling steers and at least an equal number of hogs on 20 acres. A pasture so heavily stocked, however, becomes so thickly covered with manure that the grass is made unpalatable. This is especially likely in pastures adjoining feeding yards that are used year after year. The heavy growth of grass under such conditions is not always proof that the cattle have an adequate amount of pasture. The same cattle turned onto an equal area, similar as far as growth of forage is concerned but free of objectionable odors, might eat the grass down to the roots within 2 or 3 weeks. As previously mentioned, green-chopping and feeding the forage in drylot overcomes this objection.

Legumes and legume-grass mixtures grown in a 3- to 5-year crop rotation are valued largely for the nitrogen and organic matter they add to the soil, and they should not, if pastured, be grazed too closely or they will be of little benefit to the grain crops that follow. It is much better to stock them with only enough cattle to eat 50 to 70 percent of the forage, leaving the remainder to be returned to the soil. Even permanent pastures such as bluegrass and bromegrass do not remain productive if they are grazed closely year after year. Under such unfavorable conditions the stand becomes thin and the bare spots are gradually taken over by weeds. A weedy permanent pasture is almost unmistakable evidence of

Table 206

Recommended Rate of Stocking Beef Cattle Pastures in Nonrange Areas (Acres per Head)[a]

| | Permanent Grasses Unfertilized | | Legumes or Mixed Grasses and Legumes in Rotation with Grain Crops | |
	Yearlings	2-Year-Olds	Yearlings	2-Year-Olds
Pasture only				
Entire season	1.5	2.25	1.0	1.5
Until Aug. 1, then removed				
to drylot	1.0	1.50	0.66	1.0
Full-fed on pasture				
Entire season	0.5	0.75	0.33	0.5
After July 1	1.0	1.50	0.66	1.0

[a] For soils of average fertility that will produce 75 to 90 bushels of corn or 2.5 tons of hay in an average season.

overgrazing It is believed that the stocking rates recommended in Table 206 will prevent overgrazing except during extremely dry years. As a result, permanent pastures become better with the passing of each year and rotation pastures add large amounts of both nitrogen and humus to the soil. Stocking rates for the range area vary so much with rainfall and type of vegetation that local specialists should be consulted as to recommended rates.

CATTLE GAINS PER ACRE

Gains secured from improved pastures during recent years by steers without grain have been so high as to disprove the statement that the level lands of the midwestern farm belt are too valuable to be seeded to pastures for beef cattle. (See Table 207.) It is easily shown that an acre of pasture, which puts 300 pounds of gain on a yearling or 2-year-old steer each summer, will return as much net profit per year as an acre of 80-bushel corn.

The most surprising fact disclosed by the pasture experiments summarized in Table 207 is that the productivity of pastures seems to bear little relation to the natural soil fertility. For example, pastures established on thin, eroded soil at the Dixon Springs, Illinois, station in the Ozarks

Table 207

Gains Secured per Acre of Improved Pasture When No Concentrates Are Fed[a]

Location	Kind of Forage	Period Grazed	Days Grazed	Gains per Acre (lb)
Central Missouri	Bluegrass	May 6–Sept. 27	144	216
(Columbia)	Bluegrass-lespedeza	May 6–Sept. 27	144	279
Northwest Missouri	Wheat-lespedeza	Apr. 21–June 26		
(Lathrop)		July 27–Aug. 31	91	313
	Bluegrass-sweet clover	Apr. 27–Oct. 4	160	315
	Bluegrass-ladino	Apr. 27–Oct. 4	160	359
Northwest Indiana	Alfalfa-timothy	May 5–Oct. 14	162	264
(Upland)	Bluegrass	May 5–Oct. 14	162	198
Central Illinois (Urbana)	Bluegrass			160
	Bluegrass-ladino	Apr. 15–Sept. 20	158	329
	Bromegrass-ladino	Apr. 15–Sept. 20	158	304
	Bromegrass-alfalfa	Apr. 13–Nov. 1	202	342
	Alfalfa	May 6–Sept. 23	140	383
	Sweet clover, 2nd year	May 6–Aug. 2	88	220
	Haas mixture, 2nd year	Apr. 30–Nov. 1	185	416
Southern Illinois	Basic mixture[b]	Apr. 20–Nov 25	219	277
(Dixon Springs)	Basic mixture + alta fescue	Apr. 20–Nov. 24	218	325
	Basic mixture + bromegrass	Apr. 20–Nov. 24	218	364
	Basic mixture + orchard grass	Apr. 20–Nov. 24	218	337
	Basic mixture + bluegrass	Apr. 20–Nov. 24	218	296
	Ladino, timothy, red clover, alta fescue		—	564[c]
Florida (Everglades)	St. Augustine grass		365	1,004

[a] Based on data in mimeographed reports of state agricultural experiment stations.
[b] Basic mixture (pounds per acre). ladino 1, timothy 4, redtop 3, alfalfa 4, lespedeza 5.
[c] Three-year average; combined gains of cattle and sheep.

section of the state have been somewhat more productive than those at Urbana in central Illinois on level, brown silt loam. Part of the difference may be explained by the longer grazing season, but the main difference seems to be that the legumes encounter less competition from the grasses on the poor soil and therefore constitute a larger percentage of the available forage.

Much heavier applications of fertilizers, principally limestone and phosphate, are required on the poor soil in preparing it for seeding. Also more labor is usually needed to clear the land of brush, to fill in gullies, and to construct brush dams to control erosion.

BETTER METHODS OF UTILIZING PASTURE

There are many different methods of utilizing pastures in the finishing of cattle, each of which probably has more or less merit for a given situation.

FIG. 90. Good-grade stockers on grass without grain. Such steers are ideal for a heavy feed of corn silage followed by a short feed of grain. (American Hereford Association.)

Farmers who have a considerable area of pasture land will do well to study critically their present use of pastures, because pastures can easily be a liability rather than an asset in a cattle-finishing enterpise. In fact, a cattleman of much experience and a close observer of other feeders' methods once said that "inexperienced feeders have lost more money trying to use pastures to save a little feed than pastures have ever made for those few feeders who know how to use them wisely." This is an excellent evaluation and it emphasizes the fact that summer-feeding cattle on pasture is in many respects a more difficult task than winter-feeding in the drylot, so far as profits are concerned.

Apparently one common error in pasture management is an attempt to use the available pasture in a cattle-feeding enterprise that would succeed as well or even better without it, instead of adopting a plan of feeding that seems to offer the best opportunity to use the pasture efficiently. This is merely another instance of the necessity of fitting the cattle-feeding program to the available feed supply.

Cattle that are full-fed grain on pasture throughout the spring and summer use too little grass to make this feeding plan suitable for utilizing comparatively large areas of pasture. If the pasture is stocked heavily enough to consume most of the grass—that is, 3 to 4 head per acre—the number of cattle required may be greater than the supply of corn that is available for satisfactory finishing.

Pasture is at its best in spring and early summer and should be utilized at this time if the greatest returns are to be realized from a given area. To stock a pasture in the spring with only the number of cattle that it will carry through the entire season under average weather conditions is a great waste of forage through failure to harvest the pasture when it is most valuable. Too heavy stocking at any time during the grazing season results in little or no gain, since most of the available forage is used for maintenance and little for production. Consequently, it is logical to stock the pasture with the number of cattle that will ensure the consumption of the forage at about the rate it grows at the time of year. This means stocking the field fairly heavily at the beginning, perhaps 3 or more head to the acre, and removing animals from time to time to supplementary pastures or to the drylot where they are given a full feed of grain.

This grazing system has been satisfactorily used at the Illinois station, as shown by the fact that the gains from pasture alone averaged 295 pounds per acre over a 5-year period. Harvesting a portion of the abundant early spring growth in the form of hay or silage is one practical means of balancing forage supply and cattle numbers.

Table 208
Methods of Utilizing an Alfalfa Crop Through Beef Steers[a]

	Pasture Lots		Drylot	
	5-Day Rotation Grazing	Strip Grazing	Hay	Soilage
Number of steers	9	9	10	10
Days on test	132	132	132	132
Initial weight (lb)	550	565	574	532
Average daily gain (lb)	1.79	1.64	1.45	1.73
Feed consumption per head per day (lb)				
As fed				
Alfalfa	—	—	21.4	69.0
Oat hay	4.7	4.9	—	3.9
Dry basis				
Alfalfa	—	—	18.9	14.1
Oat hay	4.2	4.4	—	3.5
Dry matter per cwt gain (lb)	—	—	1,303	1,017
Beef production per acre (lb)	689	739	856	1,080
Percentage of soiling	64	68	79	100
Percentage of rotational grazing	100	107	124	157

[a] *Journal of Animal Science*, 15:64.

STRIP GRAZING AND SOILAGE

California studies reported in Table 208 give some interesting comparative data on different methods of harvesting an alfalfa crop through beef steers. The strip grazing was controlled by an electric wire to provide a 2- or 3-foot strip across the field once or twice daily. Although oat hay was fed to minimize the danger of bloat, one steer from the soilage or green chop lot was lost from this cause. Despite the one death in the soilage lot, however, bloat was a greater problem in the pasture lots. Strip grazing of this sort is quite commonly practiced in Great Britain and northern Europe where pasture utilization is highly developed.

THE KANSAS PLAN OF UTILIZING PASTURE

The Kansas station has conducted extensive experiments on pasture utilization in finishing yearling cattle for market, with particular reference to the bluestem pastures of that state. Although this grass differs somewhat from rotation pasture, results obtained at the Missouri, Nebraska, and Illinois stations, where the same grazing methods have been used, are so similar to those obtained in Kansas as to leave little doubt that the grazing plan recommended by the Kansas station is well suited to all pasture forages that produce their heaviest growth in the spring and early summer.

The object of the Kansas experiments was to determine the best method of using pasture in the finishing of beef calves purchased in the fall and marketed approximately a year later as fed yearlings. Some of the different methods of feeding and grazing used are shown in Table 209. The following plan almost always proved most profitable:

1. Steer calves were wintered sufficiently well to gain 1.3 to 1.5 pounds a day. A ration consisting of 16 to 20 pounds of Atlas sorgo silage, 2 pounds of legume hay, 1 pound of protein concentrate, and 4 to 5 pounds of shelled corn per head daily was found to be excellent for this purpose. Omitting the shelled corn from the winter ration was advisable with heifer calves but not with steer calves, as steers wintered only on sorghum silage and hay lacked sufficient finish to sell satisfactorily the next fall.

2. Steers were grazed on pasture without grain for approximately 90 days, or from May 1 to August 1. Although these were the limiting dates in all of the Kansas experiments, results obtained at the other stations indicate that condition of the pasture should be considered in deciding when to begin and end the grazing period. Nothing is gained by leaving

Table 209

The Kansas Method of Utilizing Pasture in Finishing Yearling Cattle[a]

Comparison of Standard Plan with Variations (3-year averages)	Winter Gain 136 Days	Pasture Gain 90 Days	Gain While Full- Fed Grain	Total Gain	Margin per Head	Corn Fed per Head (bu)
Standard plan	258	98	256	612	$43.54	37
No corn fed in winter	183	123	263	569	40.54	26
Standard plan	270	91	285	646	11.76	39
Full-fed on pasture after May 1	268	—	283	551	.69	43
Full-fed on pasture after Aug. 1	269	93	271	633	4.17	39
Standard plan	240	93	259	592	4.66	37
Fed 60 days on pasture after						
Aug. 1; last 40 days in drylot	236	93	250	579	3.01	37
Standard plan	231	90	265	586	10.17	38
Winter grain ration discontinued gradually during first						
4 weeks on pasture	229	104	256	589	5.87	39
Winter grain ration continued						
on pasture	219	131	247	597	3.64	46

[a] Kansas Agricultural Experiment Station, Mimeographed Report.

cattle on a pasture that does not furnish enough feed to ensure reasonably good gains.

3. Cattle were full-fed in drylot for 100 days, or from about August 1 to November 15. Various other feeding plans were tried but none was found to be as satisfactory as this. Steers fed on pasture during the last 100 days gained less rapidly than those removed to the drylot, and they sold for a much lower price. Other plans involving various combinations of pasture and drylot feeding were also inferior to the standard method.

The main reasons for the effectiveness of the Kansas plan are: (1) the cattle are wintered in a way that makes them able to use pasture efficiently the following summer, (2) the grass is grazed at the stage of growth when it is most palatable and nutritious, and (3) when the productive pasture season is over the cattle are removed to the drylot and full-fed for the late fall market, which usually is a good time to market grain-fed yearling steers of choice quality.

THE MISSOURI GRAZING PLAN

Many valuable grazing experiments have been conducted at the Missouri station to compare different methods of utilizing the pastures of that state in cattle-finishing programs. The plan found to be most satisfactory differs from the Kansas plan mainly in the kind of pastures used and in the length of the finishing period in drylot. Fall-sown wheat, bluegrass, and lespedeza pastures, usually grazed in that order, have provided good grazing from about April 15 until October 1, or for 5.5 months. As a result of the long grazing season the period of drylot feeding has often been reduced to 60 or 70 days. Usually 550 to 600 pounds of gain are made by each steer, which is fed 20 to 30 bushels of corn, excluding the corn in the silage. About 30 percent of the total gain is made during the winter on silage and legume hay, 45 percent in the summer on pasture,

Table 210

The Missouri Plan of Utilizing Pasture in Producing Finished Yearling and 2-Year-Old Steers[a]

Sold as	Yearlings[b]			2-Year-Olds[c]		
	Length of Period (days)	Gains (lb)	Percentage of Total	Length of Period (days)	Gains (lb)	Percentage of Total Gains
Initial weight (lb)	—	(577)	—		(430)	
Gain (lb): first winter	140	167	29	127	130	15.6
first summer	170	225	39	221	282	33.8
second winter	—	—	—	125	110	13.2
second summer	—	—	—	169	201	24.1
full fed in drylot	84	185	32	51	111	13.3
Total	394	577	100	693	834	100.0
Final weight (lb)		1,154			1,264	
Average corn fed, bu.		25.3			17.2	
Weight of live cattle produced per bushel of corn fed (lb)[d]		45			73	

[a] Missouri Livestock Feeders' Day Progress Reports 7 and 10.
[b] Average of 2 years.
[d] Average of six lots.
[c] Not including corn in silage.

and 25 percent during the finishing period in drylot. The cost of gains has been remarkably low.

Some of the cattle in the Missouri experiments were carried over the second winter, grazed the following summer, and sold in the fall as 2-year-olds after a feeding period of only 6 or 8 weeks. Approximately 85 percent of the 825 to 850 pounds of total gain was made from roughages and pasture, and only 15 percent from the corn and hay fed in drylot. (See Table 210.)

THE DIXON SPRINGS GRAZING PLAN

The University of Illinois, at the Dixon Springs station located in the hilly, eroded section of the southern part of the state, compared several

Table 211
Methods of Utilizing Pastures at the Dixon Springs Experiment Station in Southern Illinois[a]

| | Sold as Yearlings | | | | Sold as 2-Year-Olds | | | |
| | Full-Fed 100 Days On Pasture | | Sold Directly Off Pasture | | Full-Fed 94 Days on Pasture | | Sold Directly Off Pasture | |
	Total	Per-centage	Total	Per-centage	Total	Per-centage	Total	Per-centage
Winter period, drylot (days)	138	39	140	44	292[b]	40	292[b]	42
Pasture, without grain	119	33	181	56	336[b]	47	405[b]	58
Full-fed on pasture	100	28	—	—	94	13	—	—
Total	357	100	321	100	722	100	697	100
Gain during winter (lb)	127	28	141	46	303[b]	38	299[b]	42
Gain on pasture, no grain	145	32	164	54	285[b]	36	418[b]	58
Gain while full-fed on pasture	184	40	—	—	206	26	—	—
Total gain	456	100	305	100	794	100	717	100
Initial weight	407		399		396		394	
Final weight	863		704		1,190		1,111	
Shrinkage	40		40		42		83	
Sale weight	823		664		1,148		1,028	
Feed per head								
Corn (bu)	23.6		—		26.9		—	
Protein concentrate (lb)	267		141		394		276	
Corn silage (lb)	3,327		3,466		9,047		9,047	
Hay (lb)	397		350		1285		1285	
Pasture (acres)	1.7		1.7		3.3		3.3	

[a] Illinois Agricultural Extension Service Mimeographed Circular.
[b] The 2-year-old steers were carried through two winters and sold at the end of the second summer.

methods of utilizing pastures with homebred calves. One plan that gave excellent results differs from the Kansas plan only with respect to the finishing period during the late summer and fall. In the Dixon Springs plan both yearling and 2-year-old steers were full-fed on late summer and fall pasture, whereas at the Kansas station they were fed in drylot. (See Table 211.) Lespedeza and ladino clover, either alone or in mixed seedings, are excellent forages during August, September, and October, and their value is unutilized if they are not grazed heavily during these months.

Yearling and 2-year-old steers that were marketed directly from pasture in October without being fed any grain were also included in the Dixon Springs tests. The younger cattle usually were sold for return to the country for feeding, but the 2-year-olds almost always were sold for slaughter. The fact that these grass-fat 2-year-old steers sold for as high as $32 per hundred when the top of the market was $35.50, and dressed as high as 59.8 percent, is evidence of the high nutritive value of the improved pastures at the Dixon Springs station.

Chapter 21

HORMONES and MISCELLANEOUS RATION ADDITIVES

Developments in genetics, physiology, ruminant nutrition, and feed preparation techniques have increased the daily gains of cattle by an estimated average of at least 25 percent during the last 20 years. As a result the feeding period is being reduced on the average by 50 to 75 days. More important, from a profit or loss standpoint, is the estimated 20 percent improvement in feed efficiency. Although the price received for the finished cattle may be unaffected, a savings in nonfeed costs due to less time required to carry cattle to the same weight can amount to $5 to $15 per head, depending on size of feedlot. (Chapter 11 explains how the number of cattle fed affects nonfeed costs more than any other item.)

To show that improved feed efficiency is more important than improved rate of gain, consider that when an average feed cost of $20 per hundredweight gain is reduced 20 percent for a yearling steer fed to gain 400 pounds, a $4 reduction in feed cost is made for each 100 pounds of gain. Thus there is a $16 reduction in the feed bill for the steer. Calves fed to gain 600 pounds at a $16 feed cost before taking advantage of recent developments would have their feed costs reduced to $12.80 per hundredweight gain by the 20 percent improvement in feed efficiency, or a saving of $3.20 per hundredweight for a total of $19.20 per head fed. Savings in both categories can, of course, be made on the same steers, but the saving in feed efficiency is in the end the more important.

Several individual additives or management practices may account for half as much improvement as the estimate made above for all of the improvements combined. Unfortunately the benefits from all of the newer improved practices are not cumulative—that is, the individual responses cannot all be added on to each other. In reality this is good, because it enables a feeder to choose, from among the wide variety, a few reliable additives and practices that are also low in cost—for, after all, some of the additives are more expensive than others and all of them must be paid for either out of savings in the nonfeed cost category or out of total feed cost savings.

The effect on the end product, the finished carcass, should not be overlooked. Improvement in performance through selection for more rapid and more efficient gains fortunately nearly always also improves carcass yield or cutability. Most feeder-accepted and government-approved additives at least do not reduce carcass merit. Only a very few improve the selling price of cattle. Thus the profits must come from the two sources mentioned above.

Progressive feeders today would not think of feeding cattle without taking advantage of the stimulus supplied by certain ration additives or treatments discovered through recent research. Some of these new developments are so recent as to require further testing before recommendations can be made concerning their use. Those that have been thoroughly tested are discussed below.

HORMONES

The greatest development in beef cattle feeding and management since the establishment of the need for protein supplementation is the development of practical methods of utilizing synthetic or manufactured hormone-like compounds to enhance growth and feed efficiency. Some of the natural sex hormones such as estradiol, progesterone, and testosterone, singly or in combination, also show promise, as do the goitrogens, but the use of these products needs further study.

Diethylstilbestrol, generally referred to as "stilbestrol" is a synthetic estrogen-like compound that has many of the physiological properties of estrogen, the female sex hormone. It is the most important of the hormone or hormone-like materials currently in use. Its action on the animal body resembles that of the natural sex hormones, both male and female, in several respects. First, it stimulates growth in immature animals. This growth, as growth nearly always is, is accompanied by economically important improved feed conversion, because it results in a corresponding increase in the protein or lean meat content of the animal or carcass, with a corresponding lessening of external fat or bark.

A second effect of diethylstilbestrol is that it causes certain side effects or secondary sex characteristics to develop that may detract from the on-foot evaluation of the live animal. These side effects are more evident in the early stages of the feeding period and may completely disappear by the time the feeder cattle are ready for market. The most common side effects are depressed loins, raised tailheads and rumps, and enlarged teats in both steers and heifers. In addition, heifers are likely to be affected with vaginal prolapse, which may occur in 5 to 10 percent of the heifers receiving stilbestrol. As carcasses do not reflect these side effects, selling on a grade and yield basis is practiced by many feeders to offset any possible discrimination for this reason, and it actually is not a serious problem today.

ORAL USE OF STILBESTROL

The oral use of stilbestrol in cattle rations was approved by the Food and Drug Administration in 1954, with the level of administration set at 10 milligrams daily. Stilbestrol has also been approved for administration by implanting pellets under the skin of the ear. The Iowa station has made extensive tests of the oral method of using stilbestrol, and at present this is the more popular of the two. The Iowa tests and others are summarized in Table 212. The data show that rate of gain was stimulated to a greater degree with high grain rations than with high roughage rations. Stimulation was greater in steers than in heifers and greater in heavy cattle than in light or young cattle. In all of the tests summarized, oral stilbestrol increased gains by an average of 16 percent and reduced feed requirements by an average of 12 percent. Carcass grades were not seriously reduced nor were dressing percentages affected.

Stilbestrol for oral use is mixed with protein supplements by commercial feed manufacturers in such amounts that 1 pound of supplement contains either 5 or 10 milligrams of stilbestrol. Thus to feed 10 milligrams of stilbestrol daily the feeder can feed either 1 pound of the 10-milligram supplement or 2 pounds of the 5-milligram supplement.

IMPLANTING STILBESTROL

The use of ear-implant stilbestrol pellets for cattle was approved in 1955. The pellets most commonly used contain 12 milligrams of stilbestrol each. For a single treatment, 12 to 36 milligrams may be administered by using 1 to 3 of the 12-milligram pellets, depending on the age and size of the animal. A 15-milligram pellet is also available. The pellets are

Table 212
Summary of Stilbestrol Cattle-Feeding Experiments Conducted at Nine Agricultural Experiment Stations[a]

Experiment Station	Kind of Cattle	Type of Ration	Number of Cattle		Average Daily Gain		Percentage Increase	Feed Saved Because of Stilbestrol (%)	Days on Experiment
			Check Lot	Stilbestrol	Check Lot	Stilbestrol			
Iowa	Steers	Grain	8	16	2.13	2.67	25	13	43
Iowa	Steers	Grain	8	24	2.23	2.72	22	12	112
Iowa	Heifers	Grain	8	16	2.03	2.29	13	13	113
Iowa	Steers	Roughage	20	20	1.10	1.21	10	10	127
Iowa	Steers	½ and ½	8	8	1.71	1.99	16	11	243
Iowa	Steers	Grain	8	14	2.36	2.53	8	7	120
Iowa	Steers	Roughage	40	40	1.00	1.08	8	8	119
Iowa	Calves	Grain	9	18	2.22	2.45	10	6	224
Colorado	Steers	Grain	9	8	2.30	2.90	26	21	84
Kansas	Calves	Roughage	20	20	1.91	1.90	0	0	140
Kansas	Calves	Roughage	5	5	1.72	1.82	6	6	140
Michigan	Steers	Grain	14	14	2.30	2.60	13	20	98
Nebraska	Steers	Grain	15	15	2.02	2.40	19	12	112
Ohio	Steers	Grain	21	21	2.17	2.47	14	13	84
Indiana	Steers	Grain	10	10	2.33	2.64	13	11	123
Indiana	Steers	Grain	9	9	2.71	3.30	21	18	98
Indiana	Calves	Grain	9	9	2.37	2.84	20	15	98
Tennessee	Steers	Grain	24	16	1.38	1.77	28	18	98
Texas	Steers	Grain	10	10	2.38	3.03	27	11	120
Average all experiments			255 total	293 total			16	12	

[a] Iowa AH 693.

553

Table 213
Comparisons Between Methods of Using Stilbestrol in Steers[a]

Rations and Stilbestrol Treatment	Number of Trials	Average Days on Trial	Total Number of Treated Animals	Average Daily Gains			Increase in Feed Efficiency (%)	Carcass Grades[b]	
				Control (lb)	Treated (lb)	Increase (%)		Control	Treated
Wintering rations									
Oral	9	117	313	1.40	1.50	7.1	5.2	—	—
Implant	10	130	212	1.25	1.57	25.6	7.0	—	—
Pasture only									
Oral	7	129	103	1.35	1.46	8.1	—	—	—
Implant	35	122	600	1.59	1.87	17.6	—	—	—
Feeding on pasture									
Oral	13	123	137	2.28	2.43	6.6	2.1	—	—
Implant	25	123	276	2.26	2.65	17.3	8.1	—	—
Finishing rations									
Oral	92	124	1,357	2.30	2.63	14.3	9.7	6.6	6.5
Implant	63	144	919	2.19	2.50	14.2	10.3	6.6	6.1

[b] South Dakota State University, Animal Husbandry Department Pamphlet No. 2, 1959.

[a] Carcass grade score based on: low prime, 10; high choice, 9; average choice, 8; low choice, 7; high good, 6; average good, 5.

implanted under the skin on the backside of the ear near the head to avoid any chance of residual or unabsorbed stilbestrol being left in the carcass proper. (In the usual packinghouse slaughter procedure the ear is left on the hide.)

South Dakota workers summarized the steer trials conducted at their station and others to compare response to oral and implanted stilbestrol. They divided the trials on the basis of nutritional plane to determine if there is any relationship between energy level and response to stilbestrol. Their summary is reported in Table 213. In direct drylot finishing comparisons between feeding 10 milligrams of stilbestrol daily and implanting at either 24- or 36-milligram levels, there was a slight advantage in feed efficiency but none in rate of gain for implanting. Carcass grade was reduced slightly, one-fifteenth of a federal grade, by implanting. When finishing rations were fed on pasture, response to implanting was considerably greater. Plant estrogens found in legumes being consumed on pasture undoubtedly account for the smaller response to oral stilbestrol in all the direct comparisons shown in the table. The response to stilbestrol was usually greater when finishing or higher-energy rations were fed.

The two methods of administering hormones approved by the Food and Drug Administration are approved for use as a single source of the hormone. In other words, both methods cannot be used on the same animal at the same time, presumably because of concern that there may be risk to consumers because of possible residual hormone in the beef carcass tissue. Calves or yearlings implanted 3 to 5 months before being placed on finishing rations in which oral stilbestrol is included presumably would not be affected by the double-use regulation, as the implanted stilbestrol would all, or almost all, have been absorbed before initiation of oral feeding. When oral stilbestrol is used, it must be removed from the ration 48 hours before slaughter.

USE OF STILBESTROL IMPLANTS IN SUCKLING CALVES

For all phases of the beef cattle industry to make the maximum use of stilbestrol its use should be begun early in the calf's life and continued to slaughter weight. This means that the cow-calf man, unless he practices creep-feeding, would use an implant, administered at some convenient time such as when the calves are castrated, dehorned, vaccinated, and branded at about 2 to 3 months of age. After the calves are weaned, either the original owner or a buyer who wishes to place them on a stocker program may reimplant at this time (7 to 8 months of age) or, if a supplement is fed, may feed 10 milligrams of stilbestrol daily during the stocker phase. Finally, the feedlot operator or farmer-feeder who finishes the calf for market also will undoubtedly wish to take advantage of this growth hormone. Thus he, too, either administers the third implant or employs the oral method of administration.

The question arises as to whether use of hormone in one phase of an animal's life affects the response to hormone use in the next or succeeding phases. The Oklahoma station studied the life-cycle use of stilbestrol rather extensively under both range and feedlot conditions. They had the unusual opportunity of having available large numbers of calves to observe from birth to slaughter, and they also were able to use both sexes simultaneously. Their range was tall grass of excellent quality, at two locations in the state, and the cattle were choice grade Herefords representative of a large portion of the cattle population in the United States.

To obtain data on the magnitude of response to stilbestrol implants, twenty different experiments were conducted, using either a 12- or 15-milli-gram implant in both steers and heifers. In eight of the trials, direct comparisons were made between steers and heifers, showing what may be expected in the way of weaning weight increases from implanting 3-to-4-month-old calves with one 12-milligram pellet of stilbestrol. Note in Table 214 that implanting increased the gains in heifers 11 pounds more than in steers. The investigators mentioned that in two of the eight trials im-planted steers averaged only 1 pound and a minus 1 pound gain while the implanted heifers averaged 34 and 41 pounds of gain in the same trials. For the other six trials, responses between the sexes were more nearly com-parable but still in favor of the heifers. Heifers showed more side effects, notably depressed loins, high tailheads, and increased udder development. These did not affect feeder grade; in fact, the implanted calves graded average choice while the controls graded low choice.

Table 214
Effect of Stilbestrol Implants in Steer and Heifer Calves (Eight Direct Comparisons)[a]

	Steers		Heifers	
	Control	Implanted	Control	Implanted
Number of calves	80	74	86	85
Average gain per calf (lb)	244	262	221	250
Increased gain (lb)		18		29

[a] Oklahoma Agricultural Experiment Station Bulletin B-620, 1964.

Table 215
Effect of Implanting versus Feeding Stilbestrol on Creep-Fed Suckling Spring Calves[a]

	Control	Feed[b]	Implant[c]
Number of calves	20	17	20
Age of calves at start (days)	96	94	92
Average gain per calf (lb)	289	287	324
Increased gain due to stilbestrol (lb)		−2	35

[a] Oklahoma Agricultural Experiment Station Bulletin B-620, 1964.
[b] Five milligrams per head daily.
[c] Implanted with 12 milligrams at start of test, reimplanted with 12 milligrams 75 days later.

Implantation of spring and fall calves was compared. The response to implantation was 28 pounds for the spring calves and 20 pounds for the fall calves, with both sexes represented about equally. Apparently the better feed conditions available to the calves and their dams under spring conditions were responsible for the additional 8 pounds of gain.

The Oklahoma workers also compared two levels of implantation in six comparisons. A 12-milligram implant improved weaning weight by 16 pounds, compared with 10 pounds for a 6-milligram implant. The side effects were more apparent in the calves receiving the 12-milligram pellet.

Table 215 summarizes the results of a trial in which implantation and oral use of stilbestrol for calves were compared. Note that the implanted calves were implanted twice, at about 3 months of age and 75 days later. The response to implantation was greater than in the other tests reported above, but both a higher-energy ration was available to the calves and twice as much hormone was implanted. Failure of oral hormone to produce a response in this one test may not be a true indication of what can always be expected. The individual responses varied from weight loss in some cases to substantial gains in others, suggesting very uneven consumption among calves, and this problem alone may be reason enough to select the implantation method for using hormones on suckling calves.

In three trials, spring calves either untreated or implanted with a 12-milligram pellet, were used in a stocker or wintering program after weaning to determine if the previous implant affects subsequent performance as stockers. The implanted calves not only weaned 24 pounds heavier but also gained an additional 9 pounds on the wintering ration, making a total of 33 pounds added weight resulting from the one implant as a suckling calf. There was no relapse or reduction in gain.

Table 216

Subsequent Feedlot Performance of Previously Implanted Fall Calves[a]

	Implant Level		
	0	6 mg	12 mg
Number of calves	16	13	13
Gain while suckling (lb)	241	247	260
Gain from weaning until start of feedlot test (lb)[b]	64	61	63
Feedlot gain, 133 days (lb)	321	335	342
Average daily gain in feedlot (lb)	2.4	2.5	2.6
Total gain (lb)	626	643	665

[a] Oklahoma Agricultural Experiment Station Bulletin B-620, 1964.
[b] All calves (steers) were implanted with 24 milligrams of stilbestrol at beginning of feedlot test.

Calves implanted with 6 or 12 milligrams of stilbestrol as suckling calves were weaned and immediately fed out on finishing rations to slaughter weights of about 850 pounds. All calves, including a control group not implanted as calves, received 24 milligrams of hormone in two pellets at the start of the finishing period. Results are summarized in Table 216. The 6- and 12-milligram implants increased weaning weights by 6 and 19 pounds, and the feedlot gains for the previously implanted calves were increased another 14 and 21 pounds for these two groups, as compared with the controls. This study again demonstrates that implantation of suckling calves does not reduce subsequent performance but rather results in some favorable carryover effect.

STILBESTROL IMPLANTS IN STOCKER PROGRAMS

Stocker calves and yearlings that are wintered on roughage rations or grazed on winter or summer range also respond to implantation with stilbestrol. Using weaner calves that had not been implanted as suckling calves, the Oklahoma workers found in three tirals that 12- and 24-milligram implants after weaning resulted in 11 and 20 pounds of additional stocker gains when calves ran on native cured range and protein supplements. These responses to stilbestrol in stocker calves are lower than in cattle receiving finishing rations but nevertheless are worthwhile and profitable.

Weanling calves not previously implanted were implanted at weaning time with either 12 or 24 milligrams of stilbestrol. They were then wintered on native cured range and protein supplements before being turned onto the same range as yearlings and grazed throughout the summer grazing season. Table 217 shows again the favorable response to implantation during wintering as stockers with increases of 14 and 24 pounds resulting from 12- and 24-milligram implants. Equally important, summer gains were not reduced by the previous implantation. Had these cattle received still another implant at the end of the wintering period, no doubt further responses would have been obtained.

EFFECT OF PREVIOUS IMPLANTATION OF YEARLINGS ON FEEDLOT PERFORMANCE

Yearling steers on grass respond to stilbestrol implantation similarly to stocker calves on winter rations. It has already been mentioned that, in calves, subsequent performance in the feedlot is not adversely but in fact is somewhat favorably influenced by the previous implantation. The Oklahoma workers were interested in both the magnitude of response to implantation by yearlings on grass and their subsequent performance in the feedlot. Results of this study are summarized in Table 218.

The summer implants all produced favorable responses, with 12 milligrams apparently being adequate. Unlike the calves, the previously unimplanted yearlings showed the highest feedlot gains. On the other hand, feed efficiency apparently was not depressed by the previous treatment and, if anything, slightly improved. Only half of the steers in each lot

Table 217

Subsequent Performance on Summer Grass of Cattle Implanted with Stilbestrol as Weanling Calves in November (Average of 2 Trials)[a]

	Implant Level		
	0	12 mg	24 mg
Number of calves	39	40	39
Winter gain (lb)	28	42	52
Summer gain (lb)	220	216	222
Total gain (lb)	248	258	274
Response to stilbestrol (lb)		10	26

[a] Oklahoma Agricultural Experiment Station Bulletin B-620, 1964.

Table 218

Effect of Previous Summer Implant on Feedlot Gains and Carcass Grades of Yearling Steers after 157 Days[a]

	Implant Level (Milligrams)			
	0	12	24	36
Number of steers	16	15	14	14
Average gains (lb)				
Summer	146	194	178	178
Response to implantation		48	32	32
Feedlot	436	378	410	426
Average daily feedlot gain	2.78	2.41	2.61	2.71
Total	582	572	588	604
Final feedlot weight (lb)	1,174	1,140	1,159	1,207
Average carcass grade	C−	C−	G+ to C−	G+ to C−
Average dressing percent	61.2	60.9	60.4	60.2
Average live value per cwt	$26.82	$26.57	$26.30	$26.24

[a] Oklahoma Agricultural Experiment Station Bulletin B-620, 1964.

were reimplanted at the start of the finishing period and, as the steers were fed together, the data are not shown separately. Therefore, one can only speculate that there were no doubt larger variations within lots that may influence the lot averages shown. In studies at other stations, when implanted steers off grass are finished on rations containing 10 milligrams of oral stilbestrol daily, the previous implantation apparently does not adversely affect feedlot gain.

EFFECT OF STILBESTROL ON CARCASS YIELD AND GRADE

Extensive tests were conducted at the Ohio station to determine the effect of stilbestrol on carcass yield and grade. Results are summarized in Table 219. In these tests dressing percentage was unaffected but carcass grade was slightly reduced in steers and improved in heifers and bulls. The reduction in grade was caused by a noticeable lack of marbling in steers fed or implanted with stilbestrol, compared with marbling in non-treated steers fed similarly.

Measurements of the area of the cut surface of the rib eye show that there is actually more lean meat in the carcasses of stilbestrol-treated cattle, but so long as marbling plays such an important part in determining carcass grade, stilbestrol-treated cattle may be slightly downgraded. A solution to

Table 219
Summary of Effect of Stilbestrol on Carcass Yield and Grade[a]

Experiment	Number of Animals Control	Stilbestrol	Dressing Percentage Control	Stilbestrol	Carcass Grade[b] Control	Stilbestrol
Bulls						
1952–1953	10	5	58.8	59.6	2.01	1.64
1953–1954	10	10	61.2	61.6	2.47	2.03
1954–1955	12	12	61.9	61.3	1.97	1.91
Total	32	27	181.9	182.5	6.45	5.58
Average			60.6	60.8	2.15	1.86
Steers						
1953–1954	10	9	61.5	61.3	1.37	1.62
1954–1955	12	12	61.7	60.6	1.20	1.25
1954–1955	7	7	60.3	62.3	1.61	1.96
		7c		59.3		1.66
	7	7	62.3	60.8	1.43	1.71
		7c		61.5		1.61
	7	7	62.3	62.7	1.34	1.49
		7c		61.4		1.23
1955–1956	7	7	61.5	59.4	1.99	2.36
	7	7	62.3	60.6	1.51	1.94
	7	7	61.9	61.3	1.57	1.89
1955–1956	10	10	61.7	61.6	1.20	1.59
		10c		61.5		1.68
1955–1956	9	8	61.5	62.5	1.39	1.55
	8	8	61.4	61.8	1.60	2.10
1954	11	10	57.7	58.9	2.43	2.45
1955	10	10	58.2	58.6	2.00	2.30
1956	12	11	56.8	56.8	1.97	2.15
Total	124	151	851.1	1,092.9	22.61	32.54
Average			60.8	60.7	1.62	1.81
Heifers						
1955–1956	8	8	60.8	62.5	1.60	1.52
	8	8	60.9	62.3	1.72	1.60
Total	16	16	121.7	124.8	3.32	3.12
Average			60.8	62.4	1.66	1.56
Grand total	172	194	1,154.7	1,400.2	32.38	41.24
Grand average			60.8	60.9	1.70	1.79

[a] Ohio Research Bulletin 802.
[b] Carcass grade factors: high, average, low good = 2.0, 2.4, 2.7; high, average, low choice = 1.0, 1.4, 1.7.
[c] Stilbestrol fed, all others implanted.

this problem is simply to feed to heavier weights with a longer feed if necessary. In other words, stilbestrol should be used to produce somewhat heavier cattle at lower feed costs, rather than to shorten the length of the feeding period.

MGA AND OTHER HORMONE-LIKE MATERIALS

Other hormones and hormone-like materials that show results comparable to those of stilbestrol are hexestrol, dienestrol, estradiol-progesterone combinations, and others. Newer materials are being tested in the experimental feedlots and others are awaiting testing.

A new hormone approved by the Food and Drug Administration in 1967 for oral use in cattle rations is one especially suited for heifers. This synthetic hormone, melengestrol acetate (MGA), suppresses estrus and thus prevents the continual disturbance that results from some heifers being in heat every day when a drove of 40 to 50 or more are fed together. It apparently also is a growth stimulant, as may be seen from the summary presented in Table 220. In trials in which both stilbestrol and MGA were used, it is interesting that MGA produced the greater response. Unfortunately no control groups were fed, hence one cannot tell how much above that of untreated heifers was the response of the two treated groups. The MGA does not produce a response in steers and it may not prove suitable when mixed droves are fed. There are apparently no ill effects from using MGA on heifers which may be put back in the herd for breeding pur-

Table 220
Response to Melengestrol Acetate (MGA)
in Feedlot Heifers (25 Trials)[a]

	Controls	MGA	Percent Improvement
Average daily gain (lb)	2.14	2.38	11.2
Feed per pound gain (lb)	9.24	8.54	7.6
	Stilbestrol	MGA	Percent Improvement
Average daily gain (lb)	2.16	2.31	6.9
Feed per pound gain (lb)	9.68	9.07	6.3

[a] U.S. Department of Agriculture, Federal Extension Service, Animal Science Report No. 4, 1968.

poses, nor does it adversely affect pregnant heifers. The present cost of MGA, at the recommended feeding rate of 0.25 to 0.5 milligram per day, is about 1 to 1.5 cents per day.

It may be safely estimated that at least 90 percent of all feeder cattle now receive hormones or hormone-like materials in one form or another.

VITAMIN A

Preformed or synthetic vitamin A is now recognized as a highly important ration additive. The requirement for this vitamin, both as the natural vitamin and as its precursor, carotene, is discussed in various appropriate chapters, as is the method of administration, whether as a natural ingredient of ration components, as an injection, or as a ration additive. The same practices and principles apply to use of the synthetic forms.

ANTIBIOTICS

The encouraging results obtained with the use of antibiotics in swine and poultry rations led investigators to study the use of these additives in beef cattle rations. Early work with antibiotics for dairy calves affected by digestive disturbances of bacterial origin appeared promising, so it was natural that tests would be made with suckling beef calves. Table 221 shows the favorable results obtained in two such experiments using aureomycin. Many tests in other herds were negative, however, making it appear that disease level and sanitation practices determine whether favorable responses may be expected from antibiotics fed to suckling calves.

Table 221
Effect of Aureomycin on Gains of Suckling Beef Calves[a]

	First Trial		Second Trial	
Aureomycin (daily)	0	24 mg/100 lb body wt.	0	20 mg/calf
Number of calves	7	6	7	8
Average initial weight (lb)	61	60	61	64
Average 80-day weight (lb)	188	203	146	157
Average daily gain (lb)	1.59	1.78	1.46	1.60
Incidence of scouring (days)	38	7	—	—

[a] Indiana Cattle Feeders Report.

Table 222

Effect of Antibiotic and Antibiotic-Hormone Combinations on the Growth and Fattening of Yearling Steers (161 Days)[a]

Treatment	No Antibiotic or Hormone	Aureomycin[b]	Aureomycin and Stilbestrol[c]	Aureomycin and Hexestrol[c]	Terramycin[b]	Terramycin and Stilbestrol	Terramycin and Hexestrol
Steers per lot	12	12	12	12	12	12	12
Average initial weight (lb)	635	630	632	634	631	637	630
Average final weight (lb)	987	1,012	1,050	1,038	1,008	1,020	1,030
Gain per steer (lb)	352	382	418	404	377	383	400
Average daily gain (lb)	2.19	2.38	2.60	2.51	2.34	2.38	2.48
Increase in gain (%)		8.6	18.8	14.7	7.2	8.8	13.6
Average Daily Feed							
Corn silage (lb)	43.2	45.4	46.7	47.4	45.1	46.8	46.1
Supplement (lb)	3.5	3.5	3.5	3.5	3.5	3.5	3.5
Mineral (lb)	0.05	0.05	0.05	0.04	0.04	0.04	0.05
Salt (lb)	0.03	0.03	0.03	0.02	0.02	0.02	0.02
Feed per Pound of Gain							
Corn silage (lb)	19.8	19.1	18.0	18.9	19.3	19.7	18.6
Supplement (lb)	1.6	1.5	1.4	1.4	1.5	1.5	1.4
Feed saved (%)		3.7	9.6	5.1	2.9	1.1	6.5
Feed cost per cwt gain ($)[d]	16.20	15.70	14.80	15.50	15.90	16.20	15.40

[a] Indiana Cattle Feeders Day Report.
[b] 80 milligrams daily.
[c] 10 milligrams daily.
[d] Feed prices: corn silage, $9.80 per ton; Supplement A, $80 per ton; stilbestrol or hexestrol, 50 cents per gram; aureomycin or terramycin, 11 cents per gram; bonemeal, $4.50 per cwt; salt, $1.60 per cwt.

Similarly, the continuous feeding of either aureomycin or terramycin in the roughage rations of stocker cattle has produced variable results. Calves infected with shipping fever after transport from range to feedlot usually will respond to continuous feeding of 60 to 80 milligrams of antibiotic daily.

Results have been even more inconsistent with finishing rations. The Indiana test summarized in Table 222 shows how gains were increased by additions to a finishing ration of either antibiotic alone or a combination of antibiotic and hormones.

Because the response to antibiotics is so inconsistent as far as feeder cattle are concerned, the decision of whether to use this ration additive must be made on an individual farm or feedlot basis.

A test conducted with yearlings by a large commercial feedlot in California, under the supervision of veterinarians, resulted in some interesting data concerning incidence of feedlot disorders, and the effect of the antibiotic aureomycin on their occurrence. The antibiotic was fed in a fortified premix at the level of 500 milligrams daily for the first 28 days, followed by 75 milligrams daily for remainder of the feeding period. Table 223 shows how the number of cases of respiratory diseases and, notably, foot rot were significantly reduced by the use of aureomycin. Average daily gain in the lot receiving the antibiotic was 0.20 pound greater, mainly because of the slower gains of the animals requiring treatment for the various diseases.

Table 223 also provides information on the relative incidence of dis-

Table 223

Effect of Aureomycin on Incidence of Disease in Feedlot Cattle[a]

	Control	Aureomycin
Number of cattle	681	684
Pneumonia	13	4
Shipping fever	26	19
Tracheitis (necrotic laryngitis)	10	1
Listerellosis	1	0
Foot rot	172	2
Photosensitization	1	0
Encephalitis (nonspecific)	0	1
Digestive disturbances	4	2
Not determined	13	7
Liver condemnation at slaughter	23	20

[a] Abstracted from *Veterinary Medicine*, 52:375.

orders normally encountered in feedlot cattle. As the steers used came
from a large ranch in New Mexico and were delivered direct to the feedlot,
there was reduced exposure to viruses normally encountered in marketing
channels, such as sale rings and the like. This undoubtedly accounts for
the low incidence of shipping or "stress" fever observed, even in the control
lot.

TRANQUILIZERS

Both natural and synthetic tranquilizing drugs have been used by the
medical and veterinary professions to reduce the symptoms associated with
hypertension and nervousness. Research workers have explored the possibility
of including these drugs at extremely low levels in the rations fed to farm
animals. Results are extremely variable but unfavorable in enough trials
to question whether to recommend tranquilizers as a feed additive. Individual cattle response varies widely and is unpredictable, causing some cattle
to become overtranquilized to the point where they go down in the trucks
during shipment, to be trampled and bruised.

GOITROGENS

Goitrogens such as thiouracil, thiourea, and methimazole are among
the compounds reported to produce a quieting effect on cattle. Goitrogens
exert a depressing action upon the thyroid gland, resulting in a "hypo-
thyroid" condition. Unfortunately, in most controlled experiments, appetite
was reduced to the extent that subnormal performance resulted. New
findings yet to come may add the compounds with goitrogenic activity
to the growing list of ration additives.

DRIED RUMEN CONTENTS

Veterinarians have used rumen contents, prepared by special drying
procedures that supposedly do not destroy the rumen microorganisms, to
bring cattle back on feed after illness, infection of the digestive tract, or
other rumen disorders. Heavy use of sulfa drugs or antibiotics in the treatment of rumen disorders reduces the normal rumen microflora to abnormally low levels. Thus the administration of specially prepared rumen contents sometimes produces favorable results by more quickly reestablishing

the normal microfloral population. Although some cattle feeders have been adding such materials to the rations of healthy cattle at considerable cost, experimental evidence does not justify this ration additive under normal circumstances.

YEAST

From time to time, yeasts of various forms are highly advertised as being valuable ration additives. Extensive research with yeast as a cattle feed, including relatively recent work by the Iowa and Ohio stations, shows that yeast serves no essential function in cattle rations.

The principal value of yeast as cattle feed lies in its high protein content, which is approximately 50 percent. Consequently, it may replace soybean or cottonseed meal or other high protein supplements on a pound-for-pound basis. However, it is usually sold to be fed in very small quantities of only 0.25 or 0.10 pound per day for the reported purpose of seeding a small quantity of the yeast germs in the paunch, where they will multiply rapidly and aid in fermentation and digestion of the feed. Feeding a small quantity of yeast for this purpose from time to time would perhaps be of value were it not that cattle in normal health, being fed a well balanced ration, already have literally billions of the yeast-type of organisms in their digestive tracts. Consequently the addition of a few more produces no noticeable results.

DRUGS

Feeding tonics or drugs to beef cattle to stimulate appetite and improve performance is occasionally practiced by professional herdsmen and fitters in conditioning cattle for the show ring. The drugs most commonly used for this purpose are Fowler's solution and nux vomica. Fowler's solution is an aqueous solution of potassium arsenite (K_3AsO_3). Nux vomica is a drug, made from the poisonous seed of an Asiatic tree, containing several alkaloids, chiefly strychnine. Both of these drugs are highly toxic, and their stimulating effect is achieved by their powerful reaction on the metabolic systems. Obviously the dosage must be gauged with great care or an acute toxemia is produced that may result in death.

Yearling heifers fed nux vomica and arsenic trioxide at the Washington station showed no greater appetites than the control heifers but made somewhat larger daily gains. When the feeding of arsenic trioxide was discontinued the rate of gain rapidly dropped from 2.25 to 1.51 pounds per day, despite the fact that the daily feed consumption did not change significantly during this period. Such behavior in cattle after stopping the feeding of these highly stimulating drugs has been noted frequently by experienced cattlemen.

It is commonly believed that feeding Fowler's solution, and perhaps nux vomica as well, to cattle in appreciable amounts or over long periods adversely affects their fertility. Consequently the use of such drugs is regarded as dishonest and unethical and is strongly condemned.

ENZYMES

Experiments with enzyme additions to barley, high-moisture corn, dry corn, and milo rations by various experiment stations have failed to demonstrate a consistent response to enzymes in the ration. Both cellulytic and proteolytic enzymes have been used and, while an occasional trial may appear to show a favorable response, it is likely that such an effect occurs merely by chance. At present, the use of enzymes in cattle rations seems unwarranted.

GRUB CONTROL AGENTS AND ANTIHELMINTHICS

Systemic or organic phosphate compounds are effective control agents for cattle grubs and certain kinds of cattle lice. They can be given orally, sprayed, or administered as a back pour-on. Other parasites, notably stomach and intestinal worms, can be effectively controlled with thiabendazole, an orally administered compound.

Cattle feeders generally are aware of the need for controlling the external and internal parasites but they expect to see results in terms of improved performance in their cattle in order to justify the expense and inconvenience caused by the use of these control compounds. The South Dakota station conducted a trial with yearling steers raised in the central part of their state, to obtain data on the use of systemic phosphate, thiabendazole, and aureomycin in finishing rations. The incidence of grubs and internal parasites was not reported, making it difficult to apply their results to all cattle. Knowing the climate prevailing in the area of origin, one would suspect that the incidence of both external and internal parasites was low.

A summary of the South Dakota data is given in Table 224. Application or dosage rates of the compounds were: aureomycin, 70 milligrams daily, in the feed; systemic phosphate (Neguvon), 0.5 ounce per 100 pounds

Table 224

Effect of Neguvon, Aureomycin, and Thiabendazole on Feedlot
Performance of Calves (142 Days)[a]

	Neguvon		Thiabendazole		Aureomycin	
	Control	Treated	Control	Treated	Control	Treated
Number of steers	72	72	72	72	72	72
Average daily gain (lb)	1.86	1.97	1.92	1.91	1.86	1.97
Percent change from control		5.9		−0.5		5.9
Average daily feed (lb)	17.3	17.3	17.2	17.3	17.2	17.4
Percent change from control		0		0.6		1.2
Feed per cwt gain (lb)	933	879	896	916	928	884
Percent change from control		−5.8		2.2		−4.7

[a] South Dakota Cattle Field Day Report, 1965.

body weight, as a back pour-on; thiabendazole, a single oral dose of 20
grams per head. The systemic phosphate and the antibiotic produced significant improvement in rate of gain and feed efficiency while the antihelminthic or "wormer" did not. The latter compound has performed favorably in heavily parasitized cattle originating in environments with higher
rainfall where worm infestation is a major problem.

FLAVOR AND COLOR

Feed intake, or level of feed consumption, has a profound effect on
performance in beef cattle, as gains can only be made from the feed consumed over and above that required for maintenance. Therefore, any addition to the ration that might improve voluntary consumption would presumably be worthwhile, cost considered.

Various natural and synthetic flavor compounds have affected intake
in some species of livestock and their effect on the acceptability of food
by man is well known. As different colors of feed, artificially applied, have
had some influence on intake in other species of farm animals, investigators
were led to explore the possibility of improving intake, hence performance,
of beef cattle by using certain colors and flavors.

Texas workers used a cafeteria system of feeding a medium-energy,
low-palatability ration to calves in two experiments, to determine whether
factors such as color and flavor influenced feed intake. The trough or

Table 225
Influence of Flavor and Color on Acceptability and Intake of Ration[a]

Flavor Test (15 Steers—24 Days) Ration treatment	Control	Sucro	Fenugreek	Sessalom	Total
Daily feed intake (lb)	3.12	4.62	3.49	5.20	16.43
First day's intake (lb)	2.60	7.00	4.10	10.30	24.00
Color Test (15 Steers—21 Days) Color used	Control	Green	Red	Blue	Total
Daily feed intake (lb)	3.97	4.16	3.24	3.14	14.51
First day's intake (lb)	7.90	5.80	4.20	3.10	21.00

[a] Adapted from Texas Agricultural Experiment Station Miscellaneous Publication MP-591, 1962.

feedbox contained four compartments, one for the control ration and the other for control ration plus the flavor or coloring agent being tested. Calves were individually fed by being given daily access to their respective stalls with their compartmented feedboxes, from 7:30 A.M. until 1 P.M. Results are shown in Table 225. Color had no effect on the choices made by the steers, but differences in intake between the flavored rations and the control suggest a selectivity by cattle for flavored feeds in general. The flavor products used contain a variety of organic and synthetic materials. The purpose of showing the data here is only to suggest that flavors may play a future role in cattle feeding, but at present no one flavor or combination appears to be outstanding.

Part V

SPECIALIZED BEEF
CATTLE PROGRAMS

Chapter 22

The BABY-BEEF and
FAT-CALF PROGRAMS

The western and southwestern range areas are usually associated with beef cows, as is mentioned in Chapter 10 in connection with sources of stocker and feeder cattle. Most of the calves and yearlings produced in the range areas are sold to cattle feeders for further development and finishing for the packer and processor. There are also beef cows to be found throughout the United States in ever-increasing numbers, and the Corn Belt and southeastern states are showing especially large increases.

Cattlemen in the nonrange areas who manage and handle their cows with the view of producing stocker and feeder calves and yearlings to sell to cattle feeders, in direct competition with the ranchers in the range states, are not making full use of all the opportunities available to the nonrange area cow-calf man. In the Corn Belt states, and in much of the remainder of the nonrange area for that matter, grain and harvested roughage are produced on every farm. Markets for fed cattle of all weights and grades are nearby and, in the Southeast and South particularly, weather conditions permit fall and winter calving with a minimum of shelter. Winter small grains, ideal pasture for lactating beef cows, are being grown more extensively.

For the cow-calf man who has considerable pasture and varying amounts of harvested roughage and grain, there is a real opportunity to increase the size of his business by feeding out his own home-raised beef calves. Two specialized programs that are proving profitable in such situations are the baby-beef and the fat-calf programs. The principal differences between them are the age and weight at which the slaughter calves are sold and the beef characteristics found in the cow herds. Both programs call for maximum growth rate from birth to market to the extent that the calves "never have a hungry day in their lives," as someone has said.

In general the baby-beef program is more suitable for the heavier grain-growing areas, whereas the fat-calf program is better adapted to the southern fringes of the Corn Belt and the whole of the Southeast, the South, and the Southwest.

DEFINITION OF BABY-BEEF AND SLAUGHTER FAT CALVES

A "baby beef" slaughter animal is usually defined as a beef calf 8 to 15 months old, weighing 650 to 950 pounds, and carrying enough condition and beef conformation to grade at least low choice, but preferably better, in the carcass. Slaughter "fat calves" are calves that weigh 500 to 750 pounds at ages of 6 to 10 months. These calves usually show considerable bloom and condition, yielding carcasses that grade in the upper end of good grade and better. The beef conformation or "beefiness" of the fat-calf carcass is less pronounced than in the baby beef. Such calves are usually sold directly from the cow—that is, without a postweaning feeding period.

OPERATION OF THE BABY-BEEF PROGRAM

It is estimated that over half of the feeder cattle fed out in the grain-growing states are of native origin—that is, they are produced in herds not too distant from the feedlots where they are eventually finished. Because grain and harvested roughage of average to good quality are generally produced either on the farm that maintains the cow herd or nearby, it seems only logical that the cow-calf man in these sections should feed out his own home-raised calves. This partially offsets the low volume of business that is one of the disadvantages of maintaining a cow herd in these areas, because at least twice as many pounds of calf can be sold.

Consumer preference studies show that the cuts from well-finished young animals of good beef breeding are always in strong demand. Animals that yield this kind of carcass are never plentiful but they are especially scarce in summer and early fall, resulting in strong prices for baby beef during these months.

QUALITY OF ANIMALS TO USE

Since baby beeves are sold at a relatively young age and yet are beyond the "milk fat" stage, the cows and bulls used in this program must have superior beef qualities. Although growth rate is important, early maturity or early finishing characteristics must be present along with beefiness and thickness. Heavy-milking cows are essential because the most profitable baby beeves make a large proportion of their total gain while nursing their dams. The extra finish or bloom on the calves at weaning time is not lost, as is the situation in the typical range cow-calf operation, because the calves immediately go onto at least a partial feed of grain, if not on full feed. Only heavy-milking cows of choice or better grade of the major British beef breeds, with pure or nearly pure breeding, should be used if the baby-beef program is to succeed. By all means, cows showing much dairy breeding should be avoided. Such cows are better utilized in a fat-calf program, as will be discussed later. Purebred beef bulls of average size with exceptional thickness, blockiness, and depth should be used and the so-called easy-keeping characteristic should be present.

BEST SEASON FOR CALVING

The smaller farm herds in the Corn Belt are usually fed and sheltered in such a way that early calving can be practiced. This practice is most satisfactory if heavyweight baby beeves are to be sold during the months of highest prices for such slaughter cattle, namely late spring, summer, and early fall.

Table 226 shows the advantage of early calving in a Missouri station test comparing winter calves with spring calves. For these calves to be considered genuine baby beeves they would have had to be fed until the spring months, but the data still show that early calving resulted in faster gains and better utilization of the concentrates fed. Note also that creep-feeding of the nursing calves did not cause the late calves to make up the deficiency in summer gains.

Farther south, where winter small-grain pastures are possible, still earlier and even fall calving is recommended, with the fed calves being sold in early fall the next year at about 10 months of age.

DEVELOPMENT OF HOME-BRED CALVES

Cow-calf men who follow the baby-beef plan have a choice among three recognized methods of developing their calves. First, the calves may be allowed to run with their mothers without grain until weaning time, when they are weaned, removed to the drylot, and started on feed. When handled in this way, home-raised calves are at least comparable to western-bred calves as far as weight, condition, and other qualities are concerned.

The second method of handling home-bred calves is to creep-feed them on a good grain mixture beginning when they are 3 to 4 months old and

Table 226
Effect of Winter or Spring Calving on Calf Gains
and Feed Requirements[a]

Calving Season	January–February	March–May
Nursing period (creep-fed)		
Number in each lot	12	14
Average initial weight (lb)	165 (Mar. 13)	157 (June 5)
Average weaning weight (lb)	583 (Oct. 3)	507 (Dec. 3)
Average gain to weaning (lb)	418	350
Average daily gain (lb)	2.05	1.89
Average daily ration (lb)		
Shelled corn	2.70	2.79
Cottonseed meal	0.34	0.35
Average total feed consumed (lb)		
Shelled corn	548 (9.7 bu)	498 (8.9 bu)
Cottonseed meal	68.7	62.4
Feed required/cwt gain (lb)		
Shelled corn	131.7	146.56
Cottonseed meal	16.48	18.35
Pestweaning period		
Average final weight (lb)	752 (Dec. 3)	658 (Feb. 12)
Average gain after weaning (lb)	169	151
Average daily gain (lb)	2.77	2.13
Average daily ration (lb)		
Shelled corn	11.73	12.09
Cottonseed meal	1.55	1.55
Alfalfa hay	3.71	2.36
Average total feed consumed (lb)		
Shelled corn	715 (12.7 bu)	859 (15.3 bu)
Cottonseed meal	94.2	108.1
Alfalfa hay	226	167.7
Feed required/cwt gain (lb)		
Shelled corn	423	568
Cottonseed meal	56	71
Alfalfa hay	134	111

[a] Missouri Agricultural Experiment Station Bulletin 652.

continuing until weaning time. This feeding may be done by the use of creeps when the calves are running with their dams, or the calves may be kept in a separate lot or pasture and hand-fed, with the cows being turned in for nursing during the night. Calves fed in this manner during the suckling period are much heavier and fatter at weaning time than western-bred calves, hence may be marketed after a much shorter feeding

FIG. 91. Home-bred calves of choice or better quality are preferred for the baby-beef program. (American Shorthorn Association.)

period. Usually such animals are in choice slaughter condition when 12 to 15 months old, when they weigh 800 to 900 pounds. This is the kind of calf that the meat retailer has in mind when he speaks of the superior quality of "native baby beef."

The third method of developing home-bred calves represents ultra-baby-beef production. This plan requires calves that are born early, preferably from September to February or at least by the middle of March, so that they are sufficiently mature and well finished to be marketed directly off the cows. If backed up with the right kind of early-maturing ancestry and if fed to the limit of their ability to utilize feed, such calves weigh 600 to 750 pounds when 9 to 11 months old. Though the beef from such animals lacks the color and flavor usually associated with more mature, choice beef, meat markets are able to dispose of it to customers whose chief considerations when purchasing meat are tenderness and absence of waste fat.

Each of the three methods of developing calves has its advantages and disadvantages. The method that is best suited to a particular farm depends largely on the type of the individuals in the cow herd, the feeds available for the cows after freshening, the equipment for taking care of young calves in cold weather, the milking qualities of the cows, and the intended date for marketing the finished calves.

VALUE OF CREEP-FEEDING

Experiments conducted at Sni-a-Bar Farms, Grain Valley, Missouri, under the supervision of the U.S. Department of Agriculture and the

Table 227

Effect of Feeding Nursing Calves on Return Realized When Calves Are Sold at Various Ages[a]

	Not Fed	Creep-Fed While Running with Cows	Kept from Cows and Fed Separately	Creep-Fed During Last 4 to 8 Weeks
Sold at weaning time				
Average weight (without shrink) (lb)	490	593	588	522
Value per cwt	$9.30	$11.30	$10.90	$9.75
Value per head	$45.60	$67.00	$64.10	$50.90
Grain fed to date (lb)	—	720	720	180
Value of grain fed to date[b]	—	$10.80	$10.80	$2.70
Gross return per head above cost of feed	$45.60	$56.20	$53.30	$48.20
Sold after 84 days of drylot feeding				
Average weight (without shrink) (lb)	665	770	760	695
Value per cwt	$10.30	$11.70	$11.25	$10.05
Value per head	$68.50	$90.10	$85.50	$69.85
Grain fed to date (lb)	760	1,760	1,700	1,050
Hay fed to date (lb)	335	265	245	290
Silage fed to date (lb)	220	135	120	200
Value of feed fed to date[b]	$14.60	$28.80	$27.70	$18.50
Gross return per head above cost of feed	$53.90	$61.30	$57.80	$51.35
Sold after 196 days of drylot feeding				
Average weight (without shrink)	927	1,007	976	947
Value per cwt	$12.75	$12.95	$12.60	$12.60
Value per head	$118.20	$130.40	$123.00	$119.30
Feed fed to date (lb)				
Grain	2,535	3,610	3,445	2,840
Hay	720	580	550	655
Silage	500	365	295	470
Value of feed fed to date[b]	$44.90	$59.60	$56.70	$48.90
Gross return per head above feed cost	$73.30	$70.80	$66.30	$70.40

[a] U.S.D.A. and Missouri Agricultural Experiment Station Reports.

[b] Feed prices used per ton: grain, $30; hay, $15; silage, $6.

University of Missouri, show conclusively that calves fed grain while suckling their dams return much more profit when sold for slaughter at weaning time than calves that are not so fed. (See Table 227.) Feeding 3 to 6 pounds of grain daily, or an average of approximately 700 pounds per head, produced calves that were 100 pounds heavier at weaning time and were valued at approximately $2 more per hundred. Thus by feeding $7 to

$22 worth of grain per head, depending on current feed prices and length of the suckling period, the average value of the calves when weaned was increased by $14.33 to $18.84 per head above the market value of the grain consumed. These figures apply to the calves that were fed while running with their mothers during the entire summer. Smaller increases resulted when the feeding was done during only the 4 to 8 weeks immediately preceding weaning.

Early calves that have been fed grain throughout the summer are sufficiently finished by late fall to be sold for slaughter. Profits from such calves have been quite satisfactory, judging from results of records made in the Kansas Beef Production Contest. (See Table 228.) When calves can be sold at about 11 months of age for $165 to $250 per head after eating 15 to 25 bushels of grain, there is little incentive for keeping them over the winter months to be sold in May or June on what is likely to be a less satisfactory market than that of the previous fall. It may be

Table 228
Results Obtained by Kansas Farmers in Creep-Fed Division of Beef Production Contest[a]

	1946	1947	1948	1949	1950	5-Year Average[b]
Number of herds	4	9	12	5	15	9
Number of cows	166	300	503	177	497	329
Percent calf crop	95	95	92	95.5	98	95
Feed cost per cow	$30.53	$37.64	$43.57	$35.36	$50.55	$39.53
Other costs[c]	$9.90	$7.74	$14.19	$12.26	$17.18	$12.25
Total costs per cow	$40.43	$45.38	$57.76	$47.62	$67.73	$51.78
Age of calves when sold (days)	347	310	351	347	334	338
Average selling weight (lb)	778	710	717	690	760	731
Average sale price per cwt	$22.80	$27.65	$25.85	$23.95	$33.15	$26.70
Gross value per calf	$177.38	$196.31	$185.34	$165.25	$251.94	$195.22
Feed fed per calf	$29.80	$53.26	$37.01	$28.99	$48.32	$39.53
Other expenses per calf[d]	$3.12	$2.93	$3.11	$4.69	$6.28	$4.03
Cow cost per calf	$42.56	$47.77	$62.78	$49.86	$67.73	$54.14
Net return per calf	$101.90	$93.35	$82.43	$81.71	$129.61	$97.80

[a] Computed from Kansas Extension Service, Mimeographed Reports, Beef Production Contests, 1946–1950.
[b] All averages are simple averages, not weighted.
[c] Bull costs, taxes, and other cash costs, but not labor, interest, depreciation.
[d] Principally marketing costs.

argued that selling calves at so young an age is economically unsound as far as the meat supply of the nation is concerned, but the individual producer can hardly be criticized for disposing of his cattle at the time when they return him the greatest profit.

Although it is highly important to feed grain to calves that are to be sold for slaughter when or soon after they are weaned, grain feeding during the suckling period does not seem especially advantageous if the calves are to be fed throughout the winter and spring and marketed sometime the following summer. Table 227 shows that the calves fed grain during the nursing period were worth about $20 more per head at weaning time than the calves that were not so fed, but this difference in value had decreased to only about $10 per head at the end of 196 days of drylot feeding. Meanwhile feed costs had mounted much more rapidly, both on the basis of cost per head and cost per hundredweight of gain, when grain was fed previous to weaning. When marketed after 6½ months of feeding, the calves that received only mother's milk during the summer were the more profitable. The practical application is this: If calves are to be sold at weaning time or after a comparatively short feed in drylot, grain-feeding during the suckling period is highly esssential in producing a satisfactory market finish. However, if the calves are to be sold as heavier baby beeves in late spring or summer the following year, creep-feeding is not recommended because the extra finish acquired results in slower and costlier gains during the long feeding period.

OPERATION OF THE FAT-CALF PROGRAM

The nonrange area cow-calf man with a herd of cows somewhat lacking in beef breeding will find that the fat-calf program offers the same opportunities for enlarging his business and for better utilizing his homegrown feed supply and pasture that the baby-beef program offers to the man with the better-bred cows discussed in the preceding section.

Returns come quickly in the fat-calf program, as calves are seldom over 8 or 9 months old when sold. There is a ready market for the 275-to-350-pound carcasses yielded by these calves because they are light, tender, and require a minimum of trimming. Many consumers in the region below a line extending from Washington, D.C., to Amarillo, Texas, are especially fond of this kind of beef. This is fortunate because the area corresponds to the region in which this specialized cow-calf program is best adapted. There is some indication that consumers in this section are gradually changing to a preference for more mature, fed beef, however. This has resulted in some softening in the demand for fat-calf beef.

BREEDING ANIMALS PREFERRED IN THE
FAT-CALF PROGRAM

In contrast to the commercial cow-calf and baby-beef programs, something less than ideal beef conformation is quite acceptable in the brood cows used in the fat-calf program. In fact, cows of mixed beef and dairy breeding are preferred because they usually are better milkers than cows of straight beef breeding, although of course there are variations within each group.

Investigators at the Kentucky station have studied this program extensively and are among its strongest proponents for mixed farming areas such as those of Kentucky. They recommend that the females selected for this program should be healthy, rugged heifers or cows of fair to poor dairy breeding or mixed dairy-beef breeding. Cows should produce 2 to 3 gallons of milk daily at freshening and should maintain fair milk production for 8 to 10 months. The better dairy cows are unsuitable because they give more milk than one calf can utilize. High-grade or purebred beef cows usually sell too high and give insufficient milk to justify their use in this plan. The dual-purpose breeds such as Milking Shorthorns and Red Polls are suitable but relatively scarce and high-priced.

The Kentucky workers concluded that the bulls best suited to the fat-calf program are purebred beef bulls, larger than average in size and thick and meaty in type. Size and the ability to sire large, growthy calves with beef conformation are more important considerations than breed itself.

FIG. 92. Well-bred Red Poll cows are exceptionally well suited to the fat-calf program. Calves may be allowed to run with their mothers or, if milk flow is heavy, it may be preferable to allow the calves to nurse twice daily. (Pinney-Purdue Farm, Wanatah, Indiana.)

Bulls answering this description sire calves that sell for $15 to $30 more per head at market time.

Unlike the recommendations made for the cow-calf programs, herd replacements for the fat-calf program should usually not be made from among the heifers raised in the herd. Continuing to save the heifer calves sired by the purebred beef bulls recommended for use in this program tends to increase the beef breeding in the cow herd with each succeeding generation, resulting in lighter calves owing to the reduced milking ability of the cows. An exception may exist when a man is striving to upgrade his herd in order to have enough quality in the herd eventually to produce high-quality feeder calves or baby beeves.

Herd replacements of the desired breeding can usually be bought in the neighborhood. Buying them at weaning time has several advantages. If heifer calves are bought directly from the breeder, the sire and dams can be evaluated, the cost is less than for older females, the health of the herd and of the heifers can be assessed, and calfhood vaccination for Bang's disease can still be practiced if the calves are not older than 8 months.

Reliable data on the economics of the two cow-calf programs just discussed are not available, but observations by the author would indicate that most of the smaller cow herds in the South, Southeast, and Southwest are more profitable when the fat-calf plan is followed.

BEST SEASON FOR CALVING

Because the fat-calf is best sold for slaughter at weaning time or "off the teat," the period of highest prices plays a larger role in determining the calving date. Figure 93 shows that, in the Southwest at least, the best prices for slaughter calves are received during May, June, and July. Therefore in this region the best time for calving is about 8 months previous to this period, or from September to December. Fortunately the weather is favorable for calving at this time of year in this region, and both cows and calves are quite adequately nourished on the winter oats and wheat pastures grown there.

Local seasonal market demands, winter feed supply, quantity and quality of the labor supply, the weather, and the shelter available are all involved in the choice of calving season but, in general, fall and winter calvings are more profitable in this program than spring calving.

FEEDING THE NURSING COWS

Cows that calve in the fall or winter present nutritional problems that differ from those previously discussed. Homegrown feeds and pasture

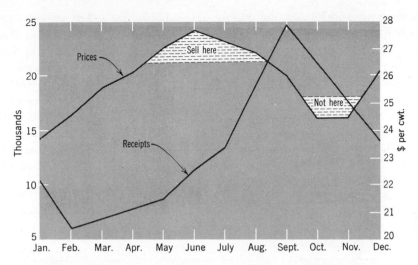

FIG. 93. Effect of season on receipts and average price received for choice slaughter calves. (Oklahoma Extension Circular 619.)

must be economically used; otherwise the cost of meeting the rather high nutrient requirements of the nursing cow soon more than absorb the profits possible with this program. If fall and winter calving is practiced, the cows will be dry in summer when pastures are usually of sufficient quality to produce good gains on such cows. Therefore, the feeding program for the nursing cows should be adequate for good milk production, but it need not contain enough energy to maintain cow weight as was advocated for the cow-calf programs in which cows were dry in winter. Loss of some weight in the nursing cow should be considered normal. A possible exception is the first-calf 2-year-old heifer that might fail to conceive for the next calf unless she is fairly well nourished.

Tables 229 and 230 show that winter pasture, in this case rye-vetch, is a satisfactory, yet cheaper ration for the nursing cow than native grass and supplement. The calves nursing the silage-fed cows were heavier and fatter at weaning and therefore sale time, but the higher cost of the cows' winter ration was not covered by the additional gross return per calf.

The use of winter pasture also reduced the amount of creep ration consumed by the nursing calves by about one-third. The carcasses were slightly less well finished, but profits were higher despite the lower selling price.

The native pasture used in the Oklahoma tests consisted of grass species that do not deteriorate so badly from weathering as do the pastures found in the Southeast and South. For this reason the native pasture results in the tests mentioned are not highly applicable in the other regions. The feeding of supplemental silage or hay is justified, in fact necessary, in these regions. In addition, winter pastures are not quite so satisfactory in these sections because the higher rainfall results in muddy pastures that are damaged by heavy grazing. A combination of winter pasture for grazing when weather permits, and silage or hay for supplemental feeding when rain and cold weather make grazing inadvisable, seems best. Tables 35, 36, and 37 give the nutrient requirements for cows that are nursing calves.

The area where the fat-calf program is recommended also happens to be the area where minerals are generally deficient in the soil (unless

Table 229

Average Cow Data from Different Feeding Systems in the Production of Fall-Dropped Calves[a]

Winter Treatment	Native Grass + Supplement	Native Grass + Silage and Supplement	Rye-Vetch Pasture
Number of cows producing calves	18	17	18
Average cow weights (lb)			
Fall	1,206	1,184	1,156
Spring	1,053	1,005	1,093
Winter weight loss	−153	−179	−63
Average daily winter ration			
Cottonseed meal (lb)	2.5	1.5	
Ground ear corn (lb)	3.0		
Alfalfa hay (lb)[b]			4.0
Oat hay (lb)[b]			12.0
Silage (lb)[c]		51	
Rye-vetch pasture (acres)			1.5
Native grass pasture (acres)	8	5	4
			(summer only)
Yearly feed cost per cow ($)			
Winter supplement	23.59	7.56	6.64
Silage		32.64	
Rye-vetch pasture			20.62
Native grass	22.50	17.50	14.00
Total	46.09	57.70	41.26

[a] Oklahoma Miscellaneous Publication MP-48.

[b] Fed during a 42-day period in January and February when it was necessary to supplement cows on rye-vetch pasture.

[c] Silage (from drought-damaged, immature corn) was available from a self-feeding pit silo, every other day. Amount consumed was estimated from silo measurements at 42 pounds of silage per cubic foot.

Table 230
Average Calf Data from Study of Systems of Management[a]
(All Calves Creep-Fed and Sold for Slaughter)[b]

Winter Treatment of Dams	Native Grass + Supplement	Native Grass + Silage and Supplement	Rye-Vetch Pasture
Number of calves marketed	18	17	18
Steers	10	9	10
Heifers	8	8	8
Average birth date, October	20th	9th	11th
Average calf weights (lb)			
Birth	77	77	69
End of winter phase, 4/19	462	474	440
Final weight, 5/5	551	574	540
Slaughter data			
Average yield (%)[c]	57.2	57.7	56.4
Average carcass grade (U.S.D.A. standard)	C−	C−	G+
Average market value per cwt ($)[d]	19.50	19.74	18.63
Total value per calf ($)	107.77	113.27	100.65
Creep-feed consumed per calf (lb)	898	856	590
Creep-feed cost per calf ($)	21.21	20.22	13.97
Cow feed cost per cow-calf unit ($)	46.09	57.70	41.26
Total cow-calf feed cost ($)	67.30	77.92	55.23
Net return over feed cost ($)	40.47	35.35	45.42

[a] Oklahoma Miscellaneous Publication MP-48.
[b] Based on average of steer and heifer data for each lot.
[c] Hot carcass weights shrunk 2.5 percent (minus hide weight). Values based on final Ft. Reno weights.
[d] On-foot market value calculated from yield, grade, and current value of carcass, based on Ft. Reno weights.

corrected by fertilization) and therefore in the plants. A mineral mixture should be available to both cows and nursing calves. A typical adequate mixture consists of equal parts of salt, limestone, and bonemeal. The bonemeal may be replaced by one-half as much dicalcium phosphate.

CREEP-FEEDING IN THE FAT-CALF PROGRAM

Creep-feeding of the calves produced in this program is highly recommended, although it is not recommended for the commercial cow-calf program mentioned earlier. Creep-feeding is more likely to be profitable if the herd consists of numbers of first-calf heifers or old cows, if drought or mud reduces the forage available as winter pasture, or if there is an appreciable spread between standard- or good-grading and choice-grading calves.

Naturally creep-feeding is successful only if the calves consume the ration offered. The quality of the pasture, location of the creep, stage of lactation of the cows, and the feeds used in the creep ration all affect the amount of creep ration consumed. The effect of the quality of the pasture has been discussed. The best location for the creep naturally varies from farm to farm, but it should be near the place where the cows spend most of their nongrazing time. This spot may be near a shade, the water supply, or the salt and mineral feeders. On farms where the pasture joins the farmstead, creep feeders can be located in a part of the barn or shed. In any case, the creep feeder should be convenient to the calves, yet protected from damaging rain and other livestock.

The creep ration should be extremely palatable, high in energy, and coarse in texture. Calves prefer either whole grains or coarsely cracked or rolled grains to finely ground feed, and a combination of grains such as corn and oats is generally preferred to a ration consisting of a single

FIG. 94. A creep feeder, ideally located near a shady nook where the the cows tend to gather. (American Angus Association.)

grain. The protein concentrate, if any, should be in pellet form, so that it will not sift out from the rest of the feed mixture. Grain sorghums or barley should always be rolled or coarsely ground, but oats and corn can be fed whole. Including a molasses feed increases feed intake but does not always result in more profit. Table 231 gives results from two tests in which various grain and supplement combinations were tested. Including a protein concentrate is usually profitable only during the late stages of lactation.

The question of the value of stilbestrol and the antibiotic terramycin in creep feeding was tested during the last 47 days of the Oklahoma test summarized in Tables 229 and 230. Table 232 shows that both the feeding of 5 milligrams of stilbestrol daily and a combination of 5 milligrams of stilbestrol and 40 milligrams of terramycin daily were beneficial. Other research referred to in Chapter 21 indicates that the earlier use of the

Table 231
Comparison of Grain Rations for Nursing Beef Calves

Rations Compared	First Series Average of 3 Years[a]			Second Series Average of 2 Years[b]		
	Shelled Corn	Shelled Corn 8 Parts, CSC[c] 1 Part	Shelled Corn 2 Parts, Oats 1 Part	Shelled Corn 8 Parts, CSC 1 Part	Ground Corn 8 Parts, CSC 1 Part	Ground Corn 8 Parts, CSC 1 Part, Ground Alfalfa, Molasses
Initial age (days)	79	81	81	77	81	85
Initial weight (lb)	222	221	220	216	217	217
Final weight (lb)	501	522	496	536	529	524
Total gain (lb)	280	301	277	320	312	307
Daily gain (lb)	2.0	2.15	1.98	2.29	2.23	2.19
Grain eaten per cwt gain (lb)	177	199	251	187	241	233
Final value per cwt ($)	11.65	12.10	11.60	6.80	6.80	6.70
Feed consumed daily (lb)						
1st 28 days	0.5	0.9	1.0	0.7	1.4	1.4
2nd 28 days	1.8	2.8	3.0	1.9	3.7	2.8
3rd 28 days	4.0	4.7	5.4	4.3	5.7	5.1
4th 28 days	5.1	5.9	6.9	6.5	6.9	7.1
5th 28 days	6.3	7.0	8.4	8.0	9.1	9.0
Total, 140 days	3.5	4.3	5.0	4.3	5.4	5.1

[a] U.S.D.A. Technical Bulletin 397.
[b] U.S.D.A. Technical Bulletin 564.
[c] Cottonseed cake.

Table 232

Effect of Stilbestrol or Stilbestrol Plus Terramycin in Creep-Feeding Beef Calves (Last 47 Days on Test)[a]

	Basal Creep Feed	Basal + 5 mg Stilbestrol per Calf	Basal + 5 mg Stilbestrol + 40 mg Terramycin
Number of calves per group[b]	15	15	15
Average calf weights (lb)			
Initial, 4/19	470	473	470
Final, 5/5	563	572	578
Average daily gain	1.97	2.10	2.30
Creep feed consumed per calf (lb)[c]	254	207	229
Creep feed per cwt gain (lb)	273	209	212
Slaughter data[d]			
Yield (%)	57.9	57.7	57.1
Carcass grade	G+	G+ to C−	G+ to C−
Marbling score	3.47	2.97	3.32
Financial results ($)			
Average cow and calf feed cost[e]	67.28	66.36	66.82
Market value per cwt[f]	19.54	19.57	19.23
Total value per calf	110.01	111.92	111.15
Net return per calf	42.73	45.56	44.33
Difference over controls		+2.83	+1.60

[a] Oklahoma Miscellaneous Publication MP-48.
[b] Nine steer calves and 6 heifers per group, 5 calves in each group from each lot of the original treatment.
[c] Concerns only the amount during this phase.
[d] Yield based on hot carcass weight shrunk 2.5 percent (hide off). Marbling score: 1 abundant, 3 moderate, 5 very slight.
[e] Costs prior to this phase based on average for original lots.
[f] Based on current value of carcass according to grade and final weight.

stilbestrol would have resulted in a still greater increase in weaning weight. Implantation of 24 milligrams of stilbestrol may be substituted if desired, without affecting results.

There is no one best combination of feeds for the creep ration. Rather, the available supply of homegrown grain should determine the ration used, with supplemental feeds being held to only the necessary minimum. Ordinarily the homegrown grain should make up at least 90 percent of the ration. Oats, because of their high fiber content, should not constitute more than half of the ration, at least in the final months before sale.

It may be necessary to make the opening into the creep feeder enclosure wide enough to permit entry of a yearling heifer in order for the nursing calves to learn to use the creep feeder. Penning a few calves in the enclosure during the day or night also is effective in teaching calves to use the creep. The addition of wheat bran to the first feed mixture placed in the feeder is also practiced. Swine, sheep, and poultry should by all means be kept away from the feeder.

Chapter 23

The PUREBRED PROGRAM

The aim of this chapter is to give information that will, it is hoped, enable a breeder or prospective breeder of beef cattle to conduct the purebred program more profitably. Most herds are, after all, operated with the intention of making a profit, contrary to the belief of some who think that most purebred herds are only a hobby or a tax write-off. It is true that a number of wealthy men are purebred breeders, but in most instances they too, like most other breeders, are seriously interested in producing the best cattle possible and in making a lasting contribution, to the beef cattle industry in particular and to agriculture in general.

Some of the requirements for success with the purebred program are the following:

1. Keen business judgment on the part of the owner or manager.
2. Location near, or within ready access of, an area with heavy beef cow population.
3. Skill in choosing herd replacements or in making outside purchases.
4. Knowledge of cattle feeding and herd management techniques.
5. Sufficient capital for the necessary investments in cattle, land, equipment, and buildings.
6. Sufficient land to produce all of the pasture and roughage and most of the grain needed.
7. A reputation for honesty.

Successful purebred herds are being operated by men who fail to meet all of these requirements, but, in general, if the herd is to be a financial success and achieve its other objectives over a period of years, the above requirements and possibly others must be met. Management and skill can be hired, and in some instances successful herds are owned by men who depend almost entirely on such assistance.

AIMS OF THE PUREBRED BREEDER

Every would-be breeder of purebred cattle should decide what his ultimate aims or goals are as a breeder. It is important also to analyze these aims in terms of the means available—that is, capital, land, climate, and skill or know-how—because these factors determine whether the aims can be achieved. It would be futile for someone living in a remote part of the country—or in a section unsuited to beef cattle because of low soil productivity or extremes in climate, for instance—to aspire to reach the top as a cattle breeder. The odds against success would be too great, regardless of how much capital such a prospective breeder might have. This is not to imply that superior cattle could not be produced, but the venture would almost certainly be a financial failure.

Purebred breeders and their operations can usually be divided into four rather broad categories. Although there is considerable overlapping and breeders are often in the process of shifting from one category to another, most breeders can be placed in one of these groups:

1. The so-called master breeders. The master breeder is one who makes a lasting contribution toward improving the cattle of the breed of his choice. He not only produces superior individuals or show-ring champions; he produces sires or families of females that will leave their influence on the breed for generations to come. The master breeder usually stays in business for a long time and often passes the intact herd on to the next generation. Although there are notable exceptions, master breeders are often men of means, and the beef cattle industry should be thankful that they are. They are usually able to obtain the cows or bulls needed in their breeding programs without undue consideration of cost. If a particularly costly individual proves unsuccessful as a producer, financial ruin does not result as may be the case with the breeder of more limited means. Master breeders usually display their cattle at the large shows which, although they may not be a paying proposition, promote the breed and serve as an educational tool for the industry as a whole.

Master breeders refuse to be swayed by the momentary popularity of temporarily fashionable pedigrees or of a type of cattle at variance with the one that will be most beneficial to both cattle producer and consumer. Master breeders have never been numerous and perhaps never will be.

2. Purebred breeders who furnish the purebred bulls for the commercial cow-calf man and the females for many new purebred breeders. Approximately only 3 percent of all beef cattle in the United States are purebred. A broader genetic base for improvement of all cattle would be established if there were more breeders of purebreds.

The quality of cattle produced by this type of breeder varies greatly, from cattle good enough to suit the master breeder, to cattle that should not be multiplied. The better breeders of this group are the backbone of the commercial beef cattle industry, because the bulls produced in their herds determine the quality of stocker and feeder cattle produced in the commercial herds. The size of herd and financial resources of purebred breeders in this group vary greatly and have little bearing on the success of the program.

One rather consistent characteristic of this program is its location, with the herds usually being situated near a market for large numbers of bulls for commercial use. Certainly not all of the calves produced in such herds should be saved as bulls or replacement females. Rather, a good percentage should be sold as stockers or feeders or fed out on the farms or ranches where they are produced. Breeders in this category often go to the master breeder as a source of herd bulls.

3. Breeders of commercial cattle who maintain a purebred herd that is large enough to produce the bulls and at least some replacement females for their own herds. This group includes the breeders who are gradually converting their herds from a commercial cow-calf program to a purebred herd. These purebred breeders are often found in the range areas where herds are comparatively large and many bulls are needed each year. Many young men use this method of growing into the purebred business and, although the process sometimes takes many years, such herds are usually founded on a sound basis because the breeders are aware of the requirements of the commercial operation.

4. Four-H and FFA projects. Members of 4-H and Future Farmers of America organizations, with projects of one to a dozen cows and heifers, comprise an extremely important category among the purebred breeders. It would be interesting to know what percentage of the present-day adult breeders got their start in one of these organizations. Knowledge and motivation gained while caring for a 4-H or FFA calf project have started many a breeder in business and, although this method of building a herd is slow, only a small initial investment is required and such herds often are very soundly built. Details concerning this program can always be obtained from the local county agricultural agent or from the vocational agriculture teacher.

GETTING A START WITH PUREBREDS

The choice of breed is discussed in connection with the commercial cow-calf program in Chapter 7, and most of the points made there apply also to the purebred program. Personal preference, adaptation to climate, and the feed and pasture supply are the most important considerations in choosing a breed.

Purebred cattle are usually purchased in a different manner than are commercial cattle. First of all, such cattle are usually bought by the head, rather than by the pound as commercial cattle are bought. Values of purebred cattle are, of course, determined by supply and demand, as is true of commercial cattle, but there is no real basis for establishing whether the prices asked are reasonable. Predicting the price that may be received for the offspring produced in a purebred program 5 years from now is all but impossible. It is true that purebred cattle prices swing up and down in sympathy with the commercial cattle market, but many other factors enter the picture. For instance, a drastic shift in demand for purebred cattle of certain bloodlines owing to an outcropping of such an inherited undesirable trait as dwarfism can quickly reduce values to the level of commercial cattle prices.

Another difference in method of purchase between purebred and commercial cattle is source. Among the more common sources of purebred cattle are the following:

1. Direct purchase from a breeder by private treaty. This purchase may apply to an entire herd or to a few head and is the best, although not the most economical, method of buying purebred cattle.

2. Production sale by larger breeders. The prospective buyer can obtain much knowledge concerning the performance of the sale cattle by studying the sires, dams, and close relatives of the offerings. It is also possible to observe the conditions under which the cattle were developed if the sale is held at the breeder's farm or ranch.

3. Consignment sale of cattle produced by the members of state, district, or local associations. This is perhaps the least desirable source of purebred cattle, because little information other than pedigree and the individuality of the animals themselves is available. Cattle sold in such sales are often highly fitted because of the competitive aspects of this method of selling. The high condition of such cattle not only serves to cover up weaknesses in conformation, but also may impair the fertility and shorten the productive life of many cattle.

4. A father's or neighbor's herd is often the source of 4-H and FFA project heifers and is to be highly recommended if the quality is sufficiently high.

Although the auction method of selling purebred cattle may be a profitable way of merchandising them, it is a questionable method of buying as far as starting a purebred herd is concerned. Sound judgment as to values is often laid aside in the competitive atmosphere of an auction sale. "Bargains" all too often turn out to be someone else's mediocre surplus females, fitted for sale to be bought by the beginner. If a large number of cattle are to be purchased, the buyer would do well to seek the advice and help of a highly reliable fellow breeder or breed association field man.

A highly trained, experienced herd manager or herdsman is often a real asset when beginning a purebred program because of his ability to select and purchase the foundation stock. A new breeder who personally buys his first breeding cattle may often pay exorbitant prices for cattle that are considerably below foundation quality. After making his initial investment he may feel the need for an experienced manager, but the financial outcome of the program is already under a real handicap. Such programs are usually doomed to failure from the outset.

Which quality of animal to buy depends largely on the breeder's goals and on the available finances. The average breeder may logically pay only moderate prices for his females but he should be willing to invest whatever is necessary to obtain a sire with superior genetic potential.

Production-testing information is even more important in the purebred program than in the commercial cow-calf programs. Many new purebred breeders are demanding performance records on animals being purchased to start new herds. Such records do not guarantee that the new herd will be a financial success but they help.

BREEDING SEASON

The purebred breeder has a few additional problems to consider in establishing his breeding season as compared with the commercial breeder. He must not only be concerned with feed supply, available shelter and labor, and marketing time. If a purebred breeder plans to show his animals, or if he wishes to attract buyers who do, he must consider the age classifications that are fairly standard in shows throughout the country. January 1, May 1, and September 1 are the principal base dates for such classifications, and the following classification is representative of those found in open class shows:

Two-year-old—calved between January 1 and August 31, two years previously.

Senior yearling—calved between September 1 and December 31, two years previously.

Junior yearling—calved between January 1 and April 30 of the previous year.

Summer yearling—calved between May 1 and August 31 of the previous year.

Senior calf—calved between September 1 and December 31 of the previous year.

Junior calf—calved after January 1 of the current year.

In several of the breed-association-sponsored shows there is a further breakdown of the three main groupings into two each, with additional base dates of March 1, July 1, and November 1. This applies especially in the yearling and calf classes. These classifications are subject to change, but the trend has been to lower the maximum ages given above and to offer a wider variety of classes for the younger ages.

It is highly desirable to bunch calvings just as soon after the base dates as is practical. Since six base dates are coming into use, this point is less important because no calf would be more than 2 months younger than its competitors. In practice each breeder must decide what his calving plan is to be. Many successful purebred herds follow a calving season that is based entirely on the same factors used in a commercial program, and still others, especially if artificial insemination is used, breed cows to calve the year around. Larger calf crops result from year-round breeding, but naturally the management of the cow herd is then less clear-cut, and overall feed and labor requirements are greater.

SUMMER FEEDING OF THE PUREBRED HERD

Most purebred cows calve in the spring or fall but, as just mentioned, some herds may calve the year around. The summer feeding program is determined by the calving season and is therefore discussed under six headings.

Spring-Calving Cows. Because an ample milk supply for the suckling calf is highly desirable in order to ensure maximum development and growth, an adequate supply of nutritious pasture is essential for the cow herd. Stocking rates slightly below those recommended for commercial cows are in order as a guarantee against the hazards of drought.

Some breeders, especially in the nonrange areas, separate the calves from the cows during the day in order to hand-feed the calves and to feed the cows a small feeding of grain when they are penned with the calves at night for nursing. First-calf heifers and old cows especially benefit from such treatment. The economy of this practice depends on method of selling, and it is doubtful whether it pays unless calves are sold at or soon after weaning or are to be fitted. Certainly it is not essential for the proper nourishment of the mature cow herself. It should be mentioned that this practice facilitates both hand mating and breeding by artificial insemination.

Creep-feeding calves that are nursing cows in summer is generally recommended for purebred calves to ensure bloom and maximum development. This feed should probably be included under the category of advertising expense because, nutritionally speaking, calves, like their mothers, do not require the added feed. Feeding the calf extra grain during the nursing period is more justifiable than feeding the cow, however. Creep rations are further discussed in Chapter 22.

Fall-Calving Cows. These cows are dry during the summer, so that there is an opportunity to effect some savings at this time. It is never justifiable to feed grain to dry bred cows on pasture in the summertime. Possible exceptions may be very old or lame cows or cows that are to be sold in an auction sale in the fall or winter. Supplementing the pasture with legume or mixed hay may be advisable, especially if pastures consist entirely of grasses and if drought reduces the total supply of forage. If the pasture is entirely legume, feeding some grassy hay reduces the danger of bloat. Cows that calve in the fall are actually in a better state of nutrition at calving time, even if pasture is short and poor in quality, than cows that calve in the spring. Problems such as poor milk production and calving difficulties caused by poor nutrition seldom occur in fall-calving cows.

Replacement Heifers. Heifers that have definitely been chosen as herd replacements need only good pasture during the summertime, whether they are being bred as yearlings or as 2-year-olds. Feeding extra feed can more easily be done during the wintering period if it is believed necessary to ensure maximum development. In the summer, economy can be practiced without fear of detracting from the herd's appearance.

Sale Heifers. Heifers that are to be sold in the fall or winter as open yearlings or as bred yearlings or two's are more favorably received if fed some grain or high-energy concentrate while they are on summer pasture. A limited feed—that is, about 1 pound per 100 pounds of body weight daily—of a bulky concentrate mixture results in gains of approximately 2 pounds per day and a definite improvement in condition. A 30-to-60-day drylot feeding period in the fall brings the heifers into sale season with adequate condition without impairing their breeding value.

Using large amounts of oats in the summer and fall rations of sale heifers is recommended. A protein concentrate at this time is of doubtful value if pasture or hay contains considerable legume.

If fall sale heifers are handled in drylot during the summer, a ration consisting of at least 50 percent roughage should be fed and, as mentioned, oats are desirable for summer feeding. Barley or ground ear corn are also bulky and therefore acceptable, but shelled or ground shelled corn or grain sorghum should be fed sparingly. A low-protein, high-molasses, vitamin-and-mineral-fortified concentrate is often used in the fitting rations fed in drylot to ensure adequate intake of total ration and to protect drylot-fed cattle against vitamin A or mineral deficiencies.

Sale Bulls. Most bull sales are held in late winter and early spring, and grain feeding of yearling or older bulls on grass the previous summer is less important than it is for fall sale heifers. If, however, the breeder wishes to attract buyers who may want to show the bulls, grain feeding from calfhood is essential. Most buyers of bulls to be used on commercial cows—and this applies especially to the ranchers in the rougher range areas—would rather have their new bulls in less than show condition. They find that such bulls are more fertile and more apt to range with the cows when turned out to pasture during breeding season. If grain is fed on pasture, bulky feeds such as oats or ground ear corn are preferred. It is doubtful whether it is economical to maintain all sale cattle, and especially bulls, in high condition at all times with the hope that the occasional visitor will make a purchase. Rather, the feeding program should usually be designed so that most of the sale cattle are in attractive sale condition only when seasonal demand is high.

The Herd Bull. The mature herd bull needs no additional feed when running with the cow herd on good summer pasture. Yearling bulls are sometimes used in the nonrange areas, and the daily feeding of 6 to 10 pounds of oats, ground ear corn, or similar bulky concentrate is recommended to promote growth and development. The same supplemental feeding may prolong by several years the useful life of old bulls that fail to graze enough to maintain their condition.

During the summer the water, salt, mineral, and shade needs of purebred cattle should be given due attention. Often, because some grain is fed, minerals are overlooked.

WINTER FEEDING OF THE PUREBRED HERD

The problem of winter feeding will also be discussed on the basis of the nutritive requirements of the various age and sex groupings. The cow herd often consists of both wet and dry cows. Dry cows can be wintered economically in the same manner as commercial dry cows—that is, on a full feed of legume hay or, better still, mixed hay, or silages made from the same crops. A partial feed of corn or sorghum silage, grass hay, or cereal straw free-choice, and a pound of protein concentrate per day is used extensively in the nonrange areas. If cured-range grass is used, supplementation with 2 pounds of a vitamin and mineral-fortified protein concentrate results in good calf production but some loss in weight on the part of the cows. This may detract from the cows' appearance to some extent, but investment in winter feed can be kept low.

Open yearling or bred yearling and 2-year-old replacement heifers attain larger mature size if 2 to 4 pounds of a farm grain or high-energy concentrate are fed with the better roughages during the winter. Again, however, it is doubtful whether the increased feed cost can be recovered in terms of larger or heavier calf crops.

Weanling heifer and bull calves should definitely be fed for at least normal growth during wintertime with possibly some improvement in condition, especially if the calves have not been sorted into sale and replacement groups. A full feed of good mixed hay or silage plus 4 to 6 pounds of almost any grain mixture is adequate. If prairie hay or other nonlegume hay or corn or sorghum silage is used, additional protein is needed. One pound per head of any of the oilseed meals is recommended in such cases.

Winter feeding and management of the herd bull are discussed in Chapter 8. If annual production sales are held during the winter, additional condition is often desired in herd bulls simply for display purposes. Although this practice cannot be advocated from a health standpoint, there is no doubt that prospective buyers like to see the sires of sale animals carrying considerable condition.

SALE CATTLE

Winter rations or, for that matter, any ration that is going to be fed to full-fed sale cattle should meet the nutrient requirements for feeder cattle on finishing rations as given in Chapter 14. However, there is a growing consensus that the purebred industry as a whole would benefit if sale cattle were sold without fitting—that is, without being fed to high condition. Someone must pay the added cost of the unnecessary fitting ration; hence it either cuts into the profits of the original breeder or increases the initial cost to the new breeder. Even more important, the reproductive lives of such animals are shortened, and many research data would indicate that fitting of young females reduces their milk production for life. Founder and cattle fed "off their feet and legs" are all too common in purebred herds. Such cattle can only be salvaged at commercial prices.

Many special fitting rations are used in purebred herds. More frequent feedings, special preparation methods such as steamrolling the grain to ensure a bulkier, dust-free ration, cooking the grain portion of the ration, using molasses and beet pulp as appetite stimulators, and many other special adaptations are used by experienced herdsmen to increase ration consumption and to maintain appetite over a long feeding period. The economy of such special treatment depends on the added price received for sale cattle. Undoubtedly the larger breeders who wish to attract buyers from among other purebred breeders are going to continue the practice, but the breeder who sells only to the commercial breeder can reduce his feed and labor investments by maintaining his cattle in less condition by use of simple and practical homegrown rations.

NURSE COWS

Nurse cows or foster mothers are used by many fitters of sale and show cattle. Certainly this practice can be recommended in the case of calves whose dams are poor milkers because of old age or for other reasons. Otherwise, in principle at least, the practice should be discouraged because inherently poor milk production on the part of the dam is apt to be obscured, whereas it would be better to recognize this fault for what it is and to cull such cows and their offspring from the herd. Furthermore, the calves on nurse cows often develop so rapidly that many become unsound in the feet and legs. Of course it is difficult to argue with the breeder who knows that he may increase the sale value of an outstanding calf many times simply by fitting the calf to the nth degree.

As to the cost of maintaining a nurse cow, it is a safe estimate that three additional mother beef cows can be fed on the feed consumed annually by a nurse cow. Nurse cows must be fed a concentrate mixture containing 15 to 20 percent protein at the rate of 10 to 20 pounds daily depending on the amount of milk produced and on the size of the cow, plus 10 to 15 pounds of good hay or its equivalent in silage or other roughage. Holstein, Brown Swiss, or other cows that produce large quantities of low-fat milk are preferred to such high-testing breeds as the Jersey or Guernsey.

The calf produced by mating a nurse cow to a beef bull usually ends up in the freezer of the owner or his employee.

Many methods are used to change a calf over from its own mother to a nurse cow. This procedure is often difficult because, in most cases, calves will not be changed over until they are 4 months old or older. Probably the most important point in making the change successfully is to be certain that the calf is hungry. This may necessitate withholding all feed and water for as long as 48 hours, although less time is usually required. It often helps to permit the calf to nurse its own mother partially, if possible, just before switching it to its foster mother.

Often a nurse cow must be hobbled or otherwise restrained to prevent her from kicking or butting the calf. Leading calves to the restrained cow in a nursing barn or stanchion is recommended, because it promotes regularity in the feeding schedule and helps in breaking calves to lead.

Dried-milk concentrates or reconstituted dried milk shows promise of replacing the nurse cow. Further research is needed, but the use of antibiotics virtually eliminates the scours often prevalent when milk substitutes are used.

SELECTING THE SHOW HERD

The selection of animals for the show herd should be made as early as possible. For most larger breeders the show season begins about August 10 and continues through the fall and winter. Because it takes 4 to 8 months to bring an animal from good breeding to show condition, the candidates for the show herd should be chosen no later than late winter or early spring. For calves that will be shown in the calf classes, selections should be made while calves are still nursing so that they can be placed on nurse cows if desired. If more than one candidate is available for a given class, two should be fitted so that a substitute will be on hand should accident, sickness, or unsatisfactory development disqualify the more promising individual. Substitute animals are especially desirable in the younger classes, where the growth rate is rapid and marked changes in conformation are common.

An important factor in selecting a string of show cattle is uniformity. If all individuals in a show herd are of the same general type and show a uniform gradation in size from the youngest to the oldest, they present the best possible appearance while stalled and when exhibited for the group prizes. Uniformity of color, although of minor importance, is not without value in the group classes.

GENERAL CARE AND MANAGEMENT

In fitting cattle for show or sale, every reasonable precaution should be taken to ensure their comfort and well-being because they make maximum gains only when they are quiet and contented. Comfort implies protection from extreme weather, from annoying parasitic insects, and from disturbance by other animals.

In hot weather the cattle should be kept in the barn during the day to prevent their hair from being sunburned and becoming dry and stiff. Moreover, the cattle will be more comfortable in the barn and will gain more rapidly than if left in the pasture throughout the day. If possible, roomy box stalls, well supplied with clean bedding or sand, should be provided, so that the cattle are inclined to lie down and rest except when they are eating. Beginning in June or July the barn should be sprayed to control flies, and the use of fans or even air-conditioning is sometimes justified in the more humid parts of the country.

Show cattle need a reasonable amount of exercise to stay sound in their feet and legs. Exercise also tends to stimulate the appetite, making for a larger consumption of feed and a higher degree of finish. Young animals usually exercise sufficiently if turned together at night in a small pasture or lot. Older cattle often stand around or lie down, especially if their feet are tender, and should be made to exercise by walking them about a mile each day.

THE FEEDING OF SHOW CATTLE

The outstanding difference between show animals and those in the breeding herd is a difference in condition. Cattle entered in an exhibition are presumed to be as nearly perfect as it is possible to make them. They are expected to show their muscling, growthiness, soundness, and breed and sex character to the best advantage. Show animals that are noticeably lacking in finish invite the criticism that they do not have the ability to finish at a reasonably young age, or that they are "hard doers." On the other hand, extreme condition is detrimental to breeding animals, as few will deny. It is a healthy sign that judges are tending to accept show cattle with less finish than traditionally has been demanded. In the end, a continuation of this trend will be for the betterment of the cattle, from the health standpoint, and for the viewing public, which can then more accurately judge the true quality of the animal.

So long as custom continues to emphasize condition in show cattle, such cattle should be fed a liberal amount of concentrates and a relatively small allowance of roughage. In general the feeding of show cattle is similar to the finishing of steers for market, except that the length of feeding period is longer for show cattle. For this reason there is much use of highly palatable feeds, such as molasses and feeds containing molasses, that tend to sustain a large intake of concentrates throughout the show season. Bran also is usually fed in rather generous amounts, not only for its protein content but also because its bulkiness exerts a regulatory effect on the digestive tract.

The grain ration of show cattle is almost always coarsely ground or rolled, particularly during the latter part of the feeding period. Such preparation has a desirable effect on grain consumption and also makes it easier to mix such feeds with other materials.

Show cattle are commonly fed three or even four times a day, particularly if they are rather thin and if rapid improvement in condition is desired. Supplying a relatively small amount of feed at frequent intervals is much better than feeding a larger quantity at one time. So far as possible, the amount fed should be no more than will be eaten within about an hour. Leftover feed soon becomes stale and unpalatable and should be promptly removed from the feedbox.

When rather large numbers of cattle are being fitted for sale, self-feeders can be used quite satisfactorily. Including more of the bulky feeds such as oats, bran, or barley makes this an entirely safe practice and saves labor just as it does in a commercial operation.

The following grain mixtures (in pounds) are examples of the many that are recommended for cattle being fitted for show:

I.	Cracked corn	50
	Crushed oats	25
	Wheat bran	15
	Linseed pellets	10
II.	Crushed barley	40
	Crushed oats	25
	Wheat bran	10
	Molasses feed	25
III.	Shelled corn	50
	Whole oats	20
	Linseed pellets	10
	Molasses feed	20

Some fitters moisten each batch of the mixed feed with just enough molasses diluted with an equal volume of water to remove dust and dampen the feed. Cooked barley is sometimes used for the same purpose, as is wet beet pulp, at the rate of about 0.25 pound per feeding.

Show cattle on a full feed of grain are sometimes turned into a short grass pasture each night during summer and fall. Animals that have the run of a clean pasture are unlikely to be troubled by foot rot. In addition to the pasture, show cattle should be fed a limited amount of bright mixed or straight grass hay.

After frost when pastures are no longer green and after winter weather arrives, the day and night procedure may be reversed with the cattle spending the day outdoors and the night inside. Having the cattle spend some time in the open promotes general health and longer, healthier haircoats. During the summer and fall, hair growth can be encouraged by dampening the hair with a sprayer each morning after the cattle have cleaned up their feed.

FINE POINTS OF FITTING, GROOMING, AND SHOWING

Such subjects as care of feet and legs, grooming, washing, breaking to lead, care of horns, clipping of heads and tails, and selection of equipment for fitting and showing are discussed in detail in excellent publications obtainable for the asking from the various breed associations, whose addresses are listed in the Appendix, and from the extension livestock specialists in each state. Generally speaking, problems encountered in developing, fitting, and showing beef cattle are quite specific in nature. It is likely that consultation with an experienced herdsman, field man, extension service representative, or vocational agriculture teacher is the quickest and best solution to individual problems.

LETTING DOWN FITTED CATTLE

Highly fitted cattle that are being retired from a show herd or that have been purchased at a sale need special attention when they are to be let down in condition. Simply turning such cattle out to pasture or placing them on an all-roughage ration is not recommended. Rather, there should be a gradual reduction in concentrate allowance. Raising the fiber content of the grain ration by increasing the oat content is also often practiced. In any case, the feeding of some grain is recommended for as long as 3 months while the animal is gradually losing its excess fat. Increased exercise such as results when cattle are penned on pasture at all times also assists in letting down fitted cattle.

THE FINANCIAL ASPECT OF PUREBRED CATTLE

The purebred breeder's financial problems are quite different from those of the man who handles only commercial cattle. In the first place, the size of his investment in animals is largely a matter of his own choosing, with "the sky as the limit." This is true even though he maintains only an average-sized herd, for what he lacks in numbers can easily be made up in quality and breeding if he is willing to pay the price.

Purebred cattle values vary greatly from herd to herd, because there are wide differences in individual merit and in the popularity of the blood-lines represented. They also vary greatly from one period to another, largely in relation to the degree of prosperity enjoyed by the men who handle market cattle. When prices of commercial cattle are high, there is a strong demand for purebred bulls to produce steers and heifers for the market. This demand results in increased prices for purebred cattle.

The close correlation usually existing between slaughter steer prices and prices for purebred cattle may be seen in Fig. 95. During the 21-year period 1900–1920, purebred prices, as determined from published reports of all important public sales of Shorthorn, Hereford, and Angus cattle, averaged approximately 3.25 times the average value per head of native beef steers on the Chicago markets. During the early years of this period some purebred sales were held at which scarcely better than beef prices were realized. Later, however, literally scores of sales averaged around $500 while the offerings from the more noted herds averaged $1000 and upward.

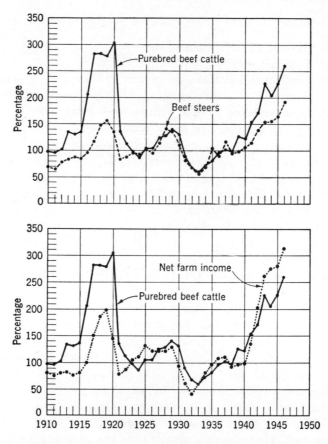

FIG. 95. The relationship of prices of purebred beef cattle to (a) prices of beef steers at Chicago and (b) net income of farm operators of the United States. Index number 1935–1939 = 100. (University of Minnesota.)

Figure 95 shows that purebred cattle prices skyrocketed during World War I but declined immediately afterward to about the same level they had occupied before 1914. A similar rise in prices occurred during World War II, but instead of receding after the war, prices climbed higher and higher.

In the 1960s purebred prices continued to rise and fall with commercial cattle prices but, because of some occasional fantastically high prices, sale averages are becoming less meaningful in terms of what the average purebred breeder may expect. If anything can be concluded it is that the previously mentioned 3.25 increase for purebred over commercial prices has been reduced to perhaps 2.50.

Some observers believe that, within some of the breeds, the domination of the show and sale rings by a few large breeders makes the small-scale breeding of purebreds of those breeds a rather discouraging prospect. In the end, if the "little" man does drop out, the resulting reduction in number of breeders could work serious harm on these breeds.

IMPORTANCE OF DEPRECIATION CHARGE IN PUREBRED HERDS

The relatively large amount of money invested in purebred cattle exerts an appreciable influence on the financial outcome of the enterprise. Interest and depreciation charges and death risks, particularly, are affected. Depreciation is an item that may be almost disregarded in the financial aspect of a herd of grade cows, because the cows are usually sold before age affects their market value. In a herd of purebreds, however, depreciation is of considerable importance, because animals bought at high prices are eventually sold for beef. Occasionally the loss suffered from the depreciation on a single animal is sufficiently large to wipe out the profits accruing from the rest of the herd for a full year. Losses from depreciation are particularly heavy during periods of financial depression when prices are falling rapidly. These, however, are to a certain extent covered by the appreciation or increase in values enjoyed during periods of prosperity when prices are advancing. Unfortunately, comparatively few breeders keep many cattle on hand when prices are rising, whereas barns and pastures are likely to be full when values are falling. Moreover, there is a pronounced tendency for new men to enter the purebred business after prices have reached a high level, and a majority of them have nothing to sell until the arrival of a period of depression, which inevitably follows inflation and overexpansion. The losses suffered on herds so founded are sometimes very great, occasionally necessitating the sale of the cattle for only a fraction

Table 233

Prices Paid for Purebred Hereford Cattle at Public Auction Sales[a]

	1960	1950	1940	1930
Number of auction sales	678	573	220	60
Number of cattle sold	48,888	38,744	17,893	3,732
Average price	$458	$604	$195	$217
Average of top sale	$4,714	$5,160	$1023	$1068
Top price for bull	$80,000[b]	$70,500	$7900	$2500
Top price for female	$21,000	$17,300	$5350	$7600
Bulls sold for $10,000 or more	16	46	33[c]	5[c]
Cows sold for $5000 or more	11	31	6[c]	3[c]

[a] Compiled from *The American Hereford Journal.*
[b] For one-fourth interest.
[c] Number sold for $2000 or more.

of their initial cost. A tax expert should be consulted by breeders of valuable animals in order to choose the best plan for computing taxable income, depreciation, and deductions.

OPERATING COSTS IN PUREBRED HERDS

Operating expenses per animal are greater for purebred than for grade cattle. The purebred breeder, to make progress and succeed, must keep his herd in fairly good condition to ensure maximum development of the young stock and to make a favorable impression on prospective purchasers. This means a larger outlay for feed and labor. Further calls for labor result from the general practice of hand mating, from the necessity of keeping the bull and heifer calves separated, from the need for more or less individual feeding and care in order to give each animal the best possible chance to develop, and from the desire of most breeders of purebred cattle to keep the barns and farmstead neat and clean and open to inspection by visitors at any time.

Usually the building and equipment charge per animal is much higher in purebred herds than in grade herds. Although a certain amount of extra equipment beyond that needed for ordinary cattle is highly desirable, elaborate barns and costly equipment are by no means essential to the raising of animals good enough to meet any requirement. A sufficient number of fairly large, roomy box stalls and well drained lots to permit proper separation of animals on the basis of age and sex constitutes the principal item of equipment needed for the satisfactory management of a herd of purebred beef cattle.

Insurance against death losses is available for highly valuable cattle. It is quite expensive but must be added as a necessary item of operating cost for breeders who cannot afford such losses. Insurance is usually used mainly for herd bulls and purchased females.

COST OF SELLING PUREBRED CATTLE

The marketing cost of purebred cattle is appreciably higher than that of grades. Because there is no established market for breeding cattle, each breeder is forced to find his own market, which he does through various kinds of advertising. Newspapers, livestock journals, billboards, and letter-heads announce to the reading public that he has stock for sale. Reproduced photographs, show herds, and public displays of prize ribbons and trophies bear witness to the excellence of his herd as a whole. Even a certain percentage of the cost of an imposing barn, the expense of painting the fences, and the cost of keeping the breeding animals in higher condition than necessary to ensure their maximum usefulness should properly be charged to advertising. Returns on money spent in this manner come from a quicker sale or from higher prices for surplus stock rather than from lower operating costs. Again, a tax expert should be consulted because many of these expense items are deductible from income under certain conditions.

The beginning breeder should be watchful lest his selling costs go out of proportion to the prices received. To consign a $300 animal to a sale is poor policy when expenses, including advertising, catalogs, auctioneer, stall rent, freight, expenses of attendant, and the like, amount to $50 or more per head. It is likely that the net return would be greater if the sale were made at home at little more than beef prices.

RECEIPTS FROM PUREBRED CATTLE

Receipts from a herd of purebred beef cattle are very irregular in both amount and season. During periods of prosperity every surplus animal is taken by eager buyers who show little tendency to haggle over prices. During less favorable times even the more noted breeders must resort to peddling tactics in order to make sales. At such times it is common for a large percentage of bull calves to be castrated, finished, and sent to market along with the plainer heifers and those cows that have shown only mediocre breeding ability. Although such sales are trying to the individual breeder, they may be of great benefit to the breed as a whole, as they eliminate the poorer animals from breeding herds.

Price fluctuations have been the downfall of a great many cattle speculators who attempted to get into the purebred business by buying a ready-made herd at peak values. However, for the man who builds his own herd from the progeny of a few carefully selected cows of proven merit, price fluctuations, although by no means welcome, are unlikely to prove disastrous.

Part VI

SPECIAL PROBLEMS
IN BEEF PRODUCTION

Chapter 24

The PREPARATION of
FEEDS and METHODS
of FEEDING

Probably no question connected with the feeding of beef cattle gives the feeder more serious concern than that of choosing the method of feed preparation. It is small wonder that he is confused, for every medium of communication is used to praise the virtues of every conceivable type of feed grinder, crusher, shredder, cooker, and the like. Many of the claims made by the manufacturers are in direct contradiction to one another, and obviously all cannot be correct in all situations. The importance of convenience too often is given more emphasis than it deserves, and the economics of a processing method is sometimes almost entirely overlooked.

Both old and new research findings indicate that altering the form in which certain feeds are fed may affect performance. The method of offering feed to cattle may also favorably affect the financial outcome of a feeding enterprise, especially if a labor saving can be made without too great a capital investment in equipment.

Common goals of any processing method or system of offering feed to cattle are:

1. Improved feed conversion or efficiency.
2. Increased feed intake, usually accompanied by faster gain.
3. Improved carcass grade, quality, and cutability.
4. Reduction in wastage of feed at the trough or feeder.
5. Lower transportation and storage costs.
6. Less labor through mechanical handling.

611

7. Improved health of cattle through reduction of digestive disorders.

8. Reduction in harvesting costs such as occurs with field shelling of corn.

In general, feeds should be fed with the least processing necessary. It is easy to overprocess feeds without necessarily enhancing their value but at the same time adding labor, equipment, and storage costs that must all be recovered from improved efficiency or lowered profits. Consultation, both with nutritionists and engineers, is indicated if an appreciable investment in a new feed processing and handling facility is contemplated. This may be even more important if remodeling an existing plant is being planned.

PREPARATION OF GRAINS

The topic of grain preparation or processing concerns most feeders more than does roughage preparation, simply because proportionately more grain is fed. An exception is the large commercial feedlot that uses complete rations in which the roughage, even though it makes up only a small portion of the ration, is blended with the grain in processed form. In this situation roughage preparation is an important factor, but the average farmer-feeder who relies on hay or silage has few decisions to make of this kind.

In discussing the many aspects of grain preparation, more emphasis will be given to preparation of corn than to other grains because, first, corn is the most important feed fed to beef cattle that requires some form of preparation and, second, more experimental work has been done with this feed.

Corn is fed to beef cattle in the following well-recognized forms: (1) shelled corn, (2) ground, crimped, or rolled shelled corn, (3) high-moisture shelled or processed corn, (4) ground ear corn, and (5) high-moisture ground ear corn.

SHELLED CORN

This form of corn is still popular with average- to small-sized farmer-feeders, especially those who feed calves instead of yearlings. Shelled corn is of course a higher-energy feed than ear corn because ear corn contains 20 percent cob by weight. This cob fraction, while adding some desirable bulk to the ration, does reduce its density or energy content and thereby slightly reduces the rate of gain. Given a supply of higher-quality roughage such as alfalfa or mixed legume-grass hay, a farmer-feeder can combine corn and hay—two homegrown feeds—with little supplementation or processing, into a very desirable economical ration on which cattle will perform well. Such smaller-scale feeders are not so concerned with the loss of the occasional whole kernel that passes through the cattle, apparently undigested, as they usually have feeder pigs following behind their cattle as scavengers.

Shelled corn, if not over 14 to 16 percent in moisture content, can be stored in rather large quantities in contrast to ground shelled and ear corn which tend to mold and grow rancid, especially at the higher moisture level. Thus the farmer-feeder who either elects to shell his entire supply of corn, or who shells a fresh batch every month or so, will probably be feeding a fresher feed at all times.

PROCESSED SHELLED CORN

Grinding, or at least some type of processing, of dry shelled corn is almost taken for granted today. The experimental evidence to support the shift away from shelled to ground corn is not extensive. Data such as those shown in Table 234, which summarizes several studies with yearling cattle, demonstrate that little is to be gained by merely grinding corn. Possibly it is the obvious presence of whole kernels in the feces that causes so many people to want to grind their shelled corn. For those feeders who still run hogs behind their steers, grinding is unjustifiable.

A processing method has been developed that shows real promise of improving the feed efficiency of rations containing fairly high levels of shelled corn. It is usually termed "flaking," but it may be called steamrolling in some parts of the country. Flaking, as most commonly used in commercial feedlots and especially in Colorado, consists of steam-cooking or heating the shelled corn in a chamber for 10 to 12 minutes at atmospheric pressure and 200°F. The moisture content is raised to about 20 percent in this process. The corn is then rolled or flaked, the thinner the better, as shown in Table 235. The Colorado workers used a constant cost figure for all three kinds of processed corn, but in reality the processing cost varies. It is fairly obvious that mill capacity is affected by fineness of grind or thinness of flake. In any case, there was no question that the bulkier, thin flake—and, for that matter, even the thick flake—were superior to finely ground corn. Undoubtedly the bulky nature of the thinner flake is involved in maintaining the normal physiological status of the rumen. The high-moisture flakes must be dried if not fed immediately. Figure 96 shows the extent to which the thin or flat flake is bulkier than the thicker flake.

Table 234
Value of Grinding Shelled Corn for Yearling Steers

| | Nebraska[a] | | | | Iowa[b] | |
| | First Trial | | Second Trial | | | |
	Shelled Corn	Coarsely Ground Shelled Corn	Shelled Corn	Coarsely Ground Shelled Corn	Shelled Corn	Coarsely Ground Shelled Corn
Initial weight (lb)	638	642	671	670	657	656
Final weight (lb)	1,067	1,075	1,063	1,076	1,121	1,105
Average daily gain (lb)	2.38	2.40	2.17	2.26	1.94	1.87
Average daily ration (lb)						
Corn (shelled basis)	16.5	17.8	14.9	15.2	12.1	11.2
Corn (cobs)	—	—	—	—	—	—
Alfalfa hay	5.3	5.9	5.4	5.3	1.5	1.5
Corn silage	—	—	—	—	11.5	11.5
Feed per cwt gain (lb)						
Corn (shelled)	692	742	686	673	625	600
Corn (cobs)	—	—	—	—	—	—
Alfalfa hay	222	247	250	235	79	82
Corn silage	—	—	—	—	593	612
Hog gains per steer (lb)	29	15	51	25	57[c]	18[c]
Dressing percent	60.4	61.2	59.7	59.5	—	—

[a] Nebraska Mimeographed Cattle Circular 143.
[b] Iowa Animal Husbandry Mimeographed Leaflet 159.
[c] Feed saved by hogs per 100 pounds of cattle gains.

The high initial cost of the equipment for processing grain by this method makes it prohibitive for the farmer-feeder with fewer than 500 head of cattle. Several manufacturers are working to develop smaller-sized units that are efficient and yet not priced out of reach of the smaller feeder. A mill with continuous grain processing is rather easily adapted for use by the very large feedlots that both process and feed for 24 hours around the clock.

Farmer-feeders who do not follow their cattle with hogs may still resort to the hammermill, burr mill, reel-type knife mill, or roller mill for processing their dry corn. Research does not indicate that one type of mill is either better or less effective than another, so the mill that grinds the corn most economically, consistent with convenience and adaptation to automation, is the one to use. If moist feeds such as silage, haylage,

Table 235

Performance of Heifers Fed Flaked Corn of Different Thickness or Finely Ground Corn (163 Days)[a]

Method of Processing	Thin Flake ($\frac{1}{32}$ in.)	Thick Flake ($\frac{1}{12}$ in.)	Finely Ground ($\frac{1}{4}$-in. screen)
Number of animals	14	13	13
Average initial weight (lb)	485	483	490
Average daily gain (lb)	2.82	2.70	2.65
Average daily ration (lb)			
Corn	12.41	12.66	12.83
Fortified supplement	0.78	0.82	0.82
Beet pulp	1.49	1.63	1.79
Chopped alfalfa hay	0.90	0.93	0.95
Corn silage	5.58	5.82	5.52
Feed consumed per pound gain (air-dry basis) (lb)	6.14	6.66	6.88
Feed cost per cwt gain	$15.60	$17.00	$17.74
Dressing percent[b]	64.21	63.71	63.25
USDA carcass grade			
Choice	13	12	12
Good	1	1	1

[a] *Colorado Farm and Home Research*, 17:4.

[b] Dressing percent = hot carcass weight less 2% cooler shrink divided by actual final weight less 4% shrink multiplied by 100.

FIG. 96. Flaked corn, showing two degrees of thickness, with the thicker flakes at left. The bulky nature of the thin flakes ensures more nearly complete digestion and greater efficiency when used in high-concentrate finishing rations. (Colorado State University.)

or green chop are not a part of the ration, it would seem that coarse processing is preferred, to provide as much bulk as possible. If on the other hand, moist feeds are fed and the grain is fed mixed or spread on top of the moist feed, then fine grinding is preferred.

HIGH-MOISTURE SHELLED OR PROCESSED CORN

The mechanical harvesting of corn with a combine, as high-moisture shelled corn, is a development that introduces some new questions for cattle feeders. As discussed in Chapter 19, corn has been harvested as high-moisture ear corn for many years, but in those circumstances it was to salvage a frosted crop of immature corn. The change to field shelling is prompted by agronomic aspects of corn production rather than for the purpose of harvesting a special kind of cattle feed.

Table 236 offers a concise explanation for one of the chief reasons for the change to field shelling. If corn is field shelled at approximately 30 percent kernel moisture, there is a harvesting loss of about 5 bushels of ears less per acre, compared with conventional harvesting at about 18 percent kernel moisture. The kernel losses due to inefficient cylinder shelling noted in the table have since been almost eliminated by more efficient cylinder and concave design in field-shelling combines. Farmers who grow corn as a cash crop also are using field shelling, but they usually prefer to wait until the moisture content of the corn is only 20 to 24 percent. The corn is then dried on the farm with forced air, either heated or unheated, in batch driers, or it may be delivered to elevators in wet form to be dried in large-capacity continuous-flow driers.

Table 236

Field Losses in Corn Harvested with a Picker-Sheller at Different Moisture Levels[a]

Observation	Percent Kernel Moisture Content When Harvested			
	35	30	25	18
Loss expressed as bushels per acre				
Ear (machine and detached)	0.8	0.8	2.7	5.4
Shelled corn (snapping rolls and separation)	2.6	1.6	2.3	2.1
Cylinder (on cobs)	5.2	2.4	2.4	.4
Total	8.6	4.8	7.4	7.9

[a] Illinois Cattle Feeders' Day Report, 1959.

Cattle feeders who have silos and who practice field shelling prefer to harvest their corn earlier or at about 25 to 30 percent moisture content, storing it as shelled corn in either airtight or conventional silos. If large pits or trenches are used, as they are by the commercial feedlots in Colorado and the Panhandle, the shelled corn is ground as it is stored, to reduce spoilage and to make it easier to mix with silage in self-unloading mixing wagons or trucks without having to process or grind at that time.

Many experiments have been conducted to compare ground or rolled high-moisture shelled corn with processed dry shelled corn. Michigan workers summarized a number of these tests, and their results may be found in Table 237. Note that feed efficiency was unimproved and gains were slightly reduced in the lots on high-moisture corn. Thus it is apparent that the rapid increase in this method of harvesting and feeding corn is the result of improvements in harvesting efficiency rather than its effect on beef production.

High-moisture corn is highly inefficient in promoting cattle gains if fed whole. The data in Table 238 offer an explanation. It is not known why, but more whole high-moisture corn kernels pass undigested through the gastrointestinal tracts of cattle than do kernels of dry corn. The same phenomenon has been observed with respect to the kernels in corn silage. Special recutter attachments for corn silage harvesters have been designed to reduce the proportion of whole kernels in corn silage. The usual way to eliminate this loss in feeding high-moisture shelled corn, however, is to follow the cattle with feeder pigs, but this practice is unfeasible in large commercial feedlots.

When a farmer-feeder contemplates changing to field-shelling his corn as high-moisture corn, he should use that opportunity to reexamine his entire feeding system so as to take full advantage of automation and thus reduce the handling and storage of his feeds to the simplest system possible. For example if high-moisture corn is to be stored in the most economical type of airtight unit, the corn must be stored as shelled corn because the bottom-unloader in the lower priced airtight silos will not handle ground grain. Placing a roller or burr mill at the mouth of the bottom-unloader and tying it in with an auger-bunk feeding system eliminates the handling of the corn entirely. Metered devices may be used to add silage and supplements into the auger system as the corn is being removed from the silo and processed, thus eliminating the need for a mixer of any kind.

For larger-scale feeders who use mobile mixing wagons or trucks to feed in fenceline bunks, a short elevator may be used to deliver the corn and other ingredients directly into the vehicle instead of into a bunk, again doing away with the need for labor.

It is possible to design literally scores of combinations of feed storage

Table 237

Summary of Eight Experiments Comparing Feeding Value of High-Moisture versus Dry Shelled Corn[a]

Station	Days on Feed	Animals per Lot	Percent Moisture		Daily Gain (lb)		Concentrate per 100 Pounds Gain (lb)		Percent Stimulation over Dry Corn	
			High-Moisture Corn	Dry Corn	High-Moisture Corn	Dry Corn	High-Moisture Corn[b]	Dry Corn	Gain	Efficiency
Michigan	147	10	30	18	2.20	2.21	696	670	0	-4
Indiana	98	40	26	18	2.22	2.38	366	375	-7	2
Illinois	112	9	37	15	1.51	1.89	832	761	-20	-9
Illinois	112	9	29	15	1.91	1.89	754	761	1	1
Illinois	112	9	24	15	1.90	1.89	761	761	0	0
Indiana	140	40	30	20	1.95	2.09	420	451	-7	7
Michigan	203	13	25	19	1.86	1.89	606	629	-2	4
Iowa	133	24	35	14	3.56	3.60	688	662	-1	-4
Average	132	19	30	17	2.14	2.23	640	634	-5	0

[a] Adapted from Michigan Beef Cattle Day Report, 1960.
[b] Corrected to same moisture content of dry corn.

Table 238
Percent High-Moisture Shelled Corn Passed
Undigested Through Yearling Steers[a]

Item	Observation
Number of animals in experiment	48
Average initial weight, February 3 (lb)	762
Average final weight, June 15 (lb)	1,234
Average daily gain (lb)	3.56
Average daily ration (lb)	
35 percent moisture shelled corn	24.3
Alfalfa hay	5.0
Supplement and minerals	1.1
Percent corn passed undigested through steers	
April 14	21
April 21	19
April 28	24
May 5	19
May 12	24
June 11	16
Average for 133-day period	21

[a] Adapted from Iowa State University AH Leaflet R14, 1960.

and handling systems. Most Land Grant universities have agricultural engineering extension specialists on their staffs who may be consulted for advice, and of course manufacturers of the components of feeding systems also provide consulting services.

GROUND EAR CORN

Although more than a score of comparisons have been made between shelled corn and ground ear corn, it appears impossible to rate one above the other without making a number of qualifications. Most of the early experiments showed that shelled corn was somewhat superior in respect to feed intake, rate of gain, and degree of finish produced, but several more recent tests have raised doubts with respect to these points. For instance, in the experiments reported in Table 239, calves and yearling cattle gained at practically the same rate on shelled corn as on ground ear corn. In all of the Ohio tests, substantially fewer bushels of corn were eaten per hundredweight gain by the steers fed ground ear corn.

Table 239

Comparative Value of Shelled and Ground Ear Corn for Feeder Cattle

	Calves				Yearlings		Yearlings	
	Ohio (Average 3 Trials)		Nebraska		Nebraska (Average 4 Trials)		Iowa	
	Shelled Corn	Ground Ear Corn	Shelled Corn	Ground Ear Corn	Shelled Corn	Ground Ear Corn	Shelled Corn, 240 Days	Ground Ear Corn, 120 Days; 25% Ground Ear Corn, 75% Ground Shelled Corn 120 Days
Average daily gain (lb)	2.00	2.10	2.52	2.48	2.59	2.53	1.94	1.98
Average daily ration (lb)								
Corn	9.7	11.2	10.9	12.8	18.3	22.3	12.1	12.0[a]
Protein concentrate	1.8	1.8	1.1	1.3	—	—	1.7	1.7
Corn silage	7.1	7.1	—	—	—	—	11.5	11.5
Hay	2.7	2.8	3.9	3.9	5.2	4.0	1.5	1.5
Feed per cwt gain								
Corn (bu)	8.7	7.6	7.7	7.4	13.5	13.2	11.2	10.9
Protein concentrate (lb)	88	83	43	51	—	—	88	86
Corn silage (lb)	267	260	—	—	—	—	593	581
Hay (lb)	280	274	155	158	212	164	79	77
Hog gains per steer (lb)	82	38	17	13	36	10	57	11
Net return over feed costs (including pork)	—	—	$8.37	$1.48	$8.52	$7.17	$17.98	$16.81

[a] Grain portion only.

In another Iowa test not reported in the table, ground ear corn produced significantly less gain than shelled corn with only 1 pound of linseed meal, but slightly larger gain with 2 pounds of protein supplement, suggesting that the ground ear corn rations need a somewhat higher level of protein supplementation.

Ground ear corn is superior to shelled corn for cattle that are full-fed on legume pastures, such as sweet clover and alfalfa, because the dry particles of cob tend to overcome the tendency of these forages to cause scours and bloat. However, shelled corn appears to be much better on bluegrass and other nonlegume pastures, because larger quantities of grain are eaten and therefore greater gains are secured. Shelled corn is less badly damaged by rain than is ground corn and is better suited for feeding outdoors and in self-feeders, which may have rain or snow blown into them during severe storms.

COMPARISON OF METHODS OF GRINDING CORN

The question of which type of mill to use for grinding either shelled or ear corn receives an unwarranted amount of discussion among cattle feeders. It is true that small differences in feed intake can be demonstrated, especially over short periods of time, if corn is coarsely ground. Such differences are not only small, but they may be more than offset by slower mill capacity and larger power requirements. Data such as those in Table 240 show that the type of mill used in grinding corn is not of great importance. In the Illinois test, where the degree of average fineness was the same for the corn ground by all three mills tested, steers tended to sort out the larger particles of cob in the hammermill-ground corn, causing slightly lower actual consumption of ground ear corn and slightly slower gains. If feedbunks are not perfectly leakproof or if they are exposed to strong winds, there is appreciable loss of the finer particles of corn.

Table 240

Comparison of Three Methods of Grinding Ear Corn for Full-Fed Yearling Steers in Drylot (56 Days)[a]

	Burr Mill	Hammermill	Reel-Type Knife Mill
Number of steers	10	10	10
Average initial weight (lb)	947	941	948
Average final weight (lb)	1,068	1,059	1,069
Average total gain (lb)	121	118	121
Average daily gain (lb)	2.17	2.11	2.17
Average daily feed consumed (lb)			
Ground ear corn	17.1	17.4	17.1
Cottonseed meal	1.49	1.49	1.49
Mixed hay	3.6	3.6	3.6
Feed consumed per cwt gain (lb)			
Ground ear corn	790.1	828.4	790.1
Cottonseed meal	68.8	70.4	68.8
Mixed hay	168.6	172.9	168.6
Feed cost per cwt gain ($)[b]	23.09	24.10	23.09
Cobs refused per day per steer (lb)	0.22	0.82	0
Average fineness moduli	4.33[c]	4.20	4.30

[a] Illinois Feeders' Day Report.
[b] Feed prices used: ground ear corn, $1.65 per bushel; cottonseed meal, $80 per ton; mixed hay, $20 per ton.
[c] Flour has a fineness modulus of 0 and particles larger than 0.375-inch are given a score of 7.

HIGH-MOISTURE EAR CORN

Many of the advantages mentioned with respect to high-moisture shelled corn also prevail when it comes to harvesting and feeding high-moisture ear corn. Most important, harvesting date is 2 to 4 weeks earlier, resulting in less ear loss. A major disadvantage is the relative cost of storing the cob portion, a low-grade roughage, in expensive silos—especially if airtight silos are used. The cob portion can supply considerable nutrients, especially when fed in ensiled form. Michigan studies showed that 7 percent fewer bushels of corn were required per hundredweight gain when high-moisture ear corn was fed in place of cracked shelled corn.

A more direct comparison may be made between dry and high-moisture ear corn. Table 241 shows that, in 14 different tests, ground high-moisture ear corn was about 10 percent more efficient than dry ground ear corn. Gains were slightly improved also, in contrast to ground high-moisture versus ground dry corn in shelled form.

High-moisture ear corn should be ground before going into the silo. Conventional silos with ordinary top-unloaders are quite satisfactory for this purpose, although the airtight silos have more built-in convenience. If the airtight bottom-unloading silo is used, the heavy-duty forage un-loader is required. This may be no inconvenience if the silo is used during part of the year for forage storage.

PREPARATION OF OATS, BARLEY, AND WHEAT

The question of processing oats, barley, and wheat for cattle is discussed in Chapter 15. As the saving effected by grinding oats for calves usually does not exceed 5 to 10 percent, grinding oats for animals of this age is unlikely to be profitable unless oats are high-priced. It is doubtful, however, that barley or wheat should ever be fed whole because so many of the hard grains are swallowed whole and are imperfectly digested.

The steam processing of barley is mentioned in Chapter 15, and results of an Arizona trial comparing dry rolling and rolling of steam-processed barley are presented in Table 126. This and other experiments showed that steam processing increases feed intake and daily gain about 10 percent, but feed efficiency is unimproved when a roughage source other than the barley hulls is included in the formula.

High-moisture barley, stored in airtight silos at 30 percent moisture, was compared with dry rolled barley by Minnesota workers. When only a small amount of hay was included in the ration, the high-moisture barley was consumed at a higher rate, resulting in an increase in gain—but, as

Table 241
Summary of 14 Experiments Comparing Feeding Value of High-Moisture versus Dry Ground Ear Corn[a]

Station	Days on Feed	Animals per Lot	Percent Moisture		Daily Gain (lb)		Concentrate per 100 Pounds Gain (lb)		Percent Stimulation over Dry Corn	
			High-Moisture Corn	Dry Corn	High-Moisture Corn	Dry Corn	High-Moisture Corn[b]	Dry Corn	Gain	Efficiency
South Dakota	97	18	40	15	2.13	1.78	790	878	19	11
Indiana	117	10	32	18	2.47	2.34	866	988	6	14
Indiana	117	10	32	18	2.56	2.33	807	951	10	17
Indiana	126	36	32	15	2.14	2.18	555	617	−2	11
Iowa	119	36	31	15	2.98	3.05	675	750	−2	11
Iowa	56	36	38	14	3.34	3.24	471	528	3	12
Indiana	133	36	37	24	1.86	1.94	634	660	−4	4
Colorado	112	8	53	14	2.41	2.52	433	483	−5	10
Michigan	147	10	31	18	1.97	1.53	754	973	29	23
Colorado	140	8	55	15	2.08	2.25	519	564	−8	8
Iowa	175	6	30	14	2.12	2.31	726	805	−8	10
Iowa	175	6	30	14	2.40	2.32	667	634	3	−5
Ohio	119	21	36	12	2.15	2.04	685	803	5	15
Michigan	203	13	23	19	1.65	1.60	734	725	3	−1
Average	131	18	36	16	2.30	2.24	663	740	3	10

[a] Adapted from Michigan Beef Cattle Day Report, 1960.
[b] Corrected to same moisture content of dry corn.

was true of steam-processed barley, there was no improvement in feed efficiency. When about 5 pounds of hay were fed per day, there was no increase in either intake or gain, and again feed efficiency was unimproved. Apparently both steam processing and rolling of high-moisture barley result in a large bulky flake that is more palatable to cattle on high-concentrate rations. The resulting increased intake steps up gain but without an improved feed efficiency or reduction in cost of gain. Therefore, processing of barley, other than grinding, dry rolling, or crimping, seems unwarranted at present.

PREPARATION OF SORGHUM GRAIN

Combined sorghum grains are usually so small and hard that they should be ground or processed for cattle of all ages. Early experiments with calves at the Texas station showed that grinding increased the value of threshed milo by 41 percent and unthreshed milo heads by 62 percent. The unground grain was poorly utilized, as indicated by the fact that the swine gains per steer were about four times as great as those of the calves fed ground milo grain.

Because milo or sorghum has assumed such an important role in the feedlots of Oklahoma, West Texas, New Mexico, and Arizona, and because it has not performed as well as its chemical composition would lead one to expect, active research programs to improve the feeding value of sorghum are under way, in the southwestern states particularly. Arizona workers have been foremost in advocating a steam-processing method, which consists of holding whole milo for 20 minutes in a 1-ton-capacity chamber into which live steam is introduced at numerous points under 20 pounds pressure. The moisture content of the grain is raised 20 percent while in the chamber, and, at the end of the tempering period as the grain drops from the chamber onto the 18- by 30-inch rollers below, it has a temperature of about 215°F. A very thin flake is preferred, hence the rollers are set at 0.003-inch spacing. Many feedlots in the Sorghum Belt now use commercially made mills comparable to the one at Arizona, but the need is for a continuous-flow mill that will achieve the same result while increasing the output and providing fresh feed throughout the 24-hour day.

A test was conducted by Oklahoma workers to compare all of the present methods of milo processing used in cattle-finishing rations. They full-fed steer calves all of the particular milo assigned to the lot plus a basal maintenance ration consisting of 35 percent chopped alfalfa, 23 percent cottonseed hulls, 40 percent cottonseed meal, and 1 percent each of salt and dicalcium phosphate. Vitamin A and antibiotic premix were also included, but no hormones were given. Five processing methods were tested. Two consisted of finely or coarsely ground milo, ground through a ⅛- and ³⁄₁₆-inch screen, respectively. Another was reconstituted milo which was prepared by placing dry whole milo in a small silo with enough water to promote fermentation, resulting in high-moisture milo with 30 percent moisture. It was rolled daily as fed. The fourth method was patterned after the Arizona steam process and flaking procedure, and the fifth consisted of reconstituted milo that was additionally steamrolled as fed. The results are summarized in Table 242.

The steam process-flaking method significantly increased feed intake and rate of gain over coarse grinding, but only very slightly improved feed efficiency. If the steers had been sold when they reached a common weight rather than all at one time, undoubtedly the faster-gaining steers on steam process-flaked feed would have been more efficient. Reconstituting milo resulted in significant improvement in efficiency without decline in gain. This same improvement in efficiency is noted in most studies with regular high-moisture milo. Fine grinding improved performance in all aspects, as it has done in previous Oklahoma tests. The coarse grind used in this test was actually finer than that used by most feedlots that grind with a hammermill.

Table 242

Feedlot Performance of Steer Calves on Milo Rations (171 Days)[a]

Processing Method	Coarsely Ground	Finely Ground	Reconstituted	Steam Process-Flaked	Reconstituted Steamrolled
Number of steers	9	9	9	9	9
Average initial weight (lb)	501	490	489	498	491
Average daily gain (lb)	2.43	2.51	2.44	2.63	2.42
Average daily ration (lb)					
Milo	10.58	9.80	9.81	11.29	9.28
Basal	8.69	8.56	8.51	8.80	8.49
Total	19.27	18.36	18.31	20.09	17.78
Feed per pound gain (lb)	7.92	7.32	7.49	7.64	7.35
Percent Change Compared with Coarsely Ground Milo					
Average daily gain (lb)		+3.3	+0.4	+8.2	−0.4
Average daily feed (lb)		−7.4	−7.3	+6.7	−12.3
Milo per pound gain (lb)		−10.1	−8.2	−1.6	−11.9

[a] Adapted from Oklahoma Agricultural Experiment Station Miscellaneous Publication MP-79, 1967.

HIGH-MOISTURE SORGHUM

High-moisture milo or sorghum may be stored in airtight silos or very well packed in conventional silos in several forms. The entire head may be removed and ground along with some stem. These heads contain about 70 to 75 percent moisture. Milo may also be combined when it contains about 30 percent moisture and ensiled whole. Finally, water may be added to dry milo to make reconstituted high-moisture milo. The Texas station has extensively studied the feeding of high-moisture milo. In a comparison between high-moisture ground heads and dry ground grain they got comparable gains and a significant improvement in feed efficiency from the high-moisture heads. They have also consistently observed improvement in feed efficiency when reconstituted grain was compared with dry grain. It appears that storing milo at about 30 percent moisture in silos offers the same potential for increasing the value of sorghums as it does for corn.

Table 243
Rolled vs. Pelleted Milo for Feeder Heifer Calves[a]

Preparation of Milo	Medium Rolled	Finely Ground and Pelleted ($\frac{3}{8}$-in.)
Number of heifers per lot	9[b]	10
Average weights (lb)		
Initial	498	497
Final	826	838
Average daily gain	2.09	2.17
Average daily ration (lb)		
Rolled milo	11.96	
Pelleted milo		11.49
Cottonseed meal + stilbestrol[c]	1.50	1.50
Dehydrated alfalfa meal pellets	1.00	1.00
Sorghum silage	11.58	10.77
2–1 mineral mix	0.10	0.06
Feed required per cwt gain (lb)		
Milo	572	529
Cottonseed meal	72	69
Dehydrated alfalfa pellets	48	46
Sorghum silage	554	496
Feed cost per cwt gain ($)[d]	18.64	18.04

[a] Oklahoma Livestock Feeders Day Report.

[b] One heifer removed from data because of founder.

[c] Cottonseed meal fed per head daily contained 10 mg stilbestrol.

[d] Charge of $0.10/cwt for rolling; $0.25/cwt for fine grinding and pelleting.

Table 244
Value of Grinding or Chaffing Hay for Feeder Cattle

	Nebraska (Two Trials)		Iowa			Minnesota	
	Shelled Corn and Alfalfa Hay		Shelled Corn and Alfalfa Hay			Ground Shelled Corn and Alfalfa Hay	
	Long Hay Fed Separately	Ground Hay Mixed with Corn	Long Hay Fed Separately	Coarsely Chopped Hay Mixed with Corn	Finely Chopped Hay Mixed with Corn	Whole Hay Fed Separately	Ground Hay Mixed with Corn
Average initial wt (lb)	611	610	692	690	690	688	689
Average daily gain (lb)	2.20	2.15	2.35	2.40	2.35	2.49	2.52
Average daily ration (lb)							
Corn	13.5	13.8	15.7	15.3	14.8	17.0	17.2
Protein supplement	—	—	1.0	1.0	1.0	—	—
Hay[a]	5.5	3.5	4.6	4.6	4.5	7.0	6.4
Feed per cwt gain (lb)							
Corn	616	642	667	636	629	682	684
Protein supplement	—	—	42	41	42	—	—
Hay	250	162	196	192	194	279	256

[a] Including hay that was wasted.

ROLLED VERSUS PELLETED GRAIN

Pelleting or cubing of the grain alone or of the entire concentrate portion of the ration has been given much attention. The Oklahoma study summarized in Table 243 shows that considerably less sorghum was required per unit of gain when the sorghum was pelleted instead of rolled, and gains were cheaper even though a much larger charge was made for pelleting than for rolling. One explanation may be that all of the smaller particles usually resulting from rolling or grinding sorghum grain were saved by being incorporated in the pellet. Another possibility is that of higher digestibility of the pelleted grain. The advantage for pelleting might be even greater if compared with whole or finely ground sorghum grain.

GRINDING AND CHAFFING HAY

The claim that the feeding value of hay is improved by grinding is not supported by most of the tests comparing long and ground hay. (See Table 244.) Instead, these experiments show that cattle often do not like ground hay, that feeding ground hay decreases the time spent in rumination, and that, because of reduced consumption and less thorough digestion of the ground roughage, gains are smaller and more costly.

The practice of grinding the hay and feeding it mixed with the grain has sometimes resulted in larger gains than when whole feeds were fed, but a number of experiments yielded opposite results. In few if any of the tests have the advantages been sufficient to pay for the expense of grinding and mixing the feeds, especially if gains made by the swine kept with the cattle are taken into account.

Apparently the principal advantage of ground hay over long hay is that less is wasted. Cattle fed long hay often pull some of it out of the manger or bunk and trample it under foot and they usually leave some of the stems uneaten, particularly if the hay is full-fed. Both of these losses are greater if the hay is coarse and poor in quality. Since only poor quality hay is fed with appreciable waste, it is this kind of hay that is most often ground. Even if the waste of the long hay were as much as 25 percent, it is doubtful whether grinding would be profitable, because the coarser portions constituting most of the waste usually are worth much less than the cost of grinding the entire supply of hay fed.

If hay is so scarce and expensive that waste must be prevented more than is possible by using properly built racks and mangers, it should be cut or chaffed by running it through a silage cutter or similar machine that will cut it into 2-to-4-inch lengths. Chaffed hay is much preferred by ruminants to ground hay and is prepared with much less labor and power.

One of the advantages of chaffed or chopped hay is the low cost of harvesting and storage. If it is chopped from the windrow with a field cutter and blown into the barn where it can be pushed down to the cattle with little or no handling, less labor is required than in harvesting and feeding long or baled hay. Also, approximately twice as many tons of chopped as of long hay can be stored in a given shed or mow.

Hay that is to be chopped in the field must be much drier when stored than ordinary hay or hay that is stored in the bale. Consequently a rather high percentage of the leaves and fine stems are reduced to such fine particles by the chopper and blower that they are lost in harvesting and feeding. Storage of chopped hay with a moisture content much above 15 percent usually results in loss of valuable feed nutrients from extensive "mow burning." There is also the risk of losing both hay and barn from spontaneous combustion. Thus the field chopping of hay can be recommended only when it is to be stored in a barn or bin equipped with forced ventilation. Stored in this manner it may be chopped while still sufficiently tough to prevent loss of valuable leaves and stems because the excess moisture is driven off by the forced ventilation before serious heating occurs.

Table 245

Effect of Method of Harvest and Storage of Alfalfa Hay on Its Value for Feeder Cattle (Average of 3 Trials)[a]

	Stacked Alfalfa Hay	Baled from Wind-row	Chopped from Wind-row	Dehydrated Alfalfa Hay	Molasses Alfalfa Silage
Average initial weight (lb)	696	692	689	694	717
Average daily gain (lb)	2.04	2.11	2.13	2.19	2.09
Average daily ration (lb)					
Grain	15.5	15.6	15.4	15.5	15.8
Soybean meal	1.0	1.0	1.0	1.0	1.0
Hay or silage	3.6	3.9	3.9	3.5[b]	9.5[b]
Feed per cwt gain (lb)					
Grain	773	756	738	724	777
Soybean meal	48	47	46	45	47
Hay or silage	177	187	183	155[c]	453[c]
Average TDN per cwt gain	712	705	689	669	702

[a] Colorado Experiment Station Progress Report.
[b] Includes approximately 0.5 pound of long stacked hay.
[c] Includes approximately 20 pounds of long stacked hay.

In the final analysis, the method of harvesting and storing roughage matters little to cattle being fed on finishing rations, as seen in Table 245. After all, roughage makes up only about 15 to 25 percent of the total ration consumed. Thus the method used in harvesting a hay crop which will be fed largely to feeder cattle is determined by the relative cost of the various harvesting methods.

EFFECT OF GRINDING AND CHOPPING ROUGHAGE ON PALATABILITY

It may seem that ground or chopped roughage should be more palatable than whole roughage because it can be eaten more easily. Also, all the ground roughage fed is usually eaten, but a noticeable percentage of long hay or stover is often refused. With coarse, stemmy hay and corn stover the refused portion may amount to 20 to 40 percent of the total amount fed.

Feeding experiments show, however, that cattle usually eat more long hay than ground or chopped hay if both kinds are fed according to appetite.

For example, in the Nebraska experiments reported in Table 244 it was necessary in both trials to limit the feeding of long hay in order to obtain the desired consumption of corn, but there was difficulty in getting the lots fed ground alfalfa to eat their hay. Also, in palatability tests conducted at the Wisconsin station with purebred beef cows and heifers a marked preference for long hay and coarsely chopped hay (1.8 inch) and a dislike for hay chopped to 0.5 and 0.25 inch was noted. In one of the Wisconsin tests, where all four kinds of hay were available in adjacent mangers, whole hay constituted 71.5 percent of all the hay consumed.

PELLETING OF BEEF CATTLE FEEDS

The pelleting or cubing of roughages or of complete rations by using pressure to extrude or force the finely ground feedstuffs through dies ranging from 0.25 to 1 inch in diameter, is one of the noteworthy developments of the last 20 years. The advantages usually named in favor of this rather costly method of preparing beef cattle feeds are the following:

1. Easier handling because of reduced bulk.
2. Increased feed intake, especially of lower-quality roughages.
3. Improved feed conversion rates.
4. Increased rate of gain on high-roughage rations.
5. Control of ratio of concentrate to roughages at the desired level.

PELLETING HAY

Of all the various components of a beef cattle ration, the roughage portion is most affected by pelleting from the standpoint of storage and labor of feeding. The bulkiness of roughage also is the limiting factor with respect to intake or level of consumption by beef cattle. Thus it appears that pelleting of roughage should offer more promise than the pelleting of concentrates. Several tests have been conducted with pelleted roughages, with conflicting results. The Illinois Dixon Springs station conducted two tests with an all-hay ration for stocker calves with results overwhelmingly in favor of pelleting, as may be seen in Table 246, which summarizes the first test. The hay fed was a first-cutting mixed hay of average to poor quality consisting of two-thirds timothy and one-third alfalfa. Before being pelleted into $\frac{3}{16}$-inch pellets, the hay was finely ground. No molasses or steam was used in the pelleting process and, as the data show, daily consumption of the ration was greatly improved by the pelleting.

The second trial produced similar results with poor-quality alfalfa, with

Table 246

**Comparison of Timothy-Alfalfa Mixture When Fed
Chopped, Pelleted, or as Long Hay (119 Days)**[a]

Method of Hay Preparation	Baled	Pelleted	Chopped
Number of steers	15	15	15
Average initial weight (lb)	421	430	423
Average final weight (lb)	496	636	497
Average daily gain (lb)	0.63	1.73	0.62
Daily feed consumption (lb)	11.0	15.7	10.7
Feed required per cwt gain (lb)	1,732	906	1,722
Feed cost per cwt gain ($)	17.32[b]	13.59	17.22
Gain per ton of feed (lb)	115.5	220.7	116.2

[a] Illinois Dixon Springs Mimeograph.
[b] Feed prices used: baled and chopped hay, $20 per ton; pelleted hay, $30 per ton.

Serecia lespedeza, and again with a timothy-alfalfa mixture. As might be expected, some of the advantage of the more rapid winter gains made by the calves on pelleted hay was lost during the subsequent summer grazing period because of compensatory gain by the lighter cattle. The roughages fed in these tests were only poor to average in quality, but much of the hay fed in any but the mountain and western states is of similar quality.

Even if further tests reveal that pelleting roughages for beef cattle is advantageous, it will be helpful if equipment manufacturers can develop a portable pelleting machine that can chop and pellet the hay out of the swath in the field. The matter of excess moisture in hay for safe storage as pellets also presents a problem that may be solved only by adding drying units to the pelleting machine or in the storage area. The cost of such a complex machine probably would be prohibitive for individual farmer-feeders. For larger commercial feedlots that buy their roughage in the neighborhood, this method of harvesting and feeding roughage shows promise, but it is doubtful whether the average or small feeder will be pelleting roughages in the near future unless he has it done on a custom basis.

WAFERS AND CUBES

Great strides are being made in developing hay-harvesting equipment that processes the hay from the windrow into extruded wafers or cubes.

It is estimated that 250,000 tons of alfalfa hay were processed by this method in 1967, much of it in California, Arizona, and New Mexico.

The wafers or cubes range in size from 1.25 inches square to 4 inches but most machines make the smaller cube. The wafering or cubing machines are self-propelled, pulling a trailer bin that holds about 4 tons of wafers. Each machine will handle 500 to 1,000 acres during the season, depending on the number of cuttings. Their rated capacity is usually about 5 tons per hour. In California economic studies it was estimated that the complete cost of wafering was $7.60 per ton as compared with $4.80 for baled hay.

The large 4-inch wafer first made did not prove especially satisfactory, and gains and feed efficiency were unimproved. There was much wastage because cattle dropped the wafers to the ground and into the manure or mud. A large commercial feedlot in California compared smaller 1.25-inch cubes with comparable hay fed in baled form to steer calves as a stocker ration and obtained gains of 1.55 and 0.95 pounds per day for cubes and baled hay respectively. The cube-fed calves required 28 percent less alfalfa. Thus it appears that when cubes make up all or most of the ration, they are more efficient, and if the cost of making the cube is no more than $3 to $4 per ton more than the cost of making hay into bales, the use of cubes may be warranted.

If cubed alfalfa is used in finishing rations, some calves tend to sort or select out the cubes, leaving other calves to eat a higher-concentrate ration. California workers were interested in the effect this selectivity might

Table 247

Influence of Form of Hay on Feedlot Performance of Beef Steers[a]

Form of Hay	Whole Cubes	Ground Cubes	Ground Hay
Number of animals	36	36	36
Average initial weight (lb)	617	616	629
Average final weight (lb)	945	934	978
Average daily weight gain (lb)	2.94	2.84	2.93
Slaughter data			
Dressing percent	60.18	60.40	60.44
Percent choice grade	64.0	69.0	72.0
Feed intake and utilization			
Daily feed intake (lb dry matter)	17.66	18.09	18.54
Feed per pound gain (as fed) (lb)	6.59	6.97	6.32

[a] Adapted from California Cattle Feeders' Day Report, 1966.

have on performance of yearlings; hence they fed hay in three forms—whole cubes, ground cubes, and chopped hay. The hay, regardless of form, was mixed with the finishing ration in the ratio of 15:85. Results are shown in Table 247. They concluded that the amount of sorting done was serious enough in the early stages of the feeding period to warrant crushing of the cubes.

Field wafering or cubing, done on a custom basis or by cooperatively owned equipment to keep down investment costs, has enough advantages with respect to handling and storage to give it a place in the already wide variety of methods of harvesting and feeding a hay crop.

PELLETING COMPLETE RATIONS

Increased mechanization of feed handling in cattle feeding has caused interest in complete mixed rations. Roughages in chopped form, either dry or as silage or green chop, are often mixed with the concentrate portion of the ration for feeding with automatic unloading wagons or in various types of conveyor or auger-bunks. Naturally the idea of pelleting the entire ration, where dry roughages are used, offers the possibility of simplifying the feeding operation considerably.

The Illinois Dixon Springs station has done some pioneer work with complete pelleted rations. Table 248 shows the result of a 130-day test comparing a complete ration fed in pelleted and in meal form. The steers fed on pellets gained faster and more efficiently, but if a $6 per ton pelleting charge is made, the economic advantage is lost.

Pelleting the complete ration may make it possible to increase the total amount of roughage in the ration without reducing performance of the steers. If this should prove to be true after more extensive testing, costs of gain could be reduced because good roughages generally are a more economical source of total digestible nutrients or energy, pound for pound, than concentrates, at least in the Midwest and in many irrigated sections of the West.

Table 249 is a summary of another Dixon Springs station experiment in which rations with various ratios of concentrate to roughage were compared when fed in pelleted form. Increasing the roughage from 25 to 35 percent did not change performance appreciably, and even when the roughage content of the ration was increased to 45 percent, reduction in rate of gain was not serious.

Pelleting the complete ration as a general farm feedlot practice awaits further developments on the part of the processing equipment manufacturers. Some commercial feedlots are pelleting their complete rations with

Table 248

Comparison of Feeding a Ration as Pellets and as a Meal to Yearling Steers (130 Days) [a]

Form of Ration	Pellets Self-Fed	Meal Self-Fed	Meal Hand-Fed
Number of steers	18	18	18
Average initial weight (lb)	651	646	653
Average final weight (lb)	1,008	981	969
Average total gain (lb)	357	335	316
Average daily gain (lb)	2.75	2.58	2.43
Average daily ration (lb)			
Concentrate and hay[b]	20.0	21.0	20.4
Corn silage	12.1	11.9	13.0
Feed eaten per cwt gain (lb)			
Concentrate and hay	729	845	840
Corn silage	442	463	534
Cost of feed ($)	60.20[c]	64.63	61.80
Cost of feed per cwt gain ($)	16.86	19.29	19.56
Sale weight (lb)	947	937	923
Shrink (lb)	61	44	46
Farm value less cost of steer and feed ($)	29.03	20.93	14.64
Dressing percent	58.5	58.5	59.0

[a] Illinois Dixon Springs Mimeograph 40-329.
[b] The pellets or meal consisted of 65% ground ear corn, 5% blackstrap molasses, 10% soybean meal, and 20% ground alfalfa hay.
[c] Feed prices: ear corn, $1.40 per bushel; soybean meal, $75 per ton; molasses, $34 per ton; hay, $24 per ton; corn silage, $10 per ton. No charge was made for pelleting.

satisfactory results, but capital investments in costly equipment can be spread over a far greater number of cattle in such operations than is possible with farm or ranch feeders.

SELF-FEEDING VERSUS HAND FEEDING

Self-feeders are commonly used in finishing cattle, especially for cattle given a short feed before being marketed. Because such animals are on a full grain ration during practically the entire feeding period, the self-fed method of feeding has many advantages. On the other hand, self-feeders are impractical for an extended feeding period using less than a full feed of grain.

The principal advantages of using self-feeders are the following:

1. There is a saving in labor per steer fed, on smaller farms at least.
2. Larger daily gains are secured.
3. The cattle are less likely to go off feed.

Of these advantages, the saving of labor is by far the most important. By using self-feeders with large bins or hoppers, it is necessary to feed cattle only once every week or 10 days. Large quantities of feed can be prepared and mechanically delivered directly into the feeders with augers or self-unloading wagons, with a minimum of time and labor. Still more convenient is the practice of grinding directly into the self-feeder, in which case almost all handling of grain is eliminated.

Hand feeding, on the other hand, requires the preparation of each feed at the time it is fed, or rehandling if the material is prepared in quantity and stored. This saving of labor is of most consequence to the farmer-feeder during the summer when he is busy in the fields with farm work. For this reason, self-feeders are more extensively used in summer than in winter feeding. Carefully conducted feeding experiments indicate that self-feeding usually results in larger gains than hand feeding. This is usually true even where the hand feeding is carefully done at regular

Table 249

Response of Yearling Feeder Steers Self-Fed Complete Pelleted Rations of Varying Ratios of Concentrate and Roughage[a]

	Lot 1	Lot 2	Lot 3
Number of steers	15	15	15
Pelleted ration (%)			
Timothy-alfalfa hay	25	35	45
Ground shelled corn	65	55	45
Soybean meal	10	10	10
Average initial weight (lb)	703	700	698
Average final weight (lb)	1,145	1,137	1,112
Average daily gain (lb)	2.89	2.85	2.71
Average daily pellet intake (lb)	21.5	22.8	23.0
Pellets required per cwt gain (lb)	745	800	851
Cost of pellets per cwt gain ($)	17.20[b]	17.49	17.58
Farm value of steer less feeder steer and feed cost ($)	15.53[c]	16.90	13.53

[a] Illinois Dixon Springs Mimeograph 40-333.
[b] Cost of feeds: ground shelled corn, $1.26 per bushel; soybean meal, $64.80 per ton; timothy-alfalfa hay, $20 per ton; grinding, mixing, and pelleting cost per ton, $6.
[c] Includes cost of pelleting ration.

intervals by experienced feeders. Under ordinary farm conditions, where there is often irregularity in the time of feeding, quantity and character of rations, and so on, the difference in favor of self-feeding is usually greater than the results obtained in experimental trials comparing these two methods.

Next in importance to the saving of labor is the value of the self-feeder in lessening the tendency of the cattle to go off feed. "Going off feed" is caused mainly by overeating. An animal's appetite is greatly affected by weather conditions and cattle that are hand-fed sometimes come to the feed trough with much keener appetites than usual. The result is that some of the larger and stronger steers eat more than their share, bolting the greater part of the feed without much chewing. On the following day these cattle are likely to be off feed. The usual symptoms are partial loss of appetite and pronounced scouring. The customary method of treatment is to cut down the feed to about one-half the usual amount, to correct the digestive disturbance. Although this is perhaps as good as any practical treatment, it is likely to cause a recurrence of the trouble unless there is great care in bringing the cattle back to their normal ration. They should be brought back as rapidly as their recovery of appetite allows, but the ration should be increased gradually to guard against repeated overeating by steers that are ravenously hungry.

Cattle that are self-fed show comparatively little tendency to go off feed. They soon learn that the feed is accessible at all times and they eat in a leisurely manner and only enough to satisfy their appetite at the moment.

Care must of course be exercised in putting cattle on a self-feeder. As a rule, the roughage content of the complete ration is high at first and gradually reduced at each filling until the cattle are accustomed to a full feed of grain. This period may vary from a week to 30 or even 50 days, depending on the length of the feeding period.

The principal disadvantages of using self-feeding are the following:

1. It is more difficult to utilize large amounts of farm-grown roughages.
2. Large amounts of high-moisture grain cannot be prepared at one time.
3. A larger investment in equipment is required.
4. There is a slight increase in cost of gains.
5. Cattle are apt to be less carefully observed.

None of these items requires any extended discussion. It is freely admitted that self-feeders have no place on farms where the grain ration is more or less limited with a view to utilizing large quantities of roughage.

If a nitrogenous concentrate is to be fed in a self-feeder along with the corn, the corn may be either shelled or ground, but a better mixture

FIG. 97. Illustrating two types of self-feeders in use on many midwestern farms; one is top-loading, the other is end-loading. The trim middles on the steers suggest that little, if any, roughage is being fed. (Hubbard Milling Company; Mankota, Minnesota.)

results with ground corn than with shelled. In all cases, of course, sorghums must be processed. If the moisture content of either ground ear corn, ground shelled corn, or sorghum is much over 15 percent in winter or 13 percent in summer, molding is apt to occur in the self-feeder, as it will if the feeder is used on pasture where a certain amount of rainwater is almost bound to get into the self-feeder.

It is rather generally believed that hand feeding results in more economical gains than can be obtained from self-feeding. This may be true where the hand feeding is carefully done but this advantage is unlikely in the ordinary methods of hand feeding. As nearly half of the experiments involving hand-fed and self-fed lots show a slightly lower feed requirement per pound of gain with self-fed lots, it may be concluded that no material difference exists between these methods of feeding as far as the cost of gains is concerned. (See Table 250.)

It is difficult to place a monetary value on good husbandry in the full feeding of cattle, but the old adage, "The eye of the master fattens his cattle," is undoubtedly an important factor in the success or failure of this cattle program.

Table 250

Comparison of Self-Feeding and Hand Feeding Steers

| | | | | Iowa Average of 2 Trials | | | |
| | Wisconsin 168 Days | | | First Trial 160 Days | | Second Trial 150 Days | |
	Fed Twice a Day	Fed Once a Day	Self-Fed Corn; Other Feeds Fed Once a Day	Self-Fed	Hand-Fed	Self-Fed	Hand-Fed
Average daily gain (lb)	2.51	2.56	2.55	3.24	3.06	2.94	2.68
Average daily ration (lb)							
Corn	12.4	12.4	12.8	15.15	14.47	16.14	15.00
Protein concentrate	1.50	1.50	1.50	2.25	2.25	2.25	2.25
Legume hay	2.0	2.0	2.0	1.44	1.44	1.28	1.28
Corn silage	17.1	17.1	17.1	32.47	31.43	29.24	28.74
Feed per pound of gain (lb)							
Corn	4.95	4.86	5.00	4.71	4.76	5.55	5.64
Protein concentrate	0.60	0.58	0.59	0.69	0.73	0.77	0.84
Legume hay	0.80	0.79	0.79	0.45	0.47	0.44	0.48
Corn silage	6.79	6.66	6.68	10.04	10.27	10.00	10.78

GROUP VERSUS STALL FEEDING

Occasionally one hears lavish claims made for the stall method of feeding practiced in some European countries. However, reports of attempts to use this method in America do not lend encouragement to its adoption here. Apparently cattle are influenced by "crowd psychology" in much the same way as human beings. Each is interested in what its companions are doing; each has a strong desire to enter into the activity engaged in by the rest. In no way is this tendency more clearly shown than in the activity of eating. No sooner does one steer get up and walk to the feed trough than others lying nearby begin to do likewise. Apparently animals with sluggish appetites are stimulated to eat through observing the behavior of their hungry mates. It is a well known fact that cattle fed together out of a common trough will eat considerably more than if they are fed separately. As a practical example, many calf club members who have been unsuccessful in putting the desired finish on a single calf fed by itself have adopted the plan of feeding two or three animals together.

DETAILS OF HAND FEEDING

There are two main considerations in the practice of hand feeding: (1) methods of feeding or combining the components of the ration and (2) when and how often to feed during a day.

Combining Feeds. Usually grain and other concentrates should be fed separately from the roughage, except when cattle are fed silage or complete pelleted rations. When silage is fed, the silage should be put into the trough first and the grain and protein concentrate poured over it. Mixing all three together with a self-unloading mixer wagon or truck is ideal. If the grain is self-fed, the protein concentrate should be spread over the silage instead of fed with the grain. When silage is not fed, all concentrates used, including grain, should be mixed together.

Number of Feedings per Day. Among practical cattle feeders there is wide variation in the number of times the different constituents of the ration are fed each day. Sometimes the cattle are fed only once daily, being given enough of every item to supply them for the next 24 hours. (Should the practice be to offer a surplus so that feed is before the cattle at all times, hand feeding no longer exists, even though the usual self-feeding equipment is not used.) Although feeding only once a day saves much time and labor, it does not have as good results as the feeding of smaller quantities at shorter intervals. This applies to the grain rather than the roughage, particularly if more than one kind of roughage is used.

Two feedings per day are the usual number, although the most careful feeders, such as those who fit show cattle, give three or even four feeds of grain. The advantage is that frequent feedings stimulate the appetites of the cattle by supplying fresh, clean feed that has not been "mussed over." Also, at each feeding time the animals' curiosity is aroused and they come to the troughs to see what they have been given.

When two feedings per day are made they should be given fairly early in the morning and late in the evening, the exact time depending on the season of the year. With three feedings, the second one should be made at noon; with four, the last one should be given about 9 o'clock at night. Four feedings are impractical except in the case of show animals or where mechanized equipment is operated with automatic timers.

Seldom are more than two feedings of roughage given per day. If more than one roughage material is used it is a common practice to feed one in the morning and the other in the evening. The less palatable material should be fed in the morning to allow a longer time for its consumption. When silage and a legume hay are fed, the former may be fed in the morning and the latter at evening, as usually more silage than hay will be used; or both silage and hay may be fed in the morning and only silage at evening.

If an auger-bunk feeding system is used and the complete system is automated, including the silo unloader when one is used, there are advantages to feeding as often as six times per day or at 4-hour intervals. Less bunk space is required, and research indicates some improvement in feed efficiency, due apparently to a more even intake of feed over the entire 24-hour period. Large commercial feedlots, equipped with self-unloading trucks and all-night lights over the feedbunks, make more efficient use of their equipment, such as trucks and feed mill, by feeding continuously for 24 hours. Each pen may be fed 2 to 4 times per day. For all practical purposes the cattle are self-fed but fresher feed, especially silage, is available if frequent feedings are made.

No firm rules can be given concerning either the best time or the best method of feeding. They necessarily must be determined largely by the location of the feed supply in relation to the cattle and by the amounts of the different feedstuffs used. The more labor involved in feeding a given material, the greater the likelihood that a single feeding per day will give the best results.

Chapter 25

BUILDINGS and EQUIPMENT for BEEF CATTLE

Reductions in capital investments in buildings and equipment, and in labor required per unit of beef produced or per head fed, have not in general kept pace with improvements in the feeding and breeding of beef cattle. An exception is found in the fast-developing feedlot areas in the Panhandle region and parts of the West, which have been able to take advantage of the experience gained in the more established feeding areas.

Recent studies by the U.S. Department of Agriculture show that crop labor requirements per unit of production have been reduced by 34 percent within the last 30 years. The reduction for all livestock enterprises has amounted to only 7 percent, and it is doubtful whether the beef production segment of the industry equals the average in this respect.

An average of 10 to 12 man-hours of labor are required to feed and care for a single feeder steer, carried through a normal feeding program in the Corn Belt area. In the commercial feedlot areas the figure is down to 2 to 3 hours per steer. The largest single labor requirement is for feeding, especially in the stocker and feeder programs; thus any appreciable reduction in labor requirements will occur only when the feeding problem is regarded as a materials-handling problem.

Efficient shed and feedlot layouts are the first step toward reducing labor requirements, followed closely by the use of carefully chosen mechanical feed processing and handling machinery. Either when building a new set of beef cattle buildings and feedlots equipped with the latest in new equipment or when remodeling older improvements, the farmer feedlot operator or the rancher must remember that most of the new capital investment must be paid for out of labor savings, and the amount that can be saved per animal unit is very small. Thus either the capacity of the

641

FIG. 98. Modest investments in buildings, silos, and small pastures are often justified for purebred herds because of the convenience afforded. The trend in design of newer barns is to provide hay storage at ground level. (American Angus Association.)

feedlot must be large, with year-round feeding operations to spread the cost over a large number of head, or the initial investment must be minimal. The great interest in mechanization, with all its possible combinations of convenient feeding systems, may lead some cattlemen to overbuild. Actually in the handling of beef cattle, comparatively little equipment is needed, especially for stockers and feeders. Purebred animals usually are allowed more pretentious shelter and equipment, although even here the investment in equipment need not be large. On the farms of the larger, wealthier breeders, however, it often totals several thousands of dollars, much of it in buildings, apparatus, and devices that make for more efficient herd management and reduce the amount of labor required for its proper feeding and care.

Following is a brief discussion of the more important articles of equipment found useful on beef cattle farms.

SHELTER

A satisfactory shelter for beef cattle is one that furnishes adequate protection from wind and rain. Protection against cold is unnecessary in all but the northern latitudes, except for young calves. Numerous experiments with full-fed steers indicate that such cattle do better in open sheds than in closed barns, provided the open sheds have a paved lot. (See Table

Table 251

Importance of Shelter for Full-Fed Steers

Kind of Shelter	Missouri (Average of Three Trials)			Pennsylvania (Average of Five Trials)	
	Barn	Open Shed	Open Lot	Barn	Open Shed
Daily gain per head (lb)	1.78	1.99	2.05	1.93	1.95
Grain per pound of gain (lb)	11.02	10.37	10.22	9.49	9.51
Dry roughage per pound of gain (lb)	4.32	3.74	4.22	6.16	6.04
Digestible matter per pound of gain (lb)	10.77	10.25	10.22	—	—
Daily ration (lb)					
Shelled corn	21.47	22.83	23.11	—	—
Dry roughage	8.42	8.24	9.55	—	—

251.) Apparently the heat generated by the digestion and metabolism of heavy grain rations is sufficient to keep the animals warm.

Comparisons between types of shelter can be misleading if they do not take into consideration the fact that a different type of lot surface is often used. Even on farms, the man with a good cattle shed will have paved lots, while the man with a poor shed or none at all will also run his cattle in unpaved muddy lots. Examples of tests in which the lot surface may have had more to do with results than type of shelter are shown in Table 252. Although the favorable results in these tests cannot entirely be ascribed to the type of shelter, the combinations tested are quite common in the states doing the testing, and thus the results can be applied to a majority of the farms. All of these states are in areas where there is either considerable snow or wintertime rain or where winter temperatures are severe, or both. This being the case, it is quite possible that the feedlot surface is the main source of differences in performance.

Cattle being fed in drylot in summertime, whether on stocker or finishing rations, are much more adversely affected by heat and humidity inside barns or even sheds if poor ventilation prevails than when simple shades or even trees are the only shelter. Fences are also a factor in air circulation, with wooden fences retarding the flow of air more than cable fences.

Although breeding cows and stocker steers on little more than maintenance rations undoubtedly are more sensitive than finishing cattle to extremely low temperatures, it is unlikely that there is any great advantage in housing them in closed barns.

Table 252

Effect of Shelter and Feedlot Surface on Cattle Performance in Wintertime[a]

Station Cattle Program	Kansas Wintering Yearling Steers		Michigan Finishing Heifer Calves		Kansas Finishing 2-Year-Old Steers		Arkansas Finishing Steer Calves	
Type of Shelter, Lot Surface	Open Shed, Concrete Lot	No Shelter, Dirt Lot	Barn, Paved Yard	Open Shed, Unpaved Yard	Open Shed, Concrete Lot	No Shelter, Dirt Lot	Open Shed	Bermuda Lot
Number of cattle per pen	20	20	26	26	20	19	39	39
Average initial weight (lb)	716	738	434	436	853	849	—	—
Average daily gain (lb)—	1.47	1.16	1.88	1.65	2.33	2.17	1.98	1.66
Average daily ration (lb)[b]	19.6	19.6	17.0	16.6	29.3	29.2	19.6	19.9
Average feed cost per cwt gain	$11.80	$15.07	$15.48	$17.34	$21.63	$23.28	$17.75	$21.50

[a] Compiled from miscellaneous Feeders' Day Reports.
[b] Air-dry basis.

FIG. 99. An example of an inexpensive pole-type cattle shed with bedding or hay storage in back, a compact adjacent concrete paved lot, a nearby, large-capacity bunker silo, and airtight high-moisture corn storage combined into an efficient cattle-feeding plant. (University of Illinois.)

PAVED FEEDLOTS

Muddy feedlots impose a severe handicap on winter-fed steers. Many a man has ordered a truck and shipped his half-finished steers to market for no other reason than to get them out of muddy lots in which they were standing up to their knees. Others have permanently abandoned late winter and spring feeding, acknowledging their defeat in the annual battle against slush and mud.

The best insurance so far devised against possible loss in cattle feeding from bad weather conditions is the paved feedlot. The three advantages of paving, briefly stated, are as follows:

1. It adds greatly to the comfort of the cattle, thus promoting greater and more efficient gains.
2. It reduces the labor involved in feeding.
3. It makes possible the saving of more manure.

The open lot used by feeder steers does not need to be large. Only enough space is required to permit the cattle to move about freely without unduly disturbing one another. Fifty square feet of floor space per mature animal, 40 for yearlings, and 30 for calves is sufficient, especially if the lot adjoins an open shed affording an equal area per head. The cost of a 5-inch concrete pavement is not excessive and soon pays for itself in the increased pork and manure credits secured, to say nothing of the greater gains made by the cattle or the labor saved in feeding and cleaning. Asphalt

FIG. 100. Several components of a modern feedlot in the making. Protected concrete water tank with enclosed electric water heater on a concrete slab with apron. In the background, poured concrete bunk and 8-foot apron; steel cable fence. (Western Livestock Journal.)

FIG. 101. Fenceline bunks and automated feeding wagons are replacing conventional feedlot bunks and the scoopshovel in the Corn Belt area. Such modern equipment is common in the larger commercial feedlots of the Southwest and West. (International Harvester Company.)

pavings are a questionable investment, especially in the extreme northern and southern states, because of the freezing and softening, respectively, that are encountered in these regions. A 6- to 10-foot concrete ramp should be provided for single-sided bunks such as those used in fencelines, and twice that width for two-sided bunks. Equally essential is a 10-foot ramp around the water tank. Reducing the feedlot concrete to these areas makes cleaning the feedlot more inconvenient, but the comfort of the cattle comes first as far as paving is concerned.

CONTROLLED-ENVIRONMENT BARNS

In the Midwest, where snow, ice, and mud prevail in winter and where high temperatures, day and night, combine with high humidity to cause problems in the summer, a few farmer-feeders are now feeding cattle in completely enclosed barns with controlled environment the year around. A portion or all of the floor is slotted, having concrete slabs 6 inches wide laid $1\frac{1}{2}$ inches apart, usually over a manure pit about 6 feet deep. The pit is equipped with a pump for emptying. Ventilation is provided by means of thermostatically controlled fans with no artificial heating or cooling being necessary. An automated feeding system is built into the barn to reduce labor to an absolute minimum. Considerably less floor space is required under these conditions than in conventional barns.

Performance of cattle housed in these plants is reportedly improved. Michigan workers report about a 5 percent improvement in gain and 8 or 9 percent reduction in feed requirements. The elimination of bedding and the reduction in labor required for feeding and barn cleanup are other positive factors. The high initial capital investment may prevent many farmer-feeders from installing this type of system.

FEEDBUNKS

Portable feedbunks used by beef cattle should be made of 2-inch material, preferably undressed cypress. The legs should be cut from 4- by 4-inch material and should be both bolted and braced to the body of the bunk. Bunks placed in the open, where cattle can eat from both sides, should be at least 30 inches wide, or preferably 36 inches for mature cattle. The floor of the bunk should be 20 inches from the ground and should have 6-to-10-inch sides, depending on the age of the cattle. Feed troughs built inside a barn or shed should be made adjustable in height to accommodate cattle of various ages. It will be necessary to raise the troughs from time to time if the manure and bedding are allowed to accumulate in the shed.

Development of the self-unloading wagon or truck has been closely

FIG. 102. Auger-bunks that deliver both roughages and concentrates to cattle at very low power costs are genuine labor savers. Another type of auger-bunk in which the auger is enclosed in a tube with openings at frequent intervals permits the separate feeding of each component of the ration. (University of Illinois.)

followed by a great increase in the use of fenceline bunks. Besides saving labor, this type of bunk, if used with self-unloading equipment, shortens travel routes, makes for easier cleaning of lots, and cuts down on damage to bunks. However, twice the linear footage of bunk space must be provided if such single-sided feeders are used.

Bunk space of 20 to 24 inches per head should be provided if all cattle are to eat at once. This may be reduced if cattle are fed two or more times daily or if bunk self-feeding is practiced. Paved, graveled, or rocked driveways are an integral part of the fenceline bunk feeding system.

CORRALS AND RESTRAINT EQUIPMENT

A set of working corrals for farm or ranch with such accessories as an unloading and loading chute, a squeeze chute with headgate, sorting chute or gate, dipping vat where needed, scales, and holding pens with sorting gate, are important tools of the beef cattle man. The size of the operation, of course, determines the most economical size of such equipment. Plans are available from the U.S. Department of Agriculture and the various agricultural experiment stations for almost any type and size of layout needed. Figure 103 is one example of a plan for an expanding operation. Note that the simplest form of the plan still contains the squeeze, headgate, and loading chute. Figure 104 shows several types of headgates, some hinged for a front-opening chute and others stationary for a side-opening chute. All such devices should combine ease of operation with strong construction, as it frequently is necessary, in confining an animal, to close and fasten a gate in the fraction of a minute.

FEEDLOT LAYOUT AND EQUIPMENT

A number of important factors must be considered in planning a feedlot of the commercial type. Some, but not all, of these principles apply in the smaller farmer-feeder operation as well. Assuming that the section of the country is already determined, there are still some local considerations involving site that must be taken into account. For instance, access to a hard-surfaced road or railroad ensures year-round shipment of cattle and feed, to and from the feedlot. Availability of an adequate supply of good-quality water is essential. Some natural drainage is desirable, and a soil type that drains well will reduce mud problems. Electrical power should be economically available. Local and state water and air pollution laws and county zoning ordinances should be studied carefully. Amount and prevailing direction of wind should be considered because of potential odor and dust problems involving the surrounding area. There should be sufficient land available for future expansion if desired.

The number of cattle to be fed has much to do with both the layout and the type of feeding equipment used. Feedlots with less than about 1,000-head capacity may well be fed and managed in pens holding 100 head each, equipped with self feeders. Auger-bunks dividing two lots so that cattle may eat from both sides may also be used. Fenceline bunks and self-unloading trucks may also be used for this size lot and are almost a necessity for lots that are larger in size.

Pen capacity may vary from as few as 50 head up to 200 as a practical

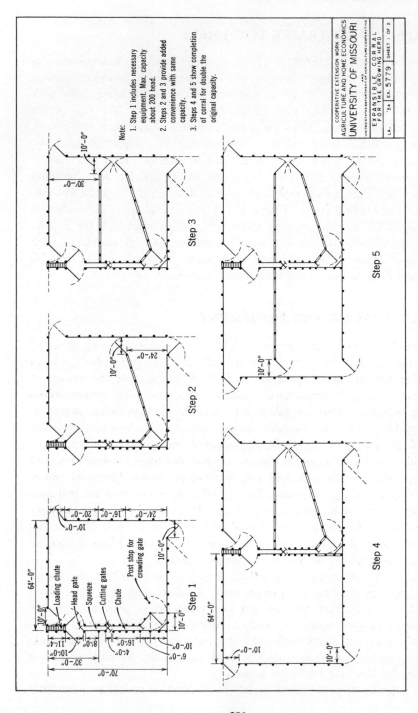

FIG. 103. An expansible corral which can be built in stages as the herd grows in size up to 200 head. Step 1 includes the essential loading shute, headgate, squeeze, and chute. Steps 2 and 3 provide added convenience with the same capacity. Steps 4 and 5 show completion of the corral, doubling the original capacity. (University of Missouri and U.S.D.A. Cooperating.)

maximum, with 100 head being about ideal for heavier cattle and 200 for calves. This number requires a pen 100 by 200 feet to provide the necessary space if the lot is unpaved.

Some additional, smaller lots are convenient for holding cattle for various reasons. A 12-foot-wide working or sorting alley and several catch pens are essential. These should be adjacent to the loading chute. Small hospital pens and perhaps a hospital shed, depending on the climate, are essential. In large lots, working alleys should be separate from feeding alleys or there will be congestion and loss of time. Feeding alleys may range from 12 to 40 feet in width, depending on size of lot. A width of 20 feet permits passage of two vehicles and in large lots this often saves time. Feeding alleys should be paved with concrete if at all possible, and in localities with 20 inches or more of rainfall annually such concrete is almost a must. Gates should be on the side of the pen facing the working alley and 12 feet is a good length, especially if it matches the width of the working alley.

Materials for the fence may be 2-inch treated lumber, $\frac{3}{8}$-to-$\frac{1}{2}$-inch new or used steel cable, or oilfield sucker rod. Cable and similar fencing allows free movement of air but provides no protection from cold winds; thus location dictates the type of material to use. Fences should be at least $5\frac{1}{2}$ feet high. Details concerning post and cable spacing, setting of posts, and so on are obtainable from builders' supply houses or the county agricultural agent. The same is true for recommendations concerning feed troughs, water systems, type of shades, and the like. Local weather conditions have so much to do with the detailed planning of such equipment that local advice should be obtained.

Research is under way to determine the value of all-night lights and windbreaks. Preliminary evidence indicates that lights are unessential but do reduce stampedes and minor disturbances as well as provide light for overnight feeding if desired. Shelters or windbreaks have proven beneficial in states such as Kansas, Nebraska, and South Dakota where snow and cold winds are common, but such structures are harmful in California, Arizona, and Texas where free circulation of air is to be desired.

In the fully equipped feedlot, arrangements must be made for office space, weighing facilities for large trucks hauling both feed and cattle, equipment storage, repair space for vehicles, and a parking lot for employees.

Manure disposal and pen cleaning must be planned for when designing a feedlot. About 0.5 ton of manure is produced per month per animal. In large dirt-surfaced lots, manure or scrapings may be bulldozed into mounds and removed only between droves of cattle fed. Concrete-surfaced lots must be hosed off or scraped more frequently.

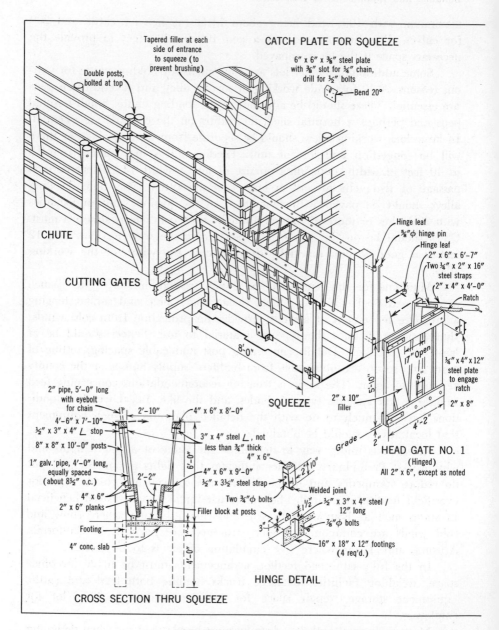

CATCH PLATE FOR SQUEEZE

6" x 6" x 3/8" steel plate with 3/8" slot for 1/4" chain, drill for 1/2" bolts

Bend 20°

Tapered filler at each side of entrance to squeeze (to prevent brushing)

Double posts, bolted at top

CHUTE

CUTTING GATES

Hinge leaf
5/8" φ hinge pin
Hinge leaf
2" x 6" x 6'-7"
Two 1/4" x 2" x 16" steel straps
2" x 4" x 4'-0"
Ratch

8'-0"

SQUEEZE

17" Open

1/4" x 4" x 12" steel plate to engage ratch

6"

2" x 8"

2" x 10" filler

Grade

2"

4'-2"

HEAD GATE NO. 1
(Hinged)
All 2" x 6", except as noted

2" pipe, 5'-0" long with eyebolt for chain

2'-10"

4" x 6" x 8'-0"

4'-6" x 7'-10"
1/2" x 3" x 4" L stop
8" x 8" x 10'-0" posts

3" x 4" steel L, not less than 3/4" thick

4" x 6"

1" galv. pipe, 4'-0" long, equally spaced (about 8 1/2" o.c.)

2'-2"

4" x 6" x 9'-0"
1/2" x 3 1/2" steel strap

4" x 6"
2" x 6" planks

13"

Footing

4" conc. slab

Filler block at posts

2'-10"

Welded joint
Two 3/4" φ bolts

1/2" x 3" x 4" steel L
12" long

3/8" φ bolts

16" x 18" x 12" footings
(4 req'd.)

4'-0"

CROSS SECTION THRU SQUEEZE

HINGE DETAIL

FIG. 104. Squeeze chute with details and three types of

652

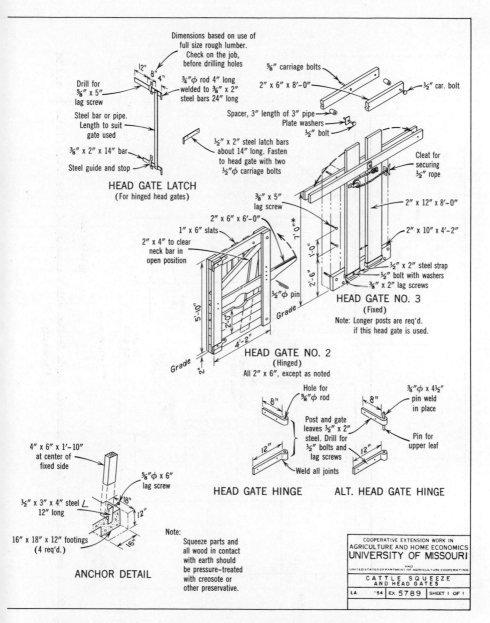

Dimensions based on use of full size rough lumber. Check on the job, before drilling holes

HEAD GATE LATCH
(For hinged head gates)

Drill for ⅝" x 5" lag screw

12" 8" 4"

¾"φ rod 4" long welded to ⅜" x 2" steel bars 24" long

Steel bar or pipe. Length to suit gate used

⅜" x 2" x 14" bar

Steel guide and stop

½" x 2" steel latch bars about 14" long. Fasten to head gate with two ½"φ carriage bolts

⅝" carriage bolts

2" x 6" x 8'-0"

½" car. bolt

Spacer, 3" length of 3" pipe
Plate washers
½" bolt

Cleat for securing ½" rope

2" x 12" x 8'-0"

2" x 10" x 4'-2"

½" x 2" steel strap
½" bolt with washers
¾" x 2" lag screws

⅜" x 5" lag screw

7'-0"*

1'-0"

2'-6"

HEAD GATE NO. 3
(Fixed)

Note: Longer posts are req'd. if this head gate is used.

2" x 6" x 6'-0"

1" x 6" slats

2" x 4" to clear neck bar in open position

½"φ pin

Grade

5'-10"

2'-0"

4'-2"

2"

Grade

HEAD GATE NO. 2
(Hinged)
All 2" x 6", except as noted

Hole for ⅝"φ rod

8"

12"

Post and gate leaves ½" x 2" steel. Drill for ½" bolts and lag screws

Weld all joints

¾"φ x 4½" pin weld in place

8"

12"

Pin for upper leaf

HEAD GATE HINGE

ALT. HEAD GATE HINGE

4" x 6" x 1'-10" at center of fixed side

½" x 3" x 4" steel ∟ 12" long

⅝"φ x 6" lag screw

18"

12"

16" x 18" x 12" footings (4 req'd.)

ANCHOR DETAIL

Note:
Squeeze parts and all wood in contact with earth should be pressure-treated with creosote or other preservative.

COOPERATIVE EXTENSION WORK IN
AGRICULTURE AND HOME ECONOMICS
UNIVERSITY OF MISSOURI
AND
UNITED STATES DEPARTMENT OF AGRICULTURE COOPERATING

CATTLE SQUEEZE
AND HEAD GATES

| LA. | '54 | EX. 5789 | SHEET 1 OF 1 |

headgates. (University of Missouri and U.S.D.A. Cooperating.)

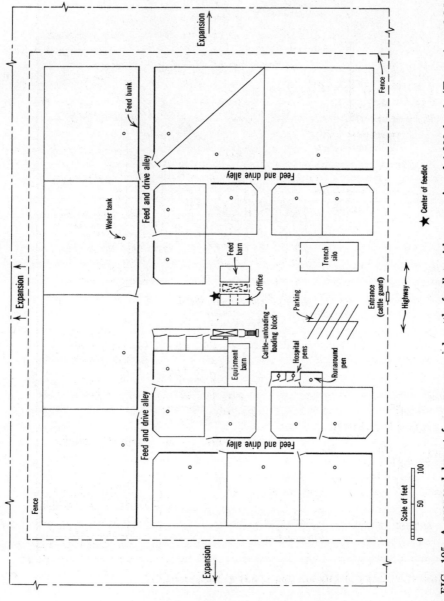

FIG. 105. A proposed layout for a commercial cattle feedlot with a capacity of 1,000 head. (Texas A. and M. University.)

FIG. 106. One man with self-unloading wagon can feed thousands of cattle daily, especially when the ration is fed in complete form—this is, roughage and concentrate combined in one mixture in the desired proportions (National Cottonseed Products Association, Inc.)

Details concerning the feed storage and feed processing facilities should be provided by an engineering service specializing in such work. Some states have Agricultural Extension Service engineers who are experienced enough to be of assistance.

CATTLE STOCKS

Cattle stocks, or bull stocks as they are often called, are essentially an item of equipment for the purebred herd. Their important features may be combined with those of an ordinary dehorning chute to make them useful with unbroken cattle, but it is more usual to build them as a separate or detached unit into which cattle are led rather than driven. This fact alone limits their use to animals that are rather easily handled.

The characteristic features of cattle stocks are (1) rollers on either side at a height of about 4 feet from the floor, made of 5-inch cedar posts or 4-inch steel pipe, for supporting a canvas sling that is placed under the animal to prevent it from lying down while in the stocks; and (2) wooden sills, 6 by 8 inches, that extend along either side at a height 15 inches from the floor, on which the feet may be rested and tied while undergoing treatment.

Cattle stocks are exceedingly useful for the restraint of animals during the performance of such minor operations as the ringing of bulls, clipping of heads, surgical treatment of lump jaw, and trimming of feet. An adjustable canvas surcingle allows the weight of the body to be taken off the legs, making it practically impossible for the animal to throw itself or lie down.

FIG. 107. Control panel for a fully automated feedmill, in the process of being installed. Large feedlots often operate their own feedmills and storage facilities adequate for a year's supply of grain. (Western Livestock Journal.)

BREEDING CRATES

As discussed in Chapter 4, breeding crates are useful in safeguarding the breeding of young heifers to old, heavy bulls. A satisfactory type of breeding crate is shown in Fig. 108. The back portion of each side of the stall is surmounted by a 2- by 12-inch plank on which cleats are nailed to prevent the bull's front feet from slipping. The gate or stanchion labeled *A* in the diagram is adjustable to accommodate cows varying in height and length of body.

LOADING CHUTES

A well-designed loading chute is a necessary item of equipment on every beef cattle farm. Stationary chutes should be located in reference to both roads and holding pens to make them accessible to heavy trucks and so that cattle may be quickly loaded and unloaded after the truck has been set. Tests conducted at the Union Stock Yards in Chicago indicate that livestock prefer low stairstep risers to a cleated ramp. Whether the advantages of such a chute are sufficient to justify the extra materials and labor required may be questioned by the farmer who handles only a load or two of cattle a year.

A portable loading chute that may be towed behind a truck is often useful in loading or unloading cattle at a distance from the farmstead where the stationary chute is built. A sketch of a portable chute with a stairstep ramp is shown in Fig. 110.

LABORSAVING EQUIPMENT

The shortage of labor on livestock farms has caused a wide interest in laborsaving devices that will shorten the time and lighten the labor required to care for the animals. Such equipment includes (1) blowers and conveyors for moving feed, (2) storage of feed and penning arrangements for animals that will require as little movement of the feed as possible, and (3) equipment for cleaning the sheds and yards. First-class equipment is fairly expensive and thus better suited for the large feeder who handles 100 or more cattle than for the farmer who feeds only a load at a time. On the other hand, the small feeder is usually able to store his feeds nearer the cattle than is possible in large operations. Often by remodeling his barns and carefully planning the location of his grain bins and silo, the small-scale feeder can feed and care for his cattle entirely under cover and with a minimum of hand labor.

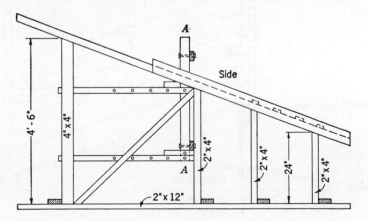

FIG. 108. Diagram of a breeding stock, showing details of construction. (University of Illinois.)

FIG. 109. A permanent loading chute designed to load cattle into single- or double-decked trucks. The 8-inch concrete step ensures more orderly loading and unloading than a slotted floor. (Western Livestock Journal.)

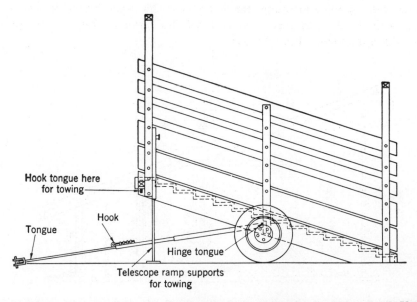

Hook tongue here for towing

Tongue

Hook

Hinge tongue

Telescope ramp supports for towing

FIG. 110. A portable step-ramp loading chute developed by the Union Stock Yards and Transit Company. (Livestock Conservation, Inc., Chicago, Illinois.)

Cleaning cattle sheds and lots and hauling and spreading the manure have always been regarded as among the hardest and most disagreeable jobs on the farm. They probably have caused many good farm boys to become grain farmers or even leave the farm entirely instead of following in the steps of their cattle-feeding fathers. However, the presence of tractors with hydraulic manure loaders on practically every farm has now eliminated most of the hand labor formerly required in cleaning sheds and feedyards on livestock farms.

Chapter 26

The MARKETING
of CATTLE

The matter of selling or buying stockers or feeders is discussed in Chapter 10; therefore, the discussion in this chapter is confined mainly to the marketing of slaughter cattle.

A rancher, farmer, or cattle feeder with cattle to sell seldom has any difficulty in disposing of them at a price near their actual value. This price may not be so high as the seller had hoped, but with the variety of marketing agencies generally available today and the competitive bidding they afford, few "steals" are made by buyers of slaughter cattle. In some instances, of course, usually owing to a low population of slaughter cattle, a locality has a limited choice of marketing facilities, and sometimes only one or no outlet is available.

In most areas with a relatively dense livestock population, most if not all of the following avenues are available for disposal of slaughter cattle:

1. Consignment to central or public terminal livestock market.
2. Use of local cooperative shipping association.
3. Consignment to local auction sale.
4. Direct on-the-farm or feedlot sale to packer buyers.
5. Sale of cattle direct to packer buyers at concentration yards or at the packing plant.
6. Consignment of cattle to a packer on the basis of carcass dressed weight and grade.
7. On-the-farm sale to cattle dealers or local packers.
8. Fulfillment of a futures contract by delivery of cattle.

The choice of the most suitable market is not a simple one, and there are no infallible rules for making such a decision. The factors that should play the most important roles in making this choice are (1) quoted market price per hundredweight, (2) transportation costs, (3) selling expenses, (4) shrinkage, (5) services rendered by the market or its marketing agencies, and (6) weighing conditions.

Still other factors that enter into the choice of a cattle market are convenience, custom, and personal preference. Many cattlemen simply prefer to deal directly with a buyer because they feel that less risk is involved. On the whole, all of the market channels listed previously offer a good market for cattle, but each man with slaughter cattle to sell owes it to himself to make comparative studies of the various markets for the particular cattle he has to sell. The market of choice varies greatly depending on locality, as is shown in Fig. 111.

Until recent times the majority of slaughter cattle eventually were sold to packers in the large central markets. Today, however, the trend is for more and more slaughter cattle to be sold direct from the feedlot to a packer buyer or at least direct to a country buyer who is buying on order for the killer. Some of the formerly large central markets—Denver, for example—have changed entirely into an auction market where, of course, fat cattle as well as stockers and feeders are sold. The overall effect is that the central market has declined in importance as a slaughter cattle market.

As already mentioned, a large percentage of slaughter cattle are purchased direct from the feedlot by the packer. The reason usually given for this practice is that it assures the packer of the necessary number of certain kinds of cattle required for a full day's kill. Many marketing experts feel that the packers follow this practice principally because buying competition is reduced, especially in areas not served by more than one packer. Another undesirable feature of this method, from the feeder's standpoint at least, is that direct-purchased cattle are usually of better quality than the cattle selling through other channels. The price at the central market is established on the poorer quality cattle offered. Because the direct-purchase price is usually calculated on the basis of central market prices, it can be seen that the overall effect is a lowering of all prices.

WHEN TO MARKET

Various rules for marketing have been stated by successful feeders, but it is doubtful whether they are taken very seriously, even by their originators. Perhaps the one most often heard is "Ship when the cattle are ready, regardless of the condition of the market." However, it would seem from the appearance of many animals received at the yards that a rule

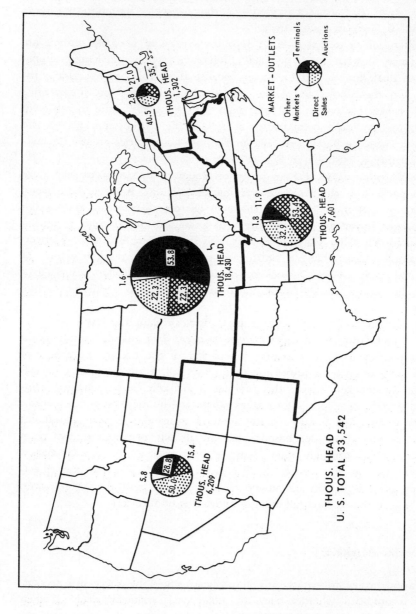

FIG. 111. The terminal market is the most important method of selling in the Corn Belt, whereas direct selling and auction selling are most important in the West and South, respectively. (U.S.D.A.)

662

more frequently followed is "Ship when the market seems right, regardless of the condition of the cattle." Neither rule is a good one if blindly followed, although each expresses an element of truth that should not be overlooked if satisfactory returns are to be realized.

The approximate time of marketing should be decided at the time the cattle are placed on feed. Only by having a definite feeding plan and length of feeding period in mind can the purchase and selling dates be planned intelligently. As the cattle begin to reach the degree of finish desired, the short-run market trends should be studied in an effort to obtain as favorable a price as possible. However, there is no way of knowing definitely which way the market will go. Even the reputation salesmen and large buyers, who have had long years of experience, occasionally make serious errors in predicting future price tendencies. Nevertheless, the judgment of such men deserves respect, and their advice regarding shipping should be carefully considered. They have at hand much information regarding total supplies, expected loadings, religious holidays affecting consumption, and conditions of the dressed beef trade that is not available to the individual shipper.

In general, the relative prices being paid for a particular weight and grade of cattle are determined by the total supply of such cattle on the major markets in the country. One of the leading livestock journals has prepared a calendar for marketing beef cattle, based on average prices generally received at the major markets throughout the year. (See Fig. 112.)

PREPARING CATTLE FOR SHIPMENT

There is much difference of opinion regarding the advisability of attempting to reduce shrinkage by changing the ration before shipping. Of course, this is not a problem when cattle are sold direct from the lot, as more and more are today.

Some feeders remove all laxative feeds such as protein concentrates and legume hay or silage a day or two before shipment and supply non-laxative feeds instead. Others withhold both feed and water on the day of shipment in the belief that the cattle fill better at the market if they arrive hungry and thirsty. Occasionally an unscrupulous shipper salts his cattle heavily on the day of shipment to obtain a heavier consumption of water at the market. Usually nothing is gained from abnormally large fills, however, because buyers quickly detect such cattle and refuse to bid on them, or they adjust their bids accordingly.

Cattle that are to be sold direct and weighed at the feedlot are usually

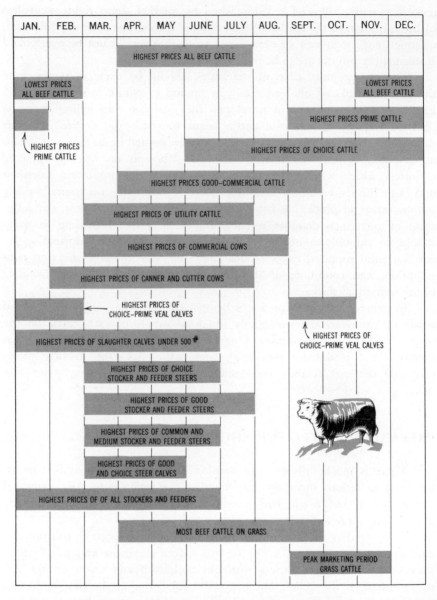

JAN.	FEB.	MAR.	APR.	MAY	JUNE	JULY	AUG.	SEPT.	OCT.	NOV.	DEC.

FIG. 112. A calendar for marketing beef cattle. Each weight and grade of slaughter cattle has its period of high and low prices during the year. There are indications that these seasonal fluctuations are narrowing, undoubtedly owing to year-round feedyard operations. (National Livestock Producers.)

fed as usual on the day of shipment. Large feedlots cannot affort to manipulate rations so as to get heavier fills because they need the repeat buyers coming to their lots weekly. These they would surely lose after only one bad experience with excessive fills.

There is little information regarding the control of shrinkage in cattle. Yet it is an extremely important subject, particularly when cattle prices are high. When cattle are worth $30 per hundredweight, the 40 pounds lost during transit amount to $12 per steer. Anything the shipper can do to reduce the loss by 5 to 10 pounds per head without noticeably impairing the slaughter value of the cattle is highly profitable. Just as with stocker cattle, some of the shrink is tissue shrink and a reduction of this loss is the goal.

Preliminary studies made at the Illinois station indicate that withholding feed and water on the day of shipment increases rather than reduces shrinkage. (See Table 253.) Apparently cattle handled in this way are greatly disturbed by the abrupt change in their feeding schedule and spend the day on their feet vainly waiting for feed and water instead of lying quietly at rest. As a result they are tired and nervous when loaded and arrive at market in a fatigued condition. Though shrinkage based on loading

Table 253
Effect of Changing Feed on Shrinkage of Slaughter Cattle During Shipment[a]

	Withholding Feed and Water on Day of Shipment		Replacing Laxative Feeds			
			Heifer Calves		Yearling Steers	
	Fed as Usual	Feed and Water Withheld	Shelled Corn, Clover Hay	Shelled Corn 50%, Oats 50%, Timothy Hay	Shelled Corn, Linseed Meal Clover Hay	Shelled Corn 50%, Oats 50%, Timothy Hay
Weight out of experiment (lb)	999	998	649	649	1,070	1,095
Length of change period	10 hr.	10 hr.	6 days	6 days	4 days	4 days
Shipping weight (lb)	999	982	666	665	1,069	1,104
Market weight (lb)	963	952	616	624	1,011	1,052.5
Shrinkage (lb)						
On final experiment weights	36	46	33	25	59	42.5
On shipping weights	36	30	50	41	58	51.5
Shrinkage (%)						
On final experiment weights	3.6	4.6	5.1	3.9	5.5	3.9
On shipping weights	3.6	3.1	7.5	6.2	5.4	4.7
Dressing percent	60.6	59.6	57.9	56.0	59.7	58.1

[a] Illinois Mimeographed Report.

weights may be in their favor, the cattle actually have lost considerable weight before they were loaded. This loss may well be 20 to 30 pounds, or the weight of the feed and water that would have been consumed if the regular feeding schedule had been followed.

Other studies made at the Illinois station indicate that substituting oats for part of the shelled corn ration, and timothy or mixed hay for alfalfa or clover hay, is a sound practice because these changes result in less shrinkage. (See Table 253.) However, these nonlaxative feeds should be introduced into the ration 4 or 5 days before shipment.

RAIL VERSUS TRUCK SHIPMENT

Trucks have almost entirely replaced rail transportation for shipping cattle from within about a 200-mile radius of the central market or slaughter plant, and the large trailer trucks are often used instead of rail for even the longer distances.

The comparative merit of rail and truck transportation for fed cattle is a disputed point. Cattle undoubtedly ride more comfortably and safely on the rails than they do in the average truck. However, the greater convenience afforded by the truck in loading the cattle at the farm or feedlot, at the hour most agreeable to the owner, and the greatly reduced shipping time have made the truck the favorite method of most feeders who are within easy trucking distance of the market. Since the heavy traffic on main highways and the poor condition of fences along secondary roads make it impractical to drive cattle to the railhead in most communities, they must be trucked from the farm or feedlot to the rail loading point even when they are shipped to market by rail. Once they are in the truck, the cattle can usually be taken directly to market in much less time and at little more expense than would be incurred if they were trucked only to the local station and reshipped by rail. (See Table 254.)

CATTLE PER CAR OR TRUCK

The standard stock car is 8 feet 6 inches wide and either 36 or 40 feet long. The shorter car is billed with a minimum weight of 22,000 pounds and the longer one with 24,400 pounds. The shipper must pay freight on this minimum weight whether his load weighs that much or not. In computing freight charges over and above that for the minimum weight just mentioned, market weights are taken, less 800 pounds deducted for fill. No maximum weights are specified, and the shipper may crowd

Table 254
Comparison of Truck and Rail Shipment of Fed Steers

	Colorado Bulletin 422		Illinois Unpublished Data	
	(3-Year Average)		(Average of 3 Shipments)	
	Truck	Rail	Truck	Rail
Number of cattle	100	100	69[a]	70[a]
Shipping weight (lb)	835	833	1,044	1,052
Market weight (lb)	806	805	1,009	1,009
Shrinkage (lb)	29	28	35	43
Shrinkage (percentage)	3.6	3.4	3.4	4.1
Bruised carcasses	5.5	4.0	10[b]	10[b]
Freight rate (cents/cwt)	17	16	25	22
Hours in transit	3.17	7.0	7.1[b]	13.0[b]
Distance shipped (miles)	70	70	135	135

[a] Total cattle in three shipments.
[b] Average of two shipments.

as many cattle into the car as he wishes. The average load for a 40-foot car is about 22 two-year-old slaughter steers or 28 yearlings. Cattle ship better if the car is comfortably filled, although overcrowding is more objectionable than underloading. If fewer than 15 mature cattle are to be shipped, they had better be partitioned off in one end of the car and the remainder of the space used for some other class of livestock. This will reduce the freight charges and will avoid undue jolting of the cattle by the sudden starting and stopping of the train.

Truckers engaged in hauling livestock have instigated more improvements in recent years than have railroads. There are a few instances where railroads give top priority in use of the lines to freight trains that are hauling large numbers of cars of live cattle and hogs from the Midwest to either coast. This makes possible a considerable reduction in the time from point of origin to destination. Livestock trucks deliver their cargo with still greater dispatch, with the time generally being saved at both ends of the haul. The net result usually is less shrink.

Truck sizes have increased immensely, with double-decked bodies not uncommon, and this has made trucks more competitive with respect to cost for hauling cattle (Table 255). Express highways with few or no stoplights also result in faster and smoother rides with lower bruise rate. On some of the very long hauls, rest stops for truck-transported cattle, especially for stocker cattle, are required by the buyer, although not by law as is the case with rail-transported cattle.

SHRINKAGE

Shrinkage refers to the loss in weight that occurs between feedlot and market scales. It may be expressed either in pounds per head or in percentage of the weight before shipment. The percentage method is preferred because the amount of weight lost is usually in direct ratio to the size of the cattle.

Shrinkage during shipment is mainly due to excretions of urine and feces and, to a lesser degree, to tissue moisture given off by the lungs in breathing. Some of this loss is regained at the market from the feed and water consumed between arrival time and the time the cattle are sold and weighed. The amount of shrinkage expressed in percentage of the home or loading weight varies greatly between different loads of cattle. The main causes of this variation are (1) length of the journey, (2) condition of the cattle at loading time, (3) the degree of comfort en route, (4) the kind of feeds used, (5) the degree of finish of the cattle, and (6) the fill at market.

Length of Journey. The longer the journey the greater the shrinkage.

Table 255

Average Capacity of Single-Deck Rail Cars and Trucks for Transporting Cattle and Calves

Length (ft)	Rail Cars[a]		Trucks[b]				
	36	40	14	18	30	36	42
Calves							
350 lb	55	62	18	24	42	50	58
450 lb	46	51	15	20	34	41	48
Cattle							
600 lb	36	40	12	16	27	33	39
800 lb	30	33	10	13	22	26	31
1,000 lb	26	28	8	11	19	22	28
1,200 lb	22	24	7	9	16	19	22
1,400 lb	19	21	6	8	14	17	20

[a] Western Weighing and Inspection Bureau, Chicago, Illinois.
[b] Livestock Conservation, Inc.

Table 256

Effect of Length of Haul on Shrinkage of Grain-Fed Cattle Transported by Trucks[a]

Weight Classes	Number of Head	Average Full Weight Out of Lot	Cumulative Percentage of Shrinkage			
			After 25 Miles	After 50 Miles	After 100 Miles	After 200 Miles
Under 1,000 lb	11	954	1.5	2.2	3.1	3.9
1,000–1,099 lb	10	1056	2.1	3.0	3.8	4.1
1,100–1,199 lb	24	1139	1.8	2.6	3.4	4.1
Over 1,200 lb	15	1263	1.9	2.4	3.1	3.6
Group average	60	1122	1.8	2.5	3.3	3.9

[a] Unnumbered report, Chicago Union Stock Yards and Transit Company.
Note: The cattle in this study were hauled in groups of five in a truck that had a scales in one end, on which they were weighed individually after covering the respective distances. No opportunity to fill was allowed before weighing. Initial weights taken in morning before the cattle were fed.

The loss in weight, however, is not in direct ratio to the distance traveled, because the greatest loss occurs during loading and the first few miles in transit. (See Table 256.)

Condition of Cattle at Loading Time. As far as possible the condition of the cattle should be normal when they are loaded. Tired, hungry, or thirsty animals are in poor physical condition to stand the trip and are slow to recover upon reaching the market. Likewise, cattle that have consumed large quantities of green grass or succulent roughage or have taken too great a fill of water just previous to loading are in poor condition for the journey.

Degree of Comfort en Route. During extremes of hot or cold weather, shrinkage runs unusually high. Badly crowded cars or trucks and slow, rough runs with frequent stops are certain to cause considerable loss in weight.

Kind of Feeds Used. Cattle that have been fed large quantities of roughage such as grass and silage commonly lose more weight than those that have received a full feed of grain. Also cattle that have been fed laxative feeds such as soybean meal and alfalfa hay shrink more than cattle fed feeds of a less laxative nature. Contrary to popular belief, silage-fed cattle do not shrink as much as cattle fed dry roughages. (See Table 257.)

Table 257
Shrinkage of Cattle During Transit[a]

Class of Cattle	Hours in Transit	Number of Shipments	Number of Cattle	Average Weight at Origin (lb)	Gross Shrinkage		Fill at Market		Net Shrinkage		
					Range (lb)	Average (lb)	Range (lb)	Average (lb)	Range (lb)	Average (lb)	Percent of Live Weight at Origin
Grain-fed											
Nonsilage	Less than 24	4	164	1,303	59–95	67	4–48	16	20–64	51	3.91
Nonsilage	24–36	59	1,853	1,157	47–128	85	19–52	37	18–88	48	4.11
Silage-fed	Less than 24	14	666	1,168	46–128	76	6–97	52	+7[b]–67	24	2.05
Silage-fed	24–36	4	169	1,204	84–121	101	50–64	58	27–75	43	3.57
Grass-finished											
Mixed range	Less than 24	21	1,511	700	19–84	37	1–56	22	+12[b]–71	15	2.14
Mixed range	24–36	17	872	848	27–118	72	–8[c]–55	18	19–114	54	6.37

Average shrinkage of grain-fed cattle in transit less than 36 hours, 3.62 percent.
Average shrinkage of grass-finished cattle in transit less than 36 hours, 3.88 percent.

[a] U.S.D.A. Bulletin No. 25.
[b] Abnormal load, market weight exceeding loading weight.
[c] Abnormal load, sale weight less than unloading weight.

670

Degree of Finish. As shrinkage is mainly due to the loss of excrement, it is more closely related to the size of the animal than to the degree of finish. In other words, thin 2-year-old steers lose about the same amount of weight per head during shipment as fat 2-year-olds that are 200 to 300 pounds heavier. However, the shrinkage per 100 pounds live weight is of course much higher for the thin cattle. Grass-finished cattle and cows are likely to have a much higher shrink, expressed as a percentage of the loading weight, than grain-fed cattle, because they have had a more laxative ration and because they are carrying less finish.

The Fill at Market. The consumption of feed and water after arrival at the market is the most important factor in determining the net amount of shrinkage. This consumption in turn is influenced by several factors, the more important being (1) weather conditions at the market on the day of sale, (2) the length of time the cattle are in the pens before they are sold and weighed, and (3) the condition of the cattle upon arrival.

Smaller fills are obtained during cold, damp weather than on bright, warm days because cattle drink little at such a time and have little appetite for hay after it becomes wet.

The most satisfactory fills result when the cattle reach the market about daylight, which permits them to be penned and fed at about their usual feeding time. As a rule the market is not under way until 9 o'clock so that the cattle have at least 2 or 3 hours in which to eat and drink. Cattle that arrive after the market has opened may be in the pen only a few minutes before being sold and are weighed almost empty. Although the price paid for such cattle is often higher than would have been bid had the cattle taken on the usual fill, the buyer rather than the seller is most likely to profit from the late arrival. After cattle have been in the pens for 4 or 5 hours they stop eating and drinking and begin to lose rather than gain in weight. Consequently, good fills are associated with brisk, active markets, rather than with slow, long-drawn-out trading that runs into the afternoon session.

Cattle that arrive at the market tired and worn out from a long hard journey, or weakened by insufficient water and feed immediately before or during shipment, frequently lie down upon being unloaded and will not eat or drink to any extent until they have obtained some rest. If sold on the day of their arrival, their shrinkage is much above the average because of their small fill. If given time to recover, or if they are only hungry or thirsty, they probably take on such a large fill that buyers will bid lower or ignore them until the effects of the fill have largely disappeared. Best results are secured when the cattle are only moderately hungry and thirsty when received. In such condition they fill to a moderate extent, but not to the point where they invite unfavorable consideration.

LOSSES SUSTAINED DURING SHIPMENT

Losses resulting from injury or death of cattle in transit are not readily apparent to the casual observer. Because of their size and strength, cattle are less likely to be injured by the rough ride in freight cars or trucks than are swine and sheep. They also withstand unfavorable weather conditions encountered en route better than do the other classes of meat animals. Probably 99 percent of all cattle shipments arrive at the market with no dead or crippled animals. Even so, there is an estimated annual loss of more than $30 million owing to bruises, crippling, and death.

Statistics kept by the Department of Agriculture of all dead and crippled livestock received at the public stockyards of the United States show that a much smaller percentage of cattle are injured during shipment than of veal calves, swine, or sheep. (See Fig. 113.) Extensive data are unavailable comparing the losses of cattle when shipped by truck and by rail.

The relatively few dead and crippled cattle that arrive at the market are not the most serious losses for which shippers must pay through lower prices received for their cattle. Unfortunately, bruised animals cannot be detected by the buyer at the time of purchase; consequently he must buy all cattle on the basis that there will be a loss of $1.50 to $2 per head

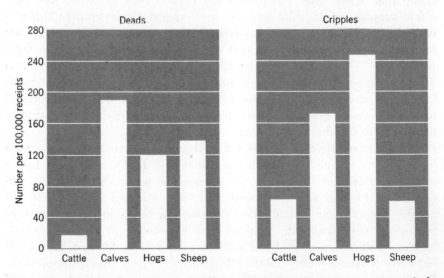

FIG. 113. Number of dead and crippled animals encountered in receipts of the different classes of livestock, as reported by the Bureau of Agricultural Economics for a typical year. (Livestock Conservation, Inc., Chicago, Illinois.)

LOCATION OF CATTLE BRUISES

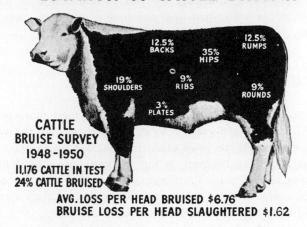

CATTLE
BRUISE SURVEY
1948-1950
11,176 CATTLE IN TEST
24% CATTLE BRUISED
AVG. LOSS PER HEAD BRUISED $6.76
BRUISE LOSS PER HEAD SLAUGHTERED $1.62

CAUSES OF CATTLE BRUISES

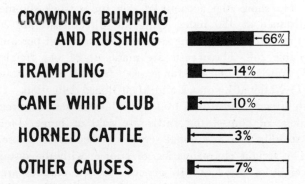

FIG. 114. Location and estimated causes of bruises observed on carcasses of beef cattle during slaughter tests conducted by Wilson and Company. (Livestock Conservation, Inc., Chicago, Illinois.)

during slaughter as the result of bruises that must be trimmed out of the carcass beef. (See Fig. 114.) This loss represents 15 to 20 cents per 100 pounds live weight, or $30 to $45 a carload.

An individual shipper can reduce losses from dead and crippled cattle by careful handling at the feedlot and during loading, but he cannot escape the losses resulting from bruising that occurs in transit or at the killer's yards. Consequently, every cattle feeder should give his enthusiastic support to the program of Livestock Conservation, Inc., a nonprofit, educational organization representing railroads, truckers, commission firms, packers, and stockyard companies, whose principal objective is to reduce losses in marketing livestock.

MARKETING COSTS

A knowledge of the items that make up the total marketing expense is highly desirable if the feeder is to estimate accurately the value of his cattle in the feedlot on the basis of current market quotations. Frequently he will need to choose between selling his cattle to a local order buyer or packer at a certain price or shipping them to the market on his own account. Obviously he must know the approximate marketing cost per hundredweight before he can make an intelligent decision.

Marketing costs are of two kinds, direct and indirect. The direct costs include the cash charges made by the transportation companies and marketing agencies. Indirect costs refer to the weight loss or shrinkage of the animals between feedlot or shipping point and the market.

Figure 115 shows the account of sale for a truck shipment of steers hauled a distance of 130 miles. Only the direct costs are considered in computing the average marketing expense per steer and per hundredweight shown in Table 258. These costs are made up of (1) freight, (2) commission, (3) yardage, (4) insurance, (5) federal transportation tax, (6) feed, and (7) National Livestock and Meat Board deduction.

Freight. Freight constitutes by far the largest item of direct marketing expense unless, of course, the cattle are sold at home. Freight charges have risen steadily, but additional services sometimes offset some of these increases. The freight charge varies, of course, with the distance traveled and the weight of the load. In general, freight charges are figured on the sale weight of the cattle, but if only a partial load is hauled the charge may be made on a mileage basis. Although it may appear that freight costs are eliminated by selling at home, the buyer in turn has to pay comparable freight costs; naturally the buyer tries to pay enough less for the cattle to offset these costs. If he is buying for a local or nearby firm he, of course, could bid more for the cattle, because his freight bill would be less in such a situation.

Commission. Cattle shipped to a central market must be consigned to a livestock commission firm, of which there are many at each of the four or five large markets in or near the Corn Belt. Employees of the commission firm receive the animals from the stockyards company, drive them to the pens, and see that they have feed and water. When the market

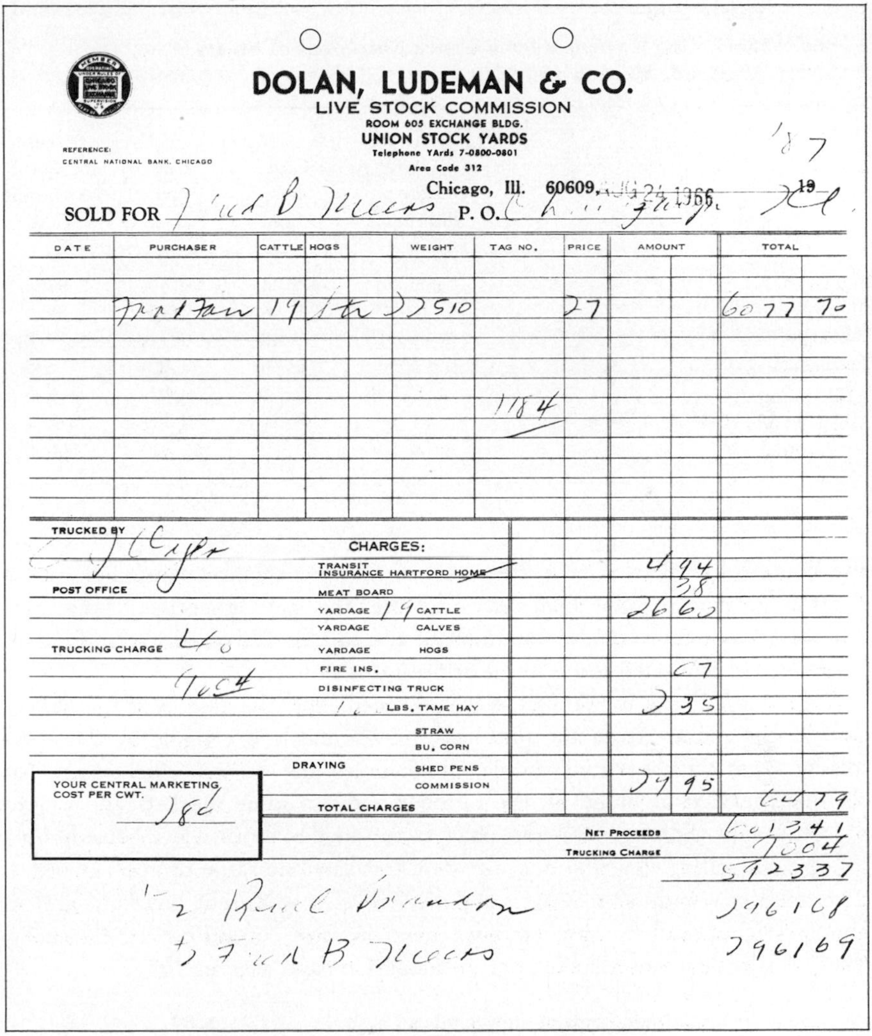

FIG. 115. Account of sale for a truckload of steers sold in Chicago. (Courtesy of Fred Meers, Champaign, Illinois.)

opens, a salesman, frequently one of the members of the firm, shows the cattle to prospective buyers and, unless otherwise instructed by the owner, finally sells them for the highest offer he has received.

Immediately after the cattle are sold they are driven to the scales where they are weighed and locked in holding pens to await the orders of the purchaser. The weigh ticket is sent by messenger to the office of

Table 258
Marketing Expenses for a Truck Shipment of Steers
(19 Head, Hauled 130 Miles)

Item of Expense	Cost, Entire Shipment	Cost per Head	Cost per Cwt	Percent of Total Marketing Cost
Freight	$90.04	$4.74	$0.400	58.3
Commission	29.95	1.58	0.133	19.5
Yardage	26.60	1.40	0.118	17.2
Insurance	5.01	0.26	0.022	3.2
Feed	2.35	0.12	0.011	1.5
National Livestock and Meat Board	0.38	0.02	0.001	0.3
Total	$154.33	$8.12	$0.685	100.0

the commission firm, where the marketing expenses are computed and a draft is drawn on the firm's account for the net proceeds of the sale. The draft, together with a statement of the sale, is delivered to the shipper, if present, or is mailed to his home address or to his bank.

All settlements for freight, yardage, feed, and the like, and the collection of the money from the purchaser of the cattle are made by the commission firm without any trouble whatever to the owner. The charge for all these services is based on the number and, on some markets, the weight of cattle sold. Formerly buying charges for feeder cattle were considerably lower than selling charges, but currently they are the same on most markets. The selling commission varies slightly between different markets and is changed from time to time to meet new business conditions. In February, 1967, the selling commission per head at Chicago was as follows:

Consignments of only one head	$1.80
First 5 head in each consignment	1.60
Next 10 head in each consignment	1.50
Each head over 15 in each consignment	1.50

Considering the multiplicity of details attended to, the value of the product sold, and the amount of responsibility assumed by the salesman individually and the firm collectively, the commission charges are remarkably low, averaging less than 0.5 to 1 percent of the gross value of fed cattle. It is doubtful whether farmers take home so high a percentage of the sale value of any other major agricultural product as they do in the case of beef cattle.

Commission charges in auction sales are generally based on the gross returns of a sale and average about 3 percent of the gross returns. At some sales the buyer reserves the right to refuse the highest bid, in which case the cattle are "passed out" with either no charge or up to 50 percent of the normal sale charge. In a few instances, commission charges are made on a straight per-head basis regardless of the age, weight, or value of the animal, but this system is relatively uncommon.

Yardage. A fee of 75 cents to $1.50 per head is charged by the stockyards company at the terminal markets for the use of the pens, watering facilities, scales, and so on. This is the main source of revenue for the stockyards company, and from it must come all funds for cleaning the pens, repairing fences, pavement, and buildings, as well as for interest, taxes, and dividends. Yardage is paid only once, regardless of how long the cattle are held before slaughter or reshipment.

Insurance. Transit insurance is required if cattle are hauled by public transportation, and it is usually a worthwhile investment. As mentioned earlier, the percentage shipping losses are small; consequently the premium rate is correspondingly low. In addition to transit insurance, at all large markets a charge of 5 to 15 cents per car is made against both shipper and buyer to provide for insurance of the cattle against fire while they are in the yards. In this way a fund is maintained which is adequate to reimburse owners for the full market value of any animals destroyed. The wisdom of providing for such a fund was fully justified in October, 1917, when a fire in the Kansas City Yards destroyed nearly 10,000 head of cattle and calves. Fortunately the insurance fund built up in the preceding years was sufficient to reimburse fully every owner involved. Checks totaling $1,733,779.99 were mailed the day after accredited appraisals had been made.

Federal Transportation Tax. All agencies that haul livestock for the public are required to pay a 3 percent federal transportation tax, based on the transportation charges. Naturally these costs are passed on to the shipper in most instances.

Feed. After being unloaded the cattle are fed hay at the rate of about 10 pounds per head, or 200 pounds per car or truckload. This hay is purchased from the stockyards company at approximately twice the price of hay on the farm. However, the price includes delivery of the hay to the pens and placing it in the mangers before the cattle.

If corn-fed cattle are held overnight before being sold, they are given a feed of shelled corn. Other feeds are usually available if the owner of the cattle wishes to use them.

National Livestock and Meat Board. A 3-cent-per-head deduction is made by commission firms in yards that cooperate with the National Livestock and Meat Board, a nonprofit organization that promotes the consumption of meat and meat products. Such promotion is conducted through all of the various mediums of communication such as demonstrations, radio, television, and newspapers. Meats research is sponsored in universities and private research organizations throughout the country to investigate subjects such as the value of meat protein in the human diet and palatability factors in meat. The deductions are voluntary and are usually matched with an equal amount by the slaughter plant that is buying the cattle.

In summary, the total direct marketing costs per hundredweight, exclusive of freight and shrink, amounted to only approximately 30 cents in the example illustrated in Table 258. It is doubtful whether so much selling and service can be obtained for so little cost in any other area of agricultural production.

EFFECT OF SHRINKAGE ON MARKETING COSTS

The weight lost by cattle between the farm or feedlot and market is as important a factor in determining their net home value per hundredweight as are the actual marketing charges that must be paid in cash. The money lost from shrinkage depends on two factors, namely, the amount of weight actually lost and the value per hundredweight of the cattle. Both items may vary considerably between different shipments. These variations make it impossible to name a figure that represents, with any degree of accuracy, the "margin" an order buyer must have in order to "break even." With 1,000-pound steers shrinking 3 percent during transit and selling for $30 at the market, a shipping margin of 90 cents per hundredweight would be necessary to cover the loss due to shrinkage. But with 1,000-pound cows shrinking 4 percent and selling for only $15 per hundredweight, a margin of only 60 cents would suffice. To this margin must be added the approximate fixed charges per hundredweight to obtain the total difference that should exist between feedlot and market values. This difference lies between $1 and $2 per hundredweight for most cattle. It must be remembered that most cattle bought direct from feedlots are bought with a 2 to 4 percent pencil shrink—that is, 2 to 4 percent of the on-farm weight is deducted when calculating the cost to the buyer. Thus at least half, or even all, of the shrink incurred in shipment to a central market or auction also occurs in direct selling.

Chapter 27

DISEASES of
BEEF CATTLE

The incidence of disease in beef cattle is low compared with the disease rate of the other important species of livestock. Nevertheless losses do occur and may be of considerable importance in individual herds. The monetary loss from cattle diseases in the United States was estimated at $300 million in a 1965 report of the Animal Health Institute. The beef cattle industry cannot afford such a loss.

Data on the incidence of specific diseases in beef cattle are limited and subject to errors in diagnosis on the part of the farmer, rancher, or veterinarian. A rather comprehensive survey covering most of the problems encountered in the beef cattle business, including diseases and parasites, was conducted by the Washington station in cooperation with the Research Committee of the American National Cattlemen's Association. Much of the Corn Belt area, the northeastern, and the upper southeastern states are not included in the survey. However, the 1,588 questionnaires, representing 502,616 head of cattle, supply information that serves as an adequate basis for discussing the incidence of disease among beef cattle.

Table 259 summarizes the data from these questionnaires with respect to incidence of the various non-nutritional diseases encountered. Figure 116 shows the five most important non-nutritional diseases, by areas surveyed. It will be noted that, with few exceptions, the same diseases are important throughout the country, and it is doubtful whether the results of the above survey would have differed much had the entire country been surveyed. An annual mortality rate of 0.59 percent from non-nutritional diseases and ailments was reported, with 60 percent of all death losses being caused by pneumonia, calf scours, shipping fever, and blackleg, in order of importance.

679

Table 259
Incidence of Beef Cattle Non-Nutritional Diseases and Ailments[a]

Disease	Herd Incidence of Disease (Cattlemen Reporting Disease)[b] (%)	Incidence of Disease (% of Total Cattle Afflicted by Disease) (%)	Herds Reporting the Disease in Which the Veterinarian:	
			Diagnosed the Disease (%)	Treated the Disease (%)
Pink eye	46.2	3.07	3.0	2.1
Calf scours	30.8	1.09	3.9	2.5
Shipping fever	11.7	0.84	2.5	2.2
Foot rot	28.3	0.74	4.8	3.8
Pneumonia	23.2	0.46	7.5	6.0
Warts	21.3	0.33	2.0	1.6
Cancer eye	36.8	0.32	7.2	8.6
Lumpy jaw and wooden tongue	25.4	0.22	5.5	6.2
Brucellosis	8.8	0.18	4.2	2.8
Leptospirosis	0.9	0.17	0.6	0.4
Calf diphtheria	8.1	0.13	3.8	2.8
Sunburned udder	3.5	0.11	0.0	0.0
Prolapse of uterus	13.0	0.10	4.0	4.9
Navel infection	7.4	0.09	1.0	1.1
Blackleg	10.2	0.04	2.6	2.1
Vaginitis	1.8	0.03	0.8	0.8
Brisket disease	0.4	0.01	0.0	0.0
Mucosal disease	0.7	0.01	0.3	0.3
Red water disease	1.3	0.01	0.4	0.4
Tetanus	0.7	0.01	0.3	0.3
Circling disease	0.5	0.00	0.2	0.3
Mastitis	0.4	0.00	0.2	0.2
Retained placenta	0.1	0.00	0.0	0.0
Anthrax	1.1	0.00	0.3	0.3
Pulmonary emphysema (grunting disease)[c]	0.06	0.00	0.0	0.0
Rabies	0.3	0.00	0.1	0.1
Vibriosis	0.06	0.00	0.1	0.1
Hardware disease	0.6	0.00	0.1	0.2
Johne's disease	0.2	0.00	0.1	0.0
Diseases not diagnosed	6.7	0.05	0.1	0.1
Other diseases	3.2	0.02	1.4	1.2
Total		8.03		

[a] Washington Experiment Station Bulletin 562.

[b] This column will total over 100% because many cattlemen reported having several diseases in their herds.

[c] This may be a nutritional disease or ailment, but proof on this point is lacking.

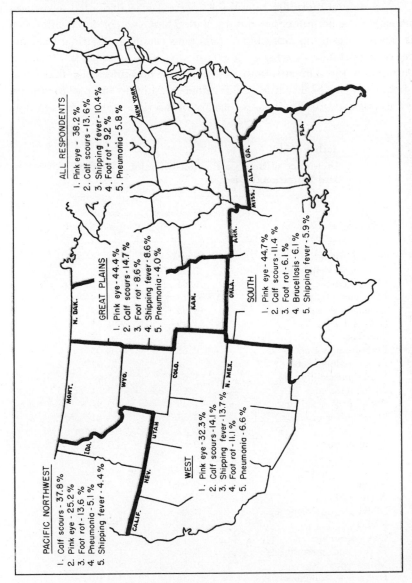

PACIFIC NORTHWEST
1. Calf scours - 37.8 %
2. Pink eye - 25.2 %
3. Foot rot - 13.6 %
4. Pneumonia - 5.1 %
5. Shipping fever - 4.4 %

WEST
1. Pink eye - 32.3 %
2. Calf scours - 14.1 %
3. Shipping fever - 13.7 %
4. Foot rot - 11.1 %
5. Pneumonia - 6.6 %

GREAT PLAINS
1. Pink eye - 44.4 %
2. Calf scours - 14.7 %
3. Foot rot - 8.6 %
4. Shipping fever - 8.6 %
5. Pneumonia - 4.0 %

SOUTH
1. Pink eye - 44.7 %
2. Calf scours - 11.4 %
3. Foot rot - 6.1 %
4. Brucellosis - 6.1 %
5. Shipping fever - 5.9 %

ALL RESPONDENTS
1. Pink eye - 38.2 %
2. Calf scours - 13.6 %
3. Shipping fever - 10.4 %
4. Foot rot - 9.2 %
5. Pneumonia - 5.8 %

FIG. 116. Five most important beef cattle diseases and ailments (non-nutritional) by areas in the United States. (Washington Experiment Station.)

681

Ailments of a non-infectious nature and nutritional diseases were tabulated separately in the survey, with results summarized in Table 260 and Fig. 117. An annual mortality rate of 0.32 percent was attributed to diseases and ailments of a noninfectious nature, with 71 percent of the death losses in this category resulting from bloat, poisonous plants, and urinary calculi, in order of importance.

Fortunately the ailments, both infectious and non-infectious, that occur most frequently in beef cattle yield rather readily to simple treatment. There are, however, a few diseases that are of a serious nature. In addition,

Table 260
Incidence of Beef Cattle Non-Infectious Ailments and Nutritional Diseases[a]

Disease or Ailment	Herd Incidence[b] (%)	Percent of Cattle Afflicted
Bloat, pasture	20.91	0.39
Vitamin A deficiency	2.58	0.35
Bloat, feedlot	10.01	0.15
Poisonous plants	10.14	0.08
Urinary calculi	10.39	0.07
Grass staggers	3.15	0.05
Salt sick	1.01	0.04
Fluorine poisoning	0.63	0.03
Pine needle abortion	1.32	0.02
White muscle disease	1.64	0.02
Iodine deficiency	2.02	0.01
Phosphorus deficiency	1.45	0.01
Sweet clover disease	0.57	0.01
X-disease	0.82	0.01
Alkali disease	0.88	0.01
Oat hay poison	0.69	0.01
Poisons, chemical	0.63	0.01
Anemia	0.94	0.01
Acetonemia	1.13	0.00
Molybdenum deficiency	0.06	0.00
Rickets	1.26	0.00
Oak poisoning	0.19	0.00
Prussic acid poisoning	0.13	0.00
Milk fever	0.38	0.00
Diseases not diagnosed	0.38	0.01
Other diseases	0.69	0.04
Total		1.33

[a] Washington Experiment Station Bulletin 562.
[b] Percent of cattlemen reporting the ailment or nutritional disease.

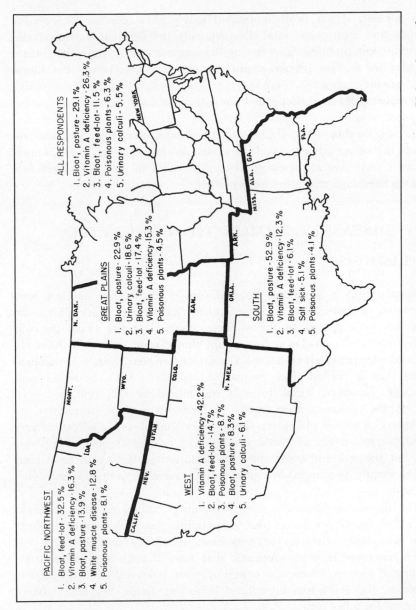

PACIFIC NORTHWEST
1. Bloat, feed-lot - 32.5 %
2. Vitamin A deficiency - 16.3 %
3. Bloat, pasture - 13.9 %
4. White muscle disease - 12.8 %
5. Poisonous plants - 8.1 %

WEST
1. Vitamin A deficiency - 42.2 %
2. Bloat, feed-lot - 14.7 %
3. Poisonous plants - 8.7 %
4. Bloat, pasture - 8.3 %
5. Urinary calculi - 6.1 %

GREAT PLAINS
1. Bloat, pasture - 22.9 %
2. Urinary calculi - 18.6 %
3. Bloat, feed-lot - 17.4 %
4. Vitamin A deficiency - 15.3 %
5. Poisonous plants - 4.5 %

SOUTH
1. Bloat, pasture - 52.9 %
2. Vitamin A deficiency - 12.3 %
3. Bloat, feed-lot - 6.1 %
4. Salt sick - 5.1 %
5. Poisoncus plants - 4.1 %

ALL RESPONDENTS
1. Bloat, pasture - 29.1 %
2. Vitamin A deficiency - 26.3 %
3. Bloat, feed-lot - 11.5 %
4. Poisonous plants - 6.3 %
5. Urinary calculi - 5.5 %

FIG. 117. Five most common beef nutritional diseases and non-infectious ailments in the United States, by area. (Washington Experiment Station.)

683

minor ailments if not properly treated may quickly develop into serious disorders that may prove fatal. Consequently the beef cattle man should want to become proficient in detecting the symptoms of the common diseases and ailments so that proper treatment may be given before the illness makes extensive progress. Perhaps even more important is a knowledge of preventive measures that can eliminate or reduce the problem in the first place.

Treatment should usually be prescribed by a competent veterinarian. Some of the minor ailments, such as scours and bloat, are of such common occurrence as to warrant keeping on hand a supply of prescribed medicines which the herdsman may administer himself.

DISEASES COMMON TO BEEF CATTLE

ANAPLASMOSIS

Anaplasmosis is a serious blood disease of cattle that has not been encountered in some sections of the United States. It is believed that the disease has been present in the southern states for many years, but was not distinguished from Texas fever until that disease had been definitely eradicated. Anaplasmosis resembles Texas fever in being caused by a blood parasite that is carried from animal to animal by ticks, horseflies, and mosquitoes. However, Texas fever is transmitted by only one species of tick, and the protozoan that causes anaplasmosis may be transmitted by more than 25 species of insects. This makes control and eradication very difficult. Anaplasmosis has been observed in nearly all of the southern states, but the few outbreaks that have occurred in the North have been confined mainly to shipped-in cattle that presumably were infected when purchased.

The disease is especially severe in mature cattle, resulting in death losses of 30 to 50 percent of the animals infected. Calves and yearlings seldom are visibly affected, but may have the disease in mild form and make a complete recovery. Animals that have the disease and recover, regardless of age, are immune to future infection but usually remain carriers throughout life, and when introduced into clean herds they are a source of infection for other cattle.

Anaplasmosis may occur in either chronic or acute form. In the acute form the first symptom is a high temperature of 103°F to 107°F. After two or three days the temperature falls to subnormal, breathing becomes difficult, the mucous membranes are pale and yellowish due to severe anemia, and the muzzle is dry. Urination is frequent, but the urine seldom is bloody. The patient usually is constipated and either blood or mucus, or both, may be voided with the feces. In acute anaplasmosis death usually occurs 1 to 5 days after the first symptoms are observed. In the chronic form the infected animals live longer and a greater percentage recover. Recovery of mature animals is usually very slow. Cows with the disease often become belligerent and hostile when moved or, at the other extreme, they may fall back to the rear of a herd and may even fail to keep up with the herd if driven for several miles. Such a crude test often sorts out most of the sick animals. Laboratory tests of blood samples should be made to verify a tentative diagnosis, and state and federal veterinarians should be notified.

No specific treatment has been perfected for anaplasmosis; consequently treatment varies among veterinarians. Good nursing and care, of course, are vital to recovery. Daily transfusions with 2 gallons of normal bovine blood usually save an infected animal but this treatment is of course expensive. Drugs that stimulate blood formation are sometimes helpful, and aureomycin has been reported beneficial in the early stages of anaplasmosis.

A vaccine for anaplasmosis was approved and released for use in 1965. Young cattle are given two treatments at 6-week intervals, and results after testing under commercial ranching conditions appear quite favorable.

ANTHRAX

Anthrax, also called splenic fever, is a highly infectious disease and often fatal. Not only is it transmissible from one animal to another, but it may spread rapidly to all species of livestock and even to man. Historically it has proved to be one of the worst scourges of animal life. This disease existed among European cattle as long ago as the early seventeenth century and has caused enormous losses. Its presence in the United States is, in general, confined to certain local areas of rather low, moist land more or less mucky in character. In such soil the spores of anthrax remain potent for years. Hay grown in such soil may carry the germs of the disease far and wide. Hides have been responsible for outbreaks of anthrax among the cattle in the vicinity of tanneries.

Death from anthrax is very sudden; finding a dead animal in the pasture or lot is commonly the first sign of the presence of the disease. In all cases of sudden death without forewarning symptoms of illness, anthrax should be suspected and the carcass handled with extreme caution until the death can be attributed to some other cause. The characteristic signs of death due to anthrax are (1) a tarry consistency of the blood, which is blackish in color and fails to clot firmly, and (2) an enlargement of the spleen of two to five times its normal size. A positive diagnosis can be made only by examining a blood sample under a microscope for the presence of the specific organism, *Bacillus anthracis.*

When anthrax is suspected a veterinarian should be called without delay. Burning or deeply burying the carcass of the dead animal, and prompt vaccination of all other animals in the herd with antianthrax serum are the best known methods of preventing spread of the disease. Following a diagnosis of anthrax, animals showing high temperatures should be separated from the others and given serum plus antibiotics, whereas others may be treated with serum and spore vaccine. Those receiving the serum and antibiotics should be revaccinated after several weeks with both serum and vaccine.

BLACKLEG

Blackleg is an acute, infectious disease attacking mainly cattle between 6 and 24 months of age. Man apparently is not susceptible to this disease. The main symptoms are a marked lameness and pronounced swellings over the shoulders and thighs, caused by the formation of gas in the subcutaneous tissues. When pressure is applied to these swellings a peculiar crackling sound is heard, which is the characteristic symptom of blackleg. The disease runs a rapid course and almost always terminates fatally in 12 to 36 hours. Like anthrax, it is more or less restricted to certain sections and even to individual fields where the soil is infested with the spores of the causative organism and where outbreaks commonly occur annually unless prevented by vaccination.

Blackleg is common in the western half of the United States and is especially prevalent throughout the Southwest and on the eastern slopes of the Rocky Mountains. Outbreaks have been reported in nearly all of the Corn Belt states, but there the disease is confined to limited areas.

Prevention rather than treatment is the practical method of combating the disease. All young cattle brought into a region known to be infested with blackleg should immediately be vaccinated with blackleg bacterin, made by treating cultures of the blackleg organism with formalin to destroy their infectious properties but not their antigenic value.

Calves born in a region where blackleg has occurred even infrequently during the preceding 10 years should be vaccinated when they are about 6 months old. Usually vaccination is performed at weaning time.

BRUCELLOSIS

Brucellosis is also called Bang's disease, contagious abortion, or simply abortion. Abortion of the infectious type is one of the worst problems for cattle breeders. It is caused, usually, by a specific organism, *Brucella abortus,* although other uterine infections of a more general character may cause a number of abortions in a herd of cows.

Apparently infection usually gains entrance through the mouth, having been scattered in the uterine discharge of an aborting animal over grass, hay, and other feed materials, or into the water source. All aborting cows should be segregated from the herd and quarantined until at least 3 weeks after all uterine discharge has ceased, or until a negative blood test has been obtained. Some authorities hold that an infected cow is never rid of the organisms, although she may eventually become sufficiently immune to carry a calf full term. Bulls as well as cows may contract the disease, and again the avenue of infection is thought to be the digestive tract. The prudent breeder, however, refrains from mating a bull that he knows is clean with cows that are suspected aborters, to prevent possible contamination from the genital tract.

The blood agglutination test furnishes a dependable basis for the control and eradication of brucellosis. In this test, which uses about 20 cubic centimeters of blood drawn from the jugular vein, various dilutions of the blood serum are mixed in small test tubes with specially prepared cultures of the *Brucella* bacillus and allowed to stand 24 to 48 hours. If the animal is negative the bacteria remain in suspension, keeping the solution cloudy, but if it is positive the bacteria are precipitated to the bottom of the tube, leaving the fluid above them clear. Samples intermediate between positive and negative are called "suspects," and animals from which these came must be rebled and tested again.

A much faster agglutination test is under study and appears promising. This test, called the card test (so named because a card with spots that are impregnated with reagents is used to make the test), requires a smaller blood sample, and the tail vein is used as the source. The need for a squeeze chute is eliminated and the cattle are not unduly disturbed. More important, results of the test can be obtained within a few minutes. Those animals that react positively—and they are either positive or negative, never "suspect" as in the older test—are retested by the older method for further verification. The big advantage of the card test is that the herd can be tested and sorted with one handling, as contrasted to the older system that called for two handlings in order to retest the suspects, with a delay of at least 2 to 3 days before definite identification of carriers could be made.

Another test known as the ring test, applied to milk samples, is impractical for beef herds but is useful in screening dairy herds for possible infected animals.

There is no known cure for brucellosis. Immunization of heifer calves between 6 and 8 months of age with a standardized live vaccine, *Brucella abortus strain 19,* is a reasonably effective immunizing agent. In older animals, immunization may effectively prevent the disease, but it interferes with blood agglutination tests, thus making later blood tests difficult to interpret. For this reason older females are seldom vaccinated. Eradication of brucellosis is usually most successful if pursued on an area plan, operated under joint federal-state arrangements. States with less than 1 percent occurrence are called "modified brucellosis-free areas." Individual herds that are tested annually by a state or federal veterinarian and found to be free of reactors in two successive years are designated as "accredited herds." Federal and state regulations relating to the transportation of breeding cattle may vary from area to area and are subject to change, as are the various approved plans for eradication of the disease. Every breeder of beef cattle, therefore, should keep abreast of the latest regulations in effect in his area.

The handling of infected animals or the drinking of milk from infected cows can spread the disease to man in a form known as undulant fever. The disease is not often fatal in man but a long, debilitating illness results.

FOOT-AND-MOUTH DISEASE

Foot-and-mouth disease is not prevalent in either the United States or Canada. It has gained access to the United States on ten different occasions since 1870, but in each instance it was promptly stamped out by the destruction of all infected and exposed animals. Such measures are justified in view of the terrible handicap imposed by this plague upon the livestock industry of the countries of Europe and South America where the disease is constantly present. Foot-and-mouth disease is not transmissible to man.

The disease was introduced into Mexico in 1947 when a shipment of Brahman bulls was smuggled into the country from Brazil, and only through heroic efforts and the expenditure of more than $100 million by the Mexican and United States governments working cooperatively was the disease brought under control. At first an attempt was made to stop the disease by the slaughter of all infected and exposed animals, but it was soon obvious that this plan was impractical, and a program of vaccination and rigid quarantine and inspection was inaugurated.

Foot-and-mouth disease is seldom fatal, but it causes enormous losses through the loss of weight suffered while the animals' mouths are so sore that they almost refuse to eat. Recovery is very slow, and many of the animals are hopelessly ruined from the standpoint of the feedlot. There is no specific cure for the disease, although vaccination is being used in many parts of the world with reported success. Federal-state agreements for prompt cooperative action in case of an outbreak of foot-and-mouth disease can quickly be put into effect. Indemnities are paid in the eradication programs that follow a diagnosis of the disease.

JOHNE'S DISEASE

This disease, known also as paratuberculosis, is believed to be on the increase. It is usually spread by the droppings of infected animals which contaminate pastures, water sources, and even feedbunks. A severe, chronic and intermittent diarrhea is responsible for the ease of contamination, and this diarrhea, accompanied by extreme loss in weight, is the principal outward symptom. Owing to the long incubation period, calves that contract Johne's disease usually do not show symptoms until they are about 2 years old or, in the cast of heifers, after they have dropped their first calf. Appetites remain good, temperature and pulse rate are unchanged, but milk production decreases. The animals soon appear unthrifty and emaciated, continuing to scour and waste away until death results in many cases. Because these symptoms are similar to the effects of internal parasites and malnutrition, Johne's disease is rather difficult to diagnose. An interdermal injection of johnin, prepared and used in the same way that tuberculin is used for the tuberculosis test, with resultant thickening of the skin within 48 hours indicates a positive reaction.

No satisfactory treatment is known and satisfactory vaccines are not yet available. A test and slaughter program with payment of indemnities by federal and state authorities has been in operation since 1927.

Johne's disease has not been reported in man.

LEPTOSPIROSIS

The incidence of this disease is not well determined, but it has been found in 48 states and seems to be spreading rapidly. Like brucellosis, this disease is spread by urine and other discharges from infected animals, and the principal point of invasion is the mucous membranes. Because the kidneys are attacked, the urine may be dark red or wine-colored. Abortions may occur when pregnant animals are infected, causing this disease to be confused with brucellosis. Leptospirosis has a short course of 3 to 10 days, with death losses running as high as one-third in calves.

Treatment is of doubtful value, although antibiotics sometimes give relief. A bacterin has been produced, but protection through its use is not always certain. A serological test has been developed for diagnostic work, and it has become a generally recommended practice to use the blood samples drawn for brucellosis testing to screen a herd at the same time for leptospirosis.

A few cases of this disease have been found in man. Until the extent of the danger to human beings is known, the usual antiseptic precautions should be taken in working with cows that abort or show other symptoms that may indicate presence of leptospirosis.

MUCOSAL DISEASE COMPLEX

Several diseases, including rhinotracheitis, mucosal disease, and virus diarrhea, are believed to be caused by several strains of the same virus. The disease, known as the mucosal disease complex, is probably much more prevalent than supposed. It is not known how these diseases are spread. Symptoms are high fever, nose and eye discharges, excessive salivation, lesions in the membranes of the respiratory and digestive tracts, and diarrhea.

Rhinotracheitis or "red nose" is the most important of the various diseases of the mucosal complex, as it has spread throughout the large feedlots of the West where large cattle numbers are congregated. This disease may often be confused with shipping fever. Weight losses are great owing to severe dehydration. The inflammation is usually confined to the respiratory system, and death often results from strangulation.

Mucosal disease itself is an infection confined largely to the digestive tract and occurs in calves more often than do the other diseases of the complex. Late winter or early spring is the season of greatest incidence. Symptoms are a rise in temperature to about 106°F and then a drop to nearly normal. There is loss of appetite, nasal discharge, and diarrhea that progresses with the disease until the feces contain much mucus and blood. Ulcers may be found in the nostrils, on the muzzle, lips, and gums, and in the mouth. Sometimes there are congested areas and hemorrhage in the colon. Death losses in a herd may amount to 20 to 50 percent of the young stock, and 100 percent losses have been reported.

Virus diarrhea, another disease of the mucosal complex, affects both the respiratory and the digestive tract. In the early stages there are nasal discharges and superficial mouth lesions, but the most characteristic symptom is diarrhea, which develops in the later stages of the infection. The watery feces contain mucus and blood, and weight losses are extremely heavy owing to the dehydration resulting both from fever, which may run as high as 108°F, and the diarrhea.

No completely successful control has been developed for any of the diseases in the mucosal complex. Antibiotics and sulfa drugs are often used but serve only to reduce the danger of secondary infections such as pneumonia. There is a vaccine for preventing the diseases, but protection is limited. Calves are usually vaccinated at branding or weaning time in the areas where the disease is quite common.

So far as is known, no danger to man exists with respect to this group of diseases.

SHIPPING FEVER

This disease has been called hemorrhagic septicemia, feedlot pneumonia, and stress fever, as well as shipping fever. Actually "stress fever" is a fitting name because any sudden change, such as weaning and shipment or change in weather, which puts a stress or strain on the cattle, particularly calves, makes them susceptible to the infective agent. The causative organism is not specifically known but is believed by some to be a mixture of viruses and bacteria.

Shipping fever spreads easily from animal to animal upon contact. Contaminated stockyards, sale barns, trucks, and cattle cars, and cattle-working equipment such as chutes and scales, are all likely sources of infection. It is not uncommon for a new shipment of stockers or feeders to spread the disease to cattle that are already on feed.

As indicated earlier, shipping fever is one of the big four in causing cattle death losses. Indirectly it may account for additional losses as calves or yearlings that have been weakened by shipping fever often fall victim to pneumonia.

Shipping fever is a respiratory infection that usually appears in cattle within 10 days of arrival after shipment or within 14 days after first exposure. A tired, hang-dog appearance is characteristically an early symptom. Appetites are dull and a mild cough is usually present, becoming more evident when the calves are moved about. Temperatures rise to 107°F in the more serious cases. Since easily recognized symptoms may not always be present, the use of a rectal thermometer to check the temperature of all calves in a drove is the only way to find those really needing treatment. In the usual outbreak of shipping fever many calves recover spontaneously but some—and this varies with the virility of the infection and the amount of stress the calves have undergone—contract pneumonia and die if left untreated.

Because shipping fever is apparently brought on by an impaired condition of the animal, preventive measures rather than a specific cure should be the first concern of the cattleman. Hard driving, overcrowding in cars, insufficient bedding, irregularity in feeding and watering en route, insufficient time to eat, drink, and rest at unloading points, improper handling, and use of unpalatable rations upon completion of the journey should be avoided.

If the cattle arrive during cold weather, especially if it is wet and stormy, adequate shelter should be provided. The cattle should have sufficient time to recover from their trip and become accustomed to their new surroundings before they are dehorned, implanted, subjected to detailed sorting, and so forth. The ration for the first few days should be of good quality, preferably a choice grade of prairie or mixed hay, with perhaps a little silage or some crushed or whole oats after the third or fourth day. During favorable weather the cattle may have the run of a rather short pasture.

The use of serums, mixed bacterins, and aggressins is not recommended for the prevention of stress or shipping fever, because their effectiveness is questionable in the absence of knowledge concerning the real cause of the disease. If an outbreak does occur in spite of all precautions or if the buyer had no control over the previous treatment of a shipment of calves, prompt diagnosis and treatment are necessary to keep both weight and death losses low. Immediate treatment with antibiotics or combinations of sulfa drugs and antibiotics on arrival of all calves is sometimes practiced, but unless the calves have had an unusually long and rough trip, such treatment probably is unwarranted. Rather, only those calves showing temperatures above 105°F need such treatment, and the remainder may be cared for in the manner previously described.

Investigations with injections of tranquilizer prior to shipment have not produced consistent results. The reduced restlessness could conceivably reduce the stress on the calves and therefore could reduce their susceptibility to stress fever. In order for this practice to work, the practical problem of sorting and handling the calves prior to shipment would have to be met. Ranchers are usually reluctant to add such operations to their established routines, mainly because they usually result in higher labor costs and lighter shipping weights. Another problem appears to be the variability in sedation effects from a given dose. Cases of "down" cattle in trucks and cars are not uncommon in tranquilized cattle.

High-level feeding of antibiotics, either aureomycin or terramycin, at the rate of 500 milligrams daily for 5 days followed by a low level of feeding (70 to 80 milligrams daily) for an additional 2 to 3 weeks, is recommended by some authorities as a preventive for shipping fever. The problem with this practice is that newly weaned calves are unaccustomed to eating concentrates unless they were previously creep-fed, which is usually not the situation. The antibiotics, usually contained in a manufactured or commercial pellet or meal, must therefore be fed with a highly palatable feed such as crushed or rolled oats, to ensure uniform consumption of the antibiotic. All too often the calves that are most apt to contract the disease owing to excessive exposure or rough treatment are also the ones that prefer to eat nothing but hay or pasture. Mixing the antibiotic-containing supplement with a palatable hay for a few days is probably the best solution. As affected cattle will usually drink water, water-soluble sulfa drugs are being used with some success.

There is no evidence that human beings contract shipping fever from exposure to infected animals.

PINK EYE

Pink eye is the term commonly applied to an infectious inflammation of the eye, technically called infectious catarrhal conjunctivitis. It is usually encountered only during the summer months and is more prevalent in cattle on pasture than in those kept in drylots. Frequently the disease persists in a herd for several months, with nearly all of the animals becoming affected in one or both eyes.

The disease is characterized by an intense inflammation of the mucous membrane of the eye accompanied by tears mixed with pus, which flow down the sides of the face. In its most aggravated form it causes a large grayish yellow ulcer to appear on the cornea, making the eye temporarily blind.

All animals in a herd in which pink eye is present should be examined carefully every day, and those that show symptoms of the disease should be segregated, if possible, in a darkened barn and supplied with plenty of fresh water and succculent feed. Their eyes should be thoroughly cleaned of dirt and pus by washing with a boric acid solution made by dissolving 1 ounce of boric acid crystals in 2 quarts of boiling water and allowing it to cool. After the eyes have been thoroughly washed, a thin coating of 5 percent sulfathiazole eye ointment should be applied to the affected parts. This ointment comes in a metal tube and can be squeeed directly on and under the eyelids. Antibiotics in fine powder form which can be "puffed" into the eye by squeezing the plastic container are also recommended. Severe cases should be treated daily, and those of greater severity at least twice a day, until all discharges cease. Eyes with ulcers sometimes require treatment for several weeks before they return to normal condition. Vaccines are available for the prevention of pink eye, but inconsistent results have been reported from their use.

PNEUMONIA

Pneumonia is one of the most common diseases affecting young cattle during the winter months. It is especially prevalent during cold, damp weather and among calves that undergo considerable exposure because of poor shelter. The characteristic symptoms are a high temperature of 105 to 107°F, quick, shallow breathing, with dilated nostrils, and a hard, pounding pulse. By applying the ear to the chest, the herdsman can hear the rasping sound made by the affected lung. The animal habitually lies on the side opposite the diseased lung in order to keep the infected organ uppermost. In advanced cases and when both lungs are involved, the animal shows little disposition to lie down. Instead, it takes an unsteady position with head down and forelegs wide apart to make for as much ease in breathing as possible.

The usual treatment for pneumonia is to administer sulfa drugs either alone or with penicillin. Whereas formerly a high percentage of pneumonia cases resulted in death, a large number are now saved by the judicious use of these wonder drugs. The animal should be kept as quiet as possible because exercise or excitement tends to aggravate the condition by increasing the pulse and respiration rate. Warmth and dry shelter are essential.

SCOURS OR DIARRHEA

An abnormal looseness of the bowels is a common disorder among young calves kept in unsanitary surroundings or fed improper rations. Calves ranging in age from a few days to 3 months are most likely to be affected, although older animals are by no means immune. The condition in reality may not be a disease in itself, but may be the result of an abnormal condition of a portion of the digestive tract that causes the feed to be improperly digested. Prompt treatment should be administered because a calf infected with a bad case of scours derives little nourishment from the milk and feed eaten and consequently loses flesh rapidly and becomes weak and thin.

Most cases of scours are caused by the presence in the digestive tract of harmful bacteria that brings about the formation of toxic products. The first step in the treatment of scours is the administration of an internal antiseptic, sulfa drug, or antibiotic, to destroy these organisms. In conjunction, a mild purgative should be given to rid the system of the objectionable toxins as soon as possible. Many biological supply houses manufacture internal antiseptics that give good results when used for scours. By giving them according to the directions on the label, along with 4 to 8 ounces of castor oil, depending on the size of the calf, the trouble is usually checked. Another remedy is 0.5 to 1 teaspoonful of formalin, diluted with 1 pint of water. In order for these medicines to bring the desired result as quickly as possible, the calf's feed intake should be somewhat reduced for 3 or 4 days.

A particularly virulent form of scours, which is highly contagious among the calves of a herd, is characterized by grayish yellow or dirty white feces with a highly offensive odor. Whenever such symptoms are observed, a veterinarian should be consulted immediately because "white scours" is one of the most serious ailments affecting young calves. White scours appears within 3 days after birth, whereas other diarrheas seldom develop before the calf is a week or 10 days old. This helps to differentiate between the two types of scours.

TUBERCULOSIS

Tuberculosis is one of the most serious diseases affecting domestic cattle. It is caused by a specific organism that attacks all parts of the body, but particularly the glands of the lymphatic system. The glands of the neck, chest, and mesentery are most likely to be affected. In advanced cases the disease spreads to all parts of the body, forming huge masses of tubercles that interfere greatly with the normal functions of the vital organs. Not only is the disease highly contagious from one cow to another, but it may be transmitted from cattle to swine through the manure, and from cattle to man through the milk.

Specific symptoms of tuberculosis are difficult to detect in the live animal. In advanced cases the hair becomes harsh and dry and the animal has a general rundown appearance. Such symptoms, however, may be caused by a variety of diseases and cannot be regarded as conclusive. The reliable test for the presence of tuberculosis is the so-called tuberculin test. Tuberculin is a laboratory product which, when injected under the skin, causes a characteristic reaction or swelling within 72 hours.

Because there is no method of treating tubercular cattle, all animals reacting to the test must be sold for slaughter. Since in most animals the disease has not progressed far enough to affect seriously the value of the carcass for meat, "reactors" should be shipped to markets where they may be slaughtered under federal inspection, to salvage those that are still usable. The law forbids the sale to local butchers of known tubercular animals. Under a plan long in force, farmers must submit all their cattle to periodic tests and may not add untested animals to their herds. The testing is done free of charge by a federal veterinarian. Any reactors that are found must be promptly disposed of.

Tuberculosis has been almost eradicated in the United States by these procedures, but constant surveillance is still necessary. The incidence of the disease is reduced to such a low status that regulations concerning free testing are in the process of change. For example, a "back tagging" system is being tested in some states wherein farmers and ranchers voluntarily tag all of their cull and older cows being marketed. This enables the federal inspectors, upon finding an animal with tubercular lesions, to identify the farm or ranch from which the animal came. Obviously the purpose is to find the herds that need testing. Unfortunately, producers seem reluctant to use this voluntary method for detecting infected herds, but in the long run it should be the most economical method for accomplishing complete eradication of this disease.

Federal and state governments cooperate in paying indemnities for cattle slaughtered after a positive reaction to the tuberculin test. Tuberculosis is transmissible to man, and it remains a health problem among human beings as long as a vestige of the disease exists in the cattle herds. Breeding cattle consigned to purebred auction sales must usually be tested for tuberculosis, brucellosis, and in some cases, leptospirosis, within 30 days before the sale. This testing is done to permit entry of animals into accredited areas or across state lines.

VESICULAR STOMATITIS

This disease is not common, but outbreaks sometimes occur in the western states. It resembles both foot-and-mouth disease and the mucosal complex. The fact that it resembles the former makes it imperative that anyone suspecting that his cattle are so affected notify the state veterinarian in the area. The disease has several forms and is caused by a virus that is transmitted by direct contact or through the drinking water.

The first symptom of vesicular stomatitis is an acute rise in temperature followed by the formation of reddened patches in the mucosa or lining of the mouth, on the tongue, inside the nostrils, between the toes, and on the teats of fresh cows. These areas quickly develop into vesicles or blisters up to an inch in diameter. They rupture, and a yellowish fluid or serum oozes from the reddened area beneath the scab that forms. Usually healing occurs in 8 to 15 days in uncomplicated cases, after much loss in condition and weight. Death seldom occurs.

Quarantines are usually imposed on all cattle in the immediate vicinity of an outbreak until it can be determined by incubation tests that the disease is not the much more serious foot-and-mouth disease. Horses may be affected; in fact, more cases are found in horses than in cattle. Treatment consists of isolation, to prevent further spread, and the provision of soft, palatable feeds such as bran, silage, or green chop. Dry hay is to be avoided if possible.

VIBRIOSIS

This infectious disease of the genital tract of cattle is fast assuming equal importance with Bang's disease and leptospirosis as a cause of abortion. It appears to be widespread, especially in the eastern parts of the Rocky Mountain states. The disease is caused by a bacterium, *Vibrio fetus,* which is spread in mating with an infected animal. Artificial insemination with semen from infected bulls can also spread the disease. Young breeding cattle seem to be most susceptible.

Infected animals may or may not conceive, but conception is followed by early abortion or, more often, resorption of the fetus. Evidence that this is happening will be the cows' returning to heat several months after they apparently were settled. Abnormally long or short heat periods are also good evidence of the presence of vibriosis. Infected cows or bulls show few other symptoms of the disease; hence any suspected animals should be tested by a veterinarian. Infected bulls should be eliminated, but sexual rest of cows for at least 3 months, plus treatment with antibiotics, will usually cure them. Only a young, virgin bull or semen from tested bulls should be used on the rested females.

A vaccine for vibriosis has been developed by Colorado workers and is commercially available, to be used on heifers in the prevention of this disease. At present, annual vaccination is required, as the vaccine does not produce permanent immunity.

NON-INFECTIOUS DISEASES AND AILMENTS

Bloat and vitamin A deficiency are two conditions coming under this heading and have already been discussed in Chapters 20 and 6, respectively. Others in this category, although not causing a large number of deaths, can be particularly trying at times. The order in which they are discussed has no bearing on their relative importance.

IMPACTION, CONSTIPATION, AND INDIGESTION

Because cattle often consume large amounts of coarse, dry roughage, they are somewhat disposed to certain digestive disorders during the fall

and winter months. Impaction is the term applied to an abnormal accumulation of material in the paunch. Constipation, on the other hand, implies
a clogging of the large intestine with hard, dry feces. Indigestion is a more
general term referring to both of these conditions, as well as to other digestive disturbances.

The first symptoms of indigestion are usually a loss of appetite and
a rise in body temperature. When such conditions are noted, the bedding
material should be examined for the amount and character of the recently
voided feces. If the feces are scanty or hard and dry, the patient should
be given a strong purgative such as 2 pounds of Epsom or Glaubers' salts
dissolved in 2 quarts of water. In the case of impaction, as much as 2
or 3 gallons of water should be given, and the left side of the animal
should be powerfully kneaded with the fist to bring about a breaking
up and softening of the impacted mass. Constipation ordinarily calls for
less drastic treatment. Drenching with 1.5 pints of castor oil, 2 quarts of
raw linseed oil, or 1.5 to 2 pounds of Epsom salts usually brings relief.
For calves these doses should be reduced by approximately one-half.

LUMPY JAW OR ACTINOMYCOSIS

Lumpy jaw is caused by a fungus that attacks the tissues of the throat,
the parotid salivary gland, and the bones of the upper and lower jaws.
Its presence is indicated by a round swelling, usually quite hard and firmly
adherent to the surrounding parts. The swelling gradually increases in size
until it finally breaks open to form an abscess that discharges thick, creamy
pus and becomes filled with raw, bleeding tissue. If not properly treated,
the fungus invades the interior of the jaw bones, causing a loosening of
the teeth. Also growths may be formed in the mouth and pharynx, sizable
enough to interfere greatly with eating and breathing.

Not all swellings in the neck region are evidence of lumpy jaw. A
diagnosis can easily be made by microscopic examination of a sample of
the abnormal growth. The fungus will appear to be made up of club-shaped
bodies, all radiating from the center of the mass to form a rosette. For
this reason the fungus is often called "ray fungus."

Cattle affected with lumpy jaw should be separated from the herd
because there is danger of spreading the disease by scattering the fungus
on feed that other animals consume. Animals of no more than market
value should be promptly sold where they will be slaughtered under proper
inspection. Only in advanced cases of the disease are the carcasses likely
to be condemned as unfit for food. Valuable breeding animals that are
suspected of having lumpy jaw should be examined by a veterinarian and,

if found diseased, should be placed under his care. The most satisfactory treatment consists in complete surgical removal of the growth, easily accomplished if the growth is not too firmly adherent to the surrounding parts. In the latter circumstance the growth may be cut open, the pus washed out, and the cavity packed with gauze saturated with tincture of iodine. If surgical aid is unavailable or if position of the growth makes an operation hazardous, the internal administration of potassium iodide over an extended period may bring about recovery. However, when surgical treatment is possible, it is much to be preferred.

FOUNDER OR LAMINITIS

Founder or laminitis in beef cattle is usually the result of overeating. It most frequently occurs among feeder or show cattle that are put on a full feed of grain too rapidly for them to become gradually accustomed to their new ration or among cattle being fed on all-concentrate or high-concentrate rations containing less than 15 to 20 percent roughages. The intake of absorbable nutrients is much greater than can be utilized by the body, necessitating the oxidation of the surplus in the tissues. In the destruction of the excess nutrients much heat is produced. Also, the amount of blood and the circulation rate are considerably increased in the effort to dispose of the heavy load of nutrients received from the lymphatic system as quickly as possible. As a result, the capillaries, especially those near the body surface and in the extremities, are gorged with blood nearly to the limit of capacity. The capillaries of the feet, surrounded as they are by the inelastic hoofs, are unable to expand to an appreciable extent. The animal is inclined to remain quiet for a considerable time after its heavy meal, and the feet do not receive the exercise necessary for vigorous blood circulation; thus there is a congestion of the blood vessels of the feet, followed by inflammation of the tissues and severe lameness.

Treatment of badly foundered cattle is seldom satisfactory, because usually the condition is not observed until the damage to the feet has been done. If it is detected in its early stages, some relief may be had by applying cold packs to the feet or compelling the animal to stand in a pond or stream of cold water. Owing to the intense pain of standing and walking, foundered animals spend a large amount of time lying down, and they go for feed and water only when driven by pains of hunger and thirst. They frequently lose weight, especially if they were carrying considerable finish when the attack occurred. Animals only mildly affected need cause little concern, but those badly involved should be disposed of as soon as it is obvious that their condition is chronic. If possible they should be sold locally, as they are likely to lie down and be trampled if shipped to a distant market. Moreover, such cattle do not command satisfactory prices at terminal markets because of their poor appearance.

FOOT ROT

Foot rot or foul foot is most likely to occur in feedlot and show cattle that are shut up in dirty barns and lots or kept on contaminated pastures during the summer. Occasionally thorns, stones, and corncobs or other bedding materials may become lodged between the toes in a way that sets up inflammation of the skin in the interdigital spaces. As this part of the foot is very sensitive and tender it is easily scratched or cut, allowing infection or attack by the fungus that causes the problem. If prompt attention is not given, the inflammation and swelling may extend well above the hoof and around to the heel.

Treatment consists of cleaning the foot thoroughly and applying a strong antiseptic that destroys the causative organism, *Actinomyces necrophorus*, which has gained access through the skin. Mild cases respond to undiluted creolin or iodine, but if the infection is well advanced, a 20 percent solution of blue vitriol or potassium permanganate is preferred. Bandaging the foot and keeping the bandage well saturated with one or the other of these solutions usually effects a cure within 3 or 4 days. Intravenous injections of sodium sulfapyridine are recommended for unbroken cattle that cannot easily be handled.

Foot rot in show cattle can be largely prevented by keeping the feet properly trimmed and by turning the cattle into dew-laden pastures at night throughout the summer and fall. A shallow footbath containing a 30 percent solution of copper sulfate, through which the animals must walk as they enter and leave the barn, is also effective in preventing the disease.

CANCER EYE

Cancer eye or epithelioma is a malignant tumor on the eyeball or eyelid of cattle. The disease is found mainly on the ranges of the Southwest where intense sunlight and irritating dust are believed to be at least indirect causes. Hereford cattle are much more susceptible to eye trouble than Shorthorns, and Angus appear to be wholly immune. Apparently the lack of pigment in the eyelid and eyelashes to shield the eye from intense sunlight contributes to the prevalence of the disease. For that reason Hereford cows with a ring of red skin around their eyes are highly regarded by many ranchers of the Southwest. Many Southwestern ranchers will not use a bull from a cow that developed cancer eye. There is strong evidence that there is a genetic basis for such discrimination against the offspring of either bulls or cows that develop cancer eye.

Cancer eye probably starts from an injury to the eye which from lack of treatment develops into a badly infected sore. Usually by the time it is observed it has affected the eyeball or the eyelids to such an extent that treatment, other than surgical, is of little avail. At any rate, if the cow or bull is of great value it should be treated by a veterinarian. Less valuable animals should be sold for beef.

URINARY CALCULI

Much trouble sometimes results from the formation of stonelike mineral deposits in the bladder in male cattle, particularly in steers that are finished on grain sorghum in the Southwest. The calculi cause no noticeable discomfort, unless they enter or block the entrance to the urethra and prevent normal discharge of urine. When this occurs the animal is extremely nervous, refuses to eat, lies down and gets up at frequent intervals, and shows other signs of acute distress. Should the bladder rupture, temporary relief may be obtained, but peritonitis caused by the urine in the abdomen soon results in death. If the bladder does not rupture, death results from uremic poisoning brought about by the continued absorption of urine by the blood.

The cause of urinary calculi is not known. There is some evidence that it is associated with mineral metabolism, but the mineral composition of the stones varies greatly, with common forms being magnesium phosphates and various silicates, indicating that a number of factors are probably involved. Drinking water that is high in mineral content does not of itself cause urinary calculi.

There is no effective treatment for urinary calculi, and feeders living in areas where it occurs should attempt to prevent the problem. Including ammonium chloride in pasture supplements or in feedlot rations appears to be one method of prevention. Details for use of this material are given in Chapter 9. Frequent salting to induce the drinking of large quantities of water is recommended in some parts of the country to reduce the incidence of calculi.

WARTS

Warts are small skin tumors that frequently appear on young cattle, especially on the neck, shoulders, and head. They are more often observed during the late winter and early spring, when the skin is in poor condition because of the low sterilizing effect of winter sunlight, and perhaps because of faulty nutrition which lowers the natural resistance of young cattle. It has been demonstrated that warts are caused by a virus that may be transmitted from one animal to another and possibly to other species, including man.

Warts are usually a temporary condition that eventually disappears of its own accord. The disappearance can be hastened by the daily application of Vaseline or castor oil, which favors their absorption. Absorption can also be accelerated and the spread of the warts to other cattle made less likely by use of a commercially prepared vaccine consisting of finely ground fresh wart tissue suspended in a salt solution to which formalin has been added to destroy the virus.

RINGWORM

This skin condition is prevalent during the winter when cattle may be rather closely confined. It is caused by a fungus and appears as circular patches of roughened, scaly skin over the body but mainly about the head and neck and at the root of the tail. Itching is a symptom easily observed as cattle rub the affected areas, thus further spreading the condition to other parts of the body or to other cattle. Curry combs and brushes can easily spread ringworm through an entire show string of cattle; consequently the condition should receive immediate attention once it is observed.

FIG. 118. A severe case of ringworm. (University of Illinois.)

Treatment for ringworm is simple, consisting of scrubbing the infected areas vigorously with soapy water, then daubing them with iodine, lime sulfur, or prepared medicants available from veterinarians. This treatment should be applied every 3 to 5 days until the condition is remedied. Gloves should always be worn when working with cattle infected with ringworm because it is highly transmissible to man.

HARDWARE DISEASE

This condition carries its descriptive name because it is caused when such objects as nails, pieces of baling wire, or parts of machinery are picked up by cattle, accidentally during eating or through curiosity, and swallowed into the reticulum or honeycomb compartment of the rumen. There the sharp edges or ends of the objects may puncture the stomach wall, passing into the pericardial cavity where they may cause immediate death by injury to the heart, or cause an accumulation of fluids that eventually leads to death.

Symptoms of hardware disease are a rapidly developing, unthrifty condition, swellings in the brisket region, a sound of fluid movement in the heart and lung regions, and obvious pain when the animal moves about. Treatment is seldom successful, and most cases are inoperable. However, when cases involving extremely valuable animals warrant the expense, objects that can be located by fluoroscopic examination are sometimes surgically removable by entry through the rumen wall.

Mineral deficiencies, especially of phosphorus, often are responsible for animals' picking up metal objects. Correction of the deficiency and careful removal of all bits of wire and the like from pastures and feedlots reduce the incidence of this problem. The use of magnetic metal arresters in feedmills removes a large part of the metal objects often responsible for the condition in feedlot cattle.

POISONOUS PLANTS AND METALS

The number of cases of sickness and death of cattle caused by eating poisonous plants and other harmful materials is difficult to determine with any degree of accuracy and it varies greatly from area to area. In the midwestern and southeastern sections of the United States the number of plants that are poisonous to livestock is small compared with the number found in the range area. Also, the feed supply is usually sufficient to cause animals to pass by poisonous plants, nearly all of which are quite unpalatable compared with good, wholesome forage. However, it occasionally happens, especially in early spring and during prolonged periods of dry weather, that pastures are so short and bare that cattle begin to eat any vegetation available, especially weeds and shrubs that are still green and succulent.

The majority of poisonous plants are found in moist, shaded regions, especially near ponds and streams. The forage of timbered and marshy pastures should be examined closely each year to see if there are plants that are poisonous to livestock. If any are found, a close watch should be maintained to detect the first tendency of the cattle to eat them.

Treatment of animals that have eaten poisonous plants should be prescribed by a veterinarian. Remedies given by mouth are often of little value in overcoming the toxic effect of poisons, owing to the great volume of the digestive tract in cattle and the mass of stomach and intestine contents that must be reached by the drugs.

Prevention of poisoning from plant sources is a matter of being able to identify the dangerous plants and conditions under which they are toxic. Limited space does not permit a detailed description of all plants that are poisonous to cattle, and only the more common ones can be described here. The local county agricultural agent usually has information relative to the plants in a given area.

COMMON CROWFOOT

This is a common weedy wild flower of the buttercup family that is found in meadows and pastures. The juices of the plant are extremely acrid, and cattle usually will not eat it except in the early spring when they are first turned onto pasture. Several varieties of crowfoot are found in the central states; nearly all are more or less poisonous in the green state but harmless after being cut and dried. The stems, which vary greatly in height according to species, are smooth, hollow, and much branched. The flowers are small, with pale yellow petals surrounding a prominent seed head.

Symptoms of crowfoot plant poisoning are gastric enteritis and diarrhea with black, foul-smelling feces, together with nervous symptoms such as difficult respiration, slow chewing of cud, and jerky movements of ears and lips, sometimes followed by convulsions and death within a few hours.

ERGOT

Poison from ergot occurs chiefly during the winter and spring when considerable amounts of grain and cured roughages are being fed. Ergot is a fungus-caused plant disease that affects the seeds of some plants, making them hard, black, somewhat curved in shape, and several times larger than natural. Of the grains, rye is most likely to be affected, and bluegrass, fescue, and redtop hays are also highly susceptible to ergot.

Ergotism in cattle is of two types, the spasmodic and gangrenous. In the former the symptoms are muscular trembling, convulsions, and delirium. Abortion is often brought about by this form of ergot poisoning. In the gangrenous type there is a mummification and sloughing off of the extremities such as ears, tail, or feet through degeneration of the small blood vessels supplying those parts. Treatment of animals with marked symptoms of ergotism is of little avail. Destruction is recommended except in the case of valuable breeding animals.

FESCUE FOOT

It is believed by some that this aptly named condition is identical with ergot poisoning, and it does resemble it in many respects. It differs principally in that it is always associated with the grazing of a heavy growth of fescue, usually in the late fall or winter. Some investigators believe that it differs from ergot poisoning because ergot is seldom found in fescue, especially after the grass drops its seed in the fall. In any case, the same crippling condition results from destruction of small blood vessels in the extremities, mainly the hind feet (therefore the name "fescue foot"). Gangrene and actual sloughing of toes, tail, and ears may occur because of the loss of circulation.

Some preventive measures that can be applied are: sowing a mixture of pasture grasses rather than straight fescue; feeding supplemental concentrate or roughage; mowing the fescue to keep it from growing tall; and rotation grazing, using the fescue no more than half the time. There is no known cure for the condition, but prompt removal of cattle from the pasture if symptoms are noticed and good nursing may prevent the loss of an animal if the condition is not far advanced. Badly affected animals should be sold for slaughter before undue loss of condition occurs, because recovery of advanced cases is seldom complete.

FERN OR BRACKEN

The common fern or bracken that grows in moist, shaded spots in woods and along streams often causes poisoning in cattle. The early symptoms are unsteady gait, loss of appetite, constipation, nervousness, and congestion of the eyes. These symptoms are followed by a spreading apart of the legs in an effort to maintain balance, extreme nervousness, and a general loss of muscular control.

JACK-IN-THE-PULPIT OR INDIAN TURNIP

This is a plant of the arum family found growing in wooded areas. It is easily recognizable by its peculiar purple-striped, vase-shaped flowers with large erect stamens and pistils. The poisonous nature of this plant is not yet fully understood, and few data are available concerning the specific symptoms in cattle that have been poisoned from this source. However, there is ample reason for classifying it among the plants dangerous to cattle.

HENBANE AND JIMSON

These weeds, which belong to the nightshade family, are coarse, smooth-stemmed, ill-scented plants that inhabit neglected fields and waste places. They prefer a rich soil and are often found in old feedlots and in fields that have been heavily manured. Poisoning is most likely to occur when good grazing is scarce or when hungry feeder cattle are turned into infested lots upon their arrival at the farm.

The common symptoms of nightshade poisoning are unsteady gait, cramps, convulsions, and loss of consciousness. Respiration is difficult and the pulse is very rapid. Fortunately cattle seldom eat these weeds except when forced to do so by extreme hunger.

WATER HEMLOCK

This plant is one of the deadliest that may be eaten by cattle. The root is the harmful part, and losses are most severe during the early spring when the ground along streams is badly eroded by high water. Also, at this season the roots may be pulled up by cattle grazing the green tops of the plants. Symptoms of hemlock poisoning are intense nervousness, frothing at the mouth and nose, and violent convulsions. Death frequently occurs within an hour after the root of a single plant is eaten.

WHITE SNAKEROOT

White snakeroot is a rather large, coarse perennial weed found in many woodland pastures throughout the central states. It can be identified rather easily by its leaves, which are opposite, from 3 to 5 inches long, broadly ovate, with serrated edges. Each leaf stalk has three main veins extending from the base of the leaf in numerous branches. These veins

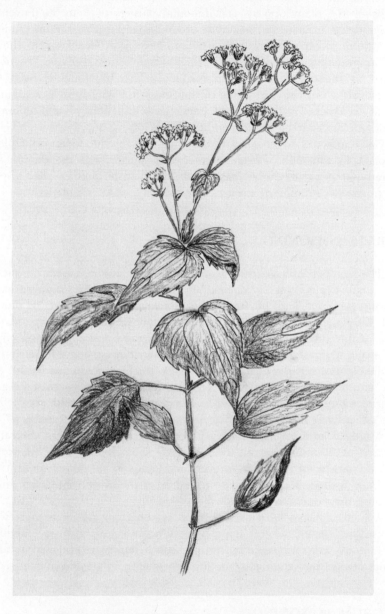

FIG 119. White snakeroot (*Eupatorium urticaefolium*).

are prominent on the underside of the leaf. In late summer the white flowers of the plant appear as compound clusters of 8 to 30 flowers.

Poisoning from white snakeroot occurs usually during the late summer and autumn when pastures are dry and brown. Affected cattle show a peculiar trembling of the muscles, which accounts for the name "trembles" applied to the disease in certain localities. This trembling is most pronounced after exercise and tends to disappear following rest. Constipation, general weakness, loss of weight, and incoordination of the voluntary muscles are associated with the condition.

Cows affected by snakeroot poisoning secrete the toxic principle of the plant in the milk. There is thus great danger that the disease may be transmitted to human beings, calves, and other animals that are fed milk from cows running in pastures infested with snakeroot plants.

NITRATE POISONING

The number of cases of nitrate poisoning has increased during the past decade, caused, it is believed, by the use of heavier application of nitrogen fertilizer, and by heavier or thicker planting rates. A number of climatic factors can be responsible for abnormal mineral metabolism in the plant, resulting in accumulation of nitrates and nitrites, mainly in the stems and leaves. Drought, excessive cloudiness, uneven distribution of rainfall, and self-shading due to a very thick stand of plants, all may cause nitrate accumulation. Corn, cut for silage or green chop, small-grain pastures, mountain meadow hay, Sudan grass, sorghum, and pigweed all cause difficulty at times.

A nitrate level higher than 1.5 percent of the dry matter of the ration, expressed as potassium nitrate or KNO_3, is cause for concern. At present the only satisfactory way to prevent difficulty is to dilute the nitrate-containing feed with a feed lower in or free of the nitrate. Molasses, mixed with or sprayed onto hay, will reduce the danger.

Animals afflicted with acute nitrate toxicity may die in a short while but usually live for several days. The difficulty is an anemia caused by a failure in the formation of normal hemoglobin, resulting in fatal or chronic anoxia brought on by insufficient oxygen in the body tissues.

SORGHUM POISONING

Under certain conditions, rapidly growing sorghum becomes a dangerous feed for cattle because of its high content of hydrocyanic or prussic acid. Severe drought and frost appear to promote the formation of this poisonous material in sorghums, especially in the regrowth that springs up after the main crop has been harvested. The grazing of sorghum stubble should be done with extreme caution so that the first signs of illness may be detected and the animals removed from the field. Feeding the harvested fodder or silage causes no trouble, even in frosted sorghum or sorghum badly fired from drought. Intravenous injections of sodium thiosulfate are used by veterinarians in treating the condition. The consumption of considerable wild cherry foliage can cause the same poisoning.

LEAD POISONING

Numerous cattle deaths have resulted from licking buildings and fences that were covered with a lead-containing paint. There have even been instances in which a single old paint bucket thrown on a rubbish pile has been responsible for the death of several valuable animals. Cattle have also been poisoned by eating silage from a silo that was filled before the lead paint on the inside walls was dry.

The symptoms of lead poisoning are general dullness, convulsive movements of the limbs, champing of the jaws, and violent bellowing, followed by prolonged stupor and death. Lead acts as a toxin that is particularly damaging to the kidneys, resulting in perforation of the glomeruli.

Use of nonlead paints has reduced the incidence of lead poisoning, but when old lead paints have only been covered over by the new, safe kinds, access to the lead-containing paint is still possible. If detected in its early stages, lead poisoning may be arrested by certain injections into the bloodstream that prevent the lead from reaching the kidneys. Prompt consultation with a veterinarian is essential in treating this condition.

RADIATION DAMAGE

Although beef cattle are unlikely to be primary targets in the event of the use of nuclear weapons, some cattle would be affected unless provided protection. Naturally, the nearer the target the greater the protection needed because of the correspondingly higher intensity of radiation and rapidity of fallout. Radiation effects are less severe in cattle than in human beings, and some cattle are more resistant than others. It is therefore likely that, although some cattle might receive external burns, most of them would survive a nuclear explosion. Temporary sterility would occur in many cases and permanent sterility in others. However, a cow that was badly burned and scarred in the first experimental atomic explosion at Trinity Site in New Mexico lived to produce 16 normal calves after she was moved to Oak Ridge, Tennessee.

The lean tissue or muscle of an animal apparently does not absorb radiation fallout. Consequently the meat from exposed animals would not become radioactive, even though burned externally, and thus would furnish a ready supply of safe food following a disaster. Such animals should not be fed contaminated feed or allowed access to contaminated water, however, if they are to be kept alive and used for food. Contamination could occur during slaughter unless precautions were taken. To ensure future generations of beef cattle, breeding stock could be fed the feeds that were subjected to the least radiation exposure. Radiation decays rapidly, and most feeds and water would be usable within a few days.

Shelters can be provided that will afford protection for cattle, but naturally in large operations it would not be feasible to attempt to protect all cattle. Priorities might be established, with valuable seedstock and milk cows having top priority. Plans for shelters, as well as estimates of feed and water requirements, are obtainable from the county agricultural agent. It is to be hoped that protection from radiation damage will never be required, but some attention and thought should be given to this potential problem.

Chapter 28

PARASITES AFFECTING
BEEF CATTLE

Internal and external parasites of beef cattle are responsible for rather serious monetary losses in individual areas, and for smaller losses in almost all situations, unless preventive and control measures are applied. Unfortunately, many of the losses are not readily apparent and thus do not seem important to many cattlemen. In 1965 the Animal Health Institute reported that cattlemen in the United States were losing $162 million annually because of cattle parasites. Undoubtedly the judicious use of insecticides is one of the best investments a cattleman can make in terms of net return per unit of investment.

The incidence of recognizable cases of harmful parasitism is difficult to assess. The Washington station survey, discussed in Chapter 27, is the most comprehensive and recent study of this problem. Figure 120 shows the percentage of farms and ranches that reported parasite problems. (The rank of each type of parasite is not necessarily its rank from the standpoint of monetary losses.) Screwworms were particularly important in the South at the time of the survey, while lice and hornflies were important in all of the areas surveyed. When cattlemen were asked which of the parasites they considered most damaging, they ranked lice, grubs, screwworms, and hornflies at the top, as shown in Fig. 121. Three of these four insects are still the most important ones, but the screwworm has almost been eliminated. Diseases such as anaplasmosis and coccidiosis were included by many cattlemen because parasites are responsible for their transmission (see Chapter 27).

The Washington survey revealed some interesting information concerning the control measures applied by cattlemen. Table 261 indicates that lice, hornfly, and grub control measures were applied by about 40 percent

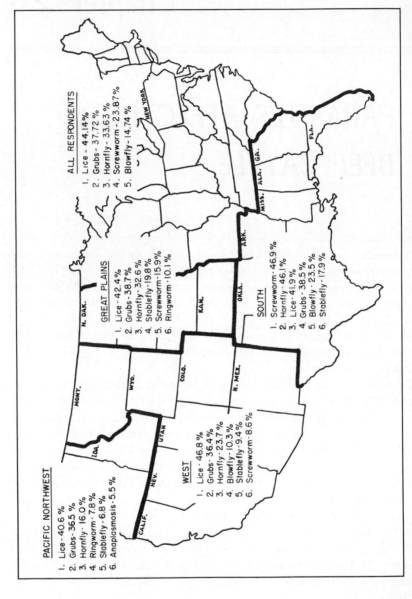

PACIFIC NORTHWEST
1. Lice - 40.6 %
2. Grubs - 36.5 %
3. Hornfly - 16.0 %
4. Ringworm - 7.8 %
5. Stablefly - 6.8 %
6. Anaplasmosis - 5.5 %

GREAT PLAINS
1. Lice - 42.4 %
2. Grubs - 38.7 %
3. Hornfly - 32.6 %
4. Stablefly - 19.8 %
5. Screwworm - 15.9 %
6. Ringworm - 10.1 %

ALL RESPONDENTS
1. Lice - 44.14 %
2. Grubs - 37.72 %
3. Hornfly - 33.63 %
4. Screwworm - 23.87 %
5. Blowfly - 14.74 %

WEST
1. Lice - 46.8 %
2. Grubs - 36.4 %
3. Hornfly - 23.7 %
4. Blowfly - 10.3 %
5. Stablefly - 9.4 %
6. Screwworm - 8.6 %

SOUTH
1. Screwworm - 46.9 %
2. Hornfly - 46.1%
3. Lice - 41.9 %
4. Grubs - 38.5 %
5. Blowfly - 23.5 %
6. Stablefly - 17.9 %

FIG. 120. Six most common beef cattle parasites in the United States, by areas. (Washington Experiment Station.)

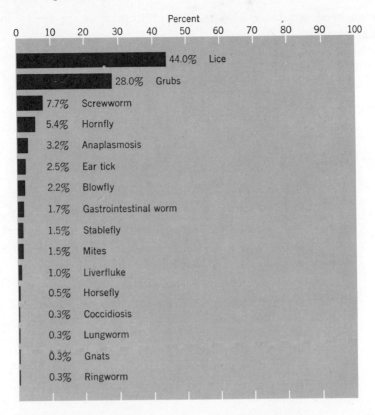

Percent

0	10	20	30	40	50	60	70	80	90	100

44.0% Lice

28.0% Grubs

7.7% Screwworm

5.4% Hornfly

3.2% Anaplasmosis

2.5% Ear tick

2.2% Blowfly

1.7% Gastrointestinal worm

1.5% Stablefly

1.5% Mites

1.0% Liverfluke

0.5% Horsefly

0.3% Coccidiosis

0.3% Lungworm

0.3% Gnats

0.3% Ringworm

FIG. 121. Most damaging beef cattle parasites in the United States, as indicated by percentage of herds affected. (Washington Experiment Station.)

or more of those answering the questionnaire. The use of control measures for intestinal parasites has increased in the South in recent years. Systemic phosphates are now being used for control of grubs, lice, intestinal parasites, and even flies.

Any discussion of control measures for parasites is apt to become out-dated because of rapid developments in this area of research. Readers would do well to follow the research work of entomologists and parasitologists through the press releases and other publications of extension specialists, experiment stations, and industrial organizations.

Many parasitic infestations are geographic in nature, and consequently only a few recommendations concerning use of certain sprays, vermifuges, and the like, are applicable in all instances throughout the country. Suppliers of insecticides, as well as county agricultural agents and vocational agriculture teachers, should be able to supply information for local conditions in a given area.

COMMON EXTERNAL PARASITES OF CATTLE

FLIES

The hornfly is the most damaging fly among those that are harmful in adult form. This fly feeds on the backs, necks, and at the base of the

Table 261

Percentage of Cattlemen Applying Beef Cattle Parasite Control Measures[a]

Parasite	Cattlemen Applying Control Measures (%)
Lice	52.7
Hornfly	41.2
Grubs	37.7
Screwworm	24.8
Stable fly	19.3
Blowfly	15.6
Mites	5.8
Ringworm	5.2
Gastrointestinal worms	4.7
Coccidiosis	2.8
Liver fluke	2.5
Anaplasmosis	2.3
Cattle tick fever	1.3
Lungworm	0.8
Bovine trichomoniasis	0.4
Others	2.5

[a] Washington Experiment Station Bulletin 562.

horns of cattle. Thousands may be found at one time on a single animal, and since each fly feeds several times daily by sucking the blood, the afflicted animal experiences much annoyance and disturbance along with loss in vitality and poor performance caused by loss of blood. Hornflies may be almost eliminated by use of any of the recommended control measures from the beginning of the grazing season and throughout the summer. Note that several chemicals are effective, and these may also be applied

FIG. 122. Hornflies can be controlled by periodic high-pressure spraying with any one of several excellent insecticides. If systemic materials are used, other insects, including grubs and even intestinal parasites, are also controlled. (Western Livestock Journal.)

by spraying or dipping or by means of back rubbers and dust bags as shown in Table 262. Preliminary investigations indicate that feeding a granular mineral mix containing 5.5 percent ronnel controls hornflies.

STABLE FLIES

Stable flies are much more difficult to control than are hornflies because they seem to become resistant to many fly-spray materials after a few years. This species of fly is most troublesome around barns and feedlots,

Table 262

Materials and Application Methods for Control and Treatment of Cattle Parasites[a]

Parasite	Treatment	Insecticide	Minimum Days from Last Application to Slaughter	Treatment Method	Precautions and Comments
Horn fly	Spray	0.5% Methoxychlor	—	Spray on 2 qt to backline of animal every 3 weeks or as needed.	Grub control can also be effected by Ruelene and Ronnel sprays if high pressures are used.
		0.5% Toxaphene	28		
		0.5% Ronnel (Korlan)	—		
		0.06–0.125% Co-Ral (Coumaphos)	—		
		0.5% Malathion			
		0.375% Ruelene	28		
		3% Ciodrin	—		
	Dust bags			Spray thoroughly, no more than 1 gal per animal. Place insecticide inside tripled burlap bag, hang so that 4–6 in. of bag touches backs of animals. Place where animals congregate or in walkways to water, mineral blocks, etc.	
	Cable rubbers	5% Methoxychlor	—	Saturate cable rubber with insecticide in No. 2 diesel oil. Recharge at 3-week intervals.	Do not use crankcase oil on cable rubbers.
		5% Toxaphene	28		
		1% Ronnel	14		
		1% Co-Ral	—		
		2% Malathion	—		
	Pour-on	8.3% Ruelene (in water) or 9.4% (ready-mix)	28	Apply 1 fl oz per 100 lb body weight up to 800 lb, no more than 8 fl oz per animal. Pour evenly along animal's backline.	Use only prepared solution in hot, humid weather. Keep sparks away from treated animals.
Grubs	Pour-on	8.3% Ruelene (in water) or 9.4% (ready-mix)	28	Apply Ruelene at 1 fl oz per 100 lb body weight up to 800 lb, no more than 8 fl oz per animal; Co-Ral and Neguvon at $\frac{1}{2}$ fl oz per 100 lb body weight up to 800 lb, maximum of 4 fl oz per animal. Pour evenly along animal's backline.	Initiate treatments 1–2 month[S] after end of heel fly season. Apply no later than Nov. 1. Also controls horn flies and lice.
		4% Co-Ral	—		
		8% Neguvon (Trichlorfon)	21		
	Spray	0.375–0.5% Co-Ral (WP)	28	Wet entire body to skin. Use nozzle pressure of 300 psi and a narrow band spray with not more than 1 gal Ruelene per animal.	Do not treat sick animals or those under 3 months old. Begin treatment after end of heel fly season. Apply no later than Nov. 1. Also controls horn flies and lice.
		or 0.375% Co-Ral (EC)	—		
		1% Neguvon	14		
	Feed additive	0.6% Ronnel	60	Mix with grain or protein supplement so that animals receive 0.3 lb treated (0.6% Ronnel) feed per 100 lb body weight daily, 7 consecutive days, or 0.3 lb treated (0.26% Ronnel) feed per 100 lb body weight daily, 14 consecutive days.	Start treatment after end of heel fly season. Apply no later than Nov. 1. Aids in control of horn flies and lice.
		0.26% Ronnel	28		
	Mineral block or granules	5.5% Ronnel	21	Feed continuously for not less than 75 days at level of 0.25 lb medicated block or granules per 100 lb body weight per month.	

Pest	Application	Insecticide	Days	Remarks
Lice	Spray	0.05% Lindane	30	Treatments also effective against horn flies. Ruelene also effective against grubs.
		0.5% Methoxychlor	—	
		0.06–0.125% Co-Ral	—	
		0.5% Malathion	28	Spray thoroughly; repeat after 3 weeks if needed. Apply with high pressure for best results. Increase methoxychlor concentration to 1.0–1.5% for tail louse.
		0.5% Ruelene	7	
		0.5% Sevin (Carbaryl)	28	
	Pour-on	9.4% Ruelene	28	Also effective against horn flies and grubs. Apply Ruelene pour-on at 1 fl oz per 100 lb body weight up to 800 lb, no more than 8 fl oz per animal. Pour evenly along animal's backline. Repeat after 3 weeks if needed.
	Dip	0.06–0.125% Co-Ral	60	Immerse animal. Repeat treatment after 2–3 weeks if needed. Stir insecticide mixture in dipping vat regularly to prevent settling out.
		0.03% Lindane	56	
		0.025% Ronnel	28	
		0.5% Toxaphene		
Ticks	Spray or dip	0.125–0.25% Co-Ral	30	Immerse or spray thoroughly. Use pressures of 300 psi for spray applications. Repeat as needed.
		0.03% Lindane (spray)	60	
		0.03% Lindane (dip)	28	
	Spray	0.5% Toxaphene	56	
		0.75% Ronnel		
Ear tick	Dust	5% Co-Ral	—	Dust lightly inside ears to reach infested areas, also adjacent head area. Repeat as needed. Use only low-pressure sprays.
	Spray	0.5% Malathion	21	Avoid ear damage by using very low pressure sprays. Handle applicators carefully.
	Aerosol	2.5% Ronnel		
Screwworm	Dust	5% Co-Ral	—	Dust wound and surrounding area thoroughly; repeat as needed.
	Spray	0.25% Co-Ral	56	Spray wound thoroughly and wet entire body; repeat as needed.
		0.5% Ronnel	21	Collect specimens before treating wound. Preserve larvae in 70% alcohol and submit to county agricultural agent for positive identification to assist in eradication program.
	Aerosol	2.5% Ronnel		Spray wound thoroughly.
Horseflies, Stable flies, Mosquitoes	Spray	1% Vapona (Dichlorvos)	—	Dilute in mineral oil. Spray 1–2 oz as a mist spray twice daily. Same as for faceflies. As a spray, 1–2 qt per animal every 2 or 3 days, or as mist spray of 1–2 oz per animal daily.
		1% Ciodrin		Only temporary control is effected. Vapona and Ciodrin not registered for control of horseflies.
		0.6–1% Pyrethrum + 0.5–1% Synergist		
Facefly	Face rubbers	1% Ciodrin	—	Dilute insecticide in No. 2 diesel oil. Saturate rubber. Construct rubber to permit animal to rub its face, or force animals to use rubber by installing it in front of mineral troughs, watering troughs, etc.
		1% Co-Ral	14	
		1% Ronnel		
	Bait	0.2% Vapona (Dichlorvos)	—	Add bait to mixture of 75% corn syrup, 25% water.

General Precautions: These control suggestions are for beef cattle only. Other sources should be consulted for control suggestions on dairy animals. Do not treat any animal with systemic insecticide (Coumaphos, Ronnel, Ruelene, Neguvon) while under treatment with another systemic. Do not treat animals under 3 months of age with Coumaphos. Spray animals 3 to 6 months old lightly. Do not treat calves less than 1 month old with Malathion. Do not spray or dip calves less than 3 months old with lindane. Do not use Coumaphos with synergized pyrethrums. Do not apply systemics in conjunction with oral drenches or other medications such as phenothiazine or with other organophosphates. Do not spray or pour-on animal for 10 days prior to shipping or weaning. Do not treat sick animals or animals under stress. Recommended levels of insecticide applications and mandatory withdrawal periods before slaughter are constantly changing. Always consult the label on the insecticide container before making applications.

[a] Prepared with the assistance of Dr. Grant Kinzer, Department of Botany and Entomology, New Mexico State University.

where it breeds in wet straw, manure, spoiled feed, and other kinds of filth. Stable flies can be greatly reduced by frequently cleaning lots and sheds and by plowing under the manure or piling it at least half a mile from the barns.

Frequent spraying of the legs of cattle with one of several effective chemicals, at levels recommended by the manufacturers, controls stable flies. The same treatment is recommended for horseflies and mosquitoes, which suck blood from the backs and ears of cattle. Various materials may be used, as recommended in Table 262.

GRUBS

The grub or ox warble is a serious cattle parasite in nearly all parts of the northern hemisphere. It has long been present throughout the United States, although in many localities it is of little economic importance. The degree of infestation appears to vary from year to year even in highly infested areas. This variation leads to the belief that the grub population is affected by climatic conditions, particularly severe cold weather, since the pupa stage of the insect is spent in the ground. The heaviest infestations are found in the Southwest, and it is possible that most of the severe outbreaks of this parasite in feedlots can be traced to feeder cattle coming from that region.

The adult fly, commonly known as the heel fly and somewhat resembling a small black bee, appears during the first hot days of summer. It deposits clusters of eggs on the hairs about the feet and legs of cattle, especially just above the heel. Although the flies can neither bite nor sting, cattle show a natural dread of them and often run from one end of the pasture or lot to the other, holding their tails high, attempting to escape from the flies. Yearlings are more likely to be attacked than calves, and calves more than adults. Because the insects do not fly over water, cattle often protect themselves by retreating into ponds or streams.

The larvae, which hatch in 4 to 6 days, immediately penetrate the skin at the base of the hairs to which they were attached. They then proceed, by no well-defined route, to the neck region where they gather about the esophagus. No plausible reason has yet been advanced to explain why the grubs congregate near the esophagus. During winter, probably in February and March in the Corn Belt but as early as October in the South or Southwest, the larvae leave the neck region and migrate to the back where they form characteristic swellings beneath the skin. There a hole is made through the hide, toward which the larva directs its posterior. In this position it increases rapidly in size, feeding on the pus and mucus caused by its presence. During this time the larva undergoes at least two metamorphoses, becoming larger and stouter until finally, with the arrival of spring, the fully grown grub works its way through the hole in the skin and falls to the ground. There it pupates before being transformed into a mature fly, when it begins the cycle over again.

The numerous holes made by grubs in the backs of cattle greatly lessen the value of the hides, and the beef in the vicinity of the grubby lesions often is slimy and has a bad color. Badly damaged carcasses require extensive trimming in the valuable loin region. Altogether the total economic loss to the country caused by the heel fly is estimated at nearly $100 million per year.

Several recommended treatments are given in Table 262. The initial treatments may begin as early as November in southern and southwestern cattle, but native cattle in the Middle West may not require attention until February. Cattle raised in the Northwest, North, and Northeast seldom are affected by grubs. Cattle shipped in from the areas where flies may overwinter are affected during the first winter but probably not thereafter.

SCREWWORMS

The screwworm, once the scourge of the southern and southwestern stockman, is all but a thing of the past. From a pest that caused untold millions of dollars in death losses and much labor and inconvenience in treating infested cattle, the screwworm has dwindled to a few hundred cases per year. The almost complete eradication of this once destructive insect pest is an outstanding example of teamwork and the widespread application of research findings.

Screwworm control was brought about and is being continued in the following manner. Each year billions of male screwworm flies are produced in the laboratory, made sterile by radiation, and then released from airplanes in areas where the female fly is found, including all of the southern United States and northern Mexico. Because the female screwworm fly mates for life with only one male, the majority of females in these areas produce infertile eggs, and the insect's life cycle is broken because only a few new young are produced. The result is that only a few isolated instances of screwworm infestation are now found in the United States.

Relatively few sterile male flies are now needed in the States, but a barrier zone several hundred miles wide is still being seeded with such flies each year across Mexico to prevent migration of fertile males back into the United States. Credit for the virtual elimination of the screwworm problem should go to the state and federal governments of both countries, to the various livestock producers' organizations, and to the individual ranchers and farmers concerned. This is one of the finest examples of how all elements of an industry have worked together and strenuously attacked a common problem.

Treatment methods for screwworm are included in Table 262, although it is doubtful whether they will often be needed.

FACEFLY

The facefly was first observed in Canada in 1952 and by 1965 had migrated into most of the United States. The facefly originated in Europe and Asia where it was more a household nuisance than a problem with livestock.

The adult facefly, slightly larger than a housefly, is distinguished by its feeding habits. The flies cluster about the eyes, nostrils, and muzzles of cattle, living off the secretions found at these sites. They do not bite or suck but are an annoyance to livestock and cause reduced gains and milk production. Faceflies are also strongly suspected of contributing to the spread of pink eye.

The adult female facefly deposits her eggs in fresh manure where they hatch within 1 day. The yellowish larvae mature in 3 or 4 days and pupate in the soil underneath the manure deposits for 3 or 4 more days, after which they emerge as adult flies, the generation having been completed in 10 to 12 days. Adult flies overwinter in warm or heated buildings, and 10 to 15 generations can be produced in one year, accounting for the large numbers that can result from only a few overwintering adults. Reasonably effective control can be obtained by following the recommendations given in Table 262.

LICE

Cattle are subject to three species of lice: long- and short-nosed blue sucking lice, and the red biting louse. Although numerous, the biting lice feed mainly on particles of hair and dead skin and are not especially troublesome. The sucking lice, however, cause the cattle much annoyance and, if numerous, sap large quantities of blood. Lice-infested cattle spend much time rubbing themselves against posts, trees, and gates in an effort to relieve the intense itching. Frequently large patches of hair are worn off and the hide becomes hard and callused.

Lice are most numerous during late winter and spring. Cattle on pasture are seldom troubled with them. Treatment should be begun in the fall by dipping or thoroughly spraying with any of the materials recommended in Table 262. Pour-on materials are also effective and easy to apply.

MANGE OR ITCH

Mange of cattle is caused by small mites that attack the skin, causing it to become thickened, covered with crusts, and devoid of hair. Usually the rump is the part most likely to be affected, although the back and shoulders may also be involved. In many respects mange of cattle resembles the disease known as scab in sheep, although it is not considered so serious. Treatment consists in liberal applications of a solution of lime sulfur or a nicotine dip. In mild cases when only a small portion of the body is affected, an ointment made of flowers of sulfur and lard often effects a cure.

TICKS

Losses from tick infestation are not so great today as they were in the preceding century when the tick that was the carrier of the dreaded Texas fever caused millions of dollars in losses every year, providing the impetus for government control programs for this and other parasites.

There are far too many species of ticks to allow a detailed discussion here, but several of the more important ones should be mentioned. The ticks found in the Southeast are often the vector or carrier for anaplasmosis in cattle. Elimination of all ticks from this area would go a long way toward eliminating this disease, but other insects also may serve as carriers, and wild animals may also carry ticks, so it is likely that the problem will never be eliminated completely. Other species of ticks serve as vectors for Rocky Mountain spotted fever and tularemia, diseases that affect man.

Spraying or dipping cattle with the preparations recommended in Table 262 at about 30-day intervals will control body ticks. The same treatments that control hornflies and lice also control most kinds of ticks.

A tick calling for special methods of control is the spinose ear tick. This parasite, more commonly called "ear tick," is rather prevalent and can be found in other species of farm animals besides cattle, and in dogs. Few death losses occur and a figure for monetary loss is difficult to estimate. Poor performance, due to the constant annoyance and irritation that cause cattle to scratch at their ears, is the real problem. The tick spends about 6 months in the ear of an animal, then drops to the ground in mature form and lays its eggs on dry debris on or near the ground surface. The eggs hatch within a few days, and the small immature ticks or larvae crawl onto grass or other plants and from there onto the animal. They then migrate through the hair to the inside of the ear, and the cycle is complete.

Several good treatment methods are available and are described in Table 262, but unfortunately all require individual treatment, usually in a squeeze chute. If cattle are being handled individually anyway for some other purpose such as Bang's testing, pregnancy testing, or weighing, this is a good time to treat.

INTERNAL PARASITES

The situation with regard to the incidence of stomach and intestinal worms in beef cattle is not clear-cut at the present time. Research has shown that cattle are hosts to many species of worms, and parasitologists say they can find worm eggs in nearly every sample of feces examined under the microscope. The problem is that differentiation between the harmful and harmless types on the basis of fecal examination alone is extremely difficult, if not impossible, with presently available methods.

The damage or loss resulting from internal parasite infestation is also difficult to assess. Few death losses occur; rather, the losses are in terms of slower gains on pasture or in the feedlot, digestive disturbances that reduce efficiency of feed conversion, or lighter calf weaning weights owing to reduced milk flow in the mother cows. In general, internal parasite damage is higher in areas of high rainfall such as the coastal regions, but other sections such as irrigated pastures or well-watered pastures in the mountain ranges may be just as hard hit.

Symptoms of severe infestation with internal parasites are diarrhea, anemia, poor performance in general, and lowered resistance to disease. Swellings under the jaw are usually seen in extremely heavy infestations, and younger animals may die if the situation is not alleviated. Worm eggs are expelled in the feces of infested cattle, contaminating the grass with the resulting larvae and thus spreading the infestation to clean cattle.

Phenothiazine, consumed daily at the low level of 1.5 to 2 grams, aids in controlling internal parasites. Such constant low levels of consumption can be maintained by mixing salt or a mineral mixture and phenothiazine in the ratio of 10:1 and feeding it free-choice. For treatment of heavily infested cattle, the oral administration of 20 grams per hundred pounds of body weight, but not over 70 or 80 grams total, as a bolus, capsule, or drench is recommended. Mixing the phenothiazine in the feed is not highly successful because it is unpalatable and because each animal might not get the correct dosage. The day following such treatment the urine is quite red in color, but this is no cause for alarm.

Thiabendazole, a new antihelminthic that is showing much promise in the southeastern part of the country, is administered orally in a single 20-gram dose. This treatment is rather expensive but is used routinely by commercial feedlots that buy feeders originating in the Gulf Coast region.

MISCELLANEOUS PARASITIC DISEASES

TRICHOMONIASIS

Trichomoniasis is a venereal disease of cattle caused by an amoeba-like organism that infects the reproductive tract of both cows and bulls. A veterinarian may diagnose the disease from the presence of the flagellate organisms in vaginal or cervical smears examined under a low-power microscope. The organism is more difficult to obtain from the sheath of the bull; consequently his infection is more readily confirmed by mating him with a virgin heifer and checking her uterine and cervical smears for the organism at her next heat period.

Infection of cows results in a mild inflammation of the vagina and uterus with an accompanying discharge, which reduces the likelihood of conception or which may cause abortion, especially during the second to fourth month of gestation. Frequently, however, the dead, macerated fetus remains in the uterus, which becomes filled with a thin, grayish white, almost odorless fluid. When this condition occurs, the cow seldom exhibits symptoms of illness, but fails to show the usual signs of approaching parturition. Should abortion occur, she may come in heat within a few days, which distinguishes this type of abortion from that caused by the brucellosis bacillus.

Bulls infected with trichomoniasis frequently show no visible signs of the disease, although usually there is a mild inflammation of the prepuce and penis during the early stages. The disease is often transitory in cows, but it usually is chronic in bulls, which reinfect the cows each time they are bred. If infected bulls are sold for slaughter and breeding operations are discontinued for at least 3 months, the disease usually disappears from the herd.

Appendix

Table A-1
Composition of Feeds[a]

Feedstuffs	Dry Matter (%)	Protein Total N × 6.25 (%)	Protein Digestible[b] (%)	Energy per Pound Digestible[b] (kcal)	Energy per Pound Metabolizable (kcal)	TDN (%)	Fat (%)	Crude Fiber (%)	Ash (%)	Calcium (%)	Phosphorus (%)	Carotene (mg/lb)	Vitamin A Equivalent (1000's IU/lb)[c]
Alfalfa hay, prebloom	89	19.1	13.3	1093	910	53	2.4	23.4	8.6	1.89	0.27	202.3	337
Alfalfa hay, early bloom	90	16.6	11.6	1048	853	51	2.0	26.8	8.5	1.12	0.21	51.9	86
Alfalfa hay, mid bloom	89	15.2	10.6	990	819	51	1.8	27.5	7.6	1.20	0.20	13.6	22
Alfalfa hay, late bloom	88	14.0	9.5	958	787	48	1.6	29.8	7.8	1.13	0.18	15.1	25
Alfalfa hay, past bloom	91	13.9	8.9	971	780	49	2.0	32.3	7.3	1.21	0.22	11.0	19
Alfalfa leaves meal, dehydrated	92	20.6	—	1140	—	57	2.9	19.6	10.9	1.60	0.23	67.4	112
Alfalfa meal, dehydrated	92	16.7	—	1080	—	54	2.3	25.8	9.5	1.09	0.29	33.4	56
Alfalfa silage	30	5.3	3.4	360	—	18	1.1	9.1	2.8	0.48	0.16	12.4	20
Alfalfa silage, wilted	36	6.4	—	400	—	20	1.2	10.9	3.1	0.51	0.12	8.5	14
Alfalfa silage with molasses preservative	32	5.6	—	360	—	18	1.1	9.3	2.8	0.63	0.13	14.2	24
Alkali sacaton grass, dormant	100[d]	3.4	0.0	891	750	35	2.2	—	12.6	0.67	0.08	0.5	1
Barley grain	89	11.7	8.4	1560	—	78	1.9	5.3	2.8	0.08	0.42	0.18	Trace
Barley grain, Pacific Coast	90	8.7	6.9	1580	—	79	—	—	—	0.06	0.33	—	—
Barley hay	87	7.7	4.3	980	—	49	1.9	23.0	6.8	0.18	0.26	—	—
Barley straw	88	3.6	0.6	820	—	41	1.6	37.3	5.8	0.30	0.08	—	—
Beans, field	90	22.9	20.2	1580	—	79	—	0.0	—	0.15	0.57	—	—
Beet mo'asses	77	6.7	3.5	1220	—	61	0.2	0.0	8.2	0.16	0.03	—	—
Beet pulp, dried	91	9.1	4.6	1744	—	62	0.6	19.4	3.6	0.68	0.10	0.1	Trace
Beet pulp with molasses, dried	92	9.1	6.0	1420	—	71	0.5	15.6	5.7	0.56	0.08	0.1	Trace

Feed													
Beet pulp, wet	10	0.9	0.5	*200*	—	10	0.2	22	0.4	0.09	0.01	—	—
Beet sugar top, silage	26	3.1	2.5	*300*	—	15	0.5	3.2	9.2	0.60	0.05	4.2	7
Bermuda grass hay	91	8.1	4.1	*880*	—	44	1.8	26.9	6.1	0.42	0.18	26.8	45
Bermuda grass coastal hay	91	8.6	5.6	*900*	—	45	2.0	27.8	4.6	0.42	0.16	—	—
Blood flour	91	82.2	78.9	*1620*	—	81	1.0	0.6	4.8	0.45	0.37	—	—
Blood meal	90	79.9	56.7	*7180*	—	59	1.6	0.8	5.6	0.28	0.22	—	—
Bone black, spent	—	—	—	—	—	—	—	—	—	22.00	13.10	—	—
Bone meal, raw	93	26.2	18.1	*360*	—	18	4.6	1.4	57.8	22.70	10.10	—	—
Bone meal, steamed	95	12.1	—	—	—	—	3.2	1.7	71.8	30.00	13.90	—	—
Brewers (see grains)	—	—	—	—	—	—	—	—	—	—	—	—	25
Bromegrass hay	90	10.6	4.4	*940*	—	47	2.3	28.8	7.7	0.39	0.25	15.0	—
Buttermilk, dried	92	32.0	28.4	*1700*	—	85	5.8	0.4	9.6	1.34	0.94	—	62
Canarygrass, reed, hay	91	8.0	5.0	*840*	—	42	2.0	31.2	6.6	0.31	0.23	37.3	—
Cane molasses	74	3.0	0.0	*1080*	—	54	0.1	0.0	8.6	0.66	0.08	—	—
Cane molasses, dried	96	10.3	—	*1260*	—	63	1.0	—	—	—	—	—	—
Carrot roots	12	1.2	0.9	*220*	—	11	—	—	—	0.05	0.04	—	Trace
Citrus pulp, dried	90	6.6	5.2	*1500*	—	75	4.6	13.0	6.0	1.96	0.12	0.1	—
Citrus pulp silage	20	1.4	0.3	*400*	—	20	1.8	3.2	1.1	—	—	—	14
Clover, alsike, hay	88	12.9	8.6	*960*	—	48	2.6	25.9	7.7	1.15	0.22	74.6	—
Clover, crimson	87	14.7	9.8	*900*	—	45	2.0	28.0	8.2	1.22	0.18	—	111
Clover, ladino, hay	91	21.0	16.1	*1200*	—	60	3.1	17.5	8.6	1.26	0.36	66.5	25
Clover, red, hay	88	13.1	8.0	*1020*	—	51	2.6	26.5	7.0	1.42	0.19	15.0	—
Coconut meal, expeller	93	20.4	16.5	*1540*	—	77	6.6	11.6	6.9	0.21	0.61	—	—
Coconut meal, solvent	92	21.3	18.1	*1380*	—	69	1.8	15.4	5.6	0.17	0.61	—	—
Corn cannery waste, silage	29	2.6	1.5	*340*	—	17	1.5	7.8	1.4	—	—	1.7	3
Corn cobs, ground	93	2.9	-0.7	*920*	—	46	0.4	34.3	1.9	—	—	—	—
Corn dent silage	26	2.2	1.5	*340*	—	20	0.8	6.7	1.5	0.10	0.06	6.1	10
Corn dent silage, milk stage	26	1.8	0.9	*337*	—	15	0.6	5.8	1.3	0.07	0.06	—	—
Corn dent silage, dough stage	29	2.3	1.5	*401*	—	19	0.8	—	1.9	—	—	4.4	7
Corn ear silage	45	2.7	1.7	*651*	—	32	0.8	2.3	0.4	—	—	—	—
Corn distillers' grains, dried	92	27.1	19.8	*1660*	—	83	9.3	11.9	2.6	0.09	0.37	1.4	2

727

Table A-1 (Continued)

Feedstuffs	Dry Matter (%)	Protein — Total N × 6.25 (%)	Protein — Digestible (%)	Energy per Pound — Digestible[b] (kcal)	Energy per Pound — Metabolizable (kcal)	TDN (%)	Fat (%)	Crude Fiber (%)	Ash (%)	Calcium (%)	Phosphorus (%)	Carotene (mg/lb)	Vitamin A Equivalent (1000's IU/lb[c])
Corn distillers' grains with solubles, dried	92	27.2	19.9	1620	—	81	9.3	9.0	4.3	—	—	—	—
Corn distillers' solubles, dried	93	26.9	21.3	1600	—	80	9.1	3.8	8.0	—	—	—	—
Corn fodder	82	7.3	4.0	1100	—	55	2.0	21.2	5.5	0.41	0.21	—	—
Corn gluten feed	90	25.3	21.8	1480	—	74	2.4	7.9	6.3	0.46	0.77	3.8	6
Corn gluten meal	91	42.9	36.9	1620	—	81	2.3	4.0	2.4	0.16	0.40	7.4	12
Corn grain and cobs meal	86	7.4	5.4	1460	—	73	—	—	1.3	0.04	0.22	—	—
Corn, white, hominy feed	90	11.1	7.8	1660	—	83	6.1	5.0	3.0	0.02	0.58	—	—
Corn stover, mature	87	5.1	3.1	960	—	48	1.0	32.3	6.2	0.40	0.07	1.4	2
Corn stover, silage	27	1.9	0.6	380	—	19	0.6	8.7	2.0	0.10	0.05	—	—
Corn yellow dent, No. 2	89	8.9	6.9	1600	—	80	3.9	2.3	1.2	0.02	0.31	0.8	1
Corn—soybean (more than 30%), silage	28	3.2	2.0	400	—	20	1.0	7.0	1.8	0.20	0.08	23.3	39
Cotton burrs	92	8.5	2.2	820	—	41	2.0	35.9	8.0	1.04	0.11	—	—
Cottonseed feed	91	39.2	31.4	1300	—	65	6.1	11.6	6.1	0.15	0.64	—	—
Cottonseed hulls	90	3.9	0.2	880	—	44	1.4	42.8	2.5	0.14	0.09	—	—
Cottonseed meal, expeller	93	41.4	33.5	1460	—	73	5.8	10.7	6.1	0.18	1.15	—	—
Cottonseed meal, solvent	91	41.6	34.5	1320	—	66	1.6	—	6.5	0.15	1.10	—	—
Cottonseed whole, pressed	90	25.1	18.1	1280	—	64	8.4	23.3	4.7	0.17	0.64	—	—
Cowpea hay	90	16.6	11.8	1020	—	51	2.6	24.6	10.4	1.21	0.29	—	—
Defluorinated phosphate	99	—	—	—	—	—	—	—	99.0	29.18	13.34	—	—
Desert molly, dormant	100[d]	9.0	5.5	978	863	50	4.1	—	24.8	2.37	0.12	8.2	14
Dicalcium phosphate	96	—	—	—	—	—	—	—	78.1	26.50	20.50	—	—

728

Feed													
Fescue, meadow, hay	88	9.2	4.9	980	—	49	2.6	27.5	7.2	0.44	0.32	28.5	47
Fish, menhaden, meal	92	61.3	49.7	1760	—	58	7.7	0.7	19.6	5.49	2.81	—	—
Flaxseed screenings	91	15.8	8.8	1760	—	58	9.4	12.4	6.7	0.37	0.43	—	—
Flaxseed screenings feed, solvent	91	24.1	13.5	1100	—	55	—	—	—	0.44	0.63	0.2	Trace
Galleta curley grass, dormant	100[d]	5.5	1.4	756	595	39	2.0	—	16.6	1.05	0.07	—	50
Grains, brewers', dried	92	25.9	20.7	1320	—	66	6.2	14.5	3.6	0.27	0.50	—	—
Grass—legume, silage	29	3.4	1.7	260	—	13	1.0	9.1	2.3	0.23	0.08	30.2	52
Grass—legume, silage with molasses present	30	3.4	2.0	300	—	15	1.0	9.3	2.2	0.28	0.08	31.3	26
Johnson grass hay	91	7.0	3.1	1000	—	50	1.9	30.3	8.0	0.74	0.28	15.3	111
Lespedeza hay, prebloom	92	16.4	8.0	1000	—	50	3.1	21.8	6.5	1.12	0.25	60.6	—
Lespedeza hay, early bloom	93	14.3	7.2	920	—	46	3.9	25.1	6.0	1.14	0.23	—	—
Lespedeza hay, mid bloom	93	13.0	6.4	920	—	46	—	—	—	1.11	0.24	—	—
Lespedeza hay, full bloom	93	12.5	5.1	760	—	38	2.9	28.8	5.0	0.97	0.21	—	—
Limestone	99	—	—	—	—	—	—	—	95.8	33.84	0.02	—	—
Linseed feed	90	33.8	28.4	1480	—	74	4.8	9.5	6.0	0.43	0.65	0.14	—
Linseed meal, expeller	91	35.3	31.1	1480	—	74	5.2	8.5	5.6	0.44	0.89	—	2
Linseed meal, solvent	91	35.1	30.9	1480	—	74	1.7	8.9	5.8	0.40	0.83	—	—
Mangel roots	9	0.8	0.6	180	—	9	0.1	0.8	1.1	0.02	0.02	—	—
Meadow hay	93	8.5	5.1	1080	—	54	2.8	28.0	7.5	0.53	0.16	—	—
Meat meal	94	53.4	43.8	1300	—	65	9.9	2.4	25.2	7.94	4.03	—	—
Milk, skimmed, dried	94	33.5	30.2	1600	—	80	0.9	0.2	7.6	1.26	1.03	—	—
Milk, whole	12	3.1	2.9	320	—	16	3.7	0.0	0.8	—	—	—	—
Milk, whole, dried	94	25.2	22.6	2380	—	119	26.4	0.2	5.4	0.89	0.68	3.2	5
Monosodium phosphate	—	—	—	—	—	—	—	—	—	—	22.40	—	—
Needle and thread grass dormant	100[d]	4.0	1.2	895	747	47	4.9	—	17.8	0.88	0.07	0.2	Trace
Oats grain	89	11.8	8.8	1300	—	65	4.5	10.7	3.6	0.10	0.35	0.0	—
Oat hay	90	6.4	3.8	900	—	45	2.5	27.5	6.1	0.23	0.21	40.4	67
Oat silage	32	3.1	1.8	380	—	19	1.3	10.0	2.7	0.12	0.10	17.2	29

Table A-1 (Continued)

Feedstuffs	Dry Matter (%)	Protein Total N × 6.25 (%)	Protein Digest-ible (%)	Energy per Pound Digest-ible[b] (kcal)	Energy per Pound Metabo-lizable (kcal)	TDN (%)	Fat (%)	Crude Fiber (%)	Ash (%)	Cal-cium (%)	Phos-phorus (%)	Caro-tene (mg/lb)	Vita-min A Equiv-alent (1000's IU/lb[c])
Oat straw	90	4.0	1.4	900	—	45	1.9	36.9	7.4	0.30	0.09	—	—
Oats grain Pacific Coast	91	9.2	7.0	1440	—	72	5.3	11.0	3.8	0.10	0.35	—	—
Oats without hulls	90	16.2	14.6	1840	—	92	—	—	—	0.08	0.46	—	—
Oats without hulls, rolled	91	16.1	14.5	1820	—	91	—	—	—	0.07	0.46	—	—
Orange pulp, dried	89	7.0	6.1	1580	—	79	1.8	9.6	4.4	0.63	0.10	—	—
Orchard grass hay	88	11.2	5.8	900	—	45	3.3	29.9	6.7	0.40	0.33	13.4	22
Oyster shells, ground	99	1.0	—	—	—	—	—	—	80.8	38.05	0.07	—	—
Pea vine silage	23	3.0	1.8	260	—	13	0.8	7.4	1.7	0.43	0.08	14.6	24
Peanut meal, expeller	92	45.8	41.2	1540	—	77	5.9	10.7	5.7	0.17	0.57	0.1	Trace
Peanut meal, solvent	92	47.4	42.7	1540	—	77	1.2	13.1	4.5	0.20	0.65	—	—
Potato meal dehydrated	90	5.9	2.1	1300	—	65	0.3	2.1	4.3	0.07	0.20	—	—
Potato tubers	21	2.2	1.3	360	—	18	0.1	0.4	1.1	0.01	0.05	—	—
Potato tubers, silage	25	2.5	1.4	320	—	16	0.2	2.1	1.5	—	—	—	—
Prairie hay, mid bloom	91	5.1	2.0	970	—	45	2.6	30.3	7.4	0.37	0.11	13.3	22
Prairie hay, late bloom	91	6.9	—	—	—	—	2.9	30.0	9.3	0.46	0.07	10.1	17
Prairie hay, past bloom	90	6.0	0.9	880	—	44	2.4	29.3	8.1	0.47	0.07	3.6	6
Rape seed	90	20.4	17.3	1360	—	68	43.6	6.6	4.2	—	—	—	—
Rice bran	91	13.5	8.5	1360	—	68	15.1	10.9	10.9	0.06	1.82	—	—
Rice polishings	90	11.8	6.7	1700	—	85	13.2	3.3	8.0	0.04	1.42	—	—
Ricegrass, Indian, dormant	100[d]	3.5	0.3	890	733	48	2.7	—	7.4	0.52	0.06	0.2	Trace
Rye, distillers' grains, dried	93	22.4	13.9	1160	—	58	6.4	13.8	2.6	0.13	0.41	—	—
Rye grain	89	11.9	9.9	1440	—	72	1.6	2.3	1.9	0.06	0.34	0.0	—

Rye middlings	90	17.1	14.4	1420	—	71	3.1	5.8	3.4	0.06	0.63	—	—
Rye straw	89	2.7	0.0	800	—	40	1.3	42.4	4.3	0.25	0.09	—	—
Sage, black, dormant	100[d]	8.5	4.4	944	510	47	9.4	—	6.2	0.60	0.16	8.0	13
Sage, bud, early leaf	100[d]	17.3	13.7	1160	911	51	4.9	—	21.4	0.97	0.33	10.8	18
Sagebrush, big, dormant	100[d]	9.4	5.4	1045	575	51	10.1	—	6.1	0.67	0.18	7.3	12
Safflower oil meal, expeller	91	19.7	16.9	1740	—	57	6.0	30.9	3.7	0.23	0.71	—	—
Safflower seeds	93	16.3	13.0	1640	—	82	29.8	26.6	2.9	—	—	—	—
Safflower with hulls, meal solvent	92	21.5	17.2	1020	—	51	6.9	32.8	3.8	—	—	—	—
Safflower, without hulls, meal solvent	91	42.5	37.4	1280	599	64	6.7	8.5	6.4	2.21	0.21	8.6	14
Saltbrush, nuttal, dormant	100[d]	7.2	3.4	677	—	36	2.2	—	21.5	0.57	0.06	0.2	Trace
Sand dropseed grass, dormant	100[d]	5.0	1.9	1093	939	59	1.4	—	6.3	2.53	0.09	8.9	15
Shadscale, dormant	100[d]	7.7	4.3	570	399	31	2.4	—	23.4	0.34	0.15	8.5	14
Sorghum fodder	86	6.8	2.6	1080	—	54	2.1	22.3	7.3	—	—	—	—
Sorghum, kafir, grain	87	11.1	9.0	1240	—	79	2.5	2.5	1.6	—	—	—	—
Sorghum, milo, grain	89	11.0	8.6	1420	—	84	2.8	2.4	1.9	—	—	—	—
Sorghum, milo, heads chop	91	9.3	7.1	1560	—	78	2.5	7.1	3.3	0.10	0.06	—	—
Sorghum, silage	29	2.3	0.6	303	—	17	0.8	7.8	2.2	0.37	0.10	4.4	7
Sorghum stover	92	4.6	—	920	—	46	1.8	29.4	9.3	0.08	0.05	2.8	5
Sorghum, sweet, silage	25	1.6	0.8	340	—	17	0.8	6.9	1.6	0.25	0.59	—	—
Soybeans, seeds	91	36.8	33.1	1640	—	82	17.4	6.4	5.0	1.15	0.20	—	24
Soybean hay	89	14.5	10.2	920	—	46	2.7	28.6	7.2	0.54	0.15	14.4	—
Soybean hulls	91	12.5	5.5	960	—	48	2.1	35.4	4.6	0.27	0.63	—	—
Soybean meal, expeller	90	43.8	39.4	1480	—	74	4.7	5.8	5.7	0.32	0.67	0.1	Trace
Soybean meal, solvent	89	45.8	43.1	1560	—	78	0.9	5.8	5.8	0.35	0.14	0.1	Trace
Soybean, silage	28	4.1	2.6	280	—	14	0.9	8.7	2.8	1.40	0.05	9.9	17
Soybean straw	88	4.8	1.1	780	—	37	1.2	38.8	5.6	—	—	—	—
Squirreltail grass, dormant	100[d]	4.5	1.1	868	732	46	2.6	—	17.1	0.67	0.07	0.5	1
Sudan grass hay	89	11.3	4.3	980	—	48	2.0	25.7	8.5	0.50	0.28	—	—

Table A-1 (Continued)

Feedstuffs	Dry Matter (%)	Protein — Total N × 6.25 (%)	Protein — Digestible (%)	Energy per Pound — Digestible[b] (kcal)	Energy per Pound — Metabolizable (kcal)	TDN (%)	Fat (%)	Crude Fiber (%)	Ash (%)	Calcium (%)	Phosphorus (%)	Carotene (mg/lb)	Vitamin A Equivalent (1000's IU/lb[c])
Sudan grass, silage	26	2.2	1.5	280	—	14	0.7	8.8	2.0	0.11	0.04	—	—
Tankage, digester	92	59.8	50.8	1320	—	66	8.1	2.0	21.4	5.94	3.17	—	—
Tankage, digester with bone	94	49.6	45.0	1300	—	65	11.9	2.5	26.3	7.34	3.73	—	—
Thistle, Russian, dormant	—	14.7	9.7	969	807	50	3.0	—	19.6	3.30	0.16	4.1	7
Timothy hay, prebloom	87	11.8	7.9	1168	963	56	3.0	31.3	7.2	0.57	0.30	—	—
Timothy hay, mid bloom	88	7.5	4.1	1020	847	53	2.4	29.5	5.1	0.36	0.16	21.0	35
Timothy hay, late bloom	87	6.9	3.4	855	685	51	2.3	29.5	4.7	0.30	0.18	—	—
Timothy silage	38	3.9	2.2	380	—	19	1.3	12.9	2.7	0.21	0.11	13.6	23
Trefoil, Birdsfoot, hay	90	14.2	9.8	960	—	48	3.0		19.6	3.30	0.16	4.1	7
Turnip roots	9	1.2	0.9	180	—	9	0.2	1.1	0.9	0.06	0.02	—	—
Vetch hay	88	17.6	11.6	1020	—	51	2.3	24.8	7.9	1.20	0.30	183.0	305
Wheat bran	89	16.0	12.2	1160	—	58	4.1	10	6.1	0.14	1.17	—	—
Wheat hay	86	6.5	3.5	880	—	44	1.7	24.0	5.9	0.10	0.14	43.5	72
Wheat germ meal	90	27.3	24.6	1600	—	80	9.1	2.6	4.8	0.07	1.04	0.0	0
Wheat, hard red spring grain	90	14.7	12.4	1620	—	81	1.9	2.5	1.9	0.05	0.47	0.0	0
Wheat, hard red winter grain	89	13.0	10.9	1600	—	80	1.6	2.7	1.8	0.05	0.40	—	—
Wheat middlings	89	18.0	15.8	1560	—	78	5.0	5.3	3.3	0.07	0.79	1.4	2
Wheat screenings	89	15.0	10.8	1380	—	69	3.0	6.5	3.2	0.08	0.39	—	—
Wheat standard middlings	90	17.2	14.3	1540	—	77	4.6	7.6	4.4	0.15	0.91	1.4	2
Wheat, soft, Pacific coast	89	9.9	8.3	1600	—	80	2.0	2.7	1.9	—	—	—	—
Wheat, soft red winter, grain	90	11.1	8.3	1560	—	78	1.6	2.2	1.8	0.09	0.30	—	—
Wheat straw	90	3.2	0.4	860	—	43	1.5	37.4	7.3	0.15	0.07	0.9	2

Feed													
Wheatgrass, beardless, dormant	100[d]	3.1	0.0	1160	—	58	4.1	—	10.6	0.49	0.06	0.5	1
Wheatgrass, western	100[d]	2.4	0.2	1280	—	64	8.3	—	10.0	0.74	0.06	0.1	Trace
Whey, whole, dried	94	13.1	11.8	1560	—	78	—	—	10.1	0.87	0.79	—	—
Wild-rye, giant, grass dormant	100[d]	3.2	−0.4	880	—	44	3.2	—	11.6	0.66	0.06	0.0	0
Winterfat	100[d]	11.0	6.9	660	—	33	2.7	—	18.6	2.14	0.12	7.6	13
Yeast, brewers', dried	93	44.6	38.4	1440	—	72	1.1	2.7	6.4	0.13	1.43	0.0	0
Yeast, torula, dried	93	48.3	41.5	1400	—	70	2.5	2.4	7.8	0.57	1.68	—	—
Yellowbrush	100[d]	6.6	3.1	1025	760	50	12.2	—	8.4	1.90	0.10	2.1	4

[a] *Nutrient Requirements of Beef Cattle*, revised 1963, National Research Council. The composition of mixed roughages can be computed as weighted means from the figures given for the pure species making up the mixtures. A dash (—) indicates that no data are available. This table was prepared by the Subcommittee of the Academy-Research Council Animal Nutrition Committee on Feed Composition—Earle W. Crampton and Lorin E. Harris.

[b] It has not been possible to obtain apparent digestible energy values for all feedstuffs; hence values have been estimated from TDN values assuming that one pound of TDN = 2000 kcal. Calculated values are in italics.

[c] Assume that 0.6 microgram of beta carotene is equal to one IU of vitamin A as determined by rat growth. When using carotene to satisfy vitamin A requirements of beef cattle, a conversion factor of 4.2 (1667 ÷ 400) should be used, since beef cattle do not convert carotene to vitamin A as efficiently as rats do.

[d] These range plants are on a 100% dry matter basis. The dry matter varies from 30% to 90% depending on weather conditions. Because of the high essential oil content of the browse species, it is recommended that metabolizable energy content be used and not TDN or DE.

Table A-2
Composition of Calcium and Phosphorus Supplements[a]

Mineral Supplement	Calcium		Phosphorus		Fluorine
	(%)	(gm/lb)	(%)	(gm/lb)	(%)
Bonemeal, raw, feeding	22.7	103	10.1	46	0.030
Bonemeal, special steamed	28.7	130	13.9	63	—
Bonemeal, steamed	30.0	136	13.9	63	0.037
Defluorinated phosphate rock a[b]	21.0	95	9.0	41	0.150 or less
Defluorinated phosphate rock b[b]	29.0	132	13.0	59	0.150
Defluorinated superphosphate	28.3	128	12.3	56	0.150
Dicalcium phosphate	26.5	120	20.5	93	0.050
Disodium phosphate	—	—	8.6	39	—
Limestone (high calcium)	38.3	174	—	—	—
Monocalcium phosphate	16.0	72	24.0	109	0.050
Monosodium phosphate	—	—	22.4	102	—
Oyster shell flour	36.9	167	—	—	—
Spent bone black	22.0	100	13.1	59	—

[a] *Nutrient Requirements of Beef Cattle*, revised 1958, National Research Council.
[b] Because of the limited number of products on the market, figures are given for two types of defluorinated rock that are being produced for livestock feeding.

Table A-3
Estimated Carotene Content of Feeds in Relation to Appearance and Methods of Conservation[a]

Feedstuff	Carotene (mg/lb)
Fresh green legumes and grasses, immature	15 to 40
Dehydrated alfalfa meal, fresh, dehydrated without field curing, very bright green color[b]	110 to 135
Dehydrated alfalfa meal after considerable time in storage, bright green color	50 to 70
Alfalfa leaf meal, bright green color	60 to 80
Legume hays, including alfalfa, very quickly cured with minimum sun exposure, bright green color, leafy	35 to 40
Legume hays, including alfalfa, good green color, leafy	18 to 27
Legume hays, including alfalfa, partly bleached, moderate amount of green color	9 to 14
Legume hays including alfalfa, badly bleached or discolored, traces of green color	4 to 8
Nonlegume hays, including timothy, cercal, and prairie hays, well cured, good green color	9 to 14
Nonlegume hays, average quality, bleached, some green color	4 to 8
Legume silage	20 to 30
Green silage	5 to 20
Corn and sorghum silages, medium to good green color	2 to 10
Grains, mill feeds, protein concentrates, and by-product concentrates, except yellow corn and its by-products	0.01 to 0.2

[a] *Nutrient Requirements of Beef Cattle*, revised 1958, National Research Council (table prepared by the late H. R. Guilbert, Davis, California).
[b] Green color is not uniformly indicative of high carotene content.

Table A-4

Breed Associations

American Angus Association, 3201 Frederick Boulevard, St. Joseph, Missouri 64500; Lloyd D. Miller, secretary.

American Belted Galloway Cattle Breeders' Association, South Fork, Missouri; Charles C. Wells, secretary.

American Brahman Breeders Association, 4815 Gulf Freeway, Houston, Texas 77000; Harry P. Gayden, secretary.

American Charbray Breeders Association, 475 Texas National Bank Building, Houston, Texas 77000; Mrs. Quinta Arrigo, secretary.

American Devon Cattle Club, Inc., Agawam, Massachusetts 01001; Kenneth Hinshaw, secretary.

American Galloway Breeders Association, P.O. Box 1424, Billings, Montana 59103; Harold Gerke, secretary.

American Hereford Association, Hereford Drive, Kansas City, Missouri 64100; W. T. Berry, Jr., secretary.

American-International Charolais Association, 923 Lincoln Liberty Life Building, Houston, Texas 77000; J. Scott Henderson, secretary.

American Polled Hereford Association, 4700 East 63rd Street, Kansas City, Missouri 64100; Orville K. Sweet, secretary.

American Red Danish Cattle Association, Marlette, Michigan 48453; Mrs. Harry Prowse, secretary.

American Shorthorn Society, 8288 Hascall Street, Omaha, Nebraska 68100; C. D. Swaffar, secretary.

International Brangus Breeders Association, 908 Livestock Exchange Building, Kansas City, Missouri 64100; Roy W. Lilley, secretary.

National Polled Cattle Club, Nicollet, Minnesota 56074; Walter Schultz, secretary.

Red Angus Association of America, P.O. Box 391, Ballinger, Texas 76821; Mrs. Sybil Parker, secretary.

Santa Gertrudis Breeders International, P.O. Box 1340, Kingsville, Texas 78363; R. P. Marshall, secretary.

Dual-Purpose Cattle

American Milking Shorthorn Society, 313 South Glenstone Avenue, Springfield, Missouri 65800; Ray R. Schooley, secretary.

Red Poll Cattle Club of America, 3275 Holdrege Street, Lincoln, Nebraska 68500; Wendell Severin, secretary.

INDEX

Abomasum: capacity and function, 133; role in protein digestion, 134

Abortion, pine needle: incidence, 682

Abortion: and reduced calf crop, 129; from ergot poisoning, 705; from trichomoniasis, 723; from vibriosis, 697; from vitamin A deficiency, 150; induced in feedlot heifers, 332; *see also* Brucellosis

Abortion, vibrionic, 697

Accredited herd, 247: in Brucellosis eradication, 688

Acetic acid, metabolism of, 133

Acetonemia, incidence, 682

Actinomyces, *see* Lumpy jaw

Adaptation, effect on performance, 62

Additives: antibiotics, 563–566; antihelminthics, 568–569; dried rumen contents, 566–567; drugs, 567–568; effect on feed efficiency, 550; enzymes, 568; flavoring and coloring agents, 569–570; goiterogens, 566; grub control agents, 568–569; tranquilizers, 566; vitamin A, 563; yeast, 567

Afterbirth, *see* Fetal membranes

Age: and capital investment, 323–324; and cost of finishing, 327; and pasture utilization, 323; and quality of feeds fed, 322–323; effect on carcass composition, 324–326; effect on economy of gains, 316–319; effect on feed consumption, 322; effect on feed requirement, 317; effect on length of feeding period, 320–321; effect on performance, 325; effect on rate of gain, 315–316, 317; effect on total gain needed to finish, 321–322; in purchase of feeder cattle, 326; in selecting bulls, 170; in selecting feeder cattle, 305; in selecting foundation

Age: (*Continued*) females, 168–170; trends in slaughter cattle, 314

Ailments, most common by area, 683

Alabama, registered cattle in, 17

Alaska, cattle numbers, 19, 20

Alfalfa: as cause of bloat, 528; as pasture in summer-feeding program, 523; as silage crop 507; compared with cottonseed hulls, 472; in rotation pastures, 214; pellets as nitrogenous roughage, 416; production related to cattle industry, 33; unidentified factors in, 136; utilizing through beef steers, 544; value of grinding and chaffing, 627; when to harvest for silage, 500; *see* also Hay, alfalfa; Pasture, alfalfa; and Silage, alfalfa

Alfalfa meal: amount fed and price, 428; as protein concentrate, 446, 447; carotene content, 735; composition and digestible nutrients, 726

Alkali disease, incidence, 682

Allantois, 115; rupture during labor, 122

All-concentrate rations: as cause of founder, 699; effect on carcass weight, 454; feeding of, 454

Altitude, effect on nutrient requirements, 196–197

American Hereford Association, studies of steer types on deferred feeding system, 345–346

American National Cattlemen's Association, survey of diseases and parasites, 679

Amides, in ruminant nutrition, 134

Amino acids, in ruminant nutrition, 134

Ammonia: as end-product of fermentation, 137; as nitrogen source, 445; in

Marketing: (*Continued*)
ing, 674–678; effect of shrinkage on
costs, 678; facilities for sale of fed
cattle, 660; factors determining selling
date, 663; increases in, of fed cattle,
294; losses due to shrinkage, 668–671;
losses during shipment, 672–674; of
fat calves, 14; of fed cattle, by states
and regions, 295, 297; rail versus truck
shipment, 666, 667; regional variations
in, 662
Mastitis, incidence of, 680
Mating, *see* Breeding
Mating, assortative: *defined,* 56; when
to practice, 58
Mating, disassortative: *defined,* 56; when
to practice, 58
Mating, random, 56
Mating systems, 56–58
Meat: and effect of radioactive fallout,
710; consumption of, in United States,
38; consumption of, worldwide, 38
Medium grade, characteristics of,
259–260
Melengestrol acetate (MGA), 562–563;
for quieting feedlot disturbance, 332
Metabolizable energy, 140
Methane, as by-product of rumen fer-
mentation, 137
Methimazole, 566
Mexico: and occurrence of foot-and-
mouth disease, 688; role in screwworm
control, 719; steers for finishing on
grass alone, 537
MGA, *see* Melengestrol acetate
Michigan, increase in cattle numbers, 24
Midwest: prevalence of poisonous plants,
703; use of controlled-environment
barns, 647
Miles, Manly: early research with silate,
477
Milk fever, incidence of, 682
Milking Shorthorn cattle, suitability of
for fat-calf program, 14, 15, 581
Mills: for grain processing, 614; for
grinding corn, 621
Milo: compared with barley, 394; feeding value of and various processed
forms, 625; steam-processed, 393–394,
624; *see also* Sorghum, grain
Minerals: and microfloral activity, 136;

Minerals: (*Continued*)
areas deficient in, 142; deficiency as
cause of hardware disease, 703; effect
of supplementation on range cattle,
227; in small-grain and ryegrass
forage, 194; suggested levels of for
high-concentration rations, 148
Minerals, toxic, 148
Minnesota: increase in cattle numbers,
24; St. Paul as source of feeder cattle,
338
Mississippi: Delta area as cattle-finishing
center, 10; registered cattle in, 17;
use of cottonseed products, 432
Missouri: East St. Louis as source of
feeder cattle, 338; grazing plan,
547–548; Kansas City as source of
stockers and feeders, 261, 338; registered cattle in, 17; studies of grain-fed
suckling calves, 577–579
Mites, efforts to control, 714
Molasses: and palatability of ration, 139;
as grain substitute, 405–406; as silage
preservative, 481; feeding value of,
404–407; for show cattle, 602; in beef
cattle rations, 34; in cattle-finishing
program, 10; kinds of, 404
Molasses, beet, 404; composition and digestible nutrients, 726
Molasses, blackstrap, 404
Molasses, cane, 404; composition and digestible nutrients, 727
Molasses, corn, 404
Molasses, holocellulose, 404
Molybdenum: deficiency, 682; effect on
copper requirement, 147; toxicity,
147; toxicity symptoms, 149
Montana: as cattle-finishing center, 10;
Miles City station breeding herd production records, 129; registered cattle
in, 17
Morea, 446
Morrison, Frank B., feeding standards:
for breeding cattle, 183; for finishing
cattle, 366; for stocker cattle, 278
Morula, 114
Mosquitoes, control methods, 717
Mothering ability: as major performance
trait, 68–69; effect of heterosis on, 59;
selection for, 79